FUNDAMENTALS AND EVOLUTION OF MPEG-2 SYSTEMS

FUNDAMENTALS AND EVOLUTION OF MPEG-2 SYSTEMS

PAVING THE MPEG ROAD

Jan van der Meer

Senior Consultant, Jan van der Meer Consultancy, The Netherlands

WILEY

This edition first published 2014
© 2014 John Wiley & Sons, Ltd

Registered office
John Wiley & Sons Ltd, The Atrium, Southern Gate, Chichester, West Sussex, PO19 8SQ, United Kingdom

For details of our global editorial offices, for customer services and for information about how to apply for permission to reuse the copyright material in this book please see our website at www.wiley.com.

The right of the author to be identified as the author of this work has been asserted in accordance with the Copyright, Designs and Patents Act 1988.

All rights reserved. No part of this publication may be reproduced, stored in a retrieval system, or transmitted, in any form or by any means, electronic, mechanical, photocopying, recording or otherwise, except as permitted by the UK Copyright, Designs and Patents Act 1988, without the prior permission of the publisher.

Wiley also publishes its books in a variety of electronic formats. Some content that appears in print may not be available in electronic books.

Designations used by companies to distinguish their products are often claimed as trademarks. All brand names and product names used in this book are trade names, service marks, trademarks or registered trademarks of their respective owners. The publisher is not associated with any product or vendor mentioned in this book.

Limit of Liability/Disclaimer of Warranty: While the publisher and author have used their best efforts in preparing this book, they make no representations or warranties with respect to the accuracy or completeness of the contents of this book and specifically disclaim any implied warranties of merchantability or fitness for a particular purpose. It is sold on the understanding that the publisher is not engaged in rendering professional services and neither the publisher nor the author shall be liable for damages arising herefrom. If professional advice or other expert assistance is required, the services of a competent professional should be sought.

Library of Congress Cataloging-in-Publication Data applied for.

ISBN: 9780470974339

Set in 10/12pt TimesLTStd-Roman by Thomson Digital, Noida, India.

1 2014

Contents

Foreword

Why do we need a book on MPEG Systems? The MPEG-2 Systems standard has been around since 1994 and is already widely deployed. Surely everyone who needs to know about it already know enough? Has it not been covered elsewhere? And is it not just a syntax of packets and time stamps? In fact, MPEG-2 Systems – the standard and designs that implement it – is significantly more involved than it first appears and there is widespread misunderstanding of some key aspects. Just as importantly, the theory, history, practice and applications are quite interesting, as this book shows. Jan van der Meer's book *Fundamentals and Evolution of MPEG-2 Systems* is the first to comprehensively and correctly explain what the MPEG-2 Systems standard is, how it works, why it works the way it does, how it relates to key video and audio standards, and the information you need to know in order to ensure that systems based on the standard perform robustly as expected. It also includes some interesting bits of history of the standard, including humorous anecdotes.

The MPEG-2 Systems standard, despite being nearly 20 years old, has succeeded wildly, with major growth in applications occurring particularly since 2000. Data from IHS quoted in the book shows that, for example, in 2014 alone over 600 million DTVs, set top boxes, BluRay and DVD disc players using MPEG-2 Systems are expected to be shipped, and that does not include the vast numbers of tablets, phones and PCs that also support MPEG-2 Systems for streaming. The MPEG-2 Systems standard has maintained its place as the systems layer for carrying video and audio, as the applications for it and the compression standards that work with it have continued to evolve. Its genesis was MPEG-1 Systems, which targeted video and audio on CDs, and the principles embodied in MPEG-1 Systems were carried over to MPEG-2 Systems. The latter was originally specified for use with MPEG-2 video and MPEG audio in both the Program Stream format targeting optical discs (DVD) and software based parsing applications and the MPEG Transport Stream format, which addressed the needs of digital terrestrial broadcasting, cable television and direct broadcast satellite. Many competing companies came together to develop, agree on, implement and promote this and related standards, kick-starting the massive revolution to digital video and audio via a range of media types. As new video and audio compression standards have been created since 1994, each gradually supplanting the ones that came before it, the MPEG-2 Systems standard has needed only minor additions to accommodate the new audio and video standards, and it has stayed with us. In addition to its use in broadcast environments, MPEG Systems is used in some popular streaming formats as well as infrastructure. The author predicts, with good reason, that it will continue to be widely used for the foreseeable future.

This book includes a suitable background of the topic of compressed video and audio, and a detailed explanation of the standard itself and in particular, what the standard really means, and how it applies in practice. For example, the System Target Decoder (STD) is not a 'standard' decoder, is not an optional part of the standard, and it is not something that one can choose to implement (or not implement). Rather, it is a minimal abstract model of an idealized hypothetical decoder that serves as a constraint on all valid MPEG Systems streams, incorporating both the unified transport, decoding and presentation timing model of the MPEG Systems standard and buffer models, with rules regarding overflow and underflow of the buffer models. If a stream fails a STD test, it is not a conforming stream, so decoder designers can rely on conforming streams always passing the STD test. The book includes an extensive set of figures illustrating buffer and buffer model fullness under a wide range of conditions.

Jan van der Meer is exceptionally well qualified to write this book. He was active in the MPEG Systems committee since its inception and remained active in it for many years. He contributed a great deal of clarity to the committee discussions and documentation and was a frequent source of ideas. As I chaired the committee through the development of the MPEG-1 and MPEG-2 Systems standards, Jan's presence, participation, clarity and humour were all invaluable in enabling the group to produce such a high quality standard. Jan's knowledge extends well beyond this standard, encompassing a wide range of related topics, as you will find evident from reading this book. Even if you are already quite familiar with the MPEG-2 Systems standard, I am confident that you will find many refreshing insights in his detailed and clever writing. I hope you enjoy it as much as I did.

Alexander (Sandy) MacInnis,
November 2013

Preface

My first name is Jan, the Dutch version of John, nothing special. My last name is van der Meer, which, with some fantasy, means 'from the Lake'. Again, nothing special, although I was born indeed very close to one of the various lakes in the north of the Netherlands, in an area considered by many Dutch people as one of the most backward regions in the Netherlands, though most of them never visited it. In fact, it is of course one of the most beautiful places in the world. Some may still recognize my origin from my articulation, as I refused to polish it; by no means do I wish to suggest coming from an unidentified region. Anyway, due to a lot of coincidences and some good reasons described elsewhere in this book, I got deeply involved in MPEG. That was special.

It is July 2007, MPEG is having its 81st meeting in Lausanne, Switzerland. I attend again, and it is a special meeting for me, not because it is the 58th MPEG meeting that I attend, but because it will be my last one. During 18 years I came to most MPEG meetings, starting with the eighth MPEG meeting in 1989. I witnessed the standardization of MPEG-1, MPEG-2, MPEG-4 and other MPEG standards. Though my initial focus was on video coding, most of my work was on MPEG-2 systems, addressing issues such as transport and synchronization of coded audio and video. Personally I am not a truly dedicated scientist, but more interested in making things work and less in the scientific considerations, though it is important to understand those.

After Sandy MacInnis's very successful chairmanship of the MPEG Systems group from 1990 to 1994, during which the MPEG-1 and MPEG-2 system standards were developed, I became MPEG Systems chair for 2 years, mainly to complete the Compliance and the Real-Time Interface specifications for MPEG-2 systems. In 1996 the main focus of MPEG moved to MPEG-4; at that time the company that I worked for, Philips, was not yet interested in MPEG-4 from a business perspective, and therefore I stopped attending every MPEG meeting, leaving the regular Philips representation to experts from Philips Research. Chairmanship of the MPEG Systems was taken over by Olivier Avaro, at that time working for France Telecom. To keep track of MPEG developments I attended one MPEG meeting in 1997, but in 1998, Philips recognized potential business in MPEG-4 technology and I was asked to attend MPEG regularly, which I continued until 2007.

Prior to attending my last MPEG meeting in Lausanne in July 2007, I informed Olivier Avaro, at that time still the MPEG Systems chair, that Lausanne was to be my last MPEG meeting. Olivier responded very disappointedly, in particular because I was one of the last MPEG-2 Systems experts still active in MPEG. During the MPEG-2 Systems development often over 100 experts attended the MPEG System meetings, but after its completion that number reduced in a significant manner.

Olivier expressed his concern that in MPEG the expertise in MPEG-2 System technology was decreasing to a level that may soon become critical, and that this may become a problem when in future MPEG-2 Systems needs to be extended, for example to support new audio and video coding specifications. Of course, I agreed with Olivier that in-depth knowledge of MPEG-2 Systems is essential for those extending the MPEG-2 Systems standard in future. Nevertheless, I explained to Olivier that it may not be good for MPEG, but that the Lausanne meeting really was to be my last one. '*Well, OK*', Olivier said, '*but if that will be your last meeting indeed, then why don't you write a book on MPEG-2 Systems to document your expertise?*'

Well, writing a book on MPEG-2 systems was already one of my popular thoughts at the time MPEG-2 systems was finished, but I was overwhelmed by other activities, as usual, and I was not the only one. A book with a comprehensive review of MPEG-2 systems still did not exist. Therefore it seems to make sense, even almost 20 years after the standardization of MPEG-2 systems, to write a book to describe the fundamentals and evolution of MPEG-2 Systems technology, and to provide the background on how and why MPEG-2 Systems became what is.

After my full retirement from Philips in 2008 I had time available and decided to work on what became this book, but the start was slow. It took until the second half of 2009 to become more productive. My desire to write a book that is accessible to people with technical expertise, not only to MPEG system experts, required the drawing of many figures. Quite a few figures needed several days of drawing, changing and changing again, and when further progress was made, sometimes the very painful conclusion was reached that another figure was needed instead.

Moreover, I wanted the book to not only explain the issues that the MPEG-2 system experts needed to resolve, but also to express how much fun it was to jointly resolve these issues. But how to explain fun . . . ? I guess people understand it is fun to write an e-mail like the one below to a MPEG-4 systems guru, who insisted that for usage of MPEG-4 technology in an MPEG-2 environment, important data structures, such as PES packets, needed to be replaced by an incompatible MPEG-4 equivalent, which means that MPEG-4 cannot be used in already existing MPEG-2 system-based applications. In an earlier e-mail I had already stated that adoption of new technology is often by building on other successful technology, rather than from scratch.

> I like your comparison. MPEG-2 a car and MPEG-4 an airplane. Quite some time ago someone told me that MPEG-4 was about a flying submarine, but that was a bad joke of course. Anyhow, airplanes fly in general. And at this very moment I travel in one and I must say, I like it far above the clouds. But back home I have to live again with my old-fashioned car. Which does a reasonable job by the way, as does MPEG-2.
>
> What I learned since I left MPEG (for the second time) was that meanwhile MPEG-4 did develop nice technology, that could be very useful in my old-fashioned MPEG-2 car. Let's say it won't let my car fly, but I would be able to navigate a lot better. But of course I would like to use the new technology in a cost-effective way as I cannot afford the costs that may be acceptable for airplanes. So I am looking for a good way to integrate. That means that I have to extend the specification of

my car (13818-1) as well as the 'safety procedure' of using the car (STD model). I hope you don't object. Of course I agree with you that my car won't fly, but still I can travel conveniently and with the new technology a lot safer. And with respect to the risk of using the new tools, I understood that they are verified extensively, so that I can apply them without risk.

At that time several MPEG-4 system people were very passionate about their design and positioned it as the only viable system for the future; and they therefore objected to compromising this approach for the 'simple purpose' of adding value to MPEG-2 system applications.

But emails are typically not very suitable to include in a book. Fortunately I met Chad Fogg when invited to attend the 100th meeting of MPEG in Geneva, in April 2012. Chad is famous in the MPEG community, not only for his expertise, but also for the jokes he regularly makes by producing MPEG quizzes, all kind of top-10 s and other funny stuff usually related to ongoing work in MPEG. Seeing Chad again did trigger the idea to include some of his material in the book, where appropriate. I was very pleased when Chad not only agreed with this proposal, but also to write the Epilogue of this book. As only a very limited subset of Chad's material was suitable, below a few examples are provided that remained unused.

e-mail Chad Fogg, 22 November 1994
Subject: Top 10 Punishments for Delinquent MPEG Delegates
Nr 9. Crime: Sleeping during plenary session
Punishment:Being woken up during plenary session
Nr 1. Crime: Laughing while saying the word "MPEG-4"
Punishment: [. . . well, none, really . . .]

While initially several jokes were made about MPEG-4, it should be noted that the MPEG-4 work item produced many very successful standards, for example on audio and video coding and on file formats.

e-mail Chad Fogg, 1 April 1995
Subject: Top 10 Resolutions of the Lausanne Meeting
Nr 7. WG11 approves conclusions of Ad-Hoc group investigation that the D-video tape recorder operates more efficiently when plugged in.
Nr 5. WG11 establishes liasonship with ski slopes to investigate 'frozen syntax' (Convenor notes observation by Swedish Head of Delegation, that like MPEG, water actually expands after freezing).

During the standardization process the draft specification is 'frozen', which means that no new technology is adopted, unless to repair a broken issue.

e-mail Chad Fogg, 30 October 1998
Subject: Leonardo's Top 10 Pet Peeves
Nr 9. Clinton gets interns, I have SC29.

A common experience when being retired is that it is hard to imagine you ever had time to work. Which explains why for writing this book I could only spend about 30% of my time. But also much more research was needed than originally anticipated; the writing of a book requires more than 'a rough idea' of the solution to a problem: you better be sure. Given that, I hugely underestimated the job; writing this book took about 4 years.

Describing MPEG-2 system technology was a challenge, but doable, though a mistake is easily made. Predicting the future is way more risky, in particular when you are no longer as deeply involved in the networks associated with MPEG-2 systems as you used to be. On the other hand, looking to developments from a certain distance also has advantages: it may provide a fresh view. Nevertheless, for the final chapter on the future of MPEG-2 systems I prepared a kind of disclaimer:

> But always keep in mind the statement similar to the one already made back in the 17th century by Mr. Renè Descartes, a French philosopher, mathematician and writer, who spent most of his adult life in the Dutch Republic: "the only thing to be sure of is doubt".

However, then I got a very positive response from the market research company IHS, who kindly provided market information on MPEG-2 system based products, not only on the past, but also on predictions for the future. This extremely valuable information provided a much firmer basis for statements on the MPEG-2 system future, so that the above disclaimer was no longer needed.

Finally, a major encouragement was provided by the National Academy of Television Arts and Sciences in New York, when MPEG was honoured with a prestigious 2013 Emmy Award for the development and standardization of MPEG-2 transport streams. This Technology and Engineering Emmy Award provides a great acknowledgement of the tremendous impact on the content distribution industry in general and on the television industry in particular of the work performed by the MPEG-2 system experts.

Jan van der Meer,
October 2013

About the Author

Jan van der Meer was born in 1947 in Burgum, a small village in the province of Fryslân in the north of the Netherlands. He received his MS in Electronic Engineering in 1978 from the University of Twente in the Netherlands. Jan has broad interests; during his study he investigated macro-economic options to improve employment by reducing labour charges and at the same time increasing value-added tax, which resulted in two published papers. But Jan decided to pursue a career in electronic engineering by joining Philips in 1979, where he became (co-)inventor of 12 patents.

Throughout his career within Philips, Jan's task has been to interface between Research and Product Development, with the objective to make new technology from Research suitable for products. From this perspective, Jan has been involved in the creation of a series of standards and products for Optical Media, Broadcast, Mobile and Internet.

Jan played a leading role in the MPEG standards committee almost from its very beginning and contributed not only to the development of the MPEG-1, MPEG-2 and MPEG-4 standards, but also to their usage in specific application areas, such as specified by 3GPP, DAVIC, DVB, IETF and ISMA. Jan is acknowledged worldwide for his contributions to MPEG and other standard bodies.

The following lists the most important activities in which Jan has been involved.

- In the early and mid-1980s, Jan developed various consumer product prototypes such as a (hand held) electronic translator, a video editor for VCRs and Camcorders and a Picture in Picture feature to extend TV sets; the latter included the development of an IC.
- The HDMAC System, the analogue High Definition TV System that was developed in Europe in the late 1980s, jointly with other companies and broadcasters in the European Community, in the context of a European Project. Here, Jan initially worked with Philips Research to develop what became the HDMAC coding algorithm, followed by the development of prototype hardware for demonstration purposes and ICs for Philips HDMAC products.
- In 1989, Jan became Manager of the so called Full Motion Video (FMV) Project with the objective to store movies and video clips on a Compact Disc. During this project, he became involved in MPEG, upon which Jan became the FMV System Architect, keeping track of ongoing product development on one hand and MPEG standardization on the other. The results of this project were the FMV extension for CD-I players (implemented on a cartridge) and the Video CD standard that has been very successful in the Far East.

- Around 1993, Jan moved to the TV Group within Philips Consumer Electronics, where he became involved in the development of Digital Broadcast Products. He was involved in defining the architecture of digital TV Set Top Boxes and in the design of ICs for such STBs. Meanwhile he represented Philips in MPEG, where he chaired the MPEG Systems Group from 1994 until 1996 during the completion of the MPEG-2 System standard.
- With a few short interruptions, Jan continued to represent Philips in MPEG until 2007. After chairing the MPEG Systems Group, he became (co-)editor of various MPEG-2 System amendments, such as:
  - Transport over MPEG-2 Systems of MPEG-4 streams;
  - Transport over MPEG-2 Systems of Metadata and MPEG-7 streams;
  - Transport over MPEG-2 Systems of AVC (H.264/MPEG-4 part 10) streams.
  - In MPEG, Jan was furthermore editor of MPEG-4 part 17 on streaming text.
- From the mid-1990s onwards, Jan discussed, promoted and defined the use of MPEG technologies in a large variety of standardization bodies, such as:
  - DVB, the Digital Video Broadcasting organization in charge of defining specifications for digital broadcast services. Jan represented Philips in DVB on issues related to coding of audio and video and on subtitling from 1994 until 2007.
  - DAVIC, a consortium to develop a complete End-to-End System for Video on Demand services. In DAVIC, Jan chaired the group in charge of defining functionalities that DAVIC compliant STBs are required to support.
  - UK DTG, the Digital Television Group in the UK in charge of defining the Digital Broadcast System for the UK; here Jan successfully promoted the use of MHEG-5 as an API.
  - W3C, the World Wide Web Consortium; from 1997 until Philip's withdrawal from W3C in 2003, Jan represented Philips in the Advisory Committee of W3C.
  - IETF, the Internet Engineering Task Force. Within the Audio and Video Transport (AVT) group in IETF, Jan got involved in transport of MPEG-4 streams over IP; he became editor of RFC 3640, which became later the basis for the ISMACryp specification.
  - 3GPP, where Jan promoted the use of MPEG-4 audio and video coding technology for use in mobile applications.
  - ISMA, the Internet Streaming Media Alliance, in charge of defining End-to-End systems for Streaming of Audio and Video over IP, where Jan represented Philips from 2001 until 2008.
  - MPEGIF, the MPEG Industry Forum, in charge of promoting the use of MPEG technology; Jan was a member of the MPEGIF Board of Directors from 2003 until 2007.
- In 2002 and 2003, Jan got involved in licensing discussions on AVC (a.k.a. H.264 and MPEG-4 part 10) to provide 'market feedback' on licensing terms, upon which he joined in 2003 the Intellectual Property and Standardization (IP&S) department in Philips, where he was Director Standardization until 1 July 2008.
- Within IP&S, Jan managed various research projects in Philips related to coding of audio and video.
- From June 2004 to June 2008, Jan chaired the OMA DRM WG, the Working Group in the Open Mobile Alliance that is responsible for the development of the OMA DRM System. Under his responsibility, the OMA DRM 2.0 specification was completed and the OMA DRM 2.1, SRM 1.0 and SCE 1.0 specifications defined. In recognition for his leadership, Jan received in June 2008 the Contributor and Achievement Award from OMA.

On 1 July 2008, Jan retired from Philips; he is currently an independent consultant, located in Heeze, near Eindhoven, the Netherlands.

Publications

- van der Meer, J. (1979) Bruto-winstnivellering ter bestrijding van werkloosheid (in Dutch), Economisch Statistische Berichten. *ESB Jaargang*, **64**(3191), 145–149.
- van der Meer, J. (1980) Arbeidsplaatsenbeleid, sociale verzekeringen en indirecte belastingen (in Dutch), Sociaal Maandblad Arbeid. *SMA Jaargang*, **35**(2), 111–118.
- Vreeswijk, F.W.P., Jonker, W., Leenen, J.R.G.M. and van der Meer, J. (1988) An HD-MAC Coding System. Proceedings of the Second International Workshop on Signal Processing of HDTV, L'Aquila, Italy.
- van der Meer, J., Carey-Smith, C.M., Rohra, K. and Vreeswijk, F.W.P. (1988) Movement Processing for an HD-MAC Coding System. Proceedings of the Second International Workshop on Signal Processing of HDTV, L'Aquila, Italy.
- van der Meer, J., Begas, H.W.A. and Vreeswijk, F.W.P. (1988) The Architecture of an HD-MAC Decoder. Proceedings of the Second International Workshop on Signal Processing of HDTV, L'Aquila, Italy.
- Sijstermans, F. and van der Meer, J. (1991) CD-I full motion video encoding on a parallel computer. *Communications of the ACM*, **34**(4), 81–91.
- van der Meer, J. (1992) The full motion system for CD-I. *IEEE Transactions on Consumer Electronics*, **38**(4), 910–920.
- van der Meer, J. (1993) A derived paper in Japanese is found in 'Data Compression and Digital Transmission', Nikkei Electronics Books, pp. 123–136.
- van der Meer, J. and Huizer, K. (1997) Interoperability between different Interactive Engines, Problems and Ways of Solutions. Symposium Record Programme Production, 20th International Television Symposium Montreux, p. 484.
- van der Meer, J. and Huizer, Cornelis(Koen) M. (1998) MHEG/JAVA Enhanced Broadcasting, the competitive edge. Tagungsband 18. Jahrestagung der FKTG (Ferhseh- und Kinotechnische Gesellschaft e.V), Erfurt, pp. 555–565.
- van der Meer, J. and Kaars, P.B. (2000) The Bridge Between Internet and Broadcast. Proceedings SCTE (Society of Cable Telecommunications Engineers) Conference on Emerging Technologies, Anaheim, CA, pp. 255–263.
- van der Meer, J. (2000) Enhanced Broadcast Services with Complimentary Delivery over IP. Technical Papers 49th Annual NTCA (National Cable Television Association) Convention, New Orleans, LA, pp. 34–39.
- Buhse, W. and van der Meer, J. (2007) The open mobile alliance digital rights management. *IEEE Signal Processing Magazine*, **24**(1), 140–143.

Acknowledgements

The person who made MPEG happen is Leonardo Chiariglione; he was the driving force behind its establishment and he remained the driving force. Leonardo is the only person who attended all MPEG meetings, totalling more than 100. He speaks many languages, including Japanese, and has an amazing capability to move MPEG experts forward in the right direction. The vision and perseverance of Leonardo made MPEG a team that has now worked for more than 25 years on standards for digital audio and video, or in more general terms, for digital media. Thousands of experts from over 25 countries and more than 300 companies participated in the joint effort to consolidate the results of company research in MPEG standards. This collaboration proved extremely efficient and productive, which applies from the perspective of the television and movie industry in particular for the standards for digital video and audio. The dedication of Leonardo Chiariglione to MPEG brought major benefits to the audio-visual industry and deserves the highest respect from everyone with a stake in digital video and audio.

MPEG-2 systems became reality through a joint effort of many excellent experts from industries all over the world, working together to create the best possible standard. When for example major company interests were involved, the discussions became sometimes heated, but were always driven by technology and remained thereby productive, with a remarkably high group comradeship. All participants should be proud of their contributions and of being part of this group. For me, working within a mix of company-political interests and technical arguments has been a great pleasure for which all participants should be acknowledged.

Unfortunately, it is totally impossible to acknowledge each person in MPEG who played an important role in the development of MPEG-2 systems. Nevertheless, with apologies to the people I forget to mention, I would like to specifically acknowledge the following MPEG system key people:

- Sandy MacInnis – chair of the MPEG systems group during the development of MPEG-1 and MPEG-2 systems. Brilliant person, both technically and as a chair. Acted as 'MPEG system conscience' by keeping track of both the 'big picture' and the details. His perfectionism and critical stance fuelled his strong desire for himself and others to be accurate.
- Bernard Szabo –another technically brilliant person; also very persistent; wrote history by continuing to actively attend a critical physical MPEG-1 systems meeting by phone all

day, though he could barely hear the experts present in the meeting room and therefore kept the phone speaker very close to his ear. However, every now and then an opinion expressed by another 'participant by phone' came through in a very loud manner, causing his ear to be 'blown away'. But he did not give up and continued to provide valuable comments.

- Matt Goldman – a highly respected participant. Made an impression on some occasions by lengthy discussions with the chair on the accuracy of his high level statements.
- Juan Pineda – the initiator and proposer of the mathematical formalism used in the STD model of MPEG systems.
- Gary Logston – the 'smart guy in purple', member of the very productive and cooperative Scientific Atlanta team that made many significant contributions to MPEG-2 systems.
- Pete Schirling – during a long period Head of Delegation of the very large United States National Body in MPEG, but also editor of the MPEG-2 system specification. During finalization of the MPEG-2 system standard, the MPEG systems group continued discussions until very late at night. When concluding the meeting somewhere between 3 and 4 a.m., he took care to be awake again, so that he could incorporate the preliminary conclusions of the meeting into the next MPEG-2 systems draft, to ensure that the discussions could be continued in an efficient manner when the group reconvened at 9 a.m.
- John Morris; a brilliant Philips Research colleague and a very nice person. Not only highly respected in systems, but also in video. Played an important role in resolving iDCT accuracy problems in video. Seems to be the source of the NNI joke discussed elsewhere in this book.
- Sam Narasimhan; excellent systems expert and the last of the MPEG-2 System Mohicans. For many MPEG-2 system amendments, Sam and I worked together pleasantly as co-editors. I also would like to acknowledge Sam for his willingness to help in answering questions during the writing of this book.

Furthermore I would like to express my special thanks to Sandy MacInnis for writing the Foreword of this book and to Chad Fogg for his contributions throughout this book, and in particular for the Epilogue he wrote. Also I am grateful to Tom Morrod and Daniel Simmons of IHS for providing the market information used in Chapter 18. I would like to thank my former Philips Research colleague Jean Gelissen for his willingness to take the role of Dutch Head of Delegation in MPEG, and for preparing the statement in the 81st MPEG meeting on my farewell when retiring from MPEG that made me feel proud:

Farewell

WG11 thanks Jan van der Meer for his longstanding and inspiring participation in the MPEG process and the many contributions to the majority of the MPEG standards. WG11 wishes Jan a very nice, joyful and exiting next phase in his life after his retirement.

For reviewing and commenting parts of this book, I would like to thank Sandy MacInnis, Ken McCann, Thomas Schierl and Ellen Mulder, as well as my former Philips colleagues Frans Vreeswijk, Leon van de Kerkhof, Fons Bruls, John Morris and Wiebe de Haan. Furthermore I thank Jeff Heynen, Leo Rozendaal and Vic Teeven for their assistance in finding important information required for this book. And last but not least I would like to thank all the people with whom it was a great pleasure to work, but who I forgot to mention here . . .

Finally, in my own family, I am very grateful to Marjolein, Marijke, Wineke, Imke and Bas who accepted my regular absence from home, taking the additional burden for granted. And, of course, I thank my entire family for their support and encouragement for writing this book.

Jan van der Meer
November 2013

Part One

Backgrounds of MPEG-2 Systems

1

Introduction

What MPEG is, the efforts that MPEG initially undertook, what MPEG-2 Systems is and how it is used by applications.

MPEG, which stands for the Moving Picture Experts Group, is the name of a group of audiovisual coding experts operating in ISO/IEC, see Note 1.1. The MPEG group is responsible for a series of well-known international standards, used for coding of audiovisual information in a digital compressed format, such as MPEG-2 video, AVC, MP3 audio and AAC. MPEG was established in 1988 and had its first meeting in May 1988 in Ottawa, Canada. The first two standards produced by MPEG were MPEG-1 and MPEG-2, published as ISO/IEC 11172 [1–6] and ISO/IEC 13818 [7–18], respectively.

Typically, each MPEG standard contains specifications for compression of audio, for compression of video and for transport and synchronization of compressed audio and video. These specifications are usually referred to as MPEG audio, MPEG video and MPEG systems, respectively, and documented in different parts of MPEG standards. For the parts contained in the MPEG-1 and MPEG-2 standards, see Note 1.1.

From the start, the objective of MPEG was to develop standards for the compression of digital video and audio. The MPEG-1 work item 'Coding of moving pictures and associated audio for digital storage media at up to about 1.5 Mbit/s' expressed a focus on Compact Disc and its bitrate. However, when it became clear that the developed technology was suitable for usage by many applications at a wide range of bitrates, the objective for MPEG-2 was broadened to 'Generic coding of moving pictures and associated audio'. As a consequence of this broadening the MPEG-3 work item on HDTV at high bitrates was dropped: MPEG-3 never happened.

The MPEG-1 standard is successfully used in Video CD, in MP3 audio devices and for coding audio in digital TV broadcast.[1] The MPEG-2 standard is almost universally used in digital cable TV, digital satellite TV, terrestrial digital TV broadcast, DVD, Blu-ray™ Disc,[2] digital camcorders and other families of products.

[1] Many digital TV broadcast systems combine MPEG-1 audio with MPEG-2 video and systems.
[2] Blu-ray™ and Blu-ray Disc™ are trademarks of the Blu-ray Disc Association.

Fundamentals and Evolution of MPEG-2 Systems: Paving the MPEG Road, First Edition. Jan van der Meer.
© 2014 John Wiley & Sons, Ltd. Published 2014 by John Wiley & Sons, Ltd.

Note 1.1 The MPEG Committee and Some of Its Standards

MPEG is a Working Group within a Sub-Committee of a Joint Technical Committee on Information Technology of ISO and IEC; more particularly, MPEG is referred to as ISO/IEC/JTC1/SC29/WG11, that is, WG 11 within SC 29 of JTC 1 of ISO and IEC. The ISO is the International Standardization Organization (see www.iso.org/). The IEC is the International Electrotechnical Commission (see http://www.iec.ch/). The first two standards produced by MPEG were MPEG-1 (1992) and MPEG-2 (1994), published as ISO/IEC 11172 and ISO/IEC 13818; each containing several parts:

ISO/IEC 11172-1	MPEG-1 systems	ISO/IEC 13818-1	MPEG-2 systems
ISO/IEC 11172-2	MPEG-1 video	ISO/IEC 13818-2	MPEG-2 video
ISO/IEC 11172-3	MPEG-1 audio	ISO/IEC 13818-3	MPEG-2 audio
ISO/IEC 11172-4	MPEG-1 compliance	ISO/IEC 13818-4	MPEG-2 compliance
ISO/IEC 11172-5	MPEG-1 software simulation	ISO/IEC 13818-5	MPEG-2 software simulation
		Various other 13818 parts, amongst others:	
		ISO/IEC 13818-9	MPEG-2 real-time interface for system decoders

Initially, MPEG was part of ISO/IEC JTC1/SC2/WG8, the same working group that was developing the JPEG standard; but in 1990, when the subgroups became too large, both MPEG and JPEG were promoted to Working Group level under SC29: JPEG became WG1, while MPEG became WG11.

The focus of this book is on MPEG-2, in particular on MPEG-2 systems. Several of the basic concepts in MPEG-2 systems were developed first for MPEG-1 systems, and therefore MPEG-1 systems also will be addressed to some extent in this book. To understand MPEG-2 systems, some basic knowledge of MPEG video and audio is needed. The audio and video parts of MPEG standards define the format of compressed audio and video streams and how to decode such MPEG audio and video streams back into uncompressed audio and video.

In the audiovisual applications addressed by MPEG-1 and MPEG-2, the MPEG audio and video streams are not transported in parallel, but instead are transported in a single stream that contains both MPEG audio and MPEG video data. Such a stream is called an MPEG-1 or MPEG-2 system stream. The format of system streams as well as the rules and conditions on their construction are specified in the MPEG-1 and MPEG-2 system specifications [2,8].

The MPEG-1 and MPEG-2 systems features include packetization of audio and video streams, their signalling, synchronization of audio and video and requirements for the decoding of audio and video from an MPEG system stream, while ensuring a high quality of service. So

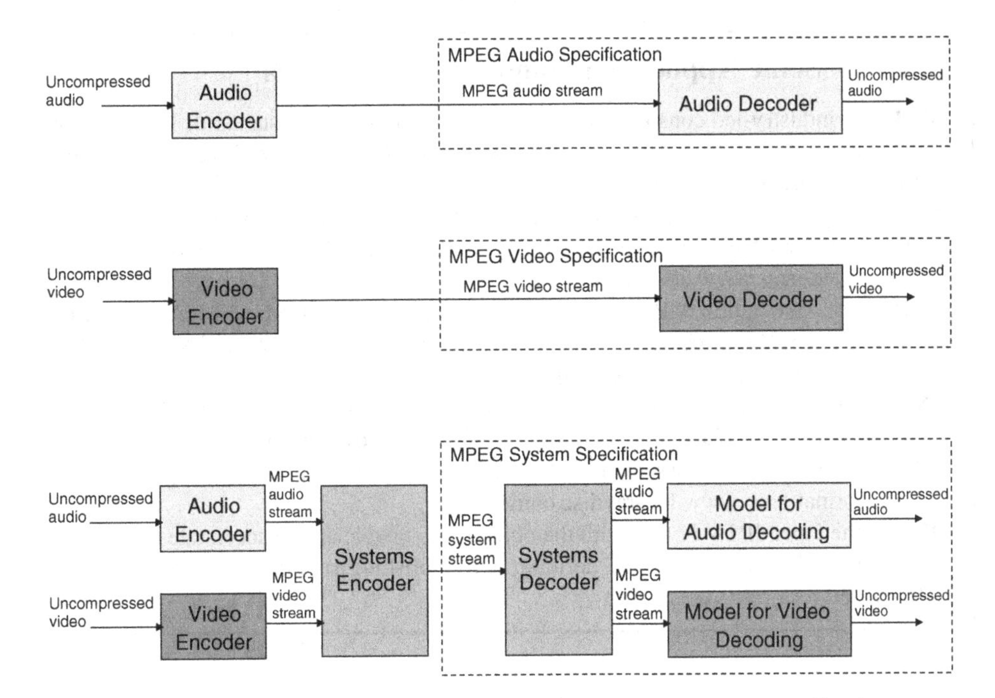

Figure 1.1 Scope of MPEG audio, MPEG video and MPEG system specifications

as to accurately define synchronization of audio and video, the system specifications include a model of audio and video decoding. MPEG does not specify how to perform audio, video and systems encoding; instead only the format of the MPEG audio, video and system streams at the output of the encoders is specified, thereby leaving to the market how to encode audio, video and systems in a most efficient and cost-effective manner (see Figure 1.1).

MPEG-2 systems provide an application independent interchange format, optimized for the target applications, so that mapping to practical transport layers can be made conveniently. For example, the MPEG-2 system specification defines for broadcast applications a transport format that is suitable for transport over terrestrial, satellite, cable and IP networks and that can also be used by recording devices. This approach allows producing and managing content independently of the delivery network to the consumer. Thereby MPEG-2 systems became the basis of an infrastructure to produce, store, exchange and transport audiovisual content.

In order to provide a generally useful interchange format, the MPEG-2 system stream format is designed so that all practical transport requirements of target applications are met. Not only MPEG audio and MPEG video streams can be carried, but also other content, such as subtitling and metadata, as well as audio or video formats defined by other standardization bodies. Moreover, when new audio and video formats evolve, MPEG-2 systems can be extended with support for these new formats, provided that a market requirement for such carriage is identified.

The MPEG committee typically only specifies carriage over MPEG-2 systems of MPEG defined streams. Support for non-MPEG defined streams, such as audio, video and subtitling

Note 1.2 Some Application Standardization Bodies

DVB is an industry-led consortium designing open interoperable standards for the global delivery of digital media services, operating from Europe. 'DVB' stands for Digital Video Broadcasting (see www.dvb.org/).

ATSC is an international organization developing standards for digital television, operating from the United States. 'ATSC' stands for Advanced Television Systems Committee (see www.atsc.org/).

ARIB is the Association of Radio Industries and Businesses, operating from Japan. ARIB aims at establishing technical standards for radio systems in the field of telecommunications and broadcasting (see http://www.arib.or.jp/english/).

DVD Forum is the international organization that defines formats for DVD (Digital Versatile Disc) products and technologies (see www.dvdforum.org/).

BDA is the Blu-ray Disc Association, dedicated to developing and promoting the Blu-ray Disc Format (see www.blu-raydisc.com/).

OIPF is the Open IPTV Forum with the objective to enable and to accelerate the creation of a mass market for IPTV by defining and publishing specifications for end-to-end IPTV services (see www.oipf.tv/).

standards evolving outside of MPEG, is usually beyond the scope of MPEG, and left to other standardization bodies or to applications. A list of important application standardization bodies is provided in Note 1.2.

The typical process for applications to adopt MPEG audio and video standards and the role of MPEG-2 systems therein is depicted in Figure 1.2. When new audio and video standards

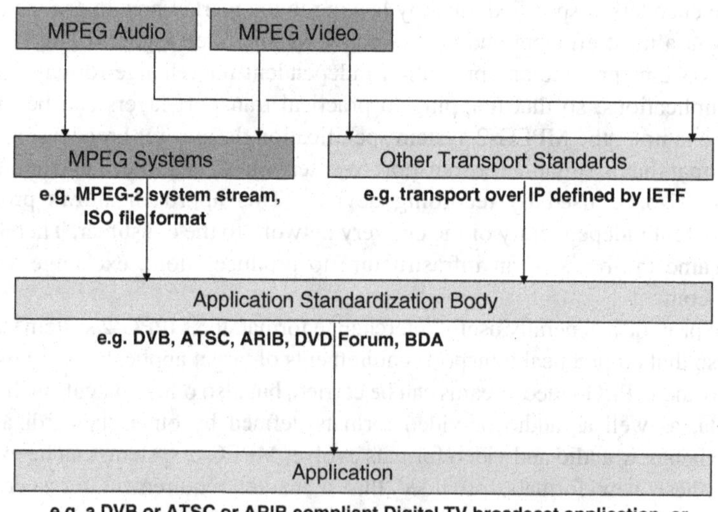

Figure 1.2 Typical process for adoption of MPEG standards

evolve in MPEG, then the MPEG-2 system standard is extended to specify carriage of the new audio or video compressed data within an MPEG system stream. Note that MPEG may also specify how to contain the new audio or video compressed data in an ISO (mp4) file and that other bodies may specify carriage over non-MPEG defined transport means; for example, IETF may specify transport over IP.

Next, an application standardization body can adopt the MPEG-2 systems extension and define guidelines for its usage within the applications governed by that application standardization body. For example, DVB may specify constraints for the use of the MPEG-2 systems extension within DVB compliant applications.

While MPEG audio and MPEG video specifications are usually succeeded by next generation audio and video compression standards, this is different for MPEG-2 systems. As long as MPEG-2 systems and the associated infrastructure are considered suitable for the applications they serve, support for new generation audio and video compression standards will be incorporated by upgrading the MPEG-2 system specification with support for these new standards. One argument here is that often new features can only be introduced in an evolutionary way, without changing the transport format. Another argument is that investing in a totally new infrastructure is usually considerably more expensive than upgrading an existing infrastructure. As a consequence, it is expected that MPEG-2 systems will remain widely used for the foreseeable future, even as video and audio standards evolve.

1.1 The Scope of This Book

This book describes the MPEG-2 system specification as developed in the early 1990s and has evolved since then into the fourth edition of the MPEG-2 systems standard. While MPEG-2 systems will continue to evolve further, this book describes the MPEG-2 system functionality as of October 2013. Also relevant background information is provided. The discussion of MPEG-2 system functionality requires knowledge of various fundamental issues, such as timing, and of supported content formats. Therefore also some basic information on video and audio coding is provided, including their evolution. Also other content formats supported in MPEG-2 systems are described, as far as needed to understand MPEG-2 systems.

Normative requirements that MPEG-2 system decoders and streams have to meet are specified in the MPEG-2 system specification. While MPEG-2 system functionality is described in this book, including clarification of requirements that apply, this book does in no way define normative MPEG-2 system requirements. In other words, this book may be used to improve the understanding of MPEG-2 systems, but not as a guideline for designing encoder and decoder implementations. Furthermore it should be taken into account that the list of requirements discussed in this book is not necessarily exhaustive.

1.2 Some Definitions

Throughout this book, data rates are expressed in units of kb/s and Mb/s, indicating 1000 (10^3) bits per second and 1 000 000 (10^6) bits per second, respectively. However, when the size of digital memory is expressed in units of KB and MB, then units of 1024 (2^{10}) bytes or 1 048 576 (2^{20}) bytes, respectively, are indicated in this book. These notations are commonly used in practice and therefore also in this book. Nevertheless, its usage may create confusion, for example when a data rate is mathematically related to a memory size.

References

1. ISO/IEC (1998) Information technology – Coding of moving pictures and associated audio for digital storage media at up to about 1.5 Mbit/s. ISO/IEC 11172. The MPEG-1 standard is published in five parts; see [2], [3], [4], [5] and [6].
2. ISO/IEC (1993) MPEG-1 Part 1: Systems. ISO/IEC 11172-1:1993. http://www.iso.org/iso/catalogue_detail.htm?csnumber=19180.
3. ISO/IEC (1993) MPEG-1 Part 2: Video. ISO/IEC 11172-2:1993. http://www.iso.org/iso/catalogue_detail.htm?csnumber=22411.
4. ISO/IEC (1993) MPEG-1 Part 3: Audio. ISO/IEC 11172-3:1993. http://www.iso.org/iso/catalogue_detail.htm?csnumber=22412.
5. ISO/IEC (1995) MPEG-1 Part 4: Compliance Testing. ISO/IEC 11172-4:1995. http://www.iso.org/iso/catalogue_detail.htm?csnumber=22691.
6. ISO/IEC (1998) MPEG-1 Part 5: Software Simulation – TR.[3] ISO/IEC 1172-5:1998. http://www.iso.org/iso/catalogue_detail.htm?csnumber=25029.
7. ISO/IEC (2013) MPEG-2 standard, published in 10 parts; see [8], [9], [10], [11], [12], [13], [14], [16], [17] and [18].
8. ISO/IEC (2007) MPEG-2 Part 1: Systems. ISO/IEC 13818-1:2013.[4] http://www.iso.org/iso/catalogue_detail.htm?csnumber=62074.
9. ISO/IEC (2013) MPEG-2 Part 2: Video. ISO/IEC 13818-2:2013. http://www.iso.org/iso/catalogue_detail.htm?csnumber=26797.
10. ISO/IEC (1998) MPEG-2 Part 3: Audio. ISO/IEC 13818-3:1998. http://www.iso.org/iso/catalogue_detail.htm?csnumber=26797.
11. ISO/IEC (2004) MPEG-2 Part 4: Compliance Testing ISO/IEC 13818-4:2004. http://www.iso.org/iso/catalogue_detail.htm?csnumber=40092.
12. ISO/IEC (2005) MPEG-2 Part 5: Software Simulation. ISO/IEC TR 13818-5:2005. http://www.iso.org/iso/catalogue_detail.htm?csnumber=39486.
13. ISO/IEC (2000) MPEG-2 Part 6: Extensions for DSM-CC. ISO/IEC 13818-6:2000. http://www.iso.org/iso/catalogue_detail.htm?csnumber=25039.
14. ISO/IEC (2006) MPEG-2 Part 7: Advanced Audio Coding (AAC). ISO/IEC 13818-7:2006. http://www.iso.org/iso/catalogue_detail.htm?csnumber=43345.
15. ISO/IEC (2007) MPEG-2 Part 8 has been withdrawn.
16. ISO/IEC (1996) MPEG-2 Part 9: Extension for real time interface for systems decoders. ISO/IEC 13818-9:1996. http://www.iso.org/iso/catalogue_detail.htm?csnumber=25434.
17. ISO/IEC (1999) MPEG-2 Part 10: Conformance extensions for Digital Storage Media Command and Control (DSM-CC). ISO/IEC 13818-10:1999. http://www.iso.org/iso/catalogue_detail.htm?csnumber=27044.
18. ISO/IEC (2004) MPEG-2 Part 11: IPMP on MPEG-2 Systems. ISO/IEC 13818-11:2004. http://www.iso.org/iso/catalogue_detail.htm?csnumber=37680.

[3] MPEG-1 part 5 is not a standard, but a technical report that provides a software implementation of the first three parts of the MPEG-1 standard. The source code is not publicly available.

[4] Approximately every 5 years the next version of the MPEG-2 System standard is published, as discussed in Chapter 8 of this book.

2

Technology Developments Around 1990

Technology developments leading to establishing MPEG and the resulting success of MPEG at the market place.

The establishment of MPEG in 1988 and its success in the market place were not coincidental, but the result of a number of important developments. From a technology perspective, digital video compression technology became feasible for widespread market introduction during the 1980s.

During the 1980s, digital video compression technology was standardized for videophone and video-conferencing applications (see Note 2.1). An important objective of this standardization effort was to resolve the critical problem of interoperability between telecommunication equipment from different manufacturers. In about the same timeframe, a multimedia desktop video standard for personal computers was developed.[1] Both standards were designed for the target application, but not for industry wide usage.

With the integration of more memory and video processing power on ICs, it became feasible at the end of the 1980s to exploit more advanced digital video compression tools in a cost-effective manner. This made it possible to achieve an acceptable picture quality at bitrates enabling various new applications, including playback of movies from Compact Disc and networks with about the same bandwidth of 1.5 Mb/s as Compact Disc. With the option to deliver comparable services across different media, the important requirement for the use of a single standard was identified.

The use of a single standard for a wide range of applications enables interoperability between services, which may be critical for a mass market to evolve. Such mass market will result in the high volume production of ICs necessary for significant cost reductions of decoders and encoders. To establish a successful standard for usage on multiple media requires commitment from the industry at large, not only from the telecommunication industry, but also from the

[1] This technology, called 'Digital Video Interactive', was developed by RCA Corporation, the Radio Corporation of America, an American electronics company. In 1986, RCA was acquired by a third party, upon which the DVI technology was sold to Intel in 1988. See for example Wikipedia http://en.wikipedia.org/wiki/Digital_Video_Interactive.

Fundamentals and Evolution of MPEG-2 Systems: Paving the MPEG Road, First Edition. Jan van der Meer.
© 2014 John Wiley & Sons, Ltd. Published 2014 by John Wiley & Sons, Ltd.

Note 2.1 Standardization for Video-conferencing and Videophone Applications

In the end of the 1980s, the telecommunication industry was heavily investing in Narrow Band ISDN (Integrated Service Digital Network). Amongst the foreseen ISDN applications were video-conferencing and videophone; these provided the motivation for the standardization of a video codec for audiovisual services at $p \times 64$ kbit/s, where p takes values from one to more than 20, corresponding to the number of used ISDN channels of 64 kbit/s each. This standardization took place in CCITT, nowadays called ITU-T [1], the Telecommunication standardization sector of the ITU, the International Telecommunication Union [2]. The resulting video coding standard H.261 [3] was ratified in 1988 and published in 1990.

consumer electronics industry, the computer industry and, last but not least, the semiconductor industry.

At the end of the 1980s, the telecommunication industry was well organized in developing and establishing standards for telecommunication applications. However, to reflect the broad industry approach, it was decided not to develop a standard in the traditional standardization bodies for the telecommunication industry, but instead to undertake an effort in ISO to develop a standard for 'video and associated audio on digital storage media', whereby the concept of digital storage media includes communication channels and networks. This effort is known by the name of the group of experts that started it in 1988: the Moving Picture Expert Group, MPEG.

For the Consumer Electronics Industry standards are also important, but they are often established de facto, rather than by committee. Successful examples of de facto standards are the Compact Cassette and the Compact Disc. However, sometimes there are competing solutions, resulting in a 'format war' that may not only be tough and expensive, but that also may fragment, and thereby hinder, the market. One such format war started in the second half of the 1970s on the video tape recorder market. After a 'bloody fight' of almost 10 years, it became clear that the VHS system would become the winner, but at that time there was also broad consensus that such a format war was (and still is) undesirable.

In MPEG, initially the telecommunication industry was present in great strength. However, also the consumer electronics industry was involved pretty much from the start of MPEG, when the so-called 'Full Motion Video' project was running inside Philips, operated in close contact with Sony and Matsushita. The aim of this project was to define a system for playback of movies from Compact Disc and the initial intent to use compression technology developed by Philips. While the developments within the 'Full Motion Video' project proceeded as scheduled, MPEG optimized its video and audio compression technology towards the final MPEG-1 standard, with active participation of Philips and its partners in the project. When the MPEG-1 standard was finalized, Philips and its partners decided to use the MPEG-1 standard in the 'Full Motion Video' project [4], which marked the important milestone for MPEG that resulted in the Video CD standard [5].

The computer industry was also an early participant in MPEG; their objective was to establish an open standard for playback of audiovisual content from hard disc. While the developments in MPEG were getting more and more attention from the market place, the semiconductor industry got involved more closely too. Opportunities for cost-effective integrated silicon solutions became evident, in particular when it became clear that MPEG-2 would offer a very challenging business case to introduce digital broadcast services. For example, in satellite applications, MPEG-2 would allow to broadcast about five digital broadcast services across the same broadcast channel that was used so far for one analogue TV service. In this way MPEG-2 promised to exploit network channels in a significantly more effective way. At more or less the same time, important players in the consumer electronics industry decided to develop a new optical disc format with a higher bitrate than available on Compact Disc. This so-called DVD format [6] could exploit MPEG-2 to distribute high-quality movies to consumers, thereby replacing the use of VHS tapes for this purpose.

All of the above resulted in a highly market driven development of the MPEG-1 and MPEG-2 standards, with a huge participation from all over the globe, representing all major industries with a stake in the audio-visual business at that time. In MPEG meetings often over 500 people participated in the various sub-groups. During the 1990s and in the following years, the MPEG-1 and MPEG-2 standards produced became extremely successful. The Video CD, based on the MPEG-1 standard mainly in the Far East, and DVD and Digital TV broadcast, based on MPEG-2 worldwide. Another major worldwide market success is the MP3 standard for music distribution; the MP3 specification is included in part 3 of the MPEG-1 standard (see Note 1.1).

References

1. ITU (2013) Information on the ITU is found at http://www.itu.int/ITU-T/.
2. ITU (2013) Information on the ITU is found at www.itu.int/.
3. ITU (2013) ITU-T Recommendation H.261, found at: http://www.itu.int/rec/T-REC-H.261/en.
4. van der Meer, J. (1992) The full motion system for CD-I. *TT IEEE Transactions on Consumer Electronics*, **38** (4), 910–920.
5. Wikipedia (2013) Information on the Video CD standard is found at: www.videohelp.com/vcd, http://en.wikipedia .org/wiki/Video_CD.
6. DVD Forum (2013) Information on the DVD standard and the DVD Forum can be accessed at: www.dvdforum .org/forum.shtml. Information on DVD is also found at: http://en.wikipedia.org/wiki/DVD.

3

Developments in Audio and Video Coding in MPEG

Why is video and audio compression needed and how much compression is required. What are the essentials of MPEG-1 and MPEG-2 video and audio compression technology. What are the various picture types used in MPEG video and what is the role of the VBV model. How MPEG-1 and MPEG-2 audio compression use a psycho-acoustic model. What are the building blocks of an MPEG audio and video decoder. How did video and audio compression technology evolve within MPEG.

3.1 The Need for Compression

New technology gets adopted by the market place once it offers added value in a cost-effective and commercially attractive manner. For audio and video compression, huge market interest evolved when it became evident that they would offer opportunities to introduce new applications and to improve the efficiency of existing applications. Two examples are given below.

- Compression of audio and video enables cost-effective distribution of movies and other audiovisual content on optical discs, such as Video CD, DVD and Blu-ray.
- Due to audio and video compression, radio and television broadcast services become more efficient. With a transmission channel that was used for broadcasting a single radio or TV programme in analogue form, now multiple programmes in digital form can be broadcast. Compression thereby offers broadcasters the option to broadcast more programmes to their customers and to reduce the transmission costs per programme. Indeed the use of audio and video compression in broadcast caused an enormous increase in the available audio and TV programmes.

  The bandwidth required for compressed audio and video depends on the conveyed quality of the audio and video. For example, to broadcast high definition video requires more bandwidth than lower resolution video. Hence, by using compression technology, broadcasters will also have the opportunity to trade off the number of transmitted programs and the audio or video quality. Both a higher and a lower quality can be selected; the latter may for

Fundamentals and Evolution of MPEG-2 Systems: Paving the MPEG Road, First Edition. Jan van der Meer.
© 2014 John Wiley & Sons, Ltd. Published 2014 by John Wiley & Sons, Ltd.

example be relevant in case the number of broadcast programs across a transmission channel is commercially more important than the conveyed video quality.

Obviously, applications benefit from compression factors that are as high as possible. However, state of the art compression technology always has its limits in achieving an acceptable quality of the compressed audio and video. At the time MPEG-1 was designed, it was considered acceptable if stereo audio could be compressed to a bitrate of about 0.2 Mb/s and video to about 1.2 Mb/s, achieving a total bitrate of about 1.4 Mb/s, the bitrate of a Compact Disc. The compression factors needed to achieve these bitrates for audio and video are discussed below.

3.1.1 Compression Factors for Audio

In its analogue form, audio is described by a continuous signal with an amplitude that varies over time, corresponding to the frequencies contained in the audio. To compress audio, the signal must be in digital form. For this purpose, digital samples can be taken from an analogue audio signal at a certain frequency. Various digital uncompressed audio formats with associated bandwidths are possible. An audio CD carries stereo (two channel) audio with a sample rate of 44.1 kHz[1] and samples of 16 bits each, consuming $2 \times 44\,100 \times 16 \approx 1.4$ Mb/s. To compress this stereo audio to a bitrate of 0.2 Mb/s needs a compression factor of about 7 (1.4/0.2).

For the MPEG-1 audio standard it proved feasible to achieve to code stereo audio at a bitrate of 192 kb/s at good quality. Due to improved state of the art technology, over time more advanced compression tools become available. Within MPEG, this is reflected in an ongoing evolution of audio standards, capable to compress stereo audio at good quality at bitrates as low as 48 kb/s, thereby achieving compression factors up to about 30.

3.1.2 Compression Factors for Video

Moving video can be thought of as a series of still images, representing scenes in motion. Apart from scene transitions, subsequent images are typically rather similar, to allow the viewer's eye to track moving objects in the scene. In a video signal, each such still image is constructed by means of a large number of horizontal lines. In the early days of television, this concept has been designed to conveniently display video on a Cathode Ray Tube, CRT, the display that was very commonly used in televisions until about 2005.

A CRT uses an electron beam that can be deflected horizontally and vertically to illuminate the fluorescent screen of the CRT, so as to draw the video images line by line. To draw one horizontal line, the intensity of the electron beam is varied corresponding to the brightness of the image to be displayed, while the deflection ensures that the beam travels across the target line. The deflection also ensures that the next line is drawn correctly below the previous one, and that at the bottom of the screen the electron beam returns to the first line at the top of the screen.

When a video system was developed for the above concept, it was found that the number of images per second had to match the frequency of the local electric power system, 60 or 50 Hz. At that time this was the only way to prevent luminance variations in video images due to

[1] Another popular audio sampling frequency is 48 kHz, often used in broadcast applications.

interference between the image rate and the power refresh rate when shooting video in environments with artificial lighting, such as a television studio and a sports stadium at night.

For the given number of images per second and the available bandwidth for the video signal, a compromise was to be found between the horizontal and vertical resolution of each image. The use of more lines per image will increase the vertical resolution, but will decrease the horizontal resolution, as the bandwidth available for the video signal is to be shared amongst more lines. The horizontal and vertical resolutions of the image should be in balance from the perspective of perception by the human eye: the perceived horizontal and vertical resolution in the image should be comparable, certainly in areas the human eye can easily focus on. To achieve a good compromise for the perceived horizontal and vertical resolution at a CRT display, video interlacing is applied.

Interlacing of video is a concept already developed around 1930. Interlace leaves the temporal resolution at 50 or 60 Hz, but increases the perceived vertical resolution by doubling the number of lines drawn on the screen as follows. Interlaced video consists of a sequence of so-called odd and even fields, whereby an odd field contains the odd lines of a picture and an even field the even lines of the picture (see Figure 3.1). Hence a picture contains two times more lines than a field. In areas with no motion, the odd and even fields carry information from the same scene. In such areas, the human visual perception system will perceive the odd and even lines of the scene as part of the same picture, so that the perceived vertical resolution increases. Interlace has no impact on the perceived vertical

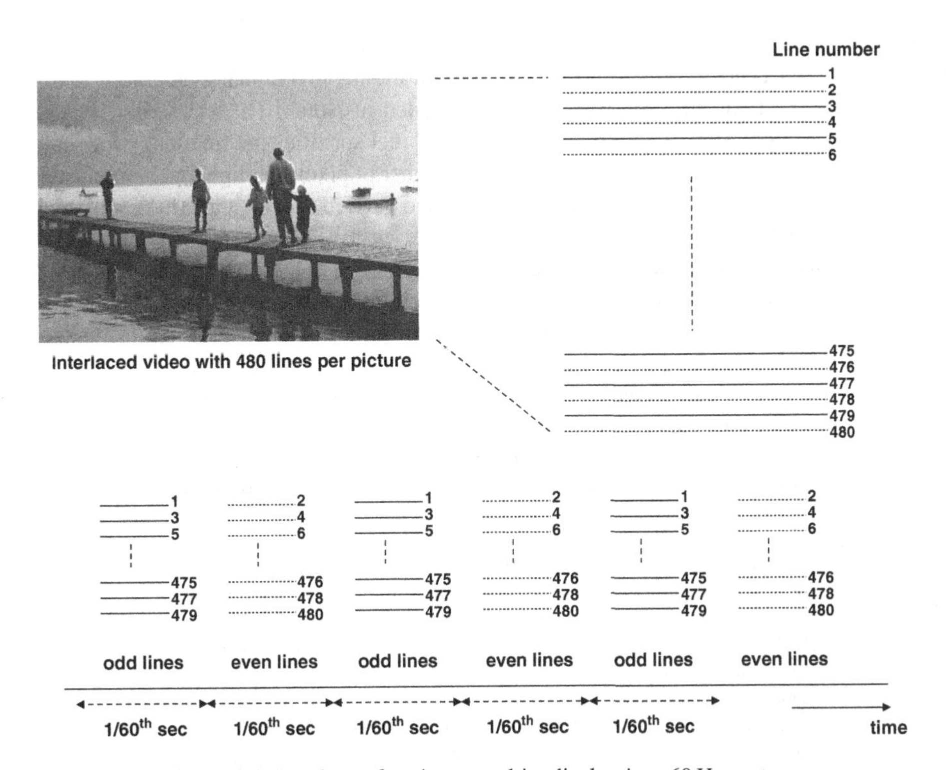

Figure 3.1 Interlace of a picture and its display in a 60 Hz system

resolution in areas with motion; in these areas, the perceived vertical resolution remains determined by the number of lines per field.

The video signal contains 50 or 60 fields per second and therefore the temporal resolution remains 50 or 60 Hz.[2] However, a picture is only complete after two fields and consequently the picture rate is equal to 25 or 30 Hz, respectively. Interlace is used in all deployed analogue television systems, both for black-and-white and for colour television.

In conclusion, two temporal resolutions evolved for television broadcast systems, both using interlace, one at 60 Hz and the other at 50 Hz. In both systems the number of drawn lines per second is approximately the same. The 50 Hz system uses pictures with 576 lines; the number of drawn lines per second equals $25 \times 576 = 14\ 400$. In the 60 Hz system each picture has 480 lines, which results in slightly less than 14 400 drawn lines per second, due to the fact that the 30 Hz picture rate is not exactly 30 Hz, but 30/1.001 Hz ≈ 29.97 Hz. Both television systems are designed so that the line generator for the deflection can operate continuously, also during the 'vertical fly back period' when the electron beam is to be moved from the end of the last line at the bottom of the picture to the beginning of the first line at the top. As a result in the 50 and 60 Hz systems there are in total 49 and 45 so-called non-active line periods per picture,[3] so that the total number of line periods per picture becomes equal to 625 lines in the 50 Hz system and equal to 525 lines in the 60 Hz system. These two television systems are therefore often referred to as 625 lines/50 Hz and 525 lines/60 Hz.

While the above choices were made for black-and-white television, there were no substantial changes to these choices when colour television was introduced. Various colour encoding schemes were defined, but each scheme was backward compatible, so as to ensure that black-and-white television receivers existing in the market could receive the colour television broadcast in black-and-white. Some background on the deployed analogue colour television standards NTSC, PAL and SECAM [1–3] is provided in Note 3.1.

For video compression, a digital format is needed that specifies the frequency for sampling the video lines. The sampling results in a certain number of pixels per line. Each pixel could for example be described by eight bit values of R, G and B (Red, Green and Blue), which would require 24 bits per pixel. From human perception perspective however, this is not an optimal choice; the human eye is much more sensitive on brightness than on colour. For broadcast therefore usually the YUV format is used, with Y being the luminance, that is the brightness, and U and V the colour information. By separating brightness from colour information, the YUV format allows anticipation on the sensitivity characteristics of the human eye; the resolution of U and V can be reduced compared to the luminance resolution, for example by means of sub-sampling the colour information.

The number of pixels per line is addressed in a specification, that was developed around the early 1980s in the radio communication sector of the ITU, ITU-R [4], then called the CCIR. The

[2] This does not apply in the case of movie material shot at the usual picture rate for movies of 24 Hz. For 60 Hz video, each movie picture is scanned during two or three field periods, the so-called 3–2 pull down mechanism, which produces the notorious jerkiness movie reproduction in 60 Hz video systems. For 50 Hz video, the speed of the movie is increased to 25 Hz, so that each movie picture can be scanned during two field periods. As a consequence, compared to a 60 Hz video, in the case of a movie, a 50 Hz video system offers consumers an improved motion portrayal and a shorter movie experience.

[3] Due to the use of interlace there are two vertical fly back periods for one picture, one for each field; the duration of each such fly back period is equal to 22.5 and 24.5 line period in the 60 and 50 Hz systems, respectively. The consequence that fields may commence or end with half a drawn line is way beyond the scope of this book.

Note 3.1 A Brief History of Analogue Colour Television Broadcast

In the 1950s, black-and-white television broadcast was introduced across the world in the consumer domain. Also in the 1950s, the first standard for colour television broadcast was defined: the, at that time, highly advanced NTSC standard, named after the National Television System Committee in the United States that developed it. Unfortunately, the NTSC standard proved weak in conveying colours to receivers in a consistent manner, due to which NTSC is sometimes jokingly referred to as 'Never The Same Colour'.

In the 1960s the PAL and SECAM standards were defined to improve the colour reproduction in receivers. PAL stands for Phase Alternating Line, the method applied in the PAL standard to ensure good colour performance. SECAM stands for SÉquentiel Couleur À Mémoire, French for 'Sequential Colour with Memory', referring to the line memory that is needed in SECAM (and also in PAL) receivers to produce colours.

Various combinations of NTSC, PAL and SECAM with 50 and 60 Hz are possible (and exist), but the most dominant are NTSC at 60 Hz, PAL at 50 Hz and SECAM at 50 Hz. For the worldwide spread of NTSC, PAL and SECAM, see below picture from Wikipedia [3].

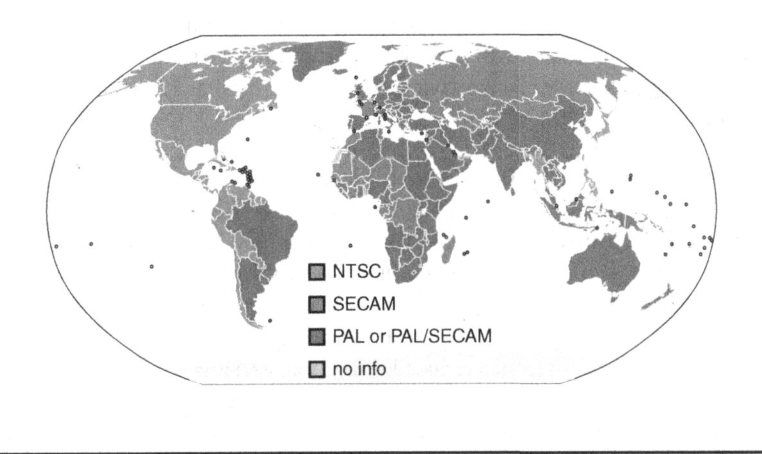

objective was to specify a YUV based digital video standard suitable for usage both in 525 line/60 Hz and in 625 line/50 Hz environments. This resulted in ITU Recommendation 601 [5,6], in which 13.5 MHz is specified as the sampling frequency for luminance Y, and 6.75 MHz as the sampling frequency for the chrominance signals U and V. With these sampling frequencies there are 720 Y samples on a drawn video line, as well as 360 U and 360 V samples.

In Rec. 601, compared to the luminance resolution, the chrominance resolution is halved in horizontal direction only. The use of vertical sub-sampling was not practical for consumer applications in 1982, the time Rec. 601 was published. However, the human visual perception characteristics allow to also halve the chrominance resolution in vertical direction without significant impact on perceived resolution. In more recent years therefore several alternatives for chrominance sub-sampling in both horizontal and vertical direction evolved.

The most relevant options for chrominance sub-sampling are listed below; each option is characterized in terms of 4:*x*:*y*: in a (conceptual) picture region of four horizontal by two vertical luminance pixels, x expresses the number of chrominance pixels[4] on the first line and y on the second line in the region [2,7].

- *4:4:4 format*: no sub-sampling is applied at all; this format is used in some professional applications. In the conceptual region there are four Y pixels, as well as four chrominance pixels at each line, hence eight Y, eight U and eight V samples, in a total of 24 samples.
- *4:2:2 format*: this is the format specified by Rec. 601 with on each line four Y pixels and two chrominance pixels; this format is often used in high-end video applications. In the conceptual region there are eight Y, four U and four V samples, in a total of 16 samples.
- *4:2:0 format*: this format is mostly used in consumer video applications; chrominance is sub-sampled by a factor two both in horizontal and in vertical direction; on the first line there are two chrominance samples, and due to the vertical sub-sampling there are no chrominance samples on the second line in the conceptual region. With this format, in the conceptual region there are eight Y, two U and two V samples, in a total of 12 samples. Thereby the 4:2:0 format reduces the bandwidth of digital video by a factor of two compared to the 4:4:4 format without any sub-sampling.
- *4:1:1 format*: this format with chrominance sub-sampling by a factor four in horizontal direction only is used in some low-end video applications; on each line in the conceptual region there are four luminance samples and one chrominance sample, which results in eight Y, two U and two V samples in the conceptual region, in total 12 samples, the same as for the 4:2:0 format.

The Rec. 601 format with 720 luminance pixels per lines is designed to convey video intended for analogue NTSC, PAL and SECAM television broadcast without loss of picture quality. This format is thereby not only very suitable for digital interfaces at production environments, but also as a format to replace analogue TV by digital TV broadcast at the same picture quality. However, for digital video, other formats are required as well. Some applications require a reduced resolution, while other applications require higher resolutions. For example, low bitrate applications benefit from a reduced resolution, as a low resolution is easier to encode than a higher one; to encode higher video resolutions obviously more bitrate is required.

In practice a limited number of spatial formats is needed. Each format is usually described by its luminance resolution; the chrominance resolution depends on the applied YUV format. In the case of the commonly used 4:2:0 YUV format, the resolution of U and V is half of the luminance resolution, both in horizontal and in vertical direction. As an example, Table 3.1 lists some popular spatial formats.[5] Note however that various other spatial formats are used as well, so that applications can trade-off conveyed picture quality and video bitrate. In Table 3.1, three resolution categories are distinguished: medium resolution, often referred to as 'VHS quality', standard definition (SD) video, equivalent with analogue NTSC, PAL and SECAM resolution, and high definition (HD) video.

To determine the data rate of uncompressed digital video, also the number of bits per sample is important. In practice, each Y, U and V sample often uses eight bits, while more high-end applications use nine or 10 bits per sample. For video compression in the target MPEG-2 applications, the use the 4:2:0 format with eight bits per sample is of most

[4] For each chrominance pixel, one U and one V sample is available.

[5] In Table 3.1 no attention is paid to other relevant video parameters, such as picture rate, picture aspect ratio and whether or not interlace is applied.

Table 3.1 Examples of popular spatial formats used for digital video

Medium ('VHS quality')	360 pixels by 240 lines 360 pixels by 288 lines
Standard definition (SD)	720 pixels by 480 lines 720 pixels by 576 lines
High definition (HD)	1920 pixels by 1080 lines 1280 pixels by 720 lines

interest. In a 525 line/60 Hz environment, digital video with pictures of 720 pixels by 480 lines based on the 4:2:0 format with eight bits per sample will contain the following per picture:

- 720×480 Y samples,
- 360×240 U samples,
- 360×240 V samples.

Hence, with eight bits per sample, each such picture consumes $720 \times 480 \times 8 + 2 \times (360 \times 240 \times 8) = 4.1472 \times 10^6$ bits. The associated picture rate is equal to 30 Hz, which results in a digital video bitrate of about 124×10^6 b/s = 124 Mb/s. Similarly, in a 625 line/50 Hz environment, digital video with pictures of 720 pixels by 576 lines based on the 4:2:0 format with eight bits per sample will consume per picture $720 \times 576 \times 8 + 2 \times (360 \times 288 \times 8) = 4.97664 \times 10^6$ bits. The associated picture rate is equal to 25 Hz, which results also in a digital video bitrate of about 124 Mb/s.

The target of video compression in MPEG-2 was that for video in the above format on average no more than about a 6 Mb/s should be needed. To achieve this bitrate or less, MPEG-2 video is required to compress the video at least with a factor of 124/6, so by a factor of roughly 20.

3.2 MPEG Video

3.2.1 Introduction

Compression of video by a factor of 20 or more means that on average each 120 bytes of uncompressed video data are represented by at most six bytes of compressed video data, while ensuring good picture quality. Some pictures are easier to compress than others, but depending on the video content, compression with a factor of 60 is proven to be achievable for MPEG-2 video, while the latest MPEG-4 video standard, AVC, also known as H.264, even provides a compression factor up to about 120. In which case 120 bytes of uncompressed video data are represented by only a single byte of compressed video data. In Note 3.2, the main targets and the achieved compression factors are discussed of the video standards developed in MPEG until 2007.

To achieve the high compression factors, MPEG video utilizes a wide variety of tools to reduce redundancy in the video signal. Significant redundancy reductions can be achieved both spatially within a picture, as well as temporally between pictures in a sequence.

Note 3.2 MPEG Video Standards, Their Main Focus and Compression Factors

The main focus of MPEG-1 video (finalized in 1992) [8] is non-interlaced medium resolution video, such as 360 pixels by 288 or 240 lines at a picture rate of 25 or 30 Hz, respectively. This format results if an SD (standard definition, see Table 3.1) video signal is sub-sampled both horizontally and vertically by a factor of two; only 360 pixels per line remain, all from one field; due to the vertical sub-sampling, only the odd or the even lines remain; the lines from the other field are no longer present and consequently there is no interlace anymore. This data rate of this medium resolution video signal is equal to 124×10^6 bit/s/4 = 31×10^6 bit/s. For such video signal, the target bitrate for MPEG-1 video was 1.2×10^6 bit/s. To achieve this, on average a compression factor is required of 31/1.2, about 25.

The MPEG-2 video specification (finalized in 1994) [9] extends MPEG-1 video with tools for the efficient compression of interlaced pictures, as required for Standard Definition (SD) and High Definition (HD) video. The compression factor of MPEG-2 video is comparable to MPEG-1 video; depending on the video content, a compression factor in the range between 20 and 60 is achieved.

MPEG-4 video, finalized as MPEG-4 part 2 in 1998 [10], is not intended as a successor of MPEG-2 video. Instead, it focuses on very low bitrates for use on the Internet; at quality levels required by MPEG-2 target applications, MPEG-4 video is slightly (10–20%) more efficient than MPEG-2 video.

The AVC (Advanced Video Coding) standard, published as MPEG-4 part 10 in ISO/IEC and as H.264 in ITU-T [11], was finalized in 2003 and is the successor of MPEG-2 video, typically providing an improvement of the compression factor by two or more compared to MPEG-2 video. AVC provides thereby a compression factor in the range between 40 and 120, depending on the video content.

3.2.2 MPEG-1 and MPEG-2 Video Essentials

To reduce spatial redundancy within a picture, the MPEG-1 and MPEG-2 video standards use a DCT, Discrete Cosine Transform, based on spatial blocks with 8 by 8 samples each. For using 8×8 blocks, each picture is subdivided into slices, that is horizontal rows of consecutive so-called macroblocks, whereby each macroblock contains 16 luminance pixels from 16 lines, as well as the spatially corresponding colour samples. Macroblocks are adjacent to each other and do not overlap. In case of the 4:2:0 YUV format, the YUV samples within each macroblock are organized in six 8×8 blocks: four 8×8 blocks with Y samples, one 8×8 block with U samples and one 8×8 block with V samples (see Figure 3.2).

A video signal can be described equally well by means of samples in the time domain and by its spectrum in the frequency domain. Like any other Fourier-related transform, a DCT converts a time domain description into a frequency domain description. Input to the DCT are samples of the video signal and at its output, the associated frequency components are available. By means of an inverse DCT (iDCT), the input samples can be reconstructed again

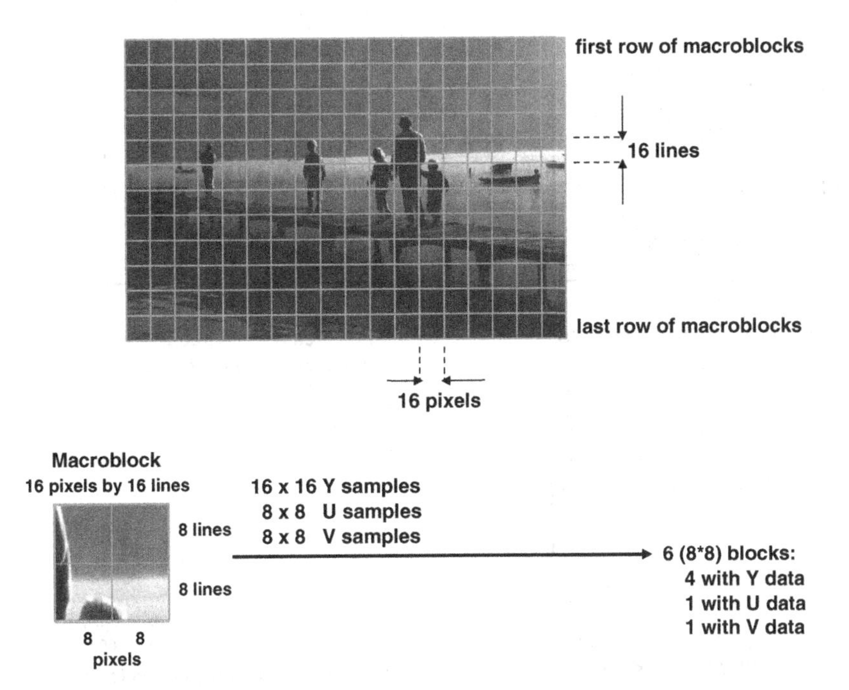

Figure 3.2 Subdividing a picture into slices with macroblocks and 8×8 blocks

from the frequency components. The DCT and iDCT operations are lossless if performed accurately; but in practice, quantization effects cause some errors.

The DCT converts the 64 samples from an 8×8 block into an 8×8 array with 64 frequency coefficients (see Figure 3.3). The first coefficient, indicated by index 0,0 in Figure 3.3 is called the DC (zero frequency) coefficient, all others are called AC coefficients. The DC coefficient represents the average value of the samples in the block; the AC coefficients with increasing vertical and horizontal index values represent higher vertical and horizontal spatial frequencies. The DCT is useful for compression because of its strong 'energy compaction' property: most of the signal information tends to be concentrated in a few low-frequency components of the DCT. A video signal can therefore more efficiently be described in the frequency domain than in the time domain. The frequency coefficients at the output of the DCT are described at an accuracy practical for implementations, for example with 12 bits per coefficient. The caused quantization errors result in some loss of picture quality; however, the quality loss would be very marginal if all 64 frequency coefficients at the output of the DCT would be provided directly to the input of an inverse DCT.

The objective is to convey the coefficients at the output of the DCT in the MPEG video stream to decoders. Each coefficient is accurately available at the output of the DCT, but for transport in the MPEG video stream per coefficient only a limited number of bits is available, depending on the available bitrate for the video stream. Therefore the next step in compressing the picture is to quantize the frequency coefficients at the DCT output. By quantizing them, the coefficients are conveyed in a less accurate manner, thereby introducing quantization errors. If a relatively high bitrate is available, then major quantization errors can be avoided, but at lower bitrates, less bits are available to convey the coefficients, upon which a more coarse quantization must be applied, resulting in larger quantization errors that may impact the picture quality.

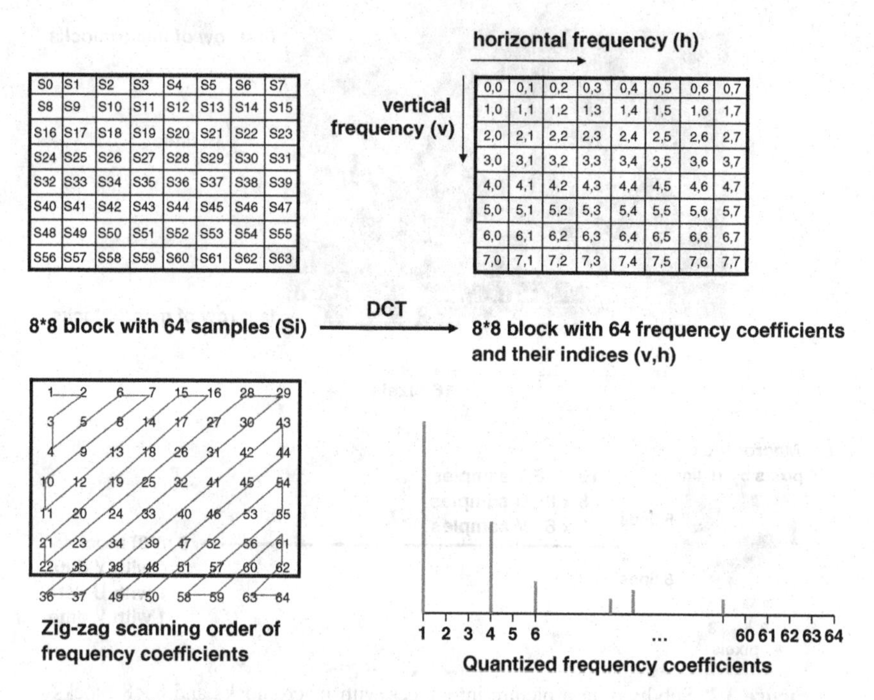

Figure 3.3 DCT transform, scanning in zigzag order and quantized coefficients

The human eye is quite sensitive to large area luminance errors and therefore the accuracy of the DC coefficient is treated very carefully. For the AC coefficients adaptive quantization is applied, depending on the sensitivity of the human eye for quantization errors. At the discretion of the encoder, in areas where the human eye is less sensitive for errors, a more coarse quantization can be applied so as to allow for a more accurate quantization in areas[6] where it is needed most.

In many 8×8 blocks, most quantized DCT coefficients have the value zero and normally, there are only a few non-zero coefficients, which is one of the main reasons why the DCT compression works as well as it does. After the adaptive quantization, the coefficients are scanned in the zigzag order indicated in Figure 3.3. As most quantized frequency coefficients have the value zero, in scan order, a non-zero coefficient is typically followed by a number of coefficients that are zero.

When considering the statistical distribution in 8×8 blocks of the number of zero coefficients prior to a non-zero coefficient, it shows that some [number of zero coefficients, non-zero coefficient value] combinations occur frequently, while other combinations are rare. This characteristic is exploited by coding the combinations with a variable number of bits, the so-called variable length coding (VLC). Frequently occurring combinations are coded with only a few bits, and rare combinations with more bits, resulting in spending on average less bits to convey the coefficient values, thereby further reducing the bitrate. The VLC code tables,

[6] The JPEG still picture coding scheme uses a very similar quantization scheme, but with a fixed quantization in the coded picture, so without the capability within one picture to take the sensitivity of the human eye for quantization errors into account. This makes the coding of still pictures in MPEG-1 and MPEG-2 video slightly more efficient than in JPEG.

expressing the value of a non-zero coefficient and the number of preceding zero coefficients are standardized by MPEG.

In summary, to compress the video the encoder takes the following steps to implement the above:

- the Y, U and V samples of a video signal are structured into horizontal slices of adjacent macroblocks;
- the 64 samples from each 8×8 block are converted by the DCT into 64 frequency coefficients;
- the 64 DCT coefficients are quantized with a variable accuracy, depending on the available bandwidth;
- the quantized DCT coefficients are VLC encoded;
- the compressed bitstream is created, containing the applied quantization levels and the VLC codes.

The video decoder reconstructs the video by processing the compressed bitstream in reverse order compared to the steps taken in the encoder. First, the VLC decoder reconstructs the quantized coefficients from the VLC tables standardized by MPEG. Next, the DCT coefficients are determined by the de-quantizer, based on the output from the VLC decoder and on the quantization level used for these coefficients, as signalled in the bitstream. Finally, the inverse-DCT, also known as i-DCT, converts the 64 DCT coefficients back into 8×8 samples, which is followed by converting the Y, U and V samples from the macroblock into a video signal. Figure 3.4 depicts the processing of the above in the video decoder.

The spatial redundancy reduction achieved by the above described combination of DCT, adaptive quantization and VLC coding achieves typically a compression factor of roughly 6 to 8. To accomplish the compression factors given in Note 3.2, also the temporal redundancy must be reduced significantly. This is practically possible because of the usual similarities in subsequent pictures in a video sequence. As an example, Figure 3.5 shows three pictures from a video sequence of several seconds, showing a ship approaching the coast; the ship is displaced in each picture, but the appearance of the ship does not change much, and neither does the background.

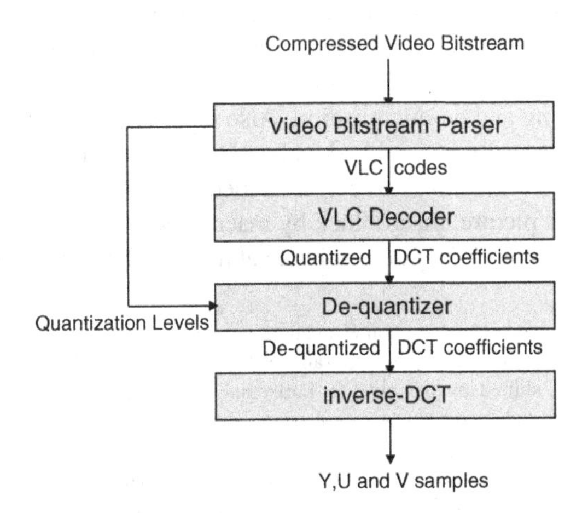

Figure 3.4 VLC decoding, de-quantization and i-DCT processing

Figure 3.5 Example of three pictures from a video sequence of several seconds

This example with little changes over a significant temporal distance may not be very typical. However, it illustrates that it is beneficial, instead of coding each picture individually, to code differences between pictures, while taking motion that may occur from picture to picture into account. In practice, video sequences are produced so that the human eye can track what happens and therefore video sequences contain a lot of temporal redundancy. As a consequence, differential coding of pictures provides a significant contribution to video compression.

To reduce the temporal redundancy between pictures, motion compensated predictive coding is used. For this purpose, the same picture segmentation into macroblock slices is applied as depicted in Figure 3.2. The objective is to predict each macroblock from a previously coded picture that serves as the reference picture for the motion compensation. Typically, a perfect prediction is not possible, and therefore the technology focuses on finding the best match and to code the difference between the best match and the actual picture. For this purpose, the encoder applies for each macroblock motion estimation, usually on the luminance,[7] so as to determine the best match, with a spatial accuracy of 0.5 pixel,[8] with a 16×16 block[9] in the reference picture (see also Figure 3.6). Next, on a sample by sample basis, the macroblock to be coded is subtracted from the best matching 16×16 block from the reference picture and instead of coding the original samples in the macroblock, the sample differences are coded, thereby typically improving the compression efficiency for such pictures by roughly a factor of two.

At the encoder, the sample differences are input to the DCT; after quantization, the associated DCT coefficients are VLC encoded and put in the video bitstream. To be able to perform the inverse operation, the decoder needs to know the motion vector applied by the encoder for the best match. Therefore the encoder also put an encoded version of that motion vector in the video bitstream. At the decoder, the sample differences are reconstructed by the iDCT after VLC decoding and de-quantization. Also the associated motion vector is decoded. Within the reference picture, the macroblock pointed to by the motion vector is retrieved by the decoder, and the samples of that macroblock are added to the reconstructed sample differences to construct the decoded picture, macroblock by macroblock. The original reference picture at the input of the encoder is obviously not available at the decoder; the decoder uses the decoded

[7] MPEG does not specify how to perform motion estimation, but leaves this as an encoder feature to the market.

[8] For this purpose not only the original 256 luminance samples from a macroblock in the reference picture are needed; also the 'in-between pixels', shifted by 0.5 pixel in horizontal and/or vertical direction, are required; to avoid mismatches between encoders and decoders, these 'in-between pixels' are calculated using an interpolation scheme defined in the MPEG video standard.

[9] This applies to non-interlaced video. In case of interlaced video, the luminance samples within a macroblock may be organized as two 8×8 blocks from one field and two 8×8 blocks from the other field; for the prediction of such macroblocks two motion vectors are used, one for each 16×8 luminance block per field.

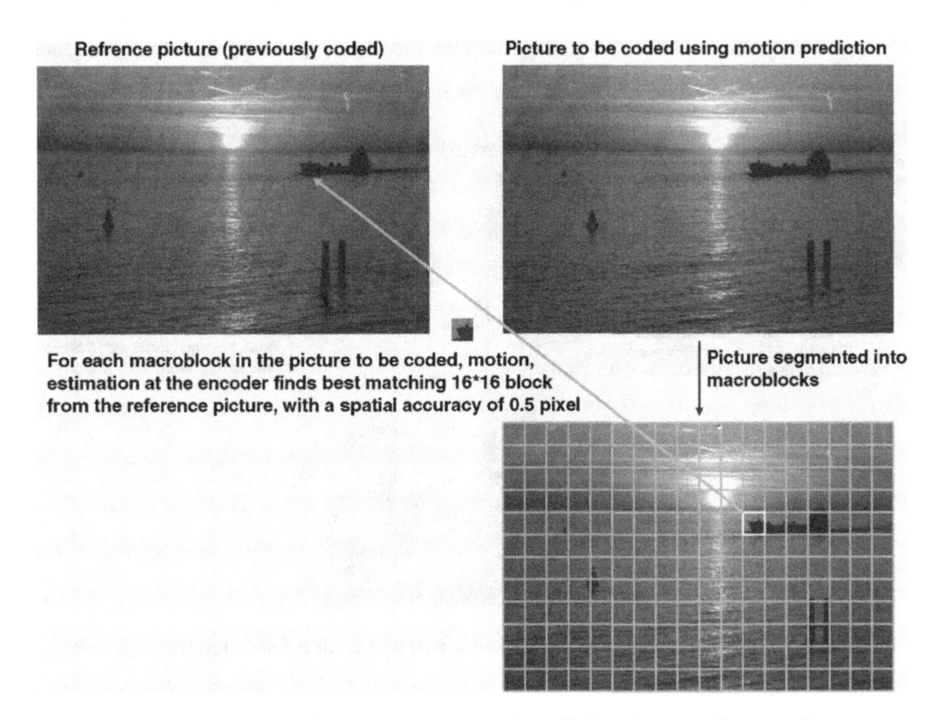

Figure 3.6 Example of motion compensated prediction from a reference picture

version. To avoid error propagating prediction mismatches between encoder and decoder, also the encoder must use decoded instead of original reference pictures. Hence, for this purpose an encoder must include a decoder as well.

A further method to reduce the temporal redundancy between pictures is the use of two reference pictures for the so-called bi-directional prediction, whereby one reference picture is from the past and the other one from the future. For example, in Figure 3.7 three pictures are depicted, in the middle the picture to be bi-directionally predicted and on the left and right side the two reference pictures, the left one from the past and the right one from the future. For each macroblock in the bi-directionally predicted picture, at the encoder the best matching 16×16 block from each reference picture is determined using motion estimation with a spatial accuracy of 0.5 pixel. To code each macroblock from the picture in the middle, the best match from the previous reference picture can be used, or the best match from the next reference picture, or the average[10] from both matches. The encoder decides which one to use. Next, the macroblock to be coded is subtracted from the 16×16 block decided by the encoder and instead of coding the original samples in the macroblock, the sample differences are coded.

To implement the above described reduction of temporal redundancy, while allowing good random access, three picture types are defined in MPEG-1 and MPEG-2. These picture types[11] are: I-pictures, P-pictures and B-pictures; they can be characterized as follows.

[10] For this purpose the samples of corresponding positions in the two 16×16 blocks are averaged.

[11] In MPEG-1 video a fourth picture type is defined: D-pictures, which are I-pictures with only DC coefficients and no AC coefficients. The D-picture concept was intended for fast search purposes, but did not prove particularly useful, as this feature can also be implemented by only decoding I-pictures. Consequently, D-pictures are not supported in other MPEG video standards.

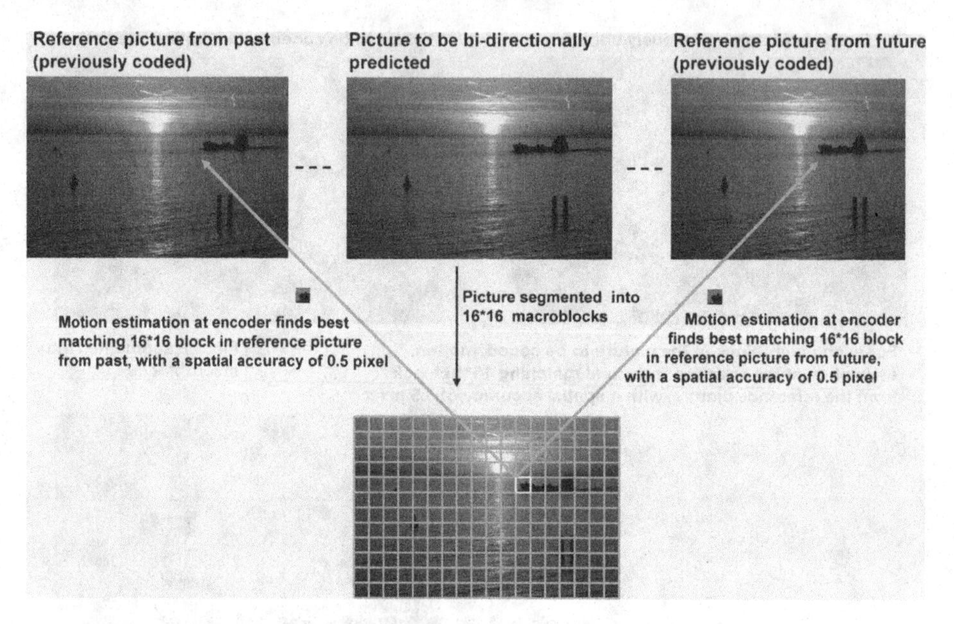

Figure 3.7 Bi-directional prediction from a past and from a future reference picture

- I-pictures, also known as intra-coded pictures; these pictures are compressed using information only from itself, so without using information from other pictures, based on the method depicted in Figures 3.2 and 3.3. In I-pictures only spatial redundancy is reduced and therefore I-pictures are only moderately compressed, typically by a factor between six and eight.
- P-pictures, also known as predictive-coded pictures; these pictures are coded using prediction from a preceding reference picture as depicted in Figure 3.6. The reference picture may be an I-picture or another (preceding) P-picture. Of course, the use of predictive coding may not be efficient for each macroblock, for instance if the macroblock contains video information not present in the reference picture. In such case that macroblock can be intra-coded in the same way as in I-pictures. Typically however, the use of predictive coding in a P-picture improves the coding efficiency significantly, roughly by a factor of two compared to an I-picture.
- B-pictures, also known as bidirectionally predictive-coded pictures in which bi-directional prediction from a past and from a future reference picture is applied, as depicted in Figure 3.7. The past as well as the future reference picture may be an I-picture or a P-picture. B-pictures are not used as a reference picture for coding of subsequent pictures and therefore the picture quality of B-pictures is less critical.[12] When more efficient, also in B-pictures macroblocks can be intra-coded. Compared to P-pictures, the use of bi-directional prediction in B-pictures roughly improves the coding efficiency with a factor of two.

[12] This is due to the absence of error propagation; errors in I-pictures and P-pictures propagate when the erroneous video is used for the prediction of subsequent P-and B-pictures; B-pictures are not used for the prediction of other pictures and therefore errors in B-pictures do not propagate.

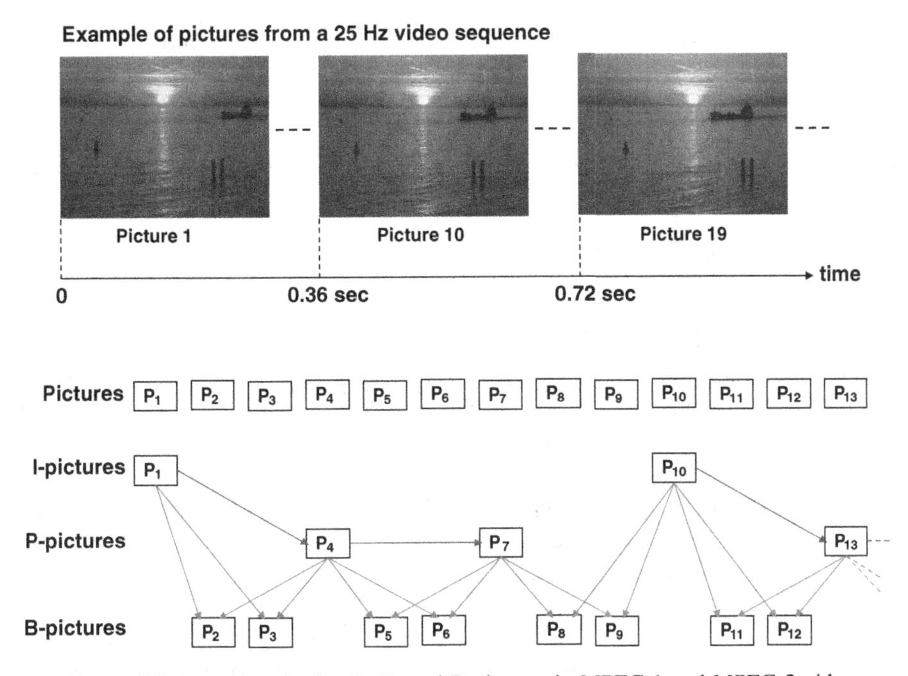

Figure 3.8 Example of using I-, P- and B-pictures in MPEG-1 and MPEG-2 video

As P- and B-pictures are reconstructed from reference pictures, the decoding of an MPEG video stream can typically not start at such pictures, but only at I-pictures.[13] For many applications, such as Digital TV broadcast, a long delay to access an MPEG video stream is not acceptable, and therefore typically at least each 0.5 s an I-picture is coded. For example, in the 25 Hz video sequence of Figure 3.8, picture 1 is coded as an I-picture, and so is picture 10; hence their temporal distance is equal to 0.36 s. The coding of the pictures is at the discretion of the encoder; in Figure 3.8, the encoder uses a scheme whereby each third picture is coded either as I-picture or as P-picture, while using these I- and P-pictures as reference pictures for the coding of both B-pictures in between.

At the decoder, the B-pictures can only be reconstructed if both reference pictures used for its compression are reconstructed first. For example, in Figure 3.9, B-pictures P_2 and P_3 are predicted from I-picture P_1 and P-picture P_4; therefore P_1 and P_4 must be decoded prior to the decoding of P_2 and P_3. For this reason, the MPEG Video stream with B-pictures does not contain the coded pictures in their natural order; the pictures must be reordered. For example, in Figure 3.9, the order of the coded pictures in the MPEG video stream is: $P_1 - P_4 - P_2 - P_3 - P_7 - P_5 - P_6 - P_{10} -$ and so on. The decoded pictures need to be presented at the output of the decoder in natural order, and therefore also at the decoder re-ordering is needed. As a consequence, a picture re-order buffer is needed, both at encoding and at decoding, as depicted in Figure 3.9.

At the encoder, pictures need to be re-ordered to ensure that a reference picture following one or more B-picture(s) is coded prior to the coding of these B-picture(s). For example, in

[13] It should be noted though that there are some strategies based on intra-coding of macroblocks in certain areas of P-pictures that enable random access without the use of I-pictures.

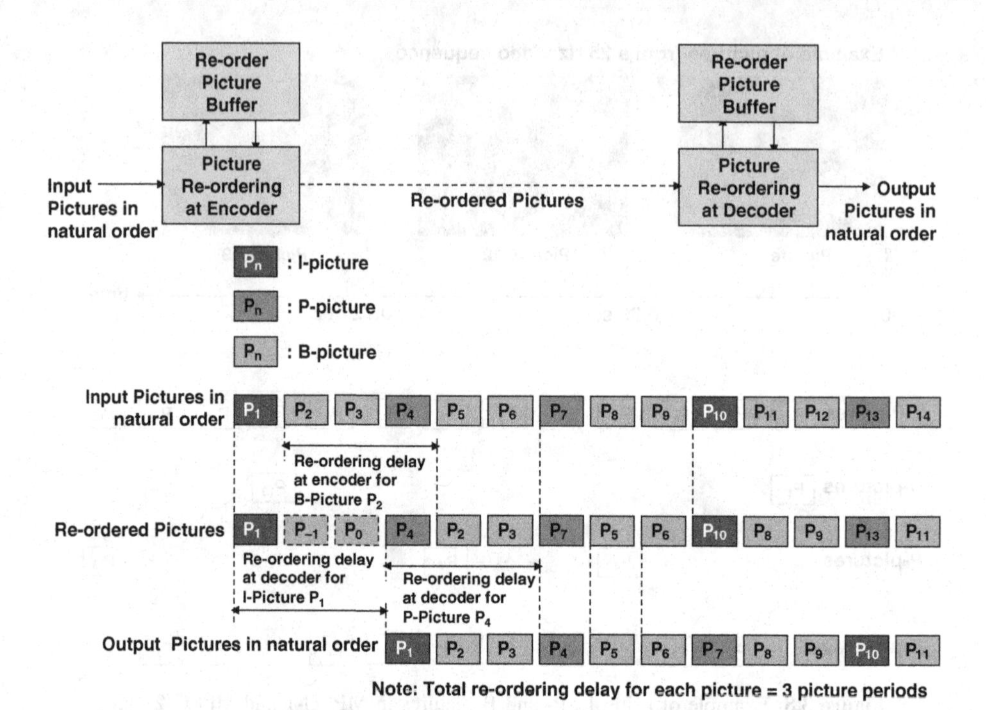

Figure 3.9 Picture re-ordering at encoder and decoder as required for MPEG Video

Figure 3.9, pictures P_2 and P_3 need to be delayed so that P_4 can be coded prior to P_2 and P_3. At the decoder, the pictures need to be re-ordered back to their natural order, in particular the reference pictures that have been moved forward by the encoder. In Figure 3.9, P_1, P_4, P_7, P_{10} and P_{13} are delayed until after the preceding B-pictures, so that the natural order is restored.

The re-ordering causes a delay for each picture; in Figure 3.9, at the encoder each B-picture is delayed by three picture periods and processed without any delay at the decoder, while the reference pictures are not delayed at the encoder, but delayed by three picture periods at the decoder. As an obvious result, the total re-ordering delay at encoding and decoding for each picture is equal to three picture periods in Figure 3.9. For a further discussion on re-ordering processing and delay see Chapter 6.

The actual number of bits for each picture depends on many variables, such as the required picture quality, the available bitrate and the characteristics of the input video. Very roughly however, the following estimate can be made:

- A coded P-picture contains about two times more bits than a coded B-picture.
- A coded I-picture contains about two times more bits than a coded P-picture.

With the above estimate, the size of a coded I-picture would be equal to twice the size of a P-picture, and four times the size of a B-picture. The compression factor for an I-picture is roughly equal to 6–8, which would mean that the compression factor for a P-picture would be about 14, and for a B-picture about 28. For random access reasons, there may be two I-pictures per second. A video sequence with 30 pictures/s can be encoded so that each I-picture and each

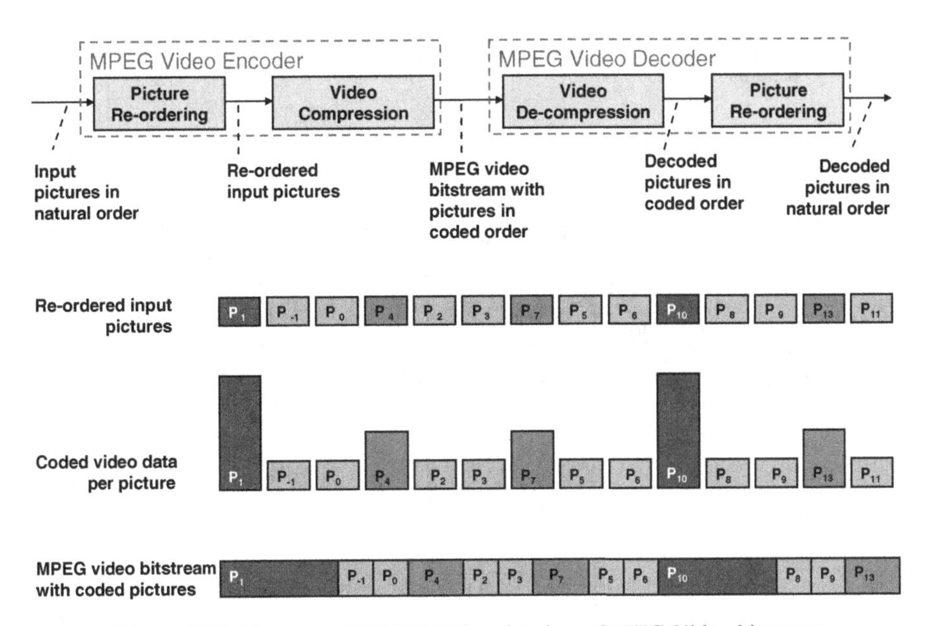

Figure 3.10 Compressed MPEG Video data in an MPEG Video bitstream

P-picture is followed by two B-pictures, as in Figure 3.9. If there are two I-pictures/s, then there would be eight P-pictures and 20 B-pictures per second. This would result in an average compression factor equal to $(2\times7+8\times14+20\times28)/30$, which is equal to about 23. While this is a reasonable typical value, in practice the achieved compression factor varies widely, roughly in the range between 20 and 60.

Due to the different sizes of the coded pictures, in an MPEG video stream with a constant bitrate, it typically takes less time to transport a coded P-picture than a coded I-picture, while a coded B-picture takes even less time. Figure 3.10 provides an example. In Figure 3.10 it is assumed for simplicity reasons that each coded I-picture is twice as large as a coded P-picture and four times as large as a coded B-picture. The MPEG video stream has a constant bitrate and therefore in this simplified case the transport in the MPEG video stream of an I-picture takes four times more time than the transport of a B-picture and two times more time as the transport of a P-picture. It should be noted however, that in practice the amount of bits allocated to a picture varies widely,[14] resulting in large variations of the 'location' of the coded picture data within an MPEG video stream.

The varying amount of data per coded picture introduces a delay, both at compression and decompression. Figure 3.11 shows the delays caused for the video stream from Figure 3.10. At the input of the compression and the output of the decompression the pictures are in coded order; see also Figure 3.10. For each picture P_n, the delay $T_E(P_n)$ at encoding is equal to the delay between the beginning of P_n at the input of the compression and the beginning of P_n in the MPEG video stream. Likewise, the delay $T_D(P_n)$ at decoding is equal to the delay between

[14] The number of bits allocated to a picture is an important factor in optimizing the picture quality; by not specifying how to allocate bits to pictures, the MPEG standard leaves the development of bit allocation strategies to the market for competing MPEG video encoders.

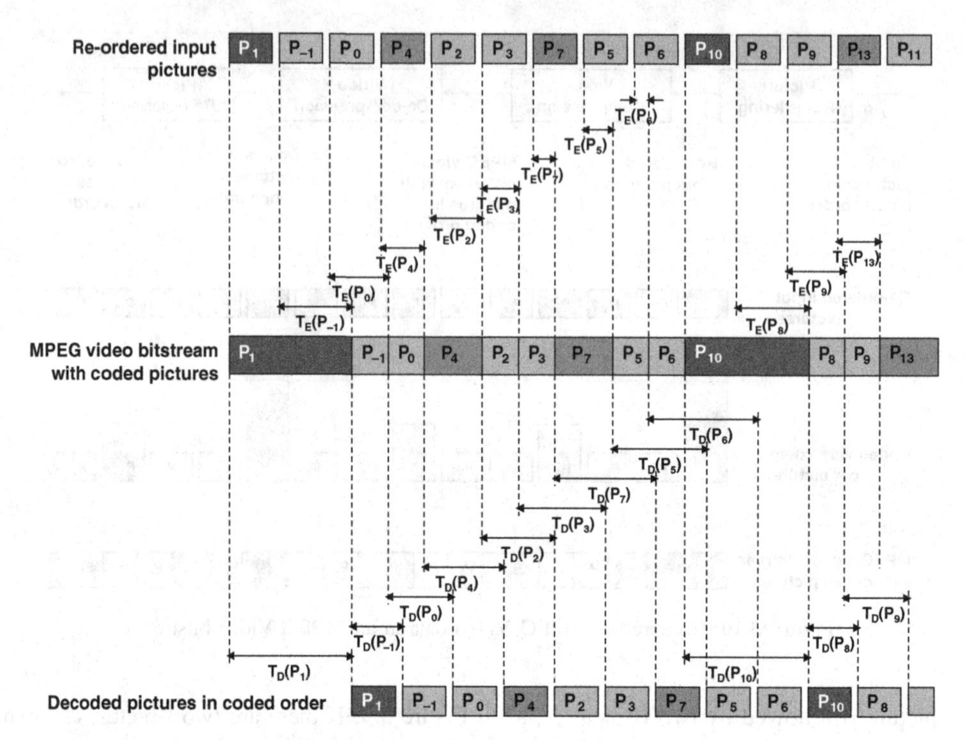

Figure 3.11 Delay caused by video compression and decompression

the beginning of P_n in the MPEG video stream and the beginning of P_n in the output of the decompression.

Figure 3.11 shows that the values of the delays at the encoder and at the decoder differ per picture, depending on its coded data size; the total delay is at least equal to the time it takes to transport the largest coded picture in the MPEG video stream. For picture P_1, the delay at compression $T_E(P_1)$ is equal to zero, because the first coded data of P_1 is immediately put in the video stream upon arrival of P_1. For picture P_{-1}, the delay $T_E(P_{-1})$ is considerable, because the compression has to wait with putting the coded data of P_{-1} in the video stream until the transport of the coded data of P_1 in the video stream is finished. On the other hand, the delay $T_D(P_{-1})$ is considerably less than $T_D(P_1)$. For each picture the total delay caused by compression and decompression is the same, so the value of $T_E(P_n) + T_D(P_n)$ is the same for each picture P_n.[15]

The above processing requires storage of coded picture data, both at the encoder and at the decoder. At the encoder the coded data of a picture need to be stored until it is transported in the video stream, while at the decoder the coded data received from the video stream need to be stored until it is decoded. It is the responsibility of an encoder to produce a compliant MPEG video stream, and the required amount of buffering to achieve that is at the encoder's discretion. To successfully play a compliant MPEG video stream requires a minimum amount of buffering

[15] Due to the fact that the re-ordered pictures at the input of the compression and the decoded pictures in coded order at the decompression output have the same picture rate.

at the decoder; this amount is determined by the encoder and must be conveyed to the decoder by means of a parameter in the MPEG video stream produced by the encoder.

Specifying the minimum size of the required buffer is only meaningful, if it is also defined how the video stream enters the buffer and how the coded picture data is removed. In the MPEG-1 and MPEG-2 video standards, for video streams with a constant bitrate, the operation of this buffer is specified by the so-called VBV, Video Buffering Verifier, model. MPEG video bitstreams with a variable bitrate are supported by the MPEG video standard too, but their timing is not specified in MPEG Video, but in MPEG Systems instead; see the STD model described in Chapters 6, 7 and 12 of this book.

The VBV model is a hypothetical video decoder that is conceptually connected to the output of a video encoder. Its input buffer is called the VBV buffer and its size is specified in the MPEG video bitstream as the VBV buffer size. The MPEG video specification defines accurately how the video stream is input to the VBV buffer as well as how the coded picture data is removed from it. As a result, the fullness of the VBV buffer is precisely specified at any point in time.

To control its operation, the video encoder uses the VBV model; in particular the encoder must ensure that each produced MPEG video stream meets the requirement specified for decoding in the VBV model; see Figure 3.12. It should be noted that the VBV does not impose any architecture of encoder implementations. The VBV also does not impose how to design a

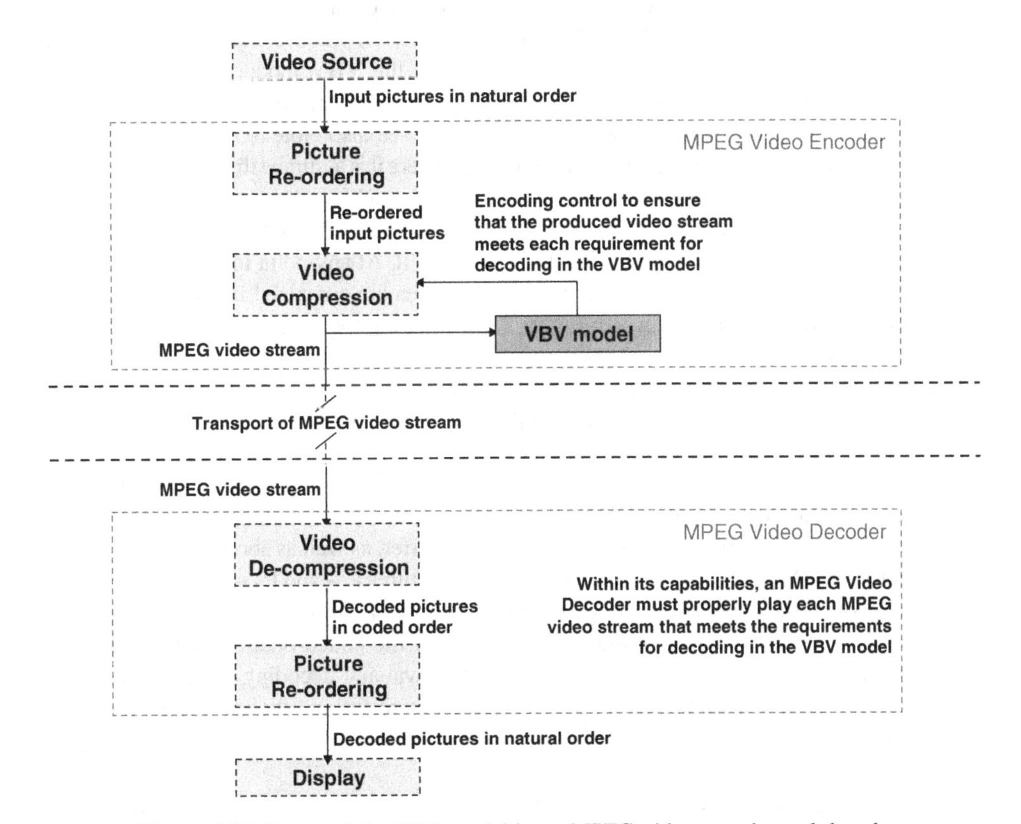

Figure 3.12 Usage of the VBV model in an MPEG video encoder and decoder

Note 3.3 Cost Effective Design of MPEG-1 and MPEG-2 Decoders

An important requirement for the MPEG-1 and MPEG-2 standards has been to allow for the cost-effective design of MPEG decoders for target applications. In practice this meant at the time of its development that it should be possible for initial players to use in their design a single chip MPEG decoder with a single memory device; the latter meant a 4 Mbit (4×2^{20} bit) memory chip for MPEG-1 and a 16 Mbit memory chip for MPEG-2. According Moore's Law [12], the complexity that can be implemented on a silicon device of a certain size doubles every 2 years. As a consequence, the initial costs of MPEG-1 and MPEG-2 decoders were more or less comparable.

video decoder; instead, for video decoder implementations, the VBV only serves as a reference decoder: within its capabilities, a physical MPEG video decoder is required to decode each video stream that meets the requirements for decoding in the VBV model.

In the VBV model, the MPEG video bitstream carrying the coded pictures, enters the VBV buffer at the bitrate of the MPEG video bitstream. At the output of the VBV buffer, the pictures are decoded at the original picture rate. In the VBV model, the pictures are decoded instantaneously: at the instant in time at which picture P_n needs to be decoded,[16] all coded data of picture P_n are removed from the input buffer and decoded instantaneously. In practice, a physical decoder will obviously need time to decode a picture, and therefore the VBV model is designed so that practical decoder designs can compensate for differences in performance between the hypothetical VBV model and the actual decoder implementation; see also Note 3.3.

The operation of the VBV is depicted in Figure 3.13, where it is assumed that the decoding starts at a random access point (an I-picture) located somewhere in an MPEG video stream, in which the same order of coded picture types is used as in Figure 3.10. Figure 3.13 depicts the delivery of the coded pictures to the VBV buffer and their removal from it. At any point in time the fullness of the VBV buffer is shown, as well as which coded pictures are contained in the VBV buffer.

Initially, the VBV buffer is empty. In Figure 3.13, at time t_0 the MPEG video bitstream starts entering the input buffer at the constant video bitrate specified in the bitstream and indicated by the slope $[F_1/(t_1 - t_0)]$. First the coded data of picture P_1 is entered; at time t_1 approximately 50% of the coded picture data of P_1 is in the buffer, in total F_1 bytes. When all coded data of P1 is in the buffer, the coded data of B-pictures P_{-1} and P_0 enters the buffer. At time t_2, all data of P_1 is in the buffer and also about three-quarters of the coded data of picture P_{-1}. At time t_3, all coded data of picture P_1 is removed instantaneously from the buffer and decoded, upon which only the coded data of pictures P_{-1} and P_0 is left in the buffer, as well as about 50% of the coded data of picture P_4. At times t_4 and t_5 the coded data of pictures P_{-1} and P_0 is removed, but these pictures cannot be decoded as one of the reference pictures used for coding P_{-1} and P_0 is not available. At t_6, t_7 and t_8, the instantaneous removal of the coded data and decoding of the pictures P_4, P_2 and P_3 takes place, followed by the removal and decoding of other subsequent pictures in the video bitstream. Figure 3.13 shows that the resulting fullness of the input buffer over time has the shape of a kind of irregular saw tooth, with the position of a tooth at the point

[16] The timing relation between the delivery of an MPEG video stream with a constant bitrate and the decoding time of a picture is provided by the MPEG video bitstream.

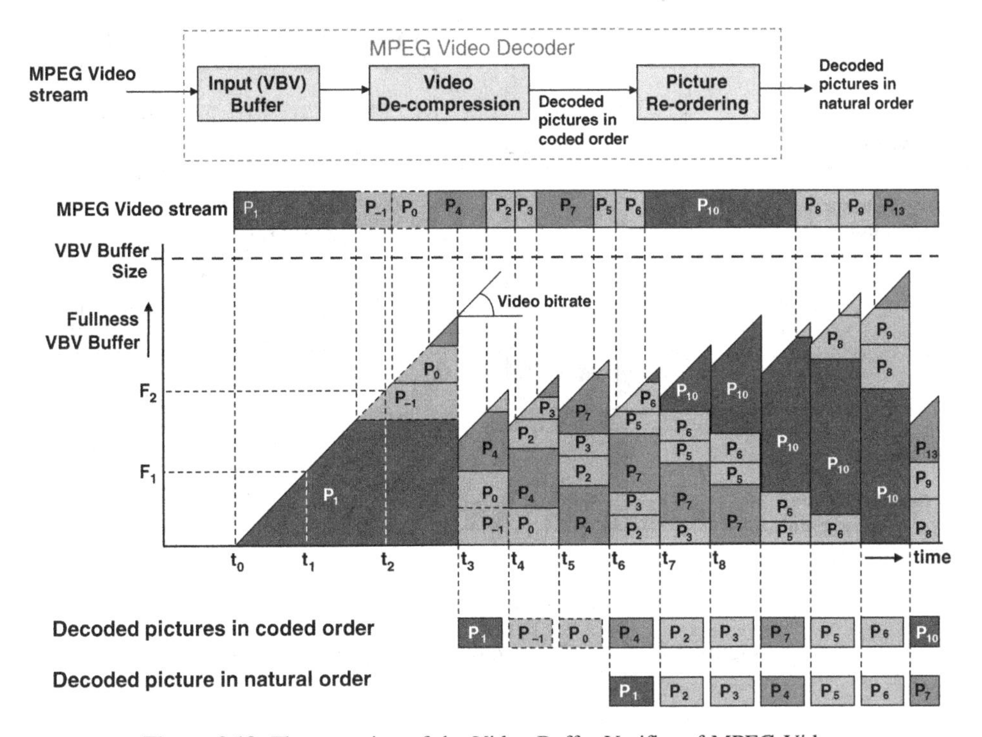

Figure 3.13 The operation of the Video Buffer Verifier of MPEG Video

in time where a picture is decoded and with a tooth size determined by the amount of coded data of the decoded picture.

For the target applications of MPEG-1 and MPEG-2 video, a high quality of service is a key requirement. One of the conditions to achieve a high quality of service, is to prevent underflow and overflow of the VBV buffer. *Underflow* occurs if the coded data of a picture is not yet completely present in the VBV buffer at the instant in time that this picture is to be decoded. For example, in Figure 3.14 underflow occurs when picture P_k needs to be decoded; at that time less than 50% of the coded picture data of P_k is in the VBV buffer. The decoder will not be able to decode picture P_k until time t_x. *Overflow* of the VBV buffer occurs if the total amount of coded picture data that is to be stored in the VBV buffer is larger than the VBV buffer size. For example, in Figure 3.14 overflow occurs during entering the coded data of picture P_n in the VBV buffer when after time t_y the total amount of the coded data of pictures P_{n-3}, P_{n-5}, P_{n-4} and P_n becomes more than the VBV buffer size. As a consequence, about 50% of the coded data of picture P_n cannot be stored in the VBV buffer and may get lost, resulting in possibly serious error propagation due to the fact that P_n is a reference picture used for the coding of subsequent B- and P-pictures.

The fullness of the VBV buffer at the decoder is controlled by the video encoder that produces the video bitstream. To prevent the severe complications of underflow and overflow of the VBV buffer, the MPEG-1 and MPEG-2 video standards require that in a compliant MPEG video bitstream the VBV buffer shall neither underflow nor overflow. There is one exception: low delay MPEG video streams are allowed to underflow the VBV buffer. Low delay streams are introduced in the MPEG video standards to accommodate real-time video

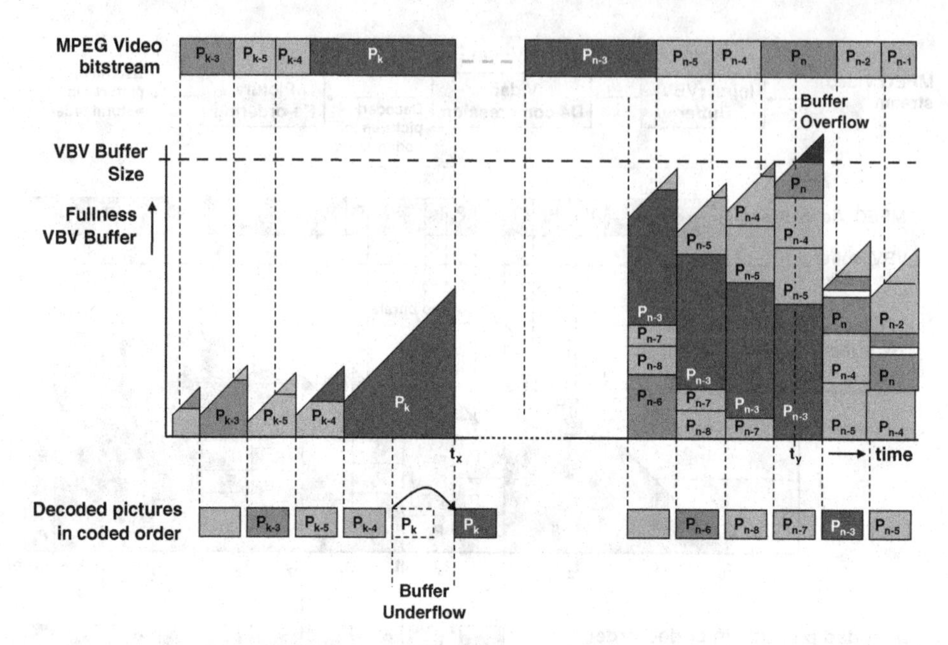

Figure 3.14 Underflow and overflow of the VBV buffer

communications such as video telephony, video conferencing and monitoring. Such low delay streams do not use B-pictures, thereby preventing the delay due to picture-re-ordering. When producing low delay streams, the encoder may decide not to code each picture at the input of the encoder, but to skip one or more picture(s). In a low delay stream, underflow of the VBV buffer indicates that the encoder did skip one or more picture(s).

The structure of an MPEG-2 video stream is depicted in Figure 3.15. The highest structure in an MPEG-2 video bitstream is the video sequence; a video stream is a concatenation of one or more video sequences. Within a video sequence, the sequence parameters are not allowed to change; examples of such parameters are the picture size, the frame rate, the aspect ratio and the size of the VBV buffer. A change of one of the sequence parameters requires to commence a new video sequence.

Each video sequence starts with a sequence header that carries the applicable values for the sequence parameters. In the course of the video sequence, the sequence header may be repeated, for example for random access purposes. The video sequence is terminated by an end of sequence code. Typically, a video sequence contains one or more Group of Pictures (GOPs). Each Group of Pictures must start with an I-picture, preceded by a group of pictures header. This I-picture is typically followed by a series of B- and P-pictures in coded order. The group of pictures header provides guideline information on the decoding of the B-pictures (if any) immediately following the first I-picture of the Group of Pictures in the case that the preceding reference picture is not available, for example due to random access or an edit. It should be noted however that the Group of Pictures structure is optional and may be absent in a video sequence,[17] in which case a video sequence is just a concatenation of I-, P- and B-pictures in coded order.

[17] The Group of Pictures structure is not used as long as the group of pictures header is absent in the stream.

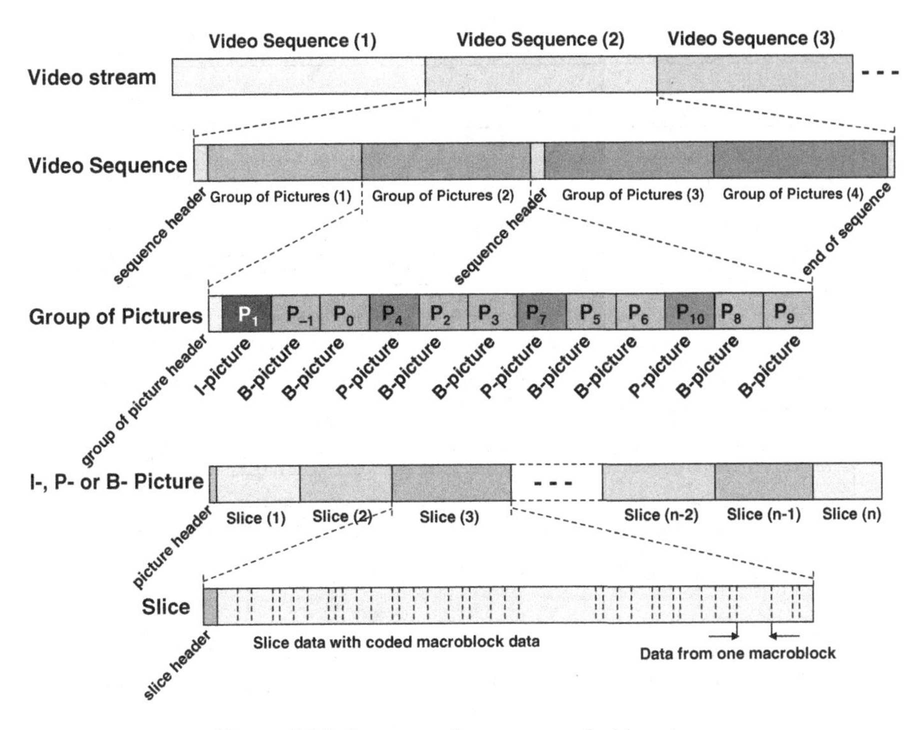

Figure 3.15 Structure of a compressed video stream

Within each I-, P- or B-picture, slices containing coded macroblock data are defined. In MPEG-2 video,[18] each slice covers one or more horizontally adjacent macroblocks: the first and the last macroblock are in the same horizontal row of macroblocks. A new slice may start at any macroblock; slices do not overlap. An example of a slice structure within an MPEG-2 video coded picture is presented in Figure 3.16.

In order to allow a video player to access a video stream randomly, the receiver needs to synchronize to the bitstream, so as to know exactly how to interpret each bit in the bitstream. For synchronization to the coded video data, MPEG defined so-called start codes. Each start code is a specific bit pattern that does not occur in the video bitstream except to indicate a specific item in the bitstream. For example, specific start codes are defined to signal the beginning of a new picture in the video stream; likewise, each new slice is signalled by a start code.

Each start code consists of a start code prefix, followed by a start code value. The start code prefix is a string of 23 bits with the value '0', followed by a single bit with the value '1', hence the bit string '0000 0000 0000 0000 0000 0001'. The start code value is an eight-bit integer that identifies the type of start code. The start codes must be byte aligned; encoders can achieve byte alignment by inserting bits with the value '0' before the start code prefix, such that the first bit of the start code prefix is the first bit of a byte.

The type of start codes specified in MPEG-2 video are given by the start code values listed in Table 3.2. The start code value 0 identifies the picture start code that signals the beginning of

[18] For MPEG-1 video slightly different slice constraints apply.

Figure 3.16 Example of a slice structure within an MPEG-2 video coded picture

the coded data of a new picture. While most start code types have a single start code value, the slice start code is represented by many values, because the start code value represents the vertical position of the slice, whereby the maximum value is equal to 175, which corresponds to a picture of 2800 lines. Though video pictures with more than 2800 lines were very rare at the time MPEG-2 was developed, MPEG defined an escape mechanism so that also larger pictures can be coded by MPEG-2 video.[19]

Table 3.2 Start code values associated with start code types

Start code type	Start code value (decimal)	Start code value (hexadecimal)
Picture start code	0	00
Slice start codes	1 through 175	01 through AF
Reserved for future use	176	B0
Reserved for future use	177	B1
User data start code	178	B2
Sequence header start code	179	B3
Sequence error code	180	B4
Extension start code	181	B5
Reserved for future use	182	B6
Sequence end code	183	B7
Group of picture header start code	184	B8
System start codes	185 through 255	B9 through FF

[19] The maximum picture size in MPEG-2 video is huge: 16 383 pixels by 16 383 lines. This reflects the approach that is generally applied in MPEG standards: an MPEG standard should be sufficiently flexible, so as to avoid constraining its own application domain.

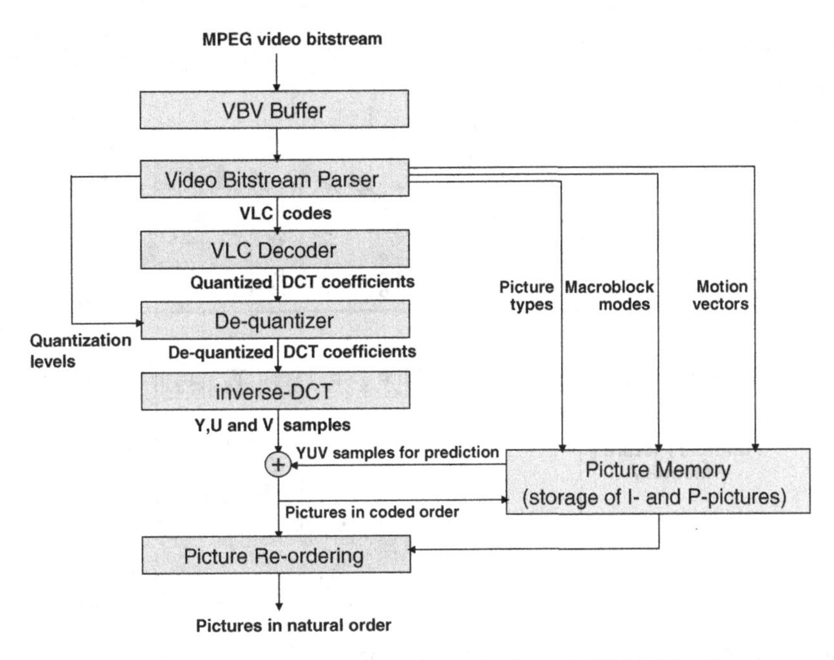

Figure 3.17 Block diagram for the processing in an MPEG Video decoder

User data is identified by the user data start code; if this start code is encountered in the video stream, the decoder ignores all (user) bytes until the next start code in the video stream. Such user bytes are private to applications and can be used for many purposes. The Extension start code signals an extension mechanism, while the Sequence error code signals that an error has been detected in the video bitstream; for example, if during transport an unrecoverable error is detected, this Sequence error code can be inserted by the transport layer. Three start code values are reserved for future use by MPEG. In total, MPEG-2 video specifies start code types for 185 start code values; the remaining values in the inclusive range between 185 and 255 are left for system usage, as will be discussed in subsequent chapters of this book.

Figure 3.17 presents a simple block diagram of the processing of an MPEG video bitstream in an MPEG video decoder. The MPEG video bitstream enters the VBV buffer, upon which it is processed by the video bitstream parser. To ensure correct synchronization to the video bitstream, the parser is continuously looking for start codes. In case of errors, either signalled by the sequence error code or identified otherwise, the decoder needs to re-synchronize to the video bitstream, while applying an error concealment strategy not specified by MPEG. Upon identifying the next start code and depending on its start code value, the decoder will be able to continue parsing and decoding the video stream.

The decoding of the video stream can commence once the picture size and other essential parameters are retrieved from the various headers. Furthermore, the parser ensures that the required information on picture types, macroblock modes and motion vectors is provided to the other building blocks of the decoder needing it. The VLC decoding, de-quantization and i-DCT processing that takes place subsequent to the parsing is the same as depicted in Figure 3.4. For the predictive and bi-directional coding, decoded I- and P-pictures are stored in a picture

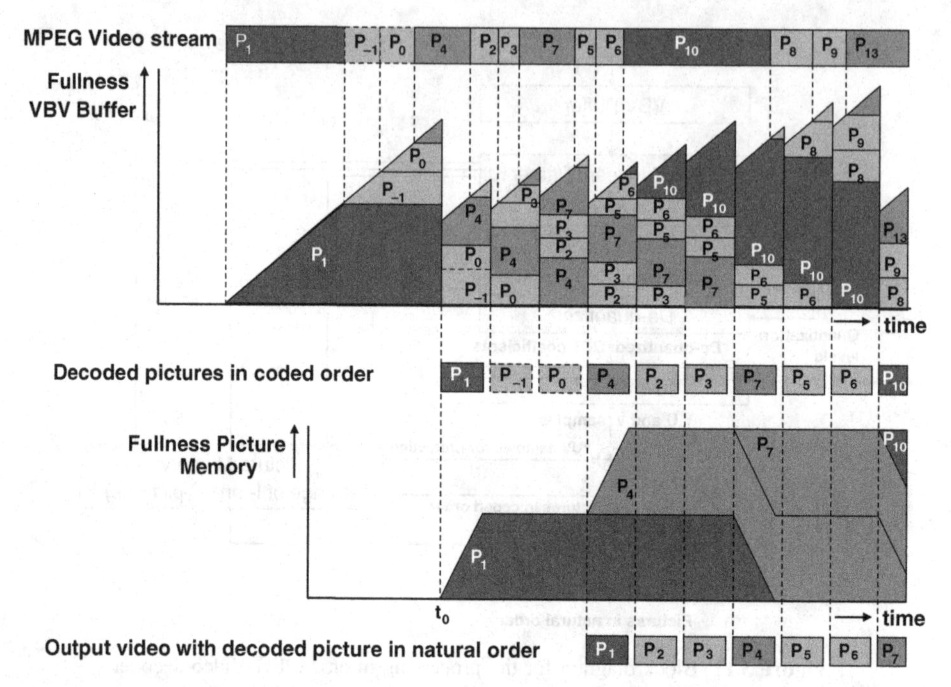

Figure 3.18 Fullness of VBV Buffer and Picture Memory in MPEG Video decoder

memory. Based on the received information from the parser, the picture memory provides the required picture data needed for the prediction, so that it can be added to the reconstructed sample differences from the i-DCT. Obviously, nothing is added to samples from the i-DCT in case of an intra-coded macroblock. Each decoded B-picture is put in the output video signal immediately, but a decoded I- or P-picture is first stored in the picture memory, so as implement the required picture re-ordering. Moreover, for bi-directional prediction of any B-pictures that may have been coded in-between two reference pictures, the availability of any required data from these reference pictures from the picture memory must be ensured.

Figure 3.18 shows a simplified example of the processing in the video decoder based on the same MPEG video stream as depicted in Figure 3.13. The fullness of the VBV buffer is the same as in Figure 3.13: for decoding and VBV buffering, the instantaneous decoding model of the VBV is assumed. However, the decoded pictures in coded order, as well as the output pictures are assumed to be video data streams at the data rate corresponding to the applied YUV format. At time t_0, the decoding of the MPEG video stream starts with the decoding of picture P_1, upon which P_1, starts to appears in the video data stream with decoded pictures in coded order. It is assumed that pictures P_{-1} and P_0 are not decoded, because the required reference picture P_{-2} is unavailable. Next, at the appropriate times, pictures P_4, P_2, P_3, P_7, P_5, P_6 and P_{10} are decoded; the B-pictures P_2, P_3, P_5 and P_6 are put straight in the output video, as opposed to the reference pictures P_4, P_7 and P_{10} that enter the picture memory. At the time P_4 starts entering the picture memory, picture P_1 starts to appear in the output video. However, because P_1 is needed for the decoding of the B-pictures P_2 and P_3, P_1 remains in the picture memory until P_3 is decoded. In Figure 3.18, P_1 is removed at the same time P_7 enters the picture memory, so that the fullness of the picture memory remains two times the size of a decoded picture. In practice a more efficient use of the picture memory is possible by means of decoder implementation specific measures.

3.2.3 Evolution of MPEG Video

The MPEG standards define the format of the coded bitstream and how to decode the bitstream, but not how encoders should create compliant bitstreams. This leaves room for the market place to improve encoders within the capabilities of the standard, without requiring any change of decoder. The result of this approach has been that over time, the compression technology applied in MPEG-2 video encoders has indeed improved significantly for several reason. First of all, the encoding algorithms could be improved, assisted by the technology progress that allowed for further integration in encoders, thereby enabling the use of more complicated tools, for example to improve motion estimation. Another reason has been the use of variable bitrate instead of constant bitrate; complicated sequences require more bits than an average sequence and by assigning no more bits than needed to obtain the required picture quality for the various sequences, the average bitrate can be reduced significantly. Finally, it was shown that by improving the applied pre-processing, such as noise reduction, the encoder efficiency could be improved. Figure 3.19 provides an indication of encoder improvements over time for MPEG-2 video for the coding of standard definition video [13]. In about 10 years the average bitrate has been reduced by more than 60%.

The MPEG-2 video standard also addresses various forms of scalability, in particular spatial scalability, SNR scalability and temporal scalability. With scalability it is possible to have multiple video layers, for example consisting of one base layer and one or more higher layers at higher (spatial or temporal) resolution that are predicted from the base layer. Scalability is from

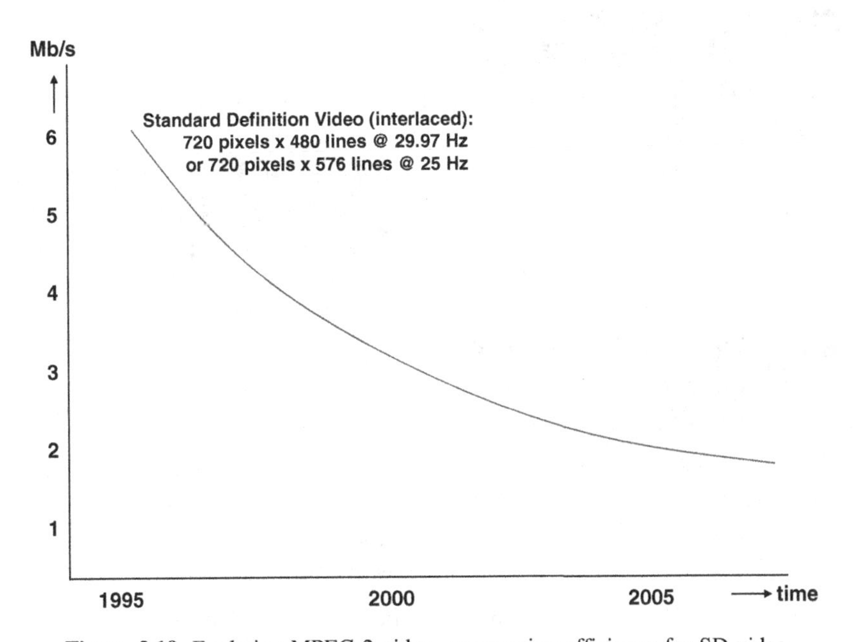

Figure 3.19 Evolution MPEG-2 video compression efficiency for SD video

technical research perspective a challenging subject, but a business model that requires scalability is hard to find. In any case, even though MPEG spent many man-years on its development, MPEG-2 scalability is not used in any commercial product or service.

During the completion of the MPEG-2 standard, MPEG started its work on MPEG-4. This MPEG-4 effort would continue for many years, during which the MPEG-4 standard was build up as a large suite of more than 25 specifications [14], one of which is MPEG-4 video, specified as part 2 of the MPEG-4 standard [10]. MPEG-4 video is largely build on top of the MPEG-2 video standard, including its scalability features, but there are also extensions, such as tools for:

- higher coding efficiency at very low bitrates, so as to improve the suitability for streaming over the internet;
- support for transparency in video;
- support for arbitrary shaped video instead of rectangular video only.

In addition, there is support in MPEG-4 video for a more general MPEG-4 feature, object orientated video. For example, one small video object can be positioned on top of a larger video at a position that may change each picture. An application of this is depicted in Figure 3.20.

The objective of the 'Buy the Ball' application in Figure 3.20 is to offer free viewing of a soccer game to users of a Pay TV service, but without showing the ball, as an attractor; for seeing the ball, payment is needed. To achieve this in a relatively simple way, in each picture it is determined whether the ball is present and if so, a rectangular area is defined around the ball. Next, for example by using information from other pictures, the background behind the ball is reconstructed and inserted in the rectangular area, so that the ball can be replaced by its background. This rectangular area is coded and broadcast to the receiver, along with its position

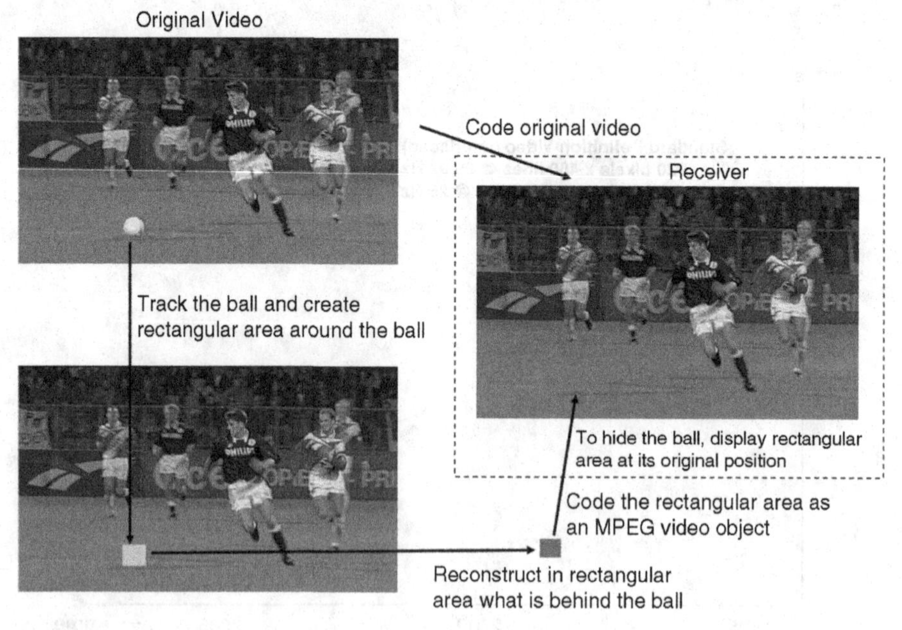

Figure 3.20 The 'Buy the Ball' example of object orientated MPEG-4 video.
Source: PSV Eindhoven

in the picture. As long as no payment is made, the receiver displays the rectangle with the background at its original position, to hide the ball. As soon as the payment is made, the rectangular with the background is no longer displayed and the ball appears.

MPEG-4 video was not intended nor suitable as a replacement of MPEG-2 video; the gain in efficiency for MPEG-2 applications is relatively minor, about 10–20%. It is applied in some internet applications, and found also some usage in applications such as camcorders. However, the MPEG-4 video adoption on the market place is far less than for MPEG-2 video. Also for object orientated video no business case was identified; so far, also the Buy the Ball application did not yet make it to the market.

During the development of the MPEG-1, MPEG-2 and MPEG-4 video standards there has been a close working relationship between MPEG and the group of video coding experts within ITU-T Study Group 16 [15], also known as VCEG (Video Coding Experts Group) [16]. The VCEG group has been responsible for the ITU-T video coding standards H.261, H.262 and H.263 [16]. As a first result of this close working relationship, MPEG-1 video is build on top of the ITU-T standard H.261. The MPEG-2 video standard is even a joint publication of ISO/IEC and ITU, published in ITU-T as H.262, while the core of the MPEG-4 video standard has a very high level of commonality with H.263.

In 2001 it was decided to form a Joint Video Team, JVT [17–19], between VCEG and MPEG to complete the next video standard, Advanced Video Coding (AVC). This joint effort resulted in the first version of the AVC standard [11] that was completed in 2003, and published as part 10 of MPEG-4 by ISO/IEC, often referred to as MPEG-4 AVC, and as H.264 by ITU-T. The AVC standard is typically at least a factor of two more efficient than MPEG-2 video, though the coding efficiency gain for high quality video may be less.

The AVC standard represents a next step in the evolution of video coding technology. At a first glance, there seems to be quite some commonality with the MPEG-2 video standard, but there is a lot of 'devil in the detail'. Indeed, the AVC standard uses block based motion compensated coding, as the MPEG-2 video standard does, but AVC differs in many details from MPEG-2 video, such as:

- Instead of a DCT, the AVC standard uses a so-called Integer Transform, which is a kind of simplified DCT.
- While MPEG-2 video uses DCT transforms on 8×8 blocks only, AVC allows the use of Integer Transforms on 4×4 and 8×8 blocks, thereby providing the option to trade-off less visible artefacts and more efficiency.[20]
- MPEG-2 video uses one reference picture for motion compensated prediction in a P-picture and equivalently two reference pictures for a B-picture. Instead, AVC allows the use of up to 16 reference pictures or, in the case of interlaced video, 32 reference fields. For example, when AVC is used in a video phone application where the user is calling in front of a fixed background, a picture of the background can be coded as the first picture of the video, to serve as a reference picture for the prediction throughout the whole video. It should be noted, however, that application standardization bodies may apply additional constraints to the maximum number and use of reference pictures in AVC.

[20] Artefacts of 4×4 blocks, such as the so-called 'ringing', are less visible than for 8×8 blocks, while the use of 8×8 blocks is more efficient, particularly in highly correlated regions.

- For prediction from a reference picture, MPEG-2 video allows one motion vector per macroblock for the entire 16 × 16 area or, in the case of interlaced video, two motion vectors for both 16 × 8 areas per field within a macroblock, while each motion vector has an accuracy of 1 or 1/2 pixel. The AVC standard is far more flexible in this respect; if so desired, each macroblock can be partitioned in non-overlapping sub-blocks of 4 × 4, 4 × 8, 8 × 4, 8 × 8, 8 × 16, 16 × 8 and 16 × 16 luminance samples. Each such partition may have two motion vectors; the motion vectors of partitions with 8 × 8 samples or more may point to different reference pictures. The motion vectors in AVC may have an accuracy of 1, 1/2, 1/4 or 1/8 pixel.
- For the coding of quantized coefficients in AVC a comparable kind of VLC coding is available as in MPEG-2 video, but there are two additional options:
  1. The first additional option is the use of CABAC (Context-Adaptive Binary Arithmetic Coding). From coding efficiency perspective, CABAC is a superior tool, but it requires complex processing involving context modelling with probability estimation, followed by arithmetic coding. Due to the associated very complex processing, arithmetic coding has a very long history (see Note 3.4).
  2. The second additional option for coding the quantized coefficients is Context Adaptive VLC (CAVLC) coding. Depending on the local context, other VLC tables may be selected.
- AVC introduces weighted prediction in motion compensation as a new tool, so as to better accommodate fade transitions, such as fade-to-black, fade-in, and cross-fade.

There are many more small differences between MPEG-2 video and AVC, but those mentioned above are the most important in practice. For an overview see also Figure 3.21. The AVC standard provides many compression tools and a lot of flexibility for encoders to use these tools, thereby providing a lot of headroom for the learning curve of AVC encoders. As with MPEG-2, it is therefore expected that the picture quality produced by AVC encoders will increase over time.

Despite various claims that have been made in the past about spectacular new video coding tools, no fundamental breakthroughs in video compression have been achieved so far. There is

Note 3.4 The History of Arithmetic Coding in MPEG

Some may consider the adoption of CABAC in AVC as a 'historical breakthrough'. Arithmetic coding for video compression was already known when the MPEG-1 video standard was developed, but despite its recognized efficiency, its use was rejected due to the huge complexity. The same happened when the MPEG-2 video and the MPEG-4 video standards were developed.

However, by the time the AVC standard was defined, IC technology was progressed to a level so that the complexity of CABAC was considered 'just acceptable'. Hence, more than 15 years after its invention, the use of arithmetic coding in video compression finally found application in real life. However, CABAC associated patents that may have been filed by the inventers were meanwhile probably not very valuable anymore, as the lifetime of those patents was likely to elapse before or soon after the AVC standard was established on the market place.

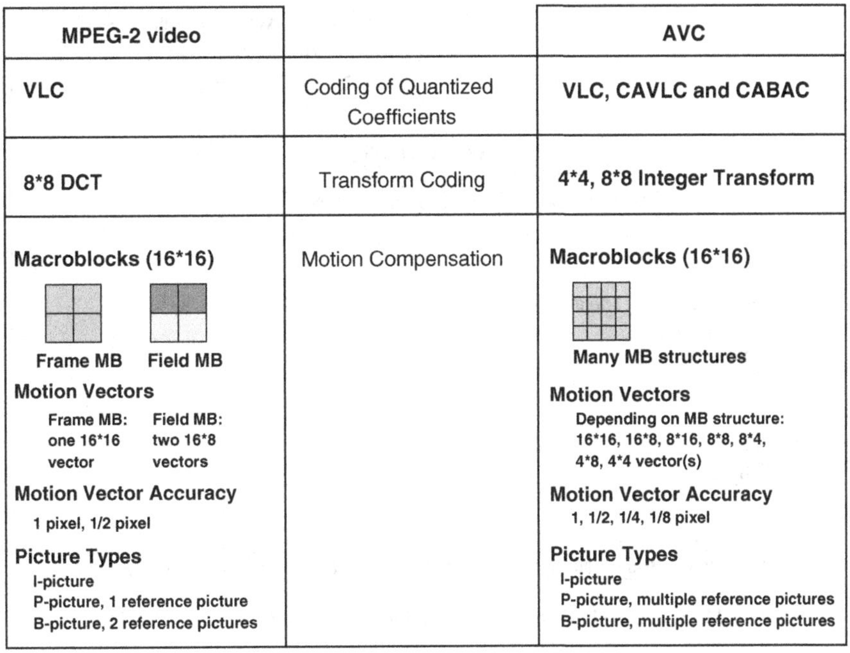

MPEG-2 video		AVC
VLC	Coding of Quantized Coefficients	VLC, CAVLC and CABAC
8*8 DCT	Transform Coding	4*4, 8*8 Integer Transform
Macroblocks (16*16) Frame MB Field MB Motion Vectors Frame MB: one 16*16 vector Field MB: two 16*8 vectors Motion Vector Accuracy 1 pixel, 1/2 pixel Picture Types I-picture P-picture, 1 reference picture B-picture, 2 reference pictures	Motion Compensation	Macroblocks (16*16) Many MB structures Motion Vectors Depending on MB structure: 16*16, 16*8, 8*16, 8*8, 8*4, 4*8, 4*4 vector(s) Motion Vector Accuracy 1, 1/2, 1/4, 1/8 pixel Picture Types I-picture P-picture, multiple reference pictures B-picture, multiple reference pictures

Figure 3.21 Important differences between MPEG-2 video and AVC

a clear evolution of video coding though, initially within existing video coding standards by improved encoding algorithms and progress in encoding and pre-processing technology without any impact on decoders. Of course, in research and standardization environments, new tools are developed and existing tools improved, but these developments will only lead to a successful new standard if at least two conditions are met. Firstly, the new standard must offer one or more features that are sufficiently interesting to the market; one such feature is gain in coding efficiency, let us say at least two times more efficient than an existing standard, so as to provide an economic justification for the introduction of a new standard. Secondly, it must be possible to design cost-effective decoders for such a new standard.

Figure 3.22 provides an overview of video coding standards introduced by MPEG between 1992 and 2007. MPEG-1 video did provide a first solution for 'VHS' picture quality and paved the way for MPEG-2 video. MPEG-4 video (part 2) offered some features, but not in a way to become a major market success. Also the scalability features developed for MPEG-2 and MPEG-4 video did not find any application; though technically very challenging, no market demand did evolve for scalability so far. This changed for AVC scalability, also known as SVC[21] (Scalable Video Coding), which found usage in video conferencing systems. However, as of 2013, opportunities in broadcast applications were not exploited yet. For example, there might have been benefits in the use of SVC when extending existing HDTV broadcast services using AVC for coding 1080 lines interlaced video in a backward compatible manner towards 1080 lines progressive video. It is questionable whether such extensions make sense from

[21] Published as Amendment 3 to the AVC standard.

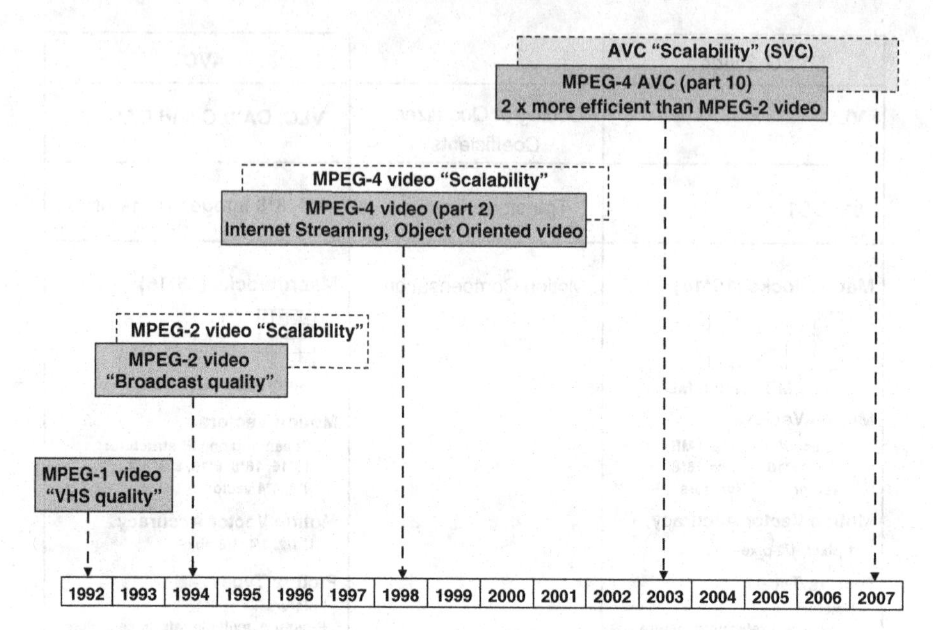

Figure 3.22 Video coding standards introduced by MPEG between 1992 and 2007

business perspective, but even if so, then the probability of future SVC usage decreases with the evolving next generation video codec HEVC.

The AVC standard will not be the last video coding standard developed by MPEG: video compression technology will continue to evolve as long as there is an economic justification. Important conditions for any new video coding standard are a sufficient gain in video compression and the availability of cheap custom silicon for decoders.[22] Experience so far implies that about every 10 years there is 'room' on the market for the next generation MPEG video standard. In 2013, MPEG published the so-called high-efficiency video coding standard (HEVC), that promises to gain a factor two or more in compression capability compared to AVC.

Video coding efficiency improvements over time are mainly driven by more complex processing becoming practical over time, as described by Moore's Law [12] (see also Note 3.3). 'McCann's Law' translates this effect into bitrate by predicting that the bitrate required to achieve a given video quality roughly halves about every 7 years [20]. Within a standard, it can be expected that the evolutionary improvements follow a rather smooth curve, and that this curve flattens in the longer term, due to limitations of the standard. To overcome these limitations, a new codec is needed; however, a new coding standard will be adopted only if justified by sufficient efficiency gain.

The experience with MPEG-2 video and AVC shows that about 10 years after the introduction of a video codec on the market, there is room for a next generation codec,

[22] It could be argued that the availability of cheap silicon becomes less relevant in future when general purpose processors find wide usage in consumer products.

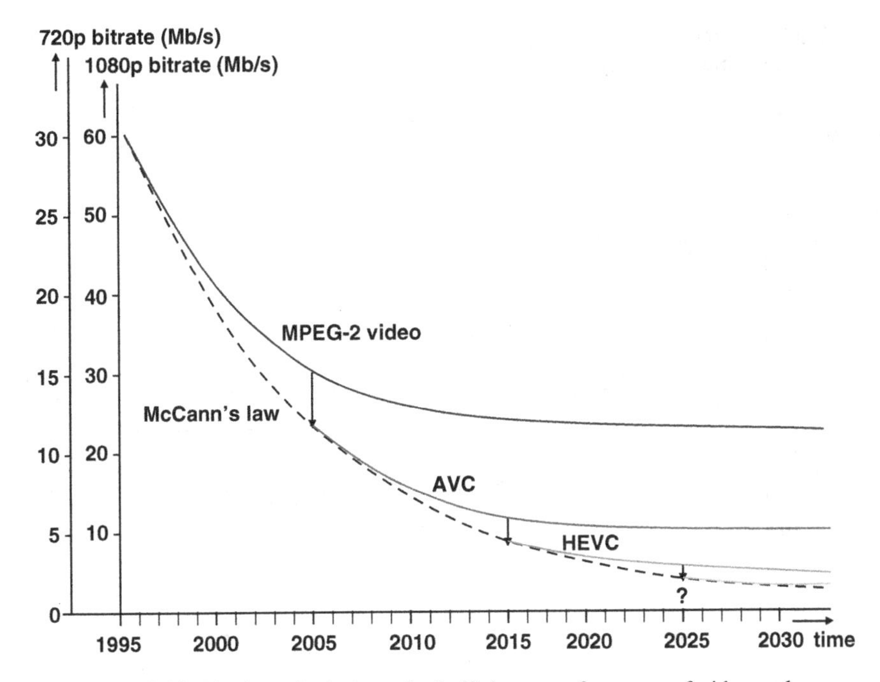

Figure 3.23 The hypothetical trend of efficiency performance of video codecs

offering an efficiency improvement by a factor of two. Indeed, in 2013, about 10 years after AVC, the HEVC standard was published. Note however that the introduction of a new codec does not necessarily mean replacement of the existing one(s); legacy issues typically prevent changing an existing service such as broadcasting. Hence, a new video codec is usually only applied for new services, while existing services continue to use the 'old' codec. As a result, MPEG-2 video, as well as AVC, may be used during several decades, even if significantly more efficient next generation codecs become available.

Figure 3.23 shows the hypothetical trend of efficiency performance of video codecs under the assumptions that McCann's Law provides a viable long term predictor of the coding efficiency and that 10 years after the introduction of a video codec a new one can be introduced with an efficiency gain with roughly a factor two. Furthermore, it is assumed that during the first 5 years of its existence, a codec follows the smooth curve of McCann's Law. After these 5 years, the efficiency improvement flattens, due to the limitations of the standard, causing a gap with the curve of McCann's Law. After 10 years, the efficiency gap becomes so large that there is room on the market to introduce a new standard with tools overcoming the above limitations. The new codec causes a stepwise improvement of the coding efficiency, back to the curve of McCann's Law. Based on the above assumptions, McCann's Law predicts that there is room on the market for HEVC from about 2015 onwards and, extrapolating further, for yet another next generation video codec in about 2025. In this context it should be noted that publication of a next generation video coding standard is an essential milestone, but prior to its adoption on the market more conditions are to be met. For example carriage of the new format over transport layers such as MPEG-2

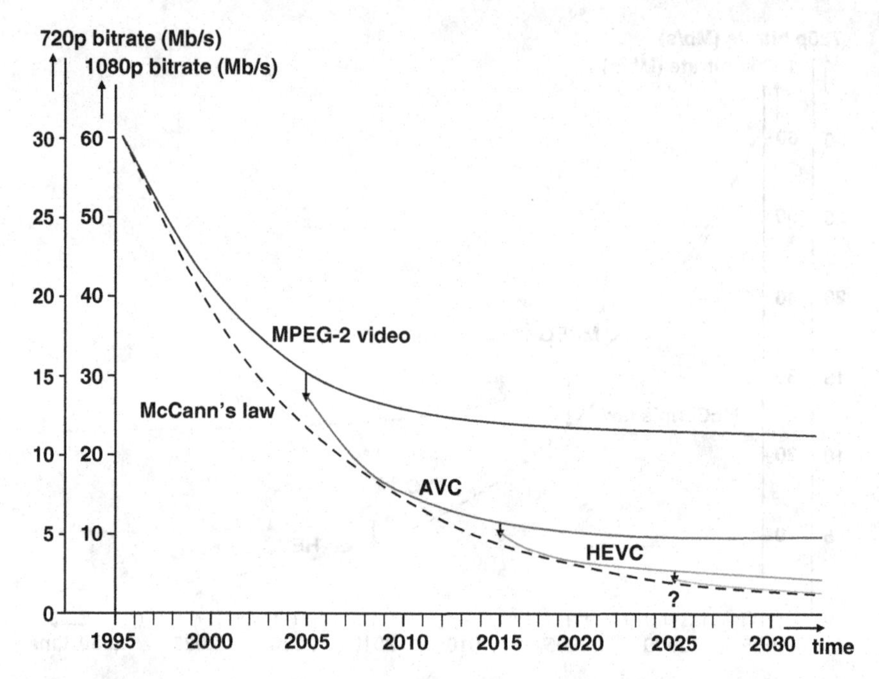

Figure 3.24 The trend of efficiency improvements of video codecs in practice. *Source:* [20] McCann 2012. Reproduced from DVB

systems and IP networks need to be specified, while application standardization bodies such as DVB may impose constraints for its usage. In other words, usage of the next generation codec is only possible when the entire end-to-end system for its usage is specified and available.

While Figure 3.23 describes the hypothetical performance, in the real world there may be an initial learning curve to get to the curve of McCann's Law during the first years of existence of a new coding standard [20]. This is illustrated in Figure 3.24, included with the kind permission of Ken McCann. Compared to the hypothetical performance in Figure 3.23, in practice the stepwise improvement when introducing a new codec may be smaller. For example, an initial encoder implementation for a new standard may be a 'quick and dirty' improved-by-software version of an existing hardware encoder designed for the preceding standard. With AVC, the encoder manufacturers initially just added more software to their MPEG-2 hardware. This provided a compliant AVC bitstream, but with very little improvement in coding efficiency over MPEG-2. However, after a few years encoders optimized for the new standard are available with a coding efficiency that approached the smooth curve of McCann's Law.

However, with respect to the adoption of a new video codec, it should be noted that coding efficiency is only one of the conditions for market adoption. Adoption of a new codec requires economical justification. For example, if an overwhelming amount of bandwidth over networks becomes available in future, it is still to be seen whether there is an economical drive towards the use of a new generation video coding standard offering more coding efficiency at higher costs.

3.3 MPEG Audio

3.3.1 MPEG-1 and MPEG-2 Audio Essentials

The inputs to an audio encoder are digital audio samples representing an analogue audio signal. Digital audio samples provide a very simple method of waveform coding, the generic name for coding an audio waveform. As an example, Figure 3.25 presents the waveform of an analogue audio signal that is sampled at a certain frequency using 16 bits for each sample, whereby each sample is usually represented by a two's complement code; see Figure 3.25. For (very) high quality audio often 20 bits are used. Popular sampling frequencies are 44.1 kHz for Compact Disc and 48 kHz for DVD, Blu-ray and various broadcast applications. For lower quality audio, in particular audio with restricted bandwidth, lower sampling frequencies are used, such as 24 kHz, while speech uses often 16 or 8 kHz as sampling frequency.[23] The choice for the sampling frequency has obviously impact on the conveyed audio bandwidth[24] and on the data rate at the input of the audio encoder. For example, stereo audio with 16 bit audio samples, a sampling rate of 44.1 kHz results in a data rate of $2 \times 16 \times 44\ 100 = 1.4112$ Mb/s at the encoder input, and a sampling rate of 8 kHz of mono audio with 16 bit samples in a data rate of $16 \times 8000 = 128$ kb/s.

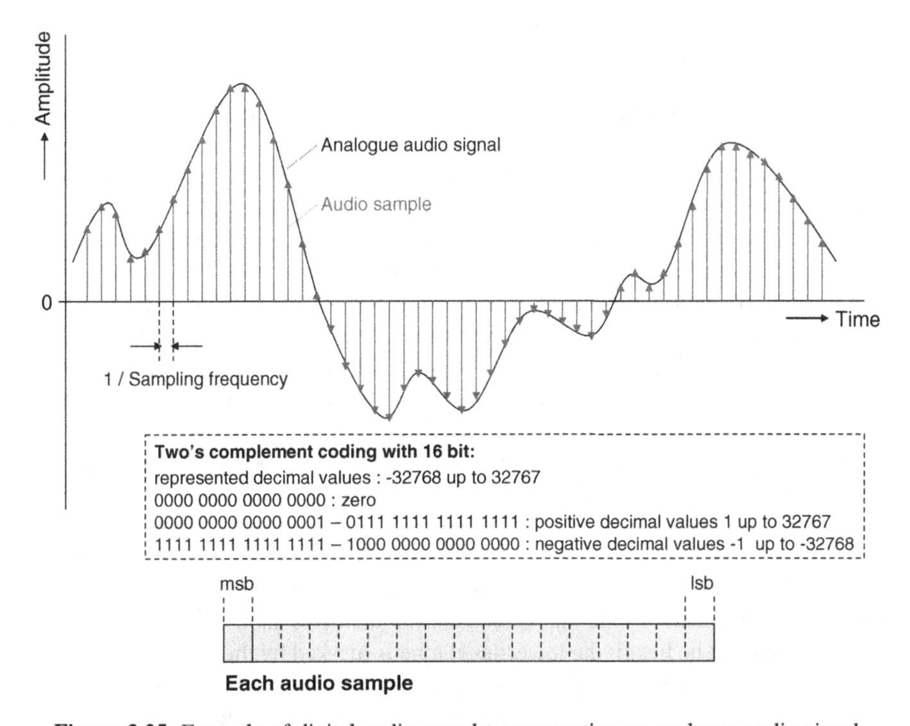

Figure 3.25 Example of digital audio samples representing an analogue audio signal

[23] In the early internet days, the 16 kHz sampling frequency was also used for (very) low quality audio.

[24] According to the so-called Nyquist-Shannon sampling theorem, only frequencies in the input signal theoretically up to half the sampling frequency can be preserved; for example on CD audio, frequencies up to 22.05 kHz. In practice, depending on filter characteristics, (slightly) less bandwidth is preserved.

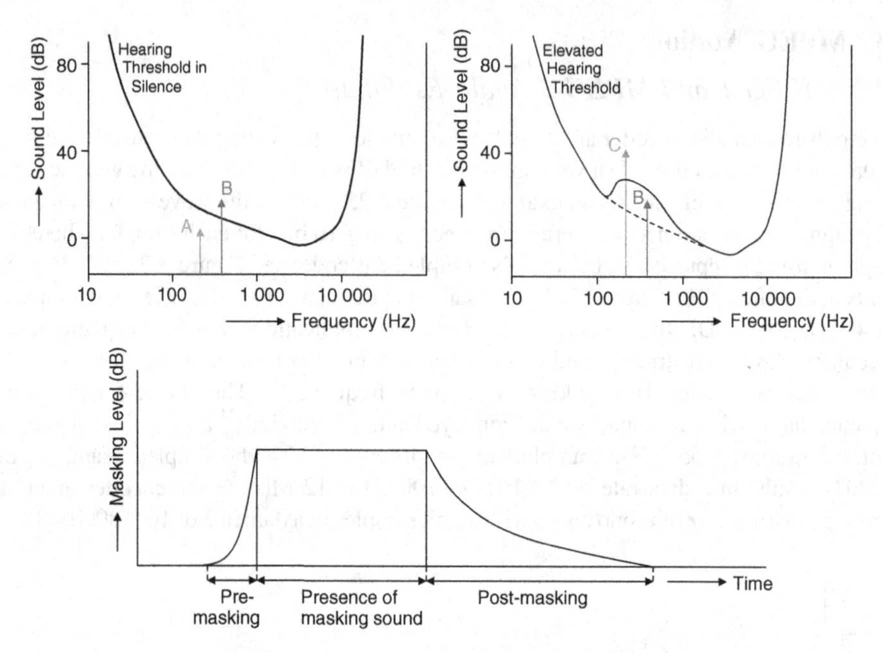

Figure 3.26 The hearing threshold and masking

Samples in the time domain are an accurate means to describe an audio signal, but in the frequency domain, the audio signal can be described equally well by means of the frequency components in the audio signal. For audio compression (as for video compression), the use of the frequency domain description is more suitable than describing the shape of the signal over time. In the case of audio the main advantage is a better match with the characteristics of the human auditory system, thereby offering opportunities to take advantage of the limitations of human perception. For this purpose, MPEG-1 audio [21] and the backward compatible multi-channel extension thereof defined in MPEG-2 audio [22] apply a technique called perceptual sub-band coding, whereby the audio input signal is split into 32 frequency bands.

A perceptual coder exploits the psycho-acoustic effect described by the hearing threshold. The hearing threshold specifies as a function of frequency the lowest level of a tone that can be sensed by the human ear. Frequency components in an audio signal falling below the hearing threshold are not relevant to the human perception and can be ignored when coding the audio signal.[25] In silence, the hearing threshold is a fixed curve, but the curve may change when audible tones are present. A tone that is close in frequency to another tone, but (sufficiently) lower in level, cannot be heard: the lower-level tone is masked by the higher-level tone. When masking occurs, the hearing threshold in silence is elevated for frequencies nearby the frequency of the masking tone during the presence of the masking tone. This masking effect includes pre- and post-masking, prior to and following, respectively, the presence of the masking signal. The hearing threshold and masking are illustrated in Figure 3.26.

[25] In the psycho-acoustical model for audio coding several complications are taken into account; for example, the hearing threshold is age dependent, in particular for frequencies above 2 kHz; furthermore, the playback level at the decoder is unknown a priori.

Figure 3.26 demonstrates the hearing threshold and masking for a few tones. In the left top graphic, the hearing threshold in silence is shown; the volume of tone A is too low to be heard, but tone B is sufficiently loud to be heard. In the right top graphic, tone C is present at the same frequency as tone A, but much louder, so that the hearing threshold is elevated for frequencies nearby the frequency of tone C. As tone B is now below the elevated hearing threshold, tone B cannot be heard anymore. In the graphic at the bottom, temporal masking is shown with pre- and post-masking. While the post-masking can be explained from the time the human auditory system needs to recover from hearing a loud signal, the (much weaker) pre-masking is a more strange phenomenon that possibly may be understood from the fact that the human auditory system needs time to build up the perception of sound, whereby louder sounds are faster processed by the human brain than softer ones.

The psycho-acoustic effect described above does not only exist for tones, but also for more complex sound, such as music and noise. Associated with every sound, there exists a hearing threshold, as a function of frequency. The hearing threshold depends on the spectrum of the sound and as this spectrum changes over time, also the associated hearing threshold changes. In a perceptual audio encoder, the hearing threshold is used to control the quantization of the frequency components; this quantization causes a noise, that may distort the decoded audio, depending on the accuracy of the applied quantization. The objective is to shape the spectrum of the caused quantization noise, so that it is kept under the hearing threshold [23].

Figure 3.27 presents a simple block diagram of the perceptual coding scheme applied in MPEG-1 and MPEG-2 audio. At the input of the encoder, audio samples are fed to a sub band

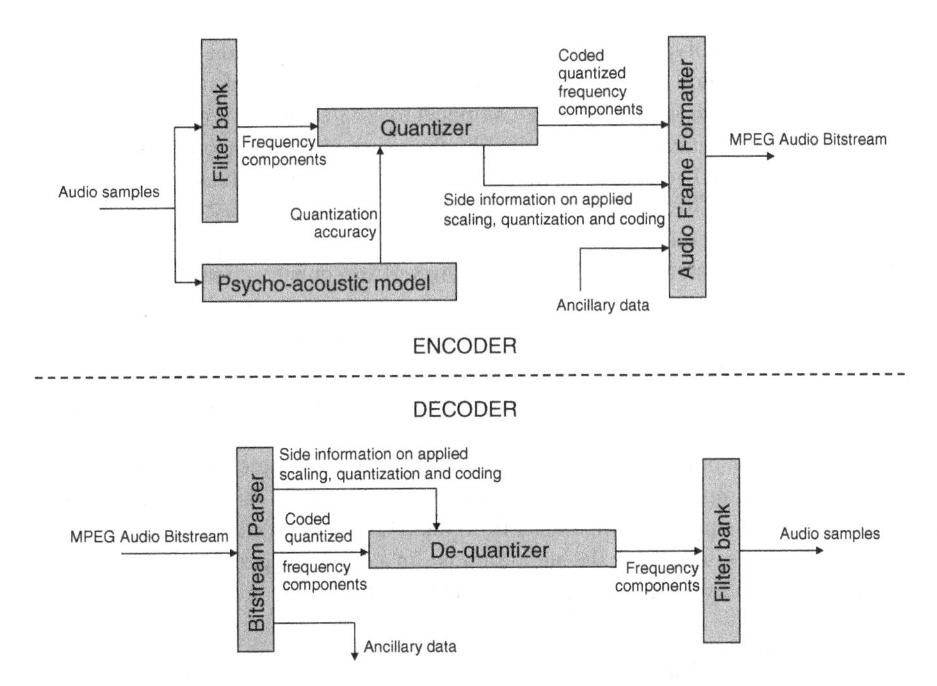

Figure 3.27 Block diagram of an MPEG audio encoder and decoder

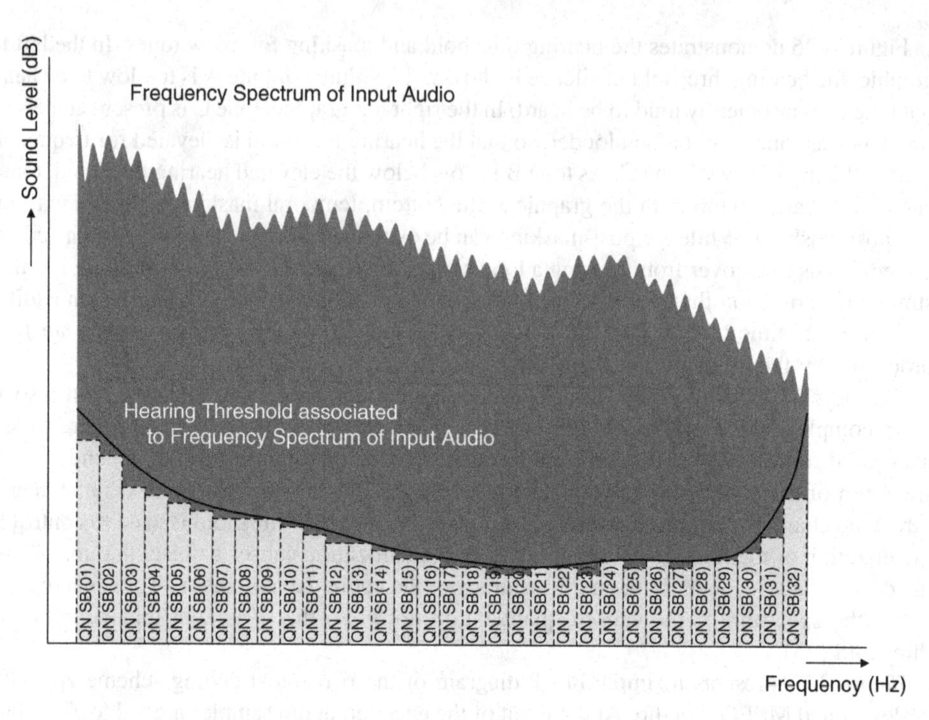

Figure 3.28 Keep quantization noise QN per sub-band SB below hearing threshold

filter consisting of a polyphase filter bank[26] with 32 frequency bands. At the output of this filter, spectral information on the input audio is available in the form of a frequency component for each sub-band. Each such frequency component is scaled and subsequently quantized by the quantizer. The quantization is controlled by the psycho-acoustic model, that continuously analyses the input signal, to determine for each sub-band the hearing threshold under which the caused quantization noise will not be audible by the human auditory system. The obvious objective is to keep the quantization noise level below the hearing threshold associated to the spectrum of the input signal. Hence, the spectrum of the quantization noise is dynamically adapted to the spectrum of the input signal. The psycho-acoustic model provides the desired quantization accuracy; in general, the achieved accuracy will be higher or lower, depending on the audio signal and the bitrate. If the audio bitrate is high enough, the noise will be masked successfully, as demonstrated in Figure 3.28, but at lower bitrates the quantization noise may produce audible distortion.

For the calculation of the frequency components, the audio samples at the input are grouped in 32 consecutive samples. Upon entering each group of 32 audio samples, the sub-band filter calculates 32 new frequency components from the 512 most recently entered audio samples. This is illustrated in Figure 3.29; the audio samples in a group are denoted as $S_{k,01}$, $S_{k,02}$, . . . , $S_{k,32}$, where k indicates the index of the group. Upon entering this group of audio samples, the

[26] A polyphase filter bank provides in terms of computational effort a very efficient implementation of a filter bank containing 32 FIR filters with 512 taps each; further details are beyond the scope of this book.

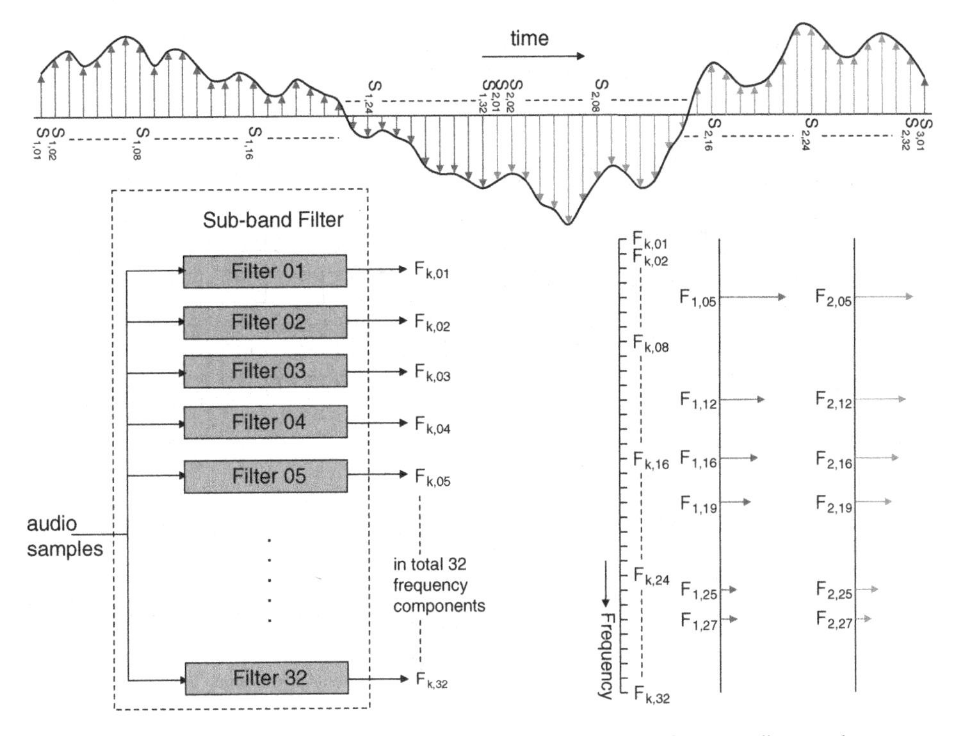

Figure 3.29 Filter bank delivers 32 frequency components for 32 audio samples

sub-band filter calculates the new frequency components $F_{k,01}, F_{k,02}, \ldots, F_{k,32}$. In Figure 3.29, the input signal carries the group of samples $S_{1,01}, S_{1,02}, \ldots, S_{1,32}$, followed by the group $S_{2,01}, S_{2,02}, \ldots, S_{2,32}$. For these audio samples, the sub-band filter delivers two frequency component groups $F_{1,01}, F_{1,02}, \ldots, F_{1,32}$, and $F_{2,01}, F_{2,02}, \ldots, F_{2,32}$. In the example depicted in Figure 3.29, most frequency components are (very) small; only the larger frequency components are shown as $F_{1,05}$, $F_{1,12}$, $F_{1,16}$, $F_{1,19}$, $F_{1,25}$ and $F_{1,27}$ and $F_{2,05}$, $F_{2,12}$, $F_{2,16}$, $F_{2,19}$, $F_{2,25}$ and $F_{2,27}$. In practice there may be less similarity between consecutive groups of frequency components than suggested in Figure 3.29.

For transport in audio frames, some structuring of the frequency components is needed; an example based on MPEG-1 audio Layer I and Layer II is given in Figure 3.30 (see Note 3.5 for a brief description of the layers in MPEG-1 audio). First, the frequency components $F_{k,01}$, $F_{k,02}, \ldots, F_{k,32}$ are organized over time in groups. At the output of the sub-band filter, 32 parallel groups are formed; each group contains 12 consecutive frequency components from the same sub-band: $F_{1,x}, F_{2,x}, \ldots, F_{12,x}$, with x as the number 1, 2, ... or 32, depending on the actual sub-band. The 32 frequency components delivered in parallel by the sub-band filter in Figure 3.30, enter the slots with the same index in the groups formed for each sub-band. Hence, the (highlighted) first slots in Figure 3.30 contain the 32 frequency components that are calculated from the same 512 audio samples at the input of the sub-band filter.

Next, the frequency component groups are arranged for constructing audio frames. For Layer I, an audio frame contains the coded data of one group from each sub-band, while a Layer II audio frame contains the coded data of three consecutive groups from each sub-band,

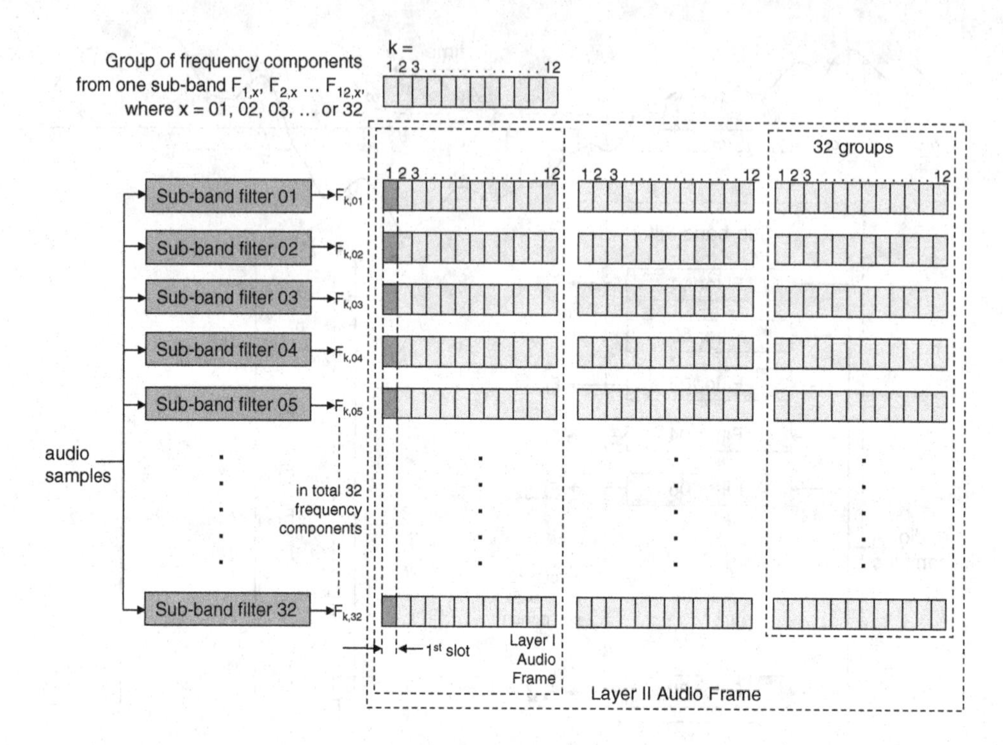

Figure 3.30 Structuring of frequency components in MPEG-1 audio Layers I and II

see Figure 3.30. Hence, an audio frame carries data representing $32 \times 12 = 384$ or $32 \times 3 \times 12 = 1152$ frequency components for Layer I and Layer II, respectively. For each 32 audio samples that enter the encoder, 32 frequency components are delivered, and therefore each frame corresponds to the same number of audio samples as frequency components. As a consequence, in the case of a sampling frequency of 48 kHz, an audio frame that contains 384 or 1152 samples, corresponds to an audio frame duration of 8 and 24 ms, respectively.

In Layer III, for grouping the frequency components, a slightly different approach is followed. Some variation is possible, but typically each such group contains 18 frequency components from a sub-band, while an audio frame contains two such subsequent groups from each sub-band. Hence, also a Layer III audio frame carries the coded data of 1152 ($32 \times 2 \times 18$) frequency components.

For efficient transport in audio frames, the quantized frequency components are coded by means of a lossless scheme to reduce redundancy. The audio frame formatter puts the coded frequency components in an audio frame, as required, along with side information on the applied scaling, quantization and coding, and concatenates the audio frames to form the MPEG audio bitstream. The decoder can decode the resulting MPEG audio bitstream without any knowledge of how the encoder performed the audio compression. This keeps the decoder complexity low, while leaving headroom for encoder improvements; when for example more efficient encoders evolve in future, the bitstreams produced will be fully compatible with all existing decoders.

The decoder parses the received bitstream and delivers the coded frequency components to the de-quantizer for decoding and reconstruction, as indicated in Figure 3.27. Finally, the reconstructed frequency components are fed to a filter bank that transforms the frequency components back into audio samples.

An audio frame may carry data from more than one audio channel, to support next to mono, dual channel (e.g. bilingual), stereo and multichannel audio. In case of multiple channels, an audio frame contains the temporally corresponding frequency components of each channel. For example, in MPEG-1 Layer II, each audio frame for a stereo signal contains the coded data of the 1152-frequency components from the left channel and the coded data of the corresponding 1152-frequency components from the right channel. Both channels can be encoded independently, but also joint stereo coding can be applied, exploiting the irrelevancy between the left and the right channel. The two tools that MPEG-1 audio provides for joint stereo coding, intensity stereo in all three layers and MS stereo in Layer II only, are beyond the scope of this book.

In MPEG-1 audio, each audio frame commences with a sync word, followed by a frame header carrying a number of parameters, such as the applied audio bitrate. As sync word, a bit string of 12 '1's is used: 1111 1111 1111'. This bit string is not unique in the audio bitstream, but the Layer I and Layer II bitstreams are designed so that this string has a very low probability of occurring. The bitstream may contain a CRC code, in which case the sync word, together with the CRC check, provides a very robust sync verification mechanism, but in addition, from the parameters provided in the frame header, the decoder can calculate after how many bits in the audio bitstream to expect the next sync word. If indeed a sync word is identified at the calculated position, then most probably the decoder is in sync with the audio stream, so that parsing can take place correctly. This synchronization mechanism can also be used by decoders in case of data loss or data error. After the error or data loss, the audio decoder will look for a sync word and verify whether the next sync word is found at the calculated position, prior to continuing the parsing and decoding of other data elements in the audio stream.

The audio bitstream may contain ancillary data. The use of ancillary data is a general mechanism that is applied in many MPEG audio specifications. Audio decoders are required to parse bitstreams that contain ancillary data, but do not need to process it. The decoder may pass it for further processing or may simply ignore it. The ancillary data is often used to create backward compatible extensions. For example, MPEG-2 audio uses it to define a multichannel extension to the stereo coding of MPEG-1 in a backward compatible manner.

3.3.2 Evolution of MPEG Audio

The evolution of audio compression is driven by progress in audio compression technology on one hand, and computing power and silicon integration technology on the other. As a result, state of the art audio compression technology changes over time when implementation of more advanced compression tools becomes feasible. An example of this phenomenon is found in the MPEG-1 audio specification, where some of the compression tools that were initially too complex for use in consumer devices, became successfully applied after several years (see Note 3.5).

The MPEG-1 audio specification was the starting point for an evolution of audio compression standards in MPEG. MPEG-1 audio supports audio in mono and stereo with sampling frequencies of 44.1 kHz, used on audio CDs, 48 kHz, often used in professional audio applications, and 32 kHz, suitable for lower quality audio. In MPEG-2 audio, three new sampling frequencies were

Note 3.5 The Evolution of MP3 and Its Feasibility

An example of audio technology becoming feasible over time is MP3. The MP3 specification is contained in the MPEG-1 standard, finalized in 1992. MPEG-1 audio, ISO/IEC 11172-3, specifies three layers: layer I, layer II and Layer III.

Layer I is the most simple one; it was intended for consumer products, such as, at that time, Digital Compact Cassette (DCC) recorders, that require both audio decoding and encoding in real-time. Layer II is more complex and more efficient than Layer I, and was intended for consumer products with only audio decoding in real-time and no encoding, such as broadcast receivers, where encoding is done by the broadcaster. MPEG-1 Layer II has been rapidly and widely adopted on the market.

Layer III is the MP3 specification, which provides the most efficient audio compression in MPEG-1, but at a complexity far beyond practical implementation for consumer products in 1992. For example, the signal adaptive hybrid filter bank (as opposed to the static one in Layer I and II), the more complicated psycho-acoustic model (note however that the simpler psycho-acoustic model from Layer I and II can be used in Layer III as well), and the variable length Huffman encoding of the quantized frequency components in Layer III were too complex for use in 1992 in consumer products.

As a result, the MP3 specification was ignored by the market place for several years. However, over time, devices became significantly more powerful, making real-time playback of MP3 files feasible. In a possibly desperate effort to safeguard the technology, MP3 encoders and players were made available on the Internet. Which turned out to be a brilliant move: in 1998, MP3 took-off as the most popular format for music distribution, initially on the Internet, but later also on a wide variety of consumers devices with MP3 capabilities.

defined as an MPEG-1 audio extension, so as to better cope with very low bitrate transmission channels: 24, 22.05 and 16 kHz, the sampling frequencies defined in MPEG-1 audio divided by two. By halving the sampling frequency, the 32 sub-bands cover each half the original bandwidth. The improved frequency resolution allows a more accurate shaping of the quantization noise and thus a lower bitrate.

In addition, MPEG-2 audio supports multichannel audio, with a special focus on the so-called 5.1 audio channel configuration, as defined in ITU-R Recommendation BS.775 [24]. In this configuration, the Left and Right speakers used for stereo are extended with one Centre speaker in the front, two Surround speakers in the side/rear and one Low Frequency Enhancement (LFE) speaker; see Figure 3.31. The location of the LFE speaker is not critical, as the human auditory system is not very capable to determine the location in a room of a speaker that only produces low frequencies.

Two multichannel audio systems are specified within the MPEG-2 standard. The first one is specified in MPEG-2 audio (part 3), defining a multichannel extension to MPEG-1 audio that is backward compatible with the MPEG-1 audio bitstream. From the multichannel audio signals, left and right signals are downmixed for coding as MPEG-1 audio Layer II stereo, while the coded data representing additional channels are transported as ancillary data within MPEG-1 audio frames. An MPEG-1 audio decoder will decode the stereo audio, while ignoring the ancillary data. An MPEG-2 audio decoder will decode the left and right downmix channels, but

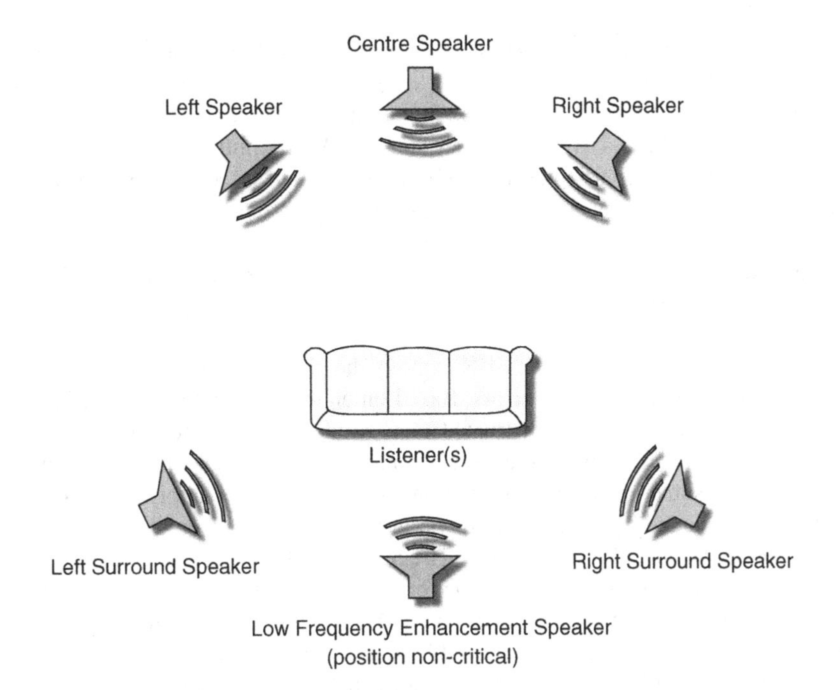

Figure 3.31 Speakers and their positions in a 5.1 multichannel audio configuration

also decodes the additional channels from the ancillary data. The multichannel audio signals are then derived from the downmixed stereo channels and the additionally transmitted channels. However, the amount of additional channel data may be more than can be carried in the ancillary data of the MPEG-1 audio frames. In such cases, the multichannel information is conveyed in two elementary streams: the base stream with MPEG-1 compatible audio frames and the extension stream with frames carrying the remaining coded data of the additional channels. Obviously, an extension stream can be decoded only if the base stream is available.

The second multichannel audio system defined within the MPEG-2 standard is MPEG-2 AAC (Advanced Audio Coding), specified in part 7 of the MPEG-2 standard [25]. MPEG-2 AAC uses all of the available advanced coding techniques feasible at the time of development, thereby providing a high quality multichannel system, but without offering backward compatibility with MPEG-1 audio.

MPEG-2 AAC does not only support consumer applications, but is also suitable for more professional applications; for example very high quality audio is supported by allowing 96 kHz as sampling frequency, which allows to convey an audio frequency spectrum way beyond what can be heard by the human auditory system according to the psycho-acoustic model.[27]

[27] On the other hand, there are audiophiles, who claim to be capable of hearing differences in sound impossible to measure, for example performance differences caused by the use of golden plated connectors. Hence, if myths from over-prized 'high-end' audio manufacturers and vendors play no role, then there may be a need to re-visit the psycho-acoustic model of the auditory system of humans with 'golden ears'.

Furthermore, a single MPEG-2 AAC audio stream is capable to carry up to 48 (!) audio channels.

In MPEG-4 audio [26], the AAC toolbox for compression is somewhat extended, while also various profiles of MPEG-4 AAC audio are defined. While initially the AAC focus was on using more advanced technologies for audio compression based on waveform coding, by introducing parametric coding as counterpart of waveform coding, MPEG-4 AAC evolved as a hybrid compression system. With parametric coding, parameters are used for the efficient modelling of an audio signal. Two methods of parametric coding are used within the MPEG-4 AAC family of standards:

- Spectral Band Replication (SBR); SBR exploits the phenomenon that human perception is less sensitive to high frequency errors. Based on an analysis of the lower frequencies, the higher frequencies can be reconstructed at a good perceptual quality, even though the waveform reconstruction is mathematically incorrect. To ensure a sufficiently adequate reconstruction, some guidance information is needed. When applying SBR, the lower frequencies are coded by MPEG-4 AAC or any other waveform coding scheme; the use of SBR is not restricted to MPEG-4 AAC. The SBR decoder reconstructs the higher frequencies by using guidance information in the form of SBR parametric information, transmitted to the decoder at a low bitrate, along with the AAC coded low frequency audio. The theoretical spectrum that can be conveyed is equal to the sampling frequency $f_s/2$; the lower frequencies are AAC coded, while the higher frequencies are SBR coded, as globally depicted in Figure 3.32.

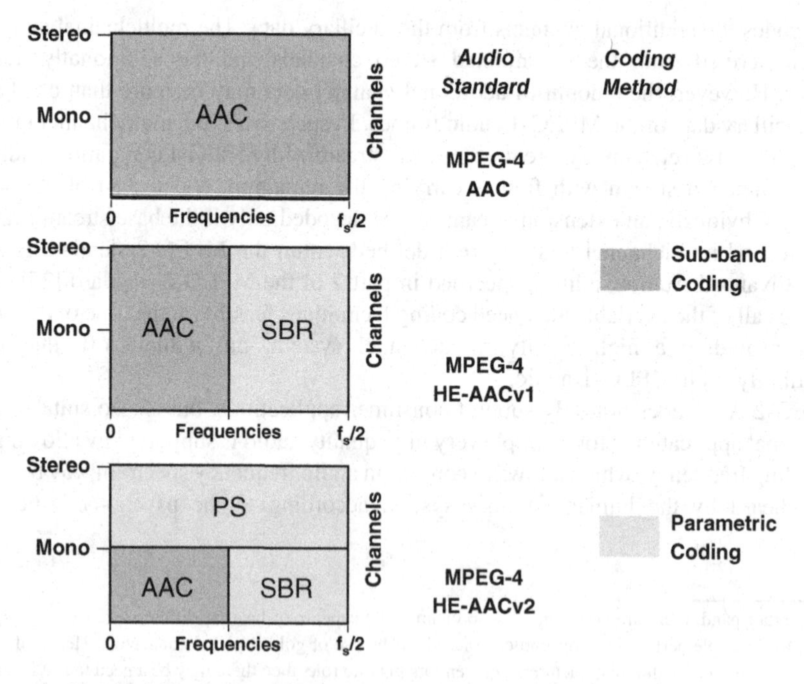

Figure 3.32 Coding Methods in AAC, HE-AAC v1 and HE-AAC v2

- Parametric Stereo (PS). With parametric stereo, the stereo image of the input signal is analysed in the encoder and represented by a set of stereo parameters. The stereo input signal is downmixed to a mono signal, and coded by for example an AAC-based encoder. Along with the AAC coded mono audio signal, the parametric stereo information is transmitted to the decoder as PS side information at a very low bitrate (2–3 kb/s). On the basis of the parametric stereo information, the decoder can reconstruct a stereo audio signal from the decoded mono audio [27] (see Figure 3.32).

The use of parametric coding is specified in the so-called MPEG-4 High Efficiency AAC (HE-AAC) standard, that is included in MPEG-4 audio (part 3).[28] In version 1 of HE-AAC (HE-AAC v1), the combination of SBR and AAC waveform coding is specified, while version 2 (HE-AAC v2) specifies the combination of SBR, PS and AAC waveform coding, as indicated in Figure 3.32.

It should be noted that parametric coding is most effective at low bitrates. Due to the fact that the reconstruction of the stereo from the mono signal and the prediction of the higher from the lower frequencies are both not always perfect, it is often difficult to reach transparent quality when the bitrate is increased. Figure 3.33 presents an impression of the resulting performance for various MPEG audio codec's, based upon a variety of listening test results. For stereo audio, bitrates for AAC start typically at 96 kb/s, for HE-AAC v1 typical bitrates are 40–96 kb/s and

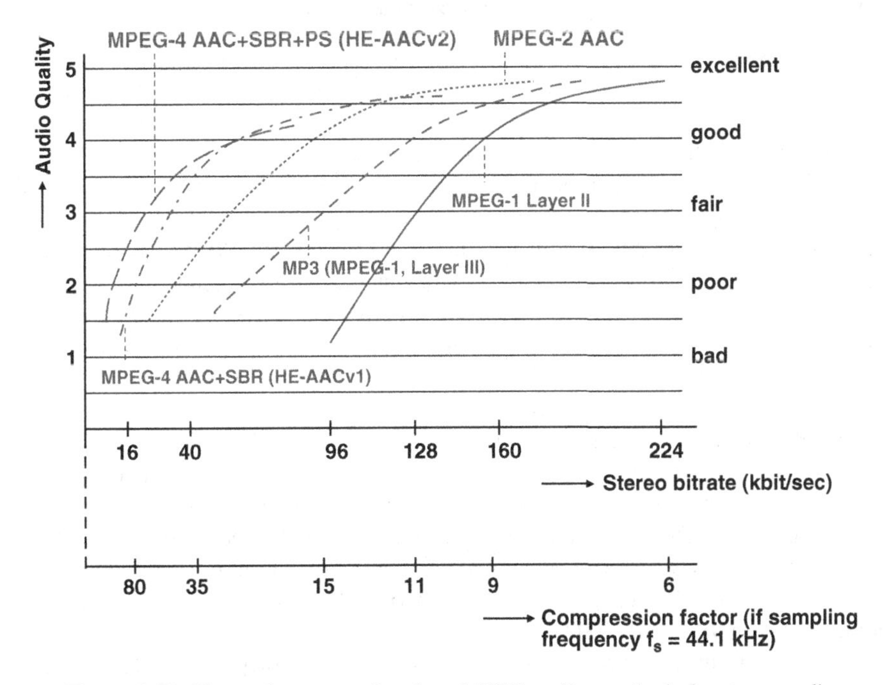

Figure 3.33 The performance of various MPEG audio standards for stereo audio

[28] MPEG-4 part 3 is not a single standard, but a suite of specifications for audio and speech coding.

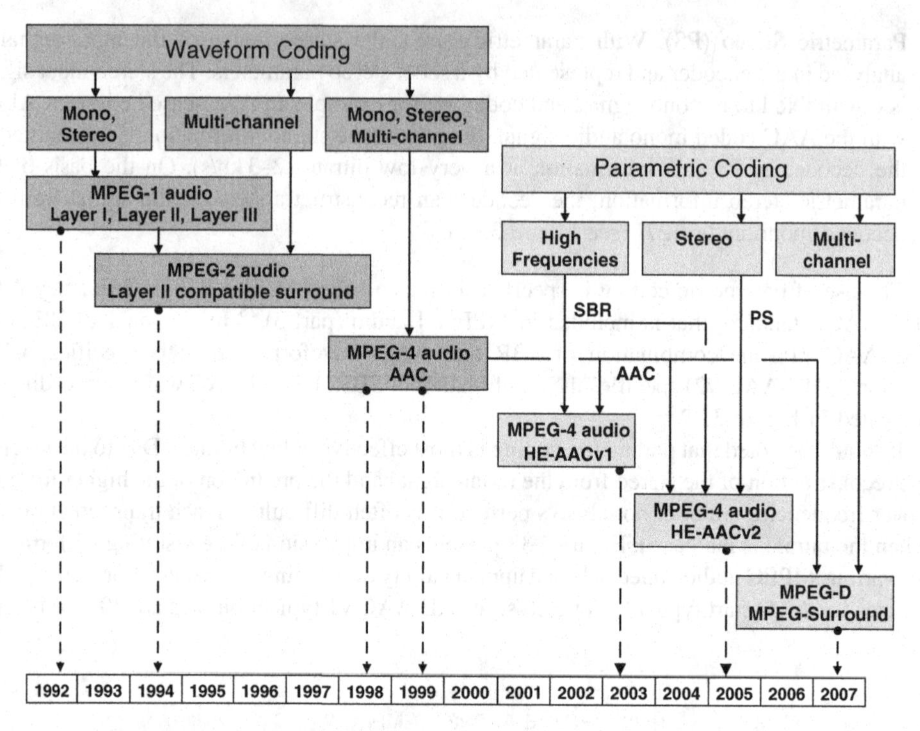

Figure 3.34 Evolution of MPEG audio standards between 1992 and 2007

for HE-AACv2 18–40 kb/s. For 5.1 multi-channel audio, the typical bitrates for AAC start at 256 kb/s; for HE-AAC v1, the typical bitrate for this channel configuration is 128–256 kb/s.

Parametric coding can also be used for surround sound multichannel audio in a manner comparable to parametric stereo. However, this is not specified in MPEG-4 audio as another HE-AAC version, but in another MPEG audio standard, the so-called MPEG Surround standard [27], published as ISO/IEC 23003-1. The MPEG Surround standard can be used to extend mono or stereo audio services to multi-channel audio in a backwards compatible manner. As the core for coding audio in mono or stereo any audio standard can be used. The total bitrates used for the (mono or stereo) core and the MPEG Surround data are typically only slightly higher than the bitrates used for coding of the core. MPEG Surround adds a side-information stream to the (mono or stereo) core bitstream, containing parametrical surround audio data. Legacy mono or stereo playback systems will ignore this side-information while players supporting MPEG Surround decoding will output the reconstructed multi-channel audio. When HE-AACv2 is used as core codec, a good quality at 64–96 kb/s for 5.1 multi-channel audio can be achieved. Figure 3.34 summarizes the evolution of the main MPEG audio standards between 1992 and 2007.

Since 2007, various further developments on audio coding took place in MPEG. In 2010, the Spatial Audio Object Coding (SAOC) standard [28] was published as ISO/IEC 23003-2. SAOC provides an audio coding algorithm which allows the distribution of multiple individual audio objects within one stream. Examples of such objects are voices, instruments and ambience; in an audio mix, the listener can adjust for instance the volume and position of a speaker or instrument. Such features may be of interest in sophisticated teleconferencing systems, games and in some other (niche) applications. Recent developments show a renewed interest in audio object coding, to support true 3D audio, but also to make loudspeaker setup independent from the transmission format. SAOC may play a role here for very low bitrate applications, but in MPEG-2 system environments no SAOC application is foreseen.

To compress speech, a dedicated state of the art speech codec, optimized for the specific characteristics of speech, is usually more efficient than the MPEG audio standards discussed so far in this chapter. To address the desire for a single audio codec, MPEG developed the Unified Speech and Audio Codec (USAC) [29–31] standard, published in 2012. USAC is capable to efficiently compress the entire spectrum of audio signals, including speech, music and speech mixed with music signals. At bitrates of 8 kb/s for mono and 96 kb/s for stereo, the quality achieved by USAC is always better and often significantly better than the quality achieved by the best state of the art codec tailored for general audio or speech. The achieved quality scales up at higher bitrates in a remarkably consistent manner, while also multi-channel audio is supported. Thereby USAC promises to find application in any area in which low-bitrate transmission or storage is necessary and audio content is an arbitrary mix of speech, speech plus music and music.

The next MPEG audio work item is on 3D audio, but to which extent the MPEG audio standards will evolve further is an interesting question. The usage of an audio codec is driven by business models; so far, a more economic use of bandwidth resources has been an important driver. However, with the achieved very low bitrates for audio, and the increase of bandwidth available over networks, it is questionable whether bandwidth will continue to drive the development of new audio codecs with improved efficiency. As shown above with SAOC, also new features may drive the evolution of audio codecs, though it should be noted that it is often easier to invent a new technical feature than to identify an associated business model providing economic justification for its usage.

References

1. ITU (2005) The NTSC, PAL and SECAM standards are published by ITU-R as Rec. ITU-R BT.470-7, referencing ITU-R BT.1700 and Rec. and Rec. ITU-R BT.1701: http://www.itu.int/rec/R-REC-BT.470-7-200502-I/en; http://www.itu.int/rec/R-REC-BT.1700-0-200502-I/en; http://www.itu.int/rec/R-REC-BT.1701-1-200508-I/en.
2. Background on the NTSC, PAL and SECAM standards, as well as on the 4:2:0 and other YUV formats is found in: Jack, Keith (2007) *Video Demystified, a Handbook for the Digital Engineer*, 5th edn, Newnes – Elsevier, ISBN: 978-0-7506-8395-1.
3. Wikipedia (2013) Further background on the NTSC, PAL and SECAM standards and their usage is found on Wikipedia: http://en.wikipedia.org/wiki/NTSC; http://en.wikipedia.org/wiki/PAL; http://en.wikipedia.org/wiki/SECAM.
4. ITU (2013) For information on the ITU-R see: http://www.itu.int/ITU-R/.
5. ITU (2013) ITU Recommendation 601: http://www.itu.int/rec/R-REC-BT.470-7-200502-I/en.
6. EBU (2013) Background on the Rec. 601 standard is provided in a paper of Staley Baron and David Wood, entitled "Rec. 601, the origins of the 4:2:2 DTV standard": http://tech.ebu.ch/docs/techreview/trev_304-rec601_wood.pdf.
7. Wikipedia (2013) The 4:2:0, 4:2:2 and other YUV formats are discussed at Wikipedia: http://en.wikipedia.org/wiki/Chroma_subsampling.
8. ISO/IEC 11172-2: (1993) Information technology – Coding of moving pictures and associated audio for digital storage media at up to about 1.5 Mb/s – Part 2: Video http://www.iso.org/iso/catalogue_detail.htm?csnumber=22412.

9. ISO/IEC 13818-2: (2000) Information technology – Generic coding of moving pictures and associated audio information – Part 2: Video http://www.iso.org/iso/catalogue_detail.htm?csnumber=31539.
10. ISO/IEC 14496-2: (2004) Information technology – Coding of audio-visual objects – Part 2: Visual http://www.iso.org/iso/catalogue_detail.htm?csnumber=39259.
11. ISO/IEC 14496-10: (2010) Information technology – Coding of audio-visual objects – Part 10: Advanced Video Coding (AVC) http://www.iso.org/iso/catalogue_detail.htm?csnumber=56538, also published as ITU-T Recommendation H.264: http://www.itu.int/rec/T-REC-H.264-201106-P.
12. Wikipedia (2013) For background on Moore's Law see for example http://en.wikipedia.org/wiki/Moore%27s_law.
13. DVB: (2003) At the DVB World conference (see www.dvbworld.org/) in 2003, Mr. Ken McCann discussed the increase of coding efficiency of MPEG 2 video encoders over time. See: http://www.zetacast.com/Assets/DVB%20World%202003.pdf.
14. ISO (2013) ISO/IEC 14496 – Coding of audio-visual objects. An overview of the complete MPEG-4 suite of specifications is found at: http://mpeg.chiariglione.org/standards/mpeg-4/mpeg-4.htm. A list of published MPEG-4 standards is found at: http://www.iso.org/iso/search.htm?qt=14496&published=on&active_tab=standards.
15. ITU (2013) Information on Study Group 16 is found at: http://www.itu.int/ITU-T/studygroups/com16/index.asp.
16. ITU (2013) Information on VCEG is found at: http://www.itu.int/ITU-T/studygroups/com16/sg16-q6.html.
17. ITU (2013) Information on JVT is found at: http://www.itu.int/en/ITU-T/studygroups/com16/video/Pages/jvt.aspx.
18. ITU (2013) A list of H.26x standards is found at: http://www.itu.int/itu-t/recommendations/index.aspx?ser=H.
19. ITU (2013) Information on JCT-VC is found at: http://www.itu.int/en/ITU-T/studygroups/com16/video/Pages/jctvc.aspx.
20. McCann, K. (2012) Progress towards High Efficiency Video Coding, DVB Scene Issue 39: http://issuu.com/dvbscene/docs/dvb_scene_issue_39/15.
21. ISO/IE C11172-3: (1993) Information technology – Coding of moving pictures and associated audio for digital storage media at up to about 1,5 Mb/s – Part 3: Audio http://www.iso.org/iso/catalogue_detail.htm?csnumber=22412.
22. ISO/IEC 13818-3: (1998) Information technology – Generic coding of moving pictures and associated audio information – Part 3: Audio http://www.iso.org/iso/catalogue_detail.htm?csnumber=26797.
23. Background on the hearing threshold and other important elements of MPEG-1 and MPEG-2 audio coding is found at: Van de Kerkhof, Leon M. and Cugnini, Aldo G. (1994) *The ISO/MPEG Audio Coding Standard*, Widescreen Review, pp. 58–61; Veldhuis, Raymond and Breeuwer, Marcel (1993) *An Introduction to Source Coding*, Prentice Hall, ISBN: 0-13-489089-2; Bosi, Marina and Goldberg, Richard E. (2003) *Introduction to Digital Audio Coding and Standards*, Kluwer Academic Publishers/Springer, ISBN: 1-4020-7357-7.
24. ITU-R (2006) The text of Recommendation ITU-R BS.775-2 (07/2006) entitled "Multichannel stereophonic sound system with and without accompanying picture" is found at: http://www.itu.int/dms_pubrec/itu-r/rec/bs/R-REC-BS.775-2-200607-I!!PDF-E.pdf.
25. ISO (2006) ISO/IEC 13818-7: Information technology – Generic coding of moving pictures and associated audio information – Part 7: Advanced Audio Coding (AAC) http://www.iso.org/iso/catalogue_detail.htm?csnumber=43345.
26. ISO (2009) ISO/IEC 14496-3: Information technology – Coding of audio-visual objects – Part 3: Audio http://www.iso.org/iso/catalogue_detail.htm?csnumber=53943.
27. Schuyers, Erik, Breebaart, Jeroen, Purnhagen, Heiko and Engdegård, Jonas (2004) Low complexity parametric stereo coding", Presented at the 116th AES Convention, 2004 May 8–11, Berlin, Germany.
28. ISO (2007) ISO/IEC 23003-1: Information technology – MPEG audio technologies – Part 1: MPEG Surround http://www.iso.org/iso/catalogue_detail.htm?csnumber=44159.
29. ISO (2010) ISO/IEC 23003-2: Information technology – MPEG audio technologies – Part 2: Spatial Audio Object Coding (SAOC) http://www.iso.org/iso/catalogue_detail.htm?csnumber=51827.
30. ISO (2009) ISO/IEC 14496-3: Transport of Unified Speech and Audio Coding (USAC; Amd 3: 2012) http://www.iso.org/iso/catalogue_detail.htm?csnumber=59635.
31. ISO (2013) ISO/IEC 23008-2: Information technology – High efficiency coding and media delivery in heterogeneous environments – Part 2: High Efficiency Video Coding (HEVC) http://www.iso.org/iso/catalogue_detail.htm?csnumber=xxxxx also published as ITU-T Recommendation H.265: http://www.itu.int/rec/T-REC-H.265-201304-I.

4

Other Important Content Formats

Which other content formats in the scope of MPEG are important for the MPEG-2 systems standard. As far as needed to understand their impact on MPEG-2 systems, metadata, timed text, lossless and scalable lossless audio, multiview video and 3D video are discussed.

During the development of the MPEG-1 and MPEG-2 standards, the focus of MPEG was very much on video and audio coding and on systems technology. However, over the years, the MPEG scope was widened significantly, as will be addressed in this book in Chapter 8, when the evolution of the MPEG-2 system standard is discussed. As a result, many of the MPEG standards produced after MPEG-2 were not relevant for MPEG-2 system applications. On the other hand, also several content formats were developed that could add value to MPEG-2 system applications, at least potentially. Sections 4.1–4.5 discuss some of those formats, as far as needed to understand the role they played in the design and evolution of MPEG-2 systems. Some content formats with impact on the MPEG-2 system standard are not discussed in this chapter, for example because the impact was minor, or because the added value for MPEG-2 system applications is expected to remain insignificant. It should be noted though that in Chapter 13 of this book each content format supported in MPEG-2 systems is discussed.

4.1 Metadata

Metadata provides information about one or more aspects of data; examples are time and date of creation, its purpose and its owner. Within an MPEG-2 system stream, metadata typically refers to a carried audiovisual work, such as a film/movie, sport event, news programme or entertainment show. For each such work, the metadata may provide the title, a summary and other information deemed useful when distributing the audiovisual work. Metadata may apply to one or more segments of the work, specified at the production of the work by means of timing parameters that are metadata format specific. If the audiovisual work is a movie, a segment could for example indicate a scene with a certain actress, in the case of a soccer game, a segment could show a goal scored by a team, and in the case of a news programme a segment could represent a news item.

Metadata associated with audiovisual works as discussed here should not be confused with information needed for the decoding and presentation of coded video and audio, such as the picture resolution and the audio sampling frequency. While such information is sometimes also

Fundamentals and Evolution of MPEG-2 Systems: Paving the MPEG Road, First Edition. Jan van der Meer.
© 2014 John Wiley & Sons, Ltd. Published 2014 by John Wiley & Sons, Ltd.

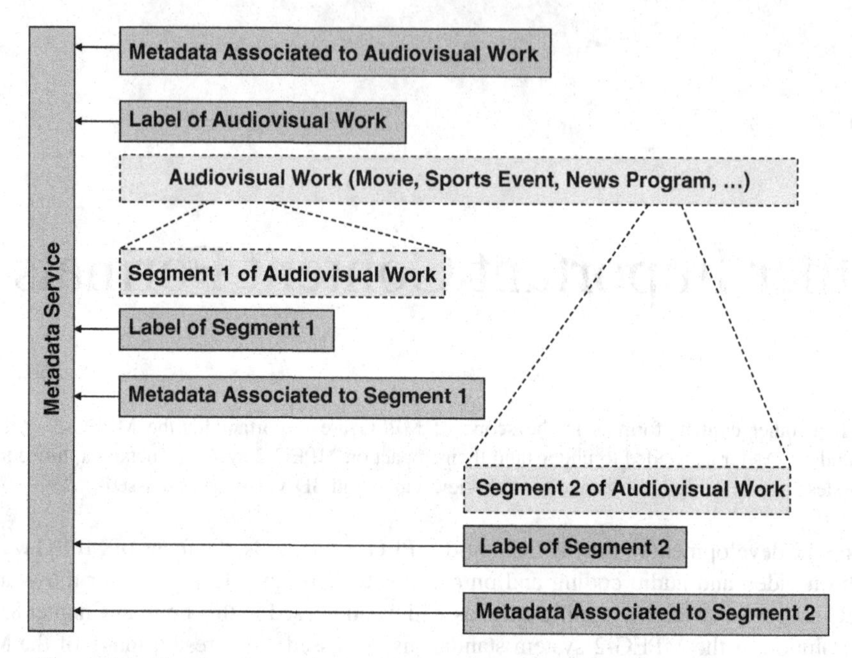

Figure 4.1 Metadata (service) associated to audiovisual work and segments thereof

referred to as metadata, it is usually carried within the coded stream or conveyed out-of-band by other means. However, in MPEG-2 system streams, the 'metadata' information associated with coded streams is carried in descriptors, as discussed in Section 9.4.4 and in more detail in Chapters 13 and 14. Many of those descriptors are specifically defined for a particular content format.

Figure 4.1 introduces the notion of a metadata service using metadata describing an audiovisual work and two segments thereof. A metadata service is a coherent set of metadata delivered to the receiver for a specific purpose, for example to indicate scenes in which a certain actress performs or to indicate the highlights of a sports event. The metadata is associated with content, and for that purpose labels are assigned to each audiovisual work or segment thereof that metadata is associated with. Figure 4.1 shows the labels assigned to the entire audiovisual work and to both segments; the metadata is linked to the associated content by referencing the label.

To uniquely label audiovisual works, an international numbering system has been standardized, within ISO,[1] but beyond MPEG. In part 1 of this standard ISO/IEC 15706 [1], the International Standard Audiovisual Number (ISAN) is specified; an ISAN uniquely identifies an audiovisual work, comparable to the ISBN number for books. Of each audiovisual work multiple versions may exist; for example, there may be differently edited versions of a movie, and a sport event may have various versions with different video resolutions and different languages. To standardize versioning of audiovisual works, part 2 of ISO/IEC 15706 [2]

[1] The work has been carried out by ISO/TC46/SC09/WG01: Working Group 1 in Sub-Committee 9 of Technical Committee 46 in ISO.

defines a version segment based on ISAN. An ISAN combined with this version segment, constitutes an ISAN version identifier, referred to as a V-ISAN. A V-ISAN is a precise and unique identifier of audiovisual works that allows for storage and retrieval of audiovisual works and associated metadata, both by automated systems and by human beings. The ISAN database contains all relevant metadata to describe the audiovisual work. This is particularly useful in audiovisual production and distribution systems, broadcasting applications and electronic programme guides.[2]

Also within MPEG work on metadata took place, but the focus of MPEG was not on labelling content but on describing content. Already in the second half of the 1990s, MPEG started its effort to establish a metadata standard, called the 'Multimedia Content Description Interface', which resulted in the MPEG-7 standard, formally known as ISO/IEC 15938 [3]. The MPEG-7 standard consists of 12 parts, providing a rich set of tools to describe multimedia content; among others, tools where developed to recognize images on their visual content, for example to retrieve all images with a colourful sunset. Furthermore, part 1 [4] defines a textual and a binary format for the transport of MPEG-7 metadata. The MPEG-7 standard has not found wide deployment on the market yet, but in future its usage may prove useful. Therefore it has been considered desirable in MPEG, when defining MPEG-2 system support for metadata, and in particular for ISAN and V-ISAN, to also support carriage over MPEG-2 systems of the textual and binary formats of MPEG-7.

Both the ISAN and V-ISAN labelling as well as the MPEG-7 descriptive metadata are defined at a generic level, without aiming at any application in particular, and fully independent of the transport of the audiovisual work. As a consequence, for carriage of the work in an MPEG-2 system stream, an interface layer is needed to specify the mapping between the audiovisual content carried in an MPEG-2 system stream and the associated metadata. This interface is specified within the MPEG-2 systems standard, and includes means for labelling ISAN and V-ISAN to content, means to transport metadata of various formats, and means to relate metadata to content.

For the use of metadata in MPEG-2 systems, a number of assumptions on the carried metadata services are made, as listed below:

- One or more metadata services are carried in one or more metadata stream(s).
- A metadata service consists of a concatenation of metadata access units; as several metadata formats are supported, the meaning of a metadata access unit depends on the applied metadata format.
- Each metadata access unit is associated to a single metadata service.
- It should be possible (but not be required) to assign a time stamp to a metadata access unit, to indicate the time the metadata access unit is intended to be decoded.
- The metadata may contain accurate timing references to the audiovisual work, for example to indicate a segment or an event within an audiovisual work. For this purpose various time clocks may be used, such as the Universal Time Clock (UTC) or SMPTE time codes. The used metadata time clock is independent of any time code that may be embedded in the audiovisual streams itself; for example, UTC may be the time clock used in the metadata, while SMPTE time codes are embedded in the video stream.

[2] A V-ISAN provides a complete solution to identify a work and its versions; therefore V-ISAN is more used in practice than the original ISAN, but the use of the latter remains possible.

- If a time clock is used in the metadata, the same clock must be used for the entire content for which the clock is used; no time discontinuities are allowed to occur. Once the times in the metadata are determined, editing can be done only if the time discontinuities caused are corrected. Furthermore, the metadata time clock and the sampling clocks of the associated content must be locked. As multiple metadata formats are supported, the above is intentionally kept abstract.

4.2 Timed Text

Soon after the completion of the MPEG-2 video, audio and systems standard, the lack of a standard for subtitling was identified in several application standardization bodies. At that time, subtitling was not explicitly addressed in MPEG; ongoing developments in the context of MPEG-4 were not expected to provide a suitable solution for a subtitling system anytime soon. Therefore several application standardization bodies decided to develop their own subtitling solution, based on coding the graphic pixels of the subtitles in the form of a bitmap with a certain colour depth, and suitable for use within the (so-called) On Screen Display (OSD) capabilities[3] available in first generation MPEG-2 video decoders for overlay on top of the decoded video.

As a result, multiple application specific subtitle standards evolved, for example in DVB, DVD and ATSC. While these standards were effective for their intended purpose, they contributed to the view that in practice there are often many standards to choose from. An important aspect of the above-mentioned application specific subtitle coding systems is that they typically allow the content distributor to fully determine the 'look and feel' of the subtitles, as the subtitle bitmap fully determines the shape, colour and other characteristics of the presentation of the subtitles. When distributing audiovisual content, not only the video and audio quality are important, but also the quality of presented subtitles, if any.

Though coding of subtitle bitmaps is not a very efficient method to convey subtitles, the required bitrate is rather marginal compared to the bitrate required for video. For example, a single language bitmap DVB subtitle stream typically requires a bitrate between 50 and 100 kb/s, depending on the size of the subtitle and the number of applied colours [5]. Compared to a bitrate of about 4 Mb/s for an MPEG-2 video stream, the DVB subtitling stream consumes roughly 2% of the video bitrate. Note however that this percentage increases with new generation video codecs that are more efficient than MPEG-2 video.

To improve the subtitle coding efficiency, instead of subtitle bitmaps, character strings can be encoded based on Unicode [6] for identifying each character. With this objective, 3GPP developed a subtitling system for multimedia services to mobile devices, resulting in the timed text standard 3GPP TS 26.245 [7]. The scope of timed text is wider than just subtitling; for example also Karaoke applications and scrolling news item applications are supported. Within TS 26.245, each subtitle is referred to as a 'text sample'.

In timed text, each text sample consists of a text string, optionally followed by one or more text modifiers (see Figure 4.2). The text string represents the characters that form the text to be

[3] These capabilities were usually based on a CLUT, a Colour Look-Up Table with, for example, four entries. This allows a bitmap with a colour depth of two bits per pixel to code the pixels of the subtitle characters. For each entry, the CLUT table is loaded with (for example) three arbitrary R, G and B values of eight bits each, so that out of the complete RGB colour space, four specific colours are available for use by the subtitle characters. Often transparency can be used as well.

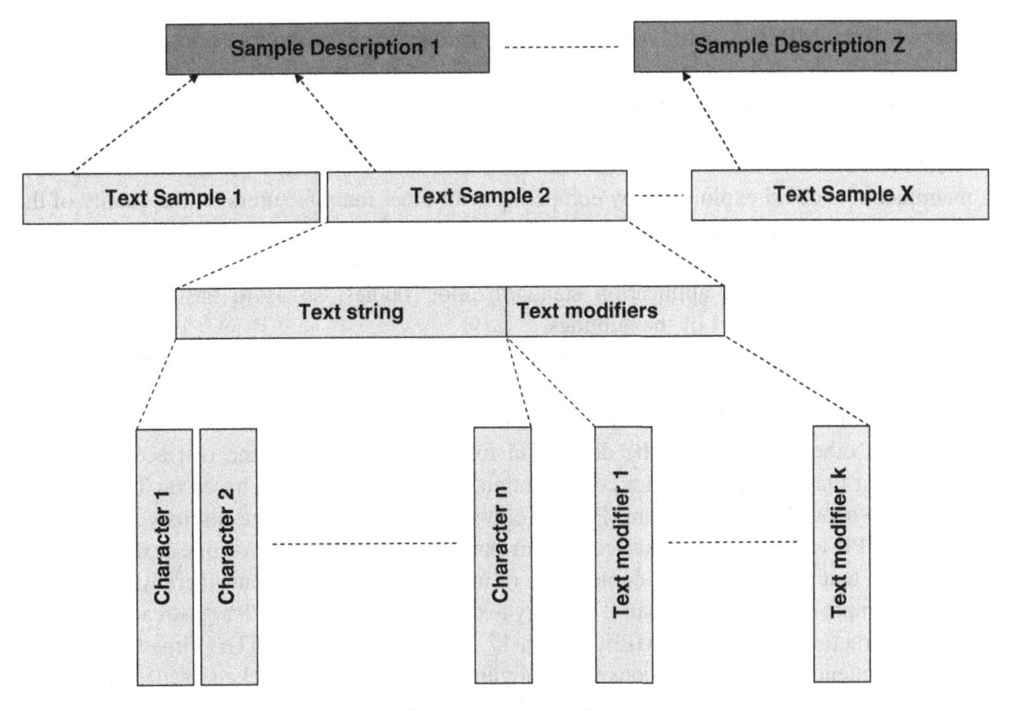

Figure 4.2 Data structure of timed text

displayed, while the text modifiers carry the changes to be applied during the presentation of the text string, such as colour changes of characters synchronized with a song for a Karaoke application or a changed position of the text string on the screen. Global information on timed text, such as the font to be used and the (default) position on the screen, is specified in sample descriptions. Multiple sample descriptions are allowed, but each text sample is associated to a single sample description. Typically, a text sample representing a subtitle is small, about 100–200 bytes, hence 800–1600 bits. If it is assumed that for a subtitling service in a certain language on average every two seconds a subtitle is to be presented, then the bitrate of a timed text subtitle stream is about 0.4–0.8 kb/s, significantly less then the subtitle bitmaps used in ATSC, DVB and DVD.

The subtitling system defined in 3GPP TS26.245, supports a wide variety of features, such as:

- Various fonts, styles, such as bold and italic, and character sizes,
- Text scrolling in vertical and horizontal direction,
- Left–right, right–left and vertical writing directions,
- Dynamic highlighting,
- Blinking,
- Hypertext.

The use of the timed text format defined in 3GPP TS 26.245, may also be of benefit to MPEG applications. To improve compatibility between MPEG and 3GPP applications, MPEG decided to specify in MPEG-4 part 17 [8] how to use 3GPP timed text streams in MPEG environments. MPEG-4 part 17 forms an intermediate layer for the transport of 3GPP timed text in MPEG defined systems, including MPEG-2 systems, thereby allowing timed text based subtitles within MPEG-2 system streams.

Compared to the subtitle bitmap coding formats, timed text offers broadcasters and other content distributors less control over the 'look and feel' of subtitle presentations. Decoder manufacturers could exploit this by competing with other manufacturers on the quality of the presented subtitles, thereby trying to offer an interesting competitive edge. However, in practice, often the use of fonts in decoders is specified too, and if not, this can be done on application level or by application standardization bodies, so as to ensure a minimum presentation quality level of the subtitles.

MPEG-4 part 17 defines timed text streams and text access units; a timed text stream is a concatenation of transported text access units with a specified format. One of the supported formats is timed text according to TS 26.245, but also private formats are allowed, while support of other formats may be defined in future. The format of the text access units is specified in the so-called decoder configuration. A timed text stream based on TS 26.245 is referred to as a 3GPP text stream. Below, only 3GPP text streams are discussed.

In a 3GPP text stream, a text access unit contains all coded data of one text sample; in addition, a text access unit may contain one or more sample descriptions, thereby allowing to convey sample descriptions in-band. To carry text access units in a 3GPP text stream, a flexible framing structure is defined in MPEG-4 part 17, based on so-called TTUs, Timed Text Units. TTUs are intended for easy and convenient alignment with transport packets used in the various transport layers. For example, by means of TTUs, small text access units can be concatenated for carriage in large transport packets, while large text access units can be fragmented for carriage in smaller transport packets.

Eight different TTU types are defined for transport of a text access unit in a 3GPP text stream, but the use of only five is actually specified; three TTU types are reserved for future use:

- TTU[1] carries a complete text sample, hence the complete text string and all modifiers associated to the text string (if any).
- TTU[2] carries one fragment of a text string.
- TTU[3] carries the first fragment of the text modifiers of a text sample.
- TTU[4] carries a non-first fragment of the text modifiers of a text sample.
- TTU[5] carries a complete sample description.
- TTU[0], TTU[6] and TTU[7] are reserved for future use

As an example, Figure 4.3 depicts the construction of a 3GPP text stream from a text access unit that contains one text sample and two sample descriptions. The text sample consist of a text string, followed by a number of text modifiers. Two cases of TTU partitioning are presented in Figure 4.3, one for large and one for small transport packets. In the first case, suitable for relatively large transport packets, the text access unit is partitioned into a single TTU[1] and two TTU[5]s; multiple text access units may be aggregated into a relatively large transport packet. To achieve alignment with relatively small transport packets, the use of small TTUs may be needed, as shown in the second case, where the text access unit is partitioned into three TTU[2]s, one TTU[3], two TTU[4]s and two TTU[5]s. The 3GPP text stream is constructed by concatenating the TTUs formed from each text access unit, whereby the text access unit data is

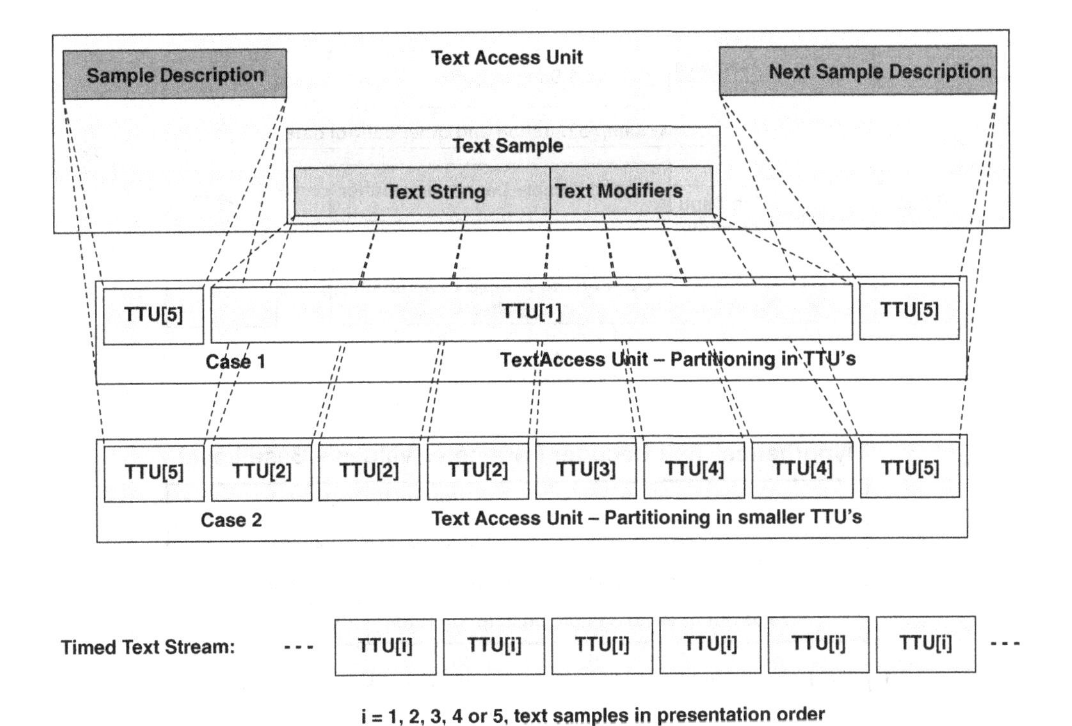

Figure 4.3 Constructing 3GPP text stream from text access units by using TTUs

in presentation order and TTUs with text sample data from one text access unit are not interleaved with TTUs with text sample data from another text access unit.

The TTU header carries information needed for the parsing and decoding of the 3GPP text stream, such as:

- the type of TTU,
- the text sample duration, the period during which the text sample is intended to be presented,
- whether the Unicode characters in the text string are UTF-8 [9] or UTF-16 [10] encoded,
- which sample description to use,
- the number of bytes of the TTU and the number of bytes in the text string, if appropriate.

For the decoding of 3GPP text streams a Hypothetical Text Decoder (HTD) is defined in MPEG-4 part 17 (see Figure 4.4). The purpose of the HTD is to define constraints a compliant 3GPP text stream needs to meet, similar to the VBV model for MPEG-1 and MPEG-2 video. Without imposing any architecture of a 3GPP text decoder, any such decoder needs to be capable of decoding a 3GPP text stream that can be decoded by the HTD; in this respect, the HTD serves as a reference for each 3GPP text decoder implementation, for example to determine the resources needed to decode a 3GPP text stream.

At the input of the HTD, the TTU decoder processes each byte of the received 3GPP text stream instantaneously. Each byte of a text sample and an in-band provided sample description is transferred to the text sample and sample description buffer, respectively. Information

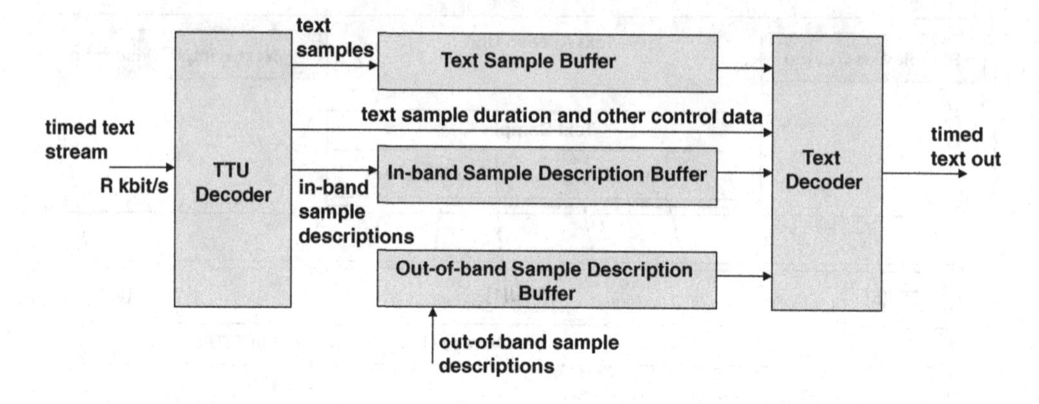

Hypothetical Text Decoder Parameter Values – Base Level

Bitrate R	10 kbit/s
Size of Text Sample Buffer	8192 Bytes
Size of In-Band Sample Description Buffer	4096 Bytes
Size of Out-of-Band Sample Description Buffer	4096 Bytes

Figure 4.4 The Hypothetical Text Decoder for timed text and its parameters

needed to control the decoding of 3GPP text, such as the text sample duration, is passed to the timed text decoder. The bytes of a 3GPP text stream enter the TTU decoder at a bitrate of R kb/s if the text sample buffer is not full; if the text sample buffer is full, no bytes enter the TTU decoder. Note that sample descriptions may also be provided out-of-band, for example, applications may require that some sample descriptions are available at decoders; sample descriptions may also be available on the internet.

At its presentation time, the text sample is removed instantaneously from the text sample buffer, decoded and presented by the timed text decoder. A 3GPP text stream must be constructed so, that at its presentation time, the complete text sample is present in the text sample buffer. Furthermore, when in-band sample descriptions are used, a 3GPP text stream must be constructed so, that each sample description is completely present in the sample description buffer at the time a text sample using data from that sample description is decoded and presented.

Figure 4.4 also provides values of HTD parameters, so as to explicitly signal the bitrate of the 3GPP text stream and the available buffer sizes for its decoding. Note, though there is a single buffer to store sample descriptions, that the available buffer size for in-band and out-of-band sample descriptions is specified separately. The HTD parameters do not apply to each 3GPP text stream; to signal whether the parameters apply, the concept of profiles and levels is used, similarly as in MPEG video standards. For MPEG video, a profile specifies a sub-set of the available coding tools, while a level indicates a maximum video resolution. For 3GPP text streams, only a single profile and a single level are defined. In the base profile each tool specified in TS 26.245 may be used, while for the base level, the parameters listed in Figure 4.4 apply. In future, there may be a need for other profiles supporting sub-sets of the tools from TS 26.245, and other levels with different HTD parameters.

4.3 Lossless and Scalable Lossless Audio

In Section 3.3, lossy audio compression is discussed. However, MPEG did not only define lossy audio coding schemes, but also lossless ones. The aim of a lossy codec is to preserve the subjective audio quality by exploiting both redundancy and irrelevancy with respect to the human auditory system, while a lossless audio codec only exploits redundancy in the signal, thereby preserving every single bit of the original audio data, though at the expense of a significantly higher bitrate. A lossless codec may be required for archiving audio that must be suitable to undergo multiple subsequent editing operations, and for distribution of audio content of superior quality. The MPEG audio group addressed the market need for a lossless audio codec, and decided to standardize three lossless audio codecs in amendments to MPEG-4 audio. These three audio codecs, called ALS, SLS and DST, are briefly discussed below.

- MPEG-4 ALS (Audio Lossless Coding), a single layer 'lossless-only' audio codec that encodes the input audio signal, while supporting a huge number of audio channels.
- MPEG-4 SLS (Scalable Lossless Coding), a lossless audio codec that can be used in a hierarchical system as a lossless enhancement layer for a lossy core codec, such as MPEG-4 AAC). Thereby SLS is suitable for use in production environments where a lossless format is needed, next to a lossy format for distribution.
- MPEG-4 DST (Direct Stream Transfer), a lossless codec of oversampled audio with an oversampling rate of 64, 128 or 256. MPEG-4 DST is used on Super Audio CD (SA-CD) [11], where a one-bit oversampled format is applied at a sampling rate of 64 times 44.1 kHz, roughly equal to 2.8 MHz, resulting in a data rate of about 2.8 Mb/s. The lossless DST codec is optionally used to compress this data rate.

The achieved compression by the lossless codecs is roughly equal to a factor two. As a consequence, there is a very significant gap between the bitrate consumed by a lossy audio codec and a lossless audio codec. For example, an MPEG Surround codec with a HE-AACv2 core codec achieves good (but not transparent) quality at 64–96 kb/s for 5.1 multichannel audio, which rougly corresponds to a compression ratio of 50–70 compared to 16 bit PCM, while a lossless codec with six channels of audio with a sampling frequency of 48 kHz and 24 bit samples would require about $(6 \times 48\,000 \times 24)/2 = 3.5$ Mb/s.

The MPEG-4 ALS, SLS and DST specifications were originally published as ISO/IEC 14496-3:2005/Amd 2:2006 [12], ISO/IEC 14496-3:2005/Amd 3:2006 [13], and ISO/IEC 14496-3:2001/Amd 6:2005 [14], respectively. All three specifications were incorporated in edition 4 of MPEG-4 audio, ISO/IEC 14496-3:2009 [15].

4.4 Multiview Video

One of the potential applications that MPEG anticipated on is multiview video. The initial objective of this activity was to offer a flexible viewpoint to viewers. For example, when watching a soccer match, the viewer may wish to watch the game from a higher position, or closer to one of the goals. With this initial approach, in principle, the user can set the viewpoint to an almost arbitrary location and direction, which can be static, change abruptly, or vary continuously, within the limits given by the available camera setup. Similarly, the audio listening point may be changed accordingly.

To encode the video streams from the various viewpoints requires a high bandwidth, but if the viewpoints are relatively close, the bandwidth can be reduced significantly by encoding these videos into a single bitstream. For this purpose, MPEG defined the Multiview Video Coding standard, MVC, as an amendment to the AVC standard [16]. When multiple cameras capture the same scene from slightly different viewpoints, a large amount of inter-view statistical dependencies occur that are exploited by MVC. To encode the pictures from a certain camera, not only temporally related pictures from the same camera can be used, but also pictures from neighbouring cameras.

A very simple multiview video application is 3D video, discussed in Section 4.5. While the MVC standard is successfully applied for 3D video, as discussed in Section 4.5.3, the market potential for more advanced multiview applications is yet to be seen.

4.5 3D Video

When a human observer views a natural scene, the right and left eye see slightly different images, due to the different positions of the eyes. When the brain combines the right- and left-eye images, depth is perceived as a third dimension of the scene.[4] This is not only the case for observing a natural scene, but also when the right- and left-eye images are presented to the viewer on a cinema or television screen or viewed in some other suitable manner. To capture both views, a stereo camera can be used with two lenses and two image sensors to simultaneously record two pictures of the same scene, as if perceived by the right and left human eye; the distance between both lenses roughly corresponds to the typical distance between the right and left human eye (about 6.5 cm) [17]. When presenting the right-eye picture to the right eye and the left-eye picture to the left eye, the viewer will perceive a three-dimensional picture. This applies both to photographic pictures and to moving video.

While this phenomenon, also known as stereoscopy, was already known for a long time, commercial exploitation did not really take off until about the 1950s, when stereo viewers and some stereo cameras for photos [17] became available on the consumer market. For moving video, commercial exploitation is more difficult, due to the need for suitable 3D display technology. For viewing 3D movies on cinema screens, polarized filters [18] can be used. The left and right views are projected superimposed onto the same screen through different polarized filters, while the viewer watches the screen through eyeglasses, also with different filters, so that the left eye only sees the left view and the right eye only the right view. This method of presenting 3D movies is based on the effect that the filter for each eye passes only the light that is similarly polarized and blocks the light polarized in the opposite direction. Both linear and circular polarization can be applied. With linear polarization, when viewers tilt their head, this impacts on their 3D experience, as both views start getting visible by each eye. This effect is improved when circular polarization is used.

Polarized filters can also be used for displaying 3D video on CRT displays, but they required a device set-up with two CRTs and one-way mirrors unsuitable for usage in the home. Therefore, commercial exploitation of 3D for moving video was difficult, but this changed when (very) high resolution flat screen displays became feasible. Next to polarized glasses, several other options are available to ensure that the left view is only seen by the left eye and the

[4] There are also other ways to perceive depth; for example, even with one eye, depth can be perceived when an observer views a scene while moving.

right view only by the right eye. While details are beyond the scope of this book, three basic approaches are briefly discussed below:

- Polarized glasses [18]: On the screen, alternate lines of the left and right view are displayed and accordingly polarized by different filters, so that the left eye only sees the left view and the right eye only the right view. Note that polarized glasses are lightweight, very cheap and do not require any battery.
- Active shutter glasses [19]: by means of an electric signal, each glass can be made either transparent or dark. For each 3D picture, a right view and a left view is available. On the screen, subsequently the right and the left view is displayed, whereby each right view is followed by a left view and vice versa. The active shutter glasses ensure by fast switching, synchronized with the screen refresh rate, that the left eye only sees the left views and the right eye only the right views. To avoid the perception of flicker by the viewer, the screen refresh rate must be sufficiently high, for example 100 or 120 Hz.
- Auto-stereoscopic displays [20]: 3D video can be viewed without any special glasses, for example by means of using lenticular lenses on the screen, but other options[5] are possible too. A lenticular lens is an array of magnifying lenses, designed so that when a viewer looks at the screen from slightly different angles, different pixels are seen. The lenses and pixels are arranged so, that within a certain range, the left eye observes pixels from a view that is left to the view observed by the right eye. With lenticular lenses multiple views can be seen from slightly different horizontal positions. An example explaining the principle of a lenticular display is depicted in Figure 4.5. In this example, the display has nine views, of which the viewer sees at most two at any time; in Figure 4.5, the left eye sees view number 5, while the right eye sees view number 6. When the right eye is moved to the original position of the left eye, then the right eye sees view number 5, while the left eye sees view number 4. Hence, a viewer positioned at the appropriate viewing distance can move to the right and left, for example to see what is behind an object in the foreground, as discussed in more detail in Section 4.5.2. Note that there is a transition area between the two views in which the perceived 3D experience may be distorted.

Typically, the depth effect on an auto-stereoscopic display is reduced compared to a 3D display with glasses. The same depth effect would be perceived if the two adjacent views seen on an auto-stereoscopic display are the same as seen through the glasses. Which would mean that the capturing positions of the adjacent views on an auto-stereoscopic display should typically differ by the nominal distance between the left and right human eye and that capturing positions of the most left and most right views in Figure 4.5 differ by eight times that nominal distance. This is undesirable. First, in practice, only one or two views may be conveyed to the receiver, in which case the receiver would be required to construct the missing views from the view(s) available at the receiver; with such large differences in capturing positions this would often be an unrealistic challenge. Second, when the user moves from one view combination to the adjacent one, the user perception of the transition should be realistic, which means that the perceived differences between the 3D scenes corresponding to adjacent view combinations should not be too large. Reducing the depth effect on auto-stereoscopic displays overcomes both problems. The reduction can be achieved by using intermediate views; for example, if the

[5] By means of parallax barriers [19] in front of the screen effects comparable to lenticular lenses can be obtained.

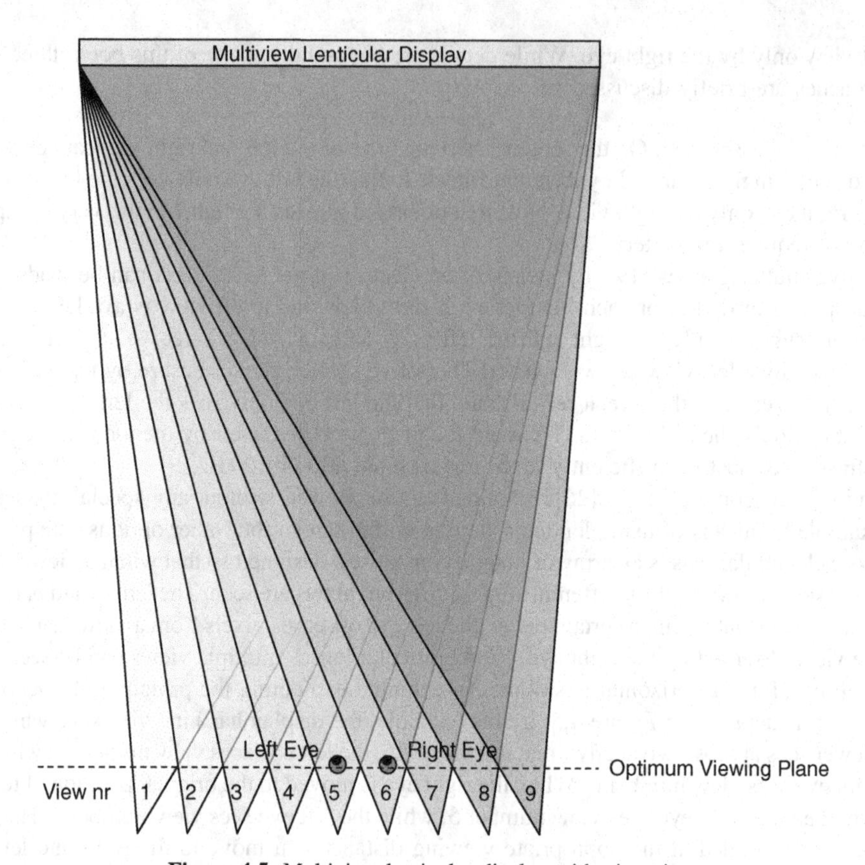

Figure 4.5 Multiview lenticular display with nine views

left and right views displayed on a 3D display with glasses are denoted as L and R, then in Figure 4.5 the lenticular display may present L and R as view number 3 and view number 7, while the other views, with capturing positions of adjacent views that differ by one-quarter of the nominal distance between the left and right human eye, are constructed by the receiver from the available view(s). For a discussion on constructing missing views see also Section 4.5.2.

In a 3D video player, independent of the type of display, in principle, freedom exists to change the depth range for the presentation of the 3D video at the screen. An example of the depth range at a 3D video screen is depicted in Figure 4.6. The depth range at the screen of observed 3D scenes is equal to the maximum depth in front of the screen plus the maximum depth behind the screen. The depth range can be increased to achieve a more spectacular 3D experience, though viewers may not appreciate watching such 3D video during a long period.[6] Manufacturers of players may offer users the choice for a depth range, so that they can choose to undergo a more spectacular 3D video experience, or to watch 3D video with a more moderate depth range, to ensure that 3D video can be conveniently watched during a longer period. It should be noted however that a change of the depth range will require processing of the views

[6] Reviewers sometimes label a movie as 'shot in headache-free 3D'.

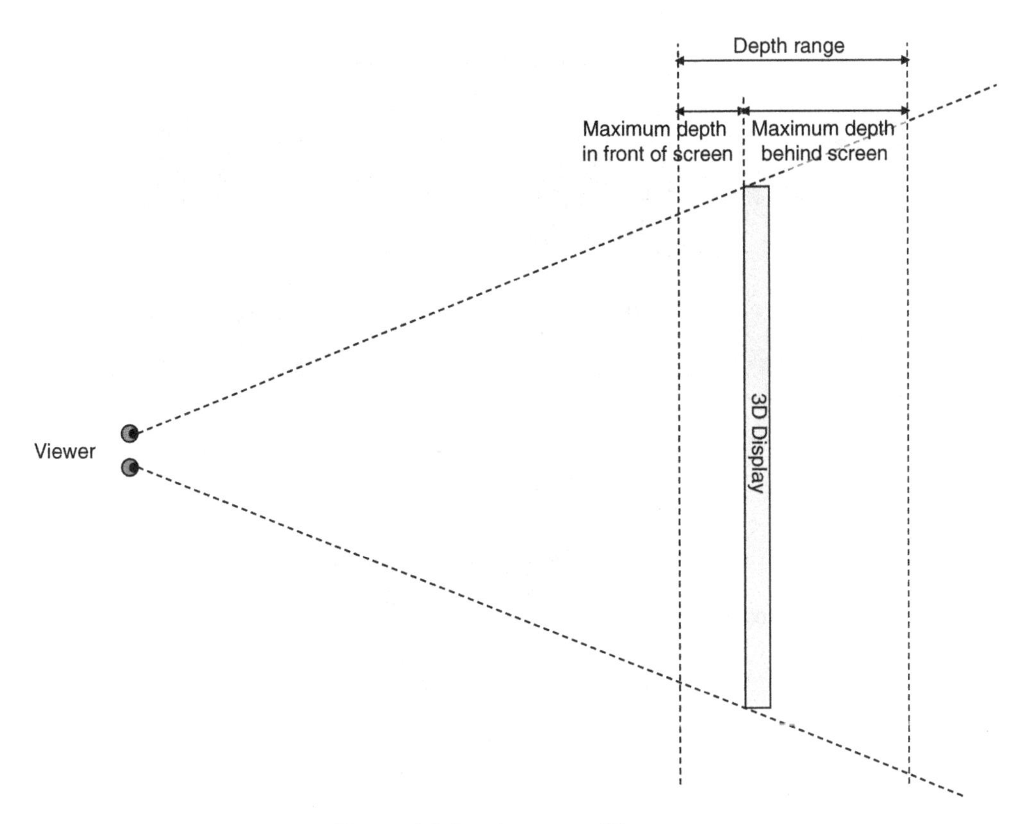

Figure 4.6 Depth range at 3D screen

presented at the screen, and depending on the available information, such processing may be complicated.

When defining support in MPEG for 3D video, the question is which format is most suitable to convey the left and right views of a 3D video signal. To this question there are multiple answers, each with specific advantages and disadvantages. The following are the options that were identified and developed further in MPEG:

- Convey the left and right views within a single video stream.
- Convey depth information associated to 2D video.
- Use MVC to convey the left and right views.
- Convey left and right views, each with associated depth information.

The above options are discussed in Sections 4.5.1–4.5.4.

4.5.1 Left and Right Views in a Single Video Stream

The left and right views can be conveyed in a single video stream by constructing a more or less regular video signal from the left and right views, in a manner so that the left and right views can be reconstructed from the decoded video at the player. To ensure correct presentation of

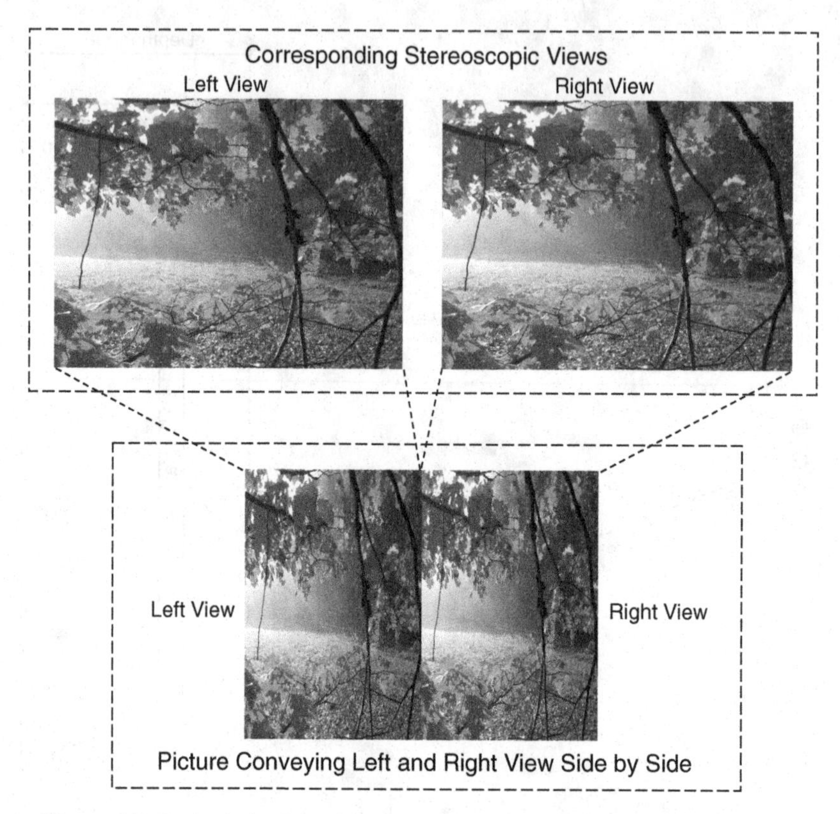

Figure 4.7 Horizontal sub-sampling of both views for side by side arrangement

corresponding left and right views, the player needs to be able to identify left view and right view pairs captured at the same temporal instant in time. A simple manner to ensure this identification, is to 'packetize' both views into the same picture; within DVB, this method is used in their so-called frame compatible plano-stereoscopic 3DTV [21]. Two packetization methods are depicted in Figure 4.7 and in Figure 4.8. In both pictures it is assumed that the left and right view both have the same format as the pictures conveying both views, for example 1280 pixels by 720 lines at 50 or 60 Hz.

In Figure 4.7, the left and right view are sub-sampled horizontally by a factor two. If the picture conveying both views contains 1280 luminance pixels per line, then the sub-sampling reduces the horizontal resolution of each view to 640 pixels. The second method is depicted in Figure 4.8, where both views are sub-sampled vertically by a factor two. If the picture conveying both views contains 720 lines, then for each view 360 lines are conveyed. The pictures conveying both views can be coded by MPEG-2 video and AVC, without making any change to these video coding standards; there is only a need for a method to signal which packetization method of the left and right views is applied, so as to enable a player to use the correct post-processing of the decoded video signal when reconstructing the left and right views.

The approaches depicted in Figures 4.7 and 4.8 are attractive for their simplicity, while there is also a nice fit with existing (broadcast) infrastructures. However, the price to pay, losing

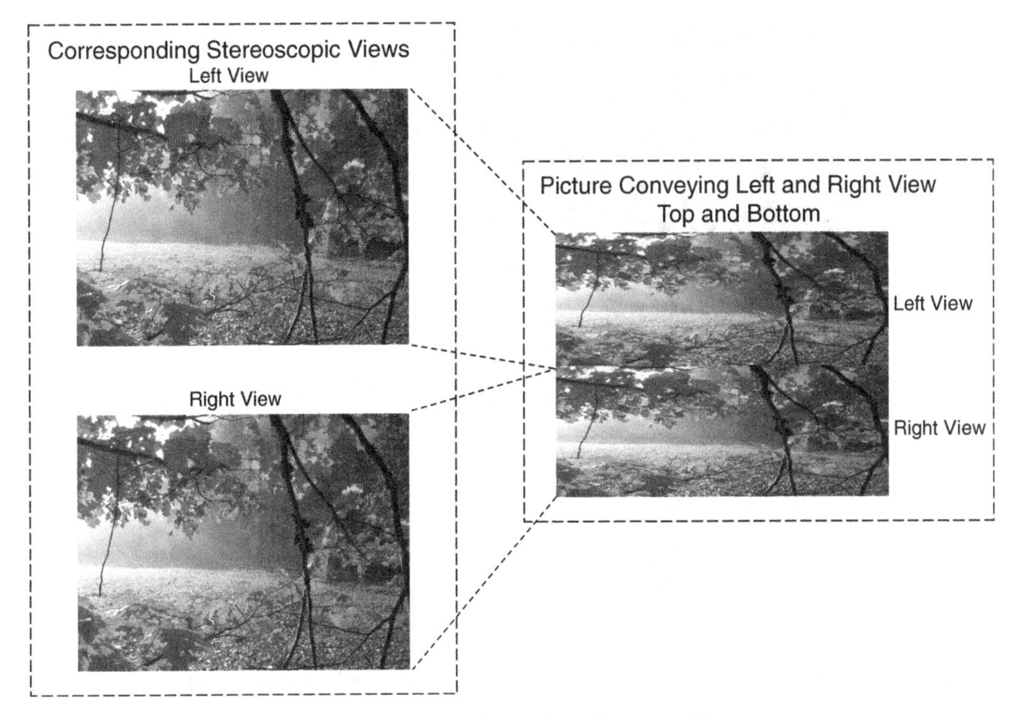

Figure 4.8 Vertical sub-sampling of both views for top and bottom arrangement

either half the horizontal or half the vertical resolution, may be considered too high. Moreover, from video compression perspective, conveying 3D video by packetization of right and left views in one picture is not particularly efficient. Therefore also other methods to convey 3D video are developed within MPEG.

4.5.2 Depth Information Associated to 2D Video

By means of depth sensing video cameras, not only 2D video pictures can be captured, but also the depth map associated to these pictures. A depth map specifies the distance between the camera and the objects within the scene. The depth map is represented by a greyscale image; the grey value of a pixel indicates a depth value; usually, more brightness indicates a more near object, less brightness a more far one. Figure 4.9 shows an example of a depth map associated to a video picture [22].

A depth map may have the same resolution as the 2D video pictures, but also a lower spatial and/or temporal resolution is possible. For example, if the 2D video has a resolution of 1280 pixels by 720 lines at 50 or 60 Hz, then the depth map may have a resolution of 640 pixels by 360 lines at 25 or 30 Hz. In any case, depth maps can be considered as regular video, but without colour and with a specific significance of the luminance pixel values. As a consequence, to compress depth maps, any video coding standard, such as MPEG-2 video and AVC, can be used. When coding depth maps, the bit rate can be reduced significantly by allowing horizontal, vertical and temporal sub-sampling. In practice, when using the same video coding standard, the bitrate for depth maps is in the range of 10–20% of the (2D) video bitrate.

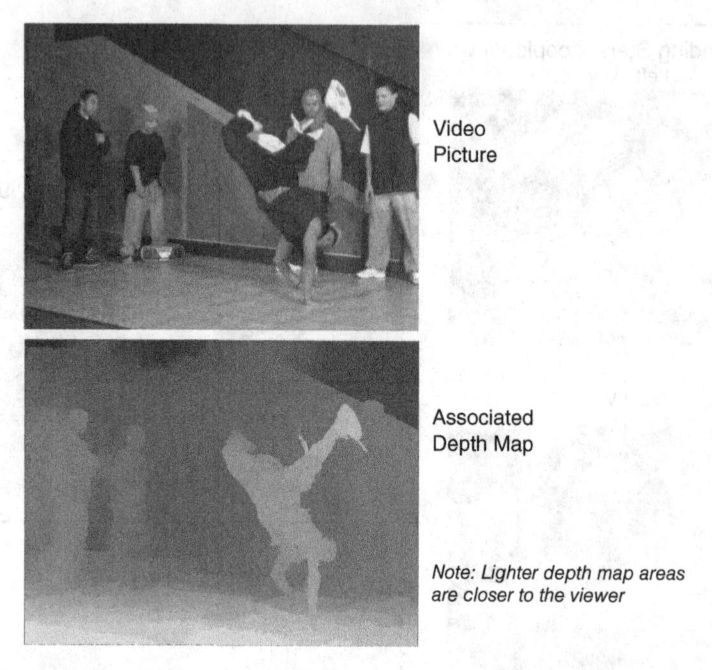

Figure 4.9 Example of a depth map associated to a video picture. *Source:* [22] Microsoft 2013. Reproduced with permission of Microsoft

How to use MPEG-2 video, AVC and other video coding standards to compress depth maps is specified by MPEG in ISO/IEC 23002-3, Representation of auxiliary video and supplemental information [23]. This includes all requirements to uniquely relate depth maps to associated 2D video pictures[7] and messages to convey parameters, such as the applied depth range. ISO/IEC 23002-3 is amazingly small: less than 10 pages of normative text; most probably the smallest MPEG standard ever!

Depth maps are important in 3D video, as the distance between the observer and an object determines the shift of that object when viewed from a (slightly) different position. For example, an object present in the right view of a stereo camera, is shifted to the right in the left view of that camera, because the left view is captured from a position slightly more to the left. The apparent shift of an object due to a change in the observing position, is called the parallax of the object. For longer distance objects, the parallax approaches zero, while the parallax increases for objects that are closer by the observer. Figure 4.10 shows parallax differences in the very simple example of a stereo camera capturing a vertical pole against a background plane. The observed parallax is depicted for the three distances D_1, D_2 and D_3 between the camera and the pole. The parallax of the pole, found by comparing the shift of the pole in the left and right view, is equal to P_1, P_2 and P_3, respectively. Figure 4.10 clearly demonstrates that the parallax of the pole decreases with a longer distance between the camera and the pole.

[7] ISO/IEC 23002-3 does not only specify coding of depth maps, but also of parallax maps. Depth and parallax are closely related, as discussed in the next paragraph.

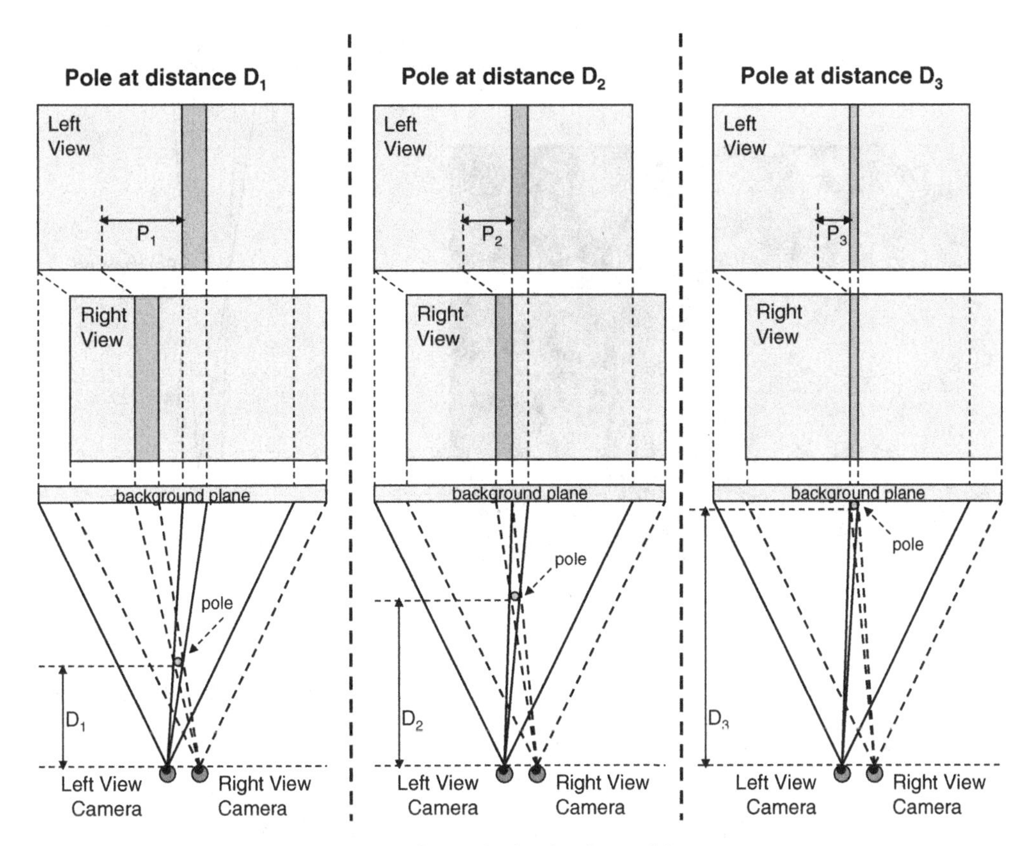

Figure 4.10 Parallax of a vertical pole observed by a stereo camera

A depth map can be used in a 3D player to calculate parallax differences at the screen for the views required for 3D video. This is very helpful when multiple views are needed for an autostereoscopic display. If a 3D video service conveys the right and left stereoscopic views, denoted as R and L, then a 3D receiver using the multiview display with nine views of Figure 4.5 must be capable of constructing the other seven views. If this player presents the conveyed views R and L as views 3 and 7, then views 1, 2, 4–6, 8 and 9 must be constructed. To assist in this construction, it is desirable that for at least one conveyed view, and preferably for both, an associated depth map is conveyed as well.

Conveying a depth map associated to one or more views is also very useful for players, when the depth range for the presentation of the 3D video at the screen is to be changed. Furthermore, depth maps are a very welcome tool to convert 2D into 3D video; for this purpose conversion algorithms can be used in 3D video players to convert regular 2D video into views needed for 3D video. These algorithms use sophisticated tools to analyse characteristics of 2D video pictures, such as foreground and background motion, to estimate the depth of 2D video pictures, with the objective to construct other 3D views, for example a left view, while using the primary 2D video as the right view (or vice versa). Errors in estimating depth by such algorithms disturb the 3D experience of users. The availability of depth maps associated with the primary 2D video can improve the reliability of 2D to 3D conversion algorithms.

Figure 4.11 Background occlusion by foreground object in left and right views

As described above, depth maps are very useful to construct additional views for 3D video, but even in the presence of depth maps, still unsolved issues in constructing additional views remain. A depth map can only indicate one depth value per pixel and has consequently difficulties with various optical phenomena such as semi-transparency and reflections. Also the problem of background information obscured by a foreground object is not solved by depth maps. An example of background occlusion is depicted in Figure 4.11, using a simplification of a picture with autumn leaves in the foreground and a relatively distant background. At the right hand side of Figure 4.11, a leave is represented as a short line in the foreground and the trees in the background by a simple background plane. In this set-up, in the right eye view, the background plane is obscured by the leave between A and C, and in the left view between B and D. In practice it is often possible to estimate obscured information from preceding or following pictures, but there are also cases where it is difficult to reconstruct the obscured information from pictures in the past or in the future, such as in the case of an obscured rotating object. Reconstruction of obscured information is a rendering issue for players, but to assist such rendering, some standardization effort may be desirable (see Section 4.5.4).

4.5.3 Use of MVC to Convey Left and Right Views

The development of MVC, discussed in Section 4.4, as an amendment to the AVC standard was initially more driven by researchers than by market requirements. However, when the MVC standard was almost finished, 3D video gained momentum in the market place, which had significant impact on the MVC market potential: all of a sudden the need for a problem to solve disappeared: MVC is the obvious solution for the coding of stereoscopic video with a left and a right view of the same scene. Note though that MVC is capable of coding more than two views.

In MVC, one of the views is used as the base video that is AVC encoded, while the encoding of the other view(s) is building on the base video. For example, when MVC is used to code stereoscopic video, the left view may be the base video, in which case pictures from the left view can be used as reference to predict right view pictures. For each view, a separate video stream is produced, hence for stereoscopic video one base video stream and one dependent video stream. This approach allows backward compatibility with existing AVC decoders. When 3D video is introduced in an already operational AVC based 2D video service by applying MVC for coding stereoscopic video, then the installed base AVC decoders for this service will be capable to decode the base video stream, while the 3D players with a MVC decoder can decode the other (dependent) stream as well.

For the coding of the dependent stream(s), MVC defines a relatively simple extension of the AVC standard: only extensions to the higher level AVC syntax are specified, without making any change to the low level AVC syntax. By making changes to the low level syntax, the compression of the dependent stream could have been made slightly more efficient, but such changes have an important drawback. In many practical implementations, the low level syntax is processed by an AVC decoding IC, while the higher level syntax is handled by control software. As a result of not changing the low level AVC syntax, an AVC decoding IC with modified control software can also be used for the processing of the dependent stream, so that for the decoding of stereoscopic video, an MVC player may use two AVC decoder ICs, one for the base video and the other for the dependent video. Furthermore, and possibly even more important, the re-use of existing building blocks for an AVC decoding IC is made possible in a single chip MVC decoding IC. Thereby, MVC players could be introduced in the market place rapidly and in a cost-effective manner, without an immediate need for huge investments in IC development. The result was that various applications rapidly adopted and applied the MVC standard to compress stereoscopic video.

4.5.4 Further 3D Video Evolution

The options discussed in Sections 4.5.1–4.5.3 to convey 3D video provide excellent opportunities to introduce 3D video in the market place, but in future, more efficient 3D video coding will become possible once new 2D video codecs such as HEVC evolve. Furthermore, some unresolved issues may need to be addressed. For example, the occlusion problem discussed in Section 4.5.2 is not very serious for moderate depth ranges at the player, but may need resolution for larger depth ranges. Within MPEG, 3D video coding will evolve further. Most probably the main, if not sole, focus will be on the coding of stereoscopic video and depth maps associated to one or both views.

A number of deliveries may be expected. One option to improve the coding of 3D video is to extend MVC with the capability to code depth maps associated with one or both stereoscopic view(s). This MVC extension may enable more cost-effective implementations of depth map decoders, due to sharing reference pictures. As long as no changes are made to the low-level AVC syntax, compatibility is achieved with MVC stereo, while the depth maps are coded with an efficiency comparable to the method discussed in Section 4.5.2 based on the use of AVC. If changes are made to low-level AVC syntax, a (slightly) better coding efficiency is achieved for coding the dependent view and depth map(s). However, the consequences, modified AVC hardware and only compatibility with the base view, may prove a too high price to pay for the achieved coding efficiency improvement.

Progress on the coding of 3D video will remain closely related to progress on 2D video coding. Based on the HEVC standardization effort, a more efficient 3D video codec is expected. This 3D video codec will probably use the HEVC standard in a way comparable to the use of AVC by MVC. If taken into account during HEVC development, HEVC may support some tools to improve the coding of dependent views and depth maps, but only if the complexity increase caused is marginal. As with MVC, probably various options for compatibility will be offered, so as to allow a trade-off between coding efficiency and market requirements.

In the long term, the evolution of 3D video coding will depend on the added value of 3D video in the market place. In this respect consumer perception of the 3D video experience will be very important; in case 3D video turns out to be no more than an interesting short term gimmick, consumers may lose interest after some years, in which case future 3D video standards will only contribute to the required size of the bookshelf for MPEG standards. However, if 3D video maintains to be of significant value on the consumer electronics market, the evolution of 3D video coding will remain to be industry driven. In any case, the focus of 3D video coding will probably remain on stereoscopic video and associated depth map(s).

References

1. ISO (2002) ISO 15706-1: Information and documentation – International Standard Audiovisual Number (ISAN) – Part 1: Audiovisual work identifier http://www.iso.org/iso/catalogue_detail.htm?csnumber=28779.
2. ISO (2007) ISO 15706-2: Information and documentation – International Standard Audiovisual Number (ISAN) – Part 2: Version identifier (V-ISAN) http://www.iso.org/iso/catalogue_detail.htm?csnumber=35581.
3. ISO (2012) ISO/IEC 15938: Information technology – Multimedia content description interface. An overview of the MPEG-7 suite of specifications is found at: http://mpeg.chiariglione.org/standards/mpeg-7/mpeg-7.htm A list of published standards from ISO/IEC/JTC1/SC29, including all parts of the MPEG 7 standard, is found at: http://www.iso.org/iso/catalogue_tc/catalogue_tc_browse.htm?commid=45316&published=on.
4. ISO/IEC 15938-1: (2002) Information technology – Multimedia content description interface – Part 1: Systems http://www.iso.org/iso/catalogue_detail.htm?csnumber=34228.
5. Acessibilidade (2013) Background information on the DVB subtitling standard is found at: http://www.acessibilidade.net/tdt/DVB_Subtitling_FAQ.pdf.
6. ISO (2013) The Unicode standard – The Unicode Consortium – Published as ISO/IEC 10646; the latest version is found via: http://www.iso.org/iso/catalogue_detail.htm?csnumber=51273 An overview of the various Unicode versions is found at Wikipedia: http://en.wikipedia.org/wiki/Unicode.
7. 3GPP (2009) 3GPP TS 26.245,[8] 3rd Generation Partnership Project; Transparent end-to-end packet switched streaming service (PSS); Timed text format (Release 6) http://www.3gpp.org/ftp/Specs/html-info/26245.htm.
8. ISO (2006) ISO/IEC 14496-17: Information technology – Coding of audio-visual objects – Part 17: Streaming Text Format http://www.iso.org/iso/catalogue_detail.htm?csnumber=39478.
9. Yergeau, F. (2003) "UTF-8, a transformation format of ISO 10646", STD 63, RFC 3629, November 2003, found at: http://datatracker.ietf.org/doc/rfc3629/.
10. Hoffman, P. and Yergeau, F. (2000) "UTF-16, an encoding of ISO 10646", RFC 2781, February 2000, found at: http://datatracker.ietf.org/doc/rfc2781/.
11. Wikipedia (2013) Information on the Compact Disc and on the Super Audio Compact Disc is found at: http://en.wikipedia.org/wiki/Super_Audio_CD, http://en.wikipedia.org/wiki/Compact_Disc.

[8] This specification has been updated (unchanged) for subsequent releases (release 7, release 8, etc.)

12. ISO (2005) ISO/IEC 14496-3: Information technology – Coding of audio-visual objects – Part 3: Audio, 3rd Edition, Amendment 2 (2006): Audio Lossless Coding (ALS), new audio profiles and BSAC extensions. The specification of MPEG-4 ALS is included in edition 4 of MPEG-4 audio, ISO/IEC 14496-3:2009 [15].
13. ISO (2005) ISO/IEC 14496-3: Information technology – Coding of audio-visual objects – Part 3: Audio, 3rd Edition, Amendment 3 (2006): Scalable Lossless Coding (SLS). The specification of MPEG-4 SLS is included in edition 4 of MPEG-4 audio, ISO/IEC 14496-3:2009 [15].
14. ISO (2001) ISO/IEC 14496-3: Information technology – Coding of audio-visual objects – Part 3: Audio, 2nd Edition, Amendment 6 (2005): Lossless coding of oversampled audio. The specification of MPEG-4 DST is included in edition 4 of MPEG-4 audio, ISO/IEC 14496-3:2009 [15].
15. ISO (2009) ISO/IEC 14496-3: Information technology – Coding of audio-visual objects – Part 3: Audio, 4th Edition: http://www.iso.org/iso/catalogue_detail.htm?csnumber=53943.
16. ISO (2010) ISO/IEC 14496-10: Information technology – Coding of audio-visual objects – Part 10: Advanced Video Coding (AVC) http://www.iso.org/iso/catalogue_detail.htm?csnumber=56538, also published as ITU-T Recommendation H 264: http://www.itu.int/rec/T-REC-H.264-201106-P.
17. Wikipedia (2013) Information on Stereo Cameras and on 3D imaging in general is found at: http://en.wikipedia.org/wiki/Stereo_camera, http://en.wikipedia.org/wiki/Stereoscopy.
18. Wikipedia (2013) Information on the use of polarized glasses for 3D movies and video is found at: http://en.wikipedia.org/wiki/Polarized_3D_glasses.
19. Wikipedia (2013) Information on the use of shutter glasses for 3D video is found at: http://en.wikipedia.org/wiki/Shutter_glasses.
20. Wikipedia (2013) Information on auto-stereoscopic 3D video displays is found at: http://en.wikipedia.org/wiki/Autostereoscopy.
21. DVB (2013) In DVB, frame compatible plano-stereoscopic 3DTV is specified in three parts of ETSI TS 101 547, Plano-stereoscopic 3DTV, Part 1: Overview of the multipart, Part 2: Frame Compatible Plano-stereoscopic 3DTV and Part 3: HDTV Service Compatible Plano-stereoscopic 3DTV, found at: www.dvb.org/standards.
22. Wikipedia (2013) This picture and other information on 2D-plus-depth is found at: http://en.wikipedia.org/wiki/2D-plus-depth.
23. ISO (2007) ISO/IEC 23002-3: Information technology – MPEG video technologies – Part 3: Representation of auxiliary video and supplemental information: http://www.iso.org/iso/catalogue_detail.htm?csnumber=44354.

5

Motivation for a Systems Standard

Why an MPEG systems standard is needed to packetize video and audio and what kind of problems such a standard is required to meet.

The need for systems technology for audiovisual applications arises from the wish to combine multiple elementary audio and/or video streams into a single system stream in a manner independent of the transport medium. Though it is possible to provide audio and video as parallel streams, it is in many situations far more practical to provide a single stream that contains audio and video. For example:

- from optical disc a single serial data stream is delivered;
- a digital broadcast stream is a single serial data stream;
- for movie playback from hard disc, it is easier to deliver a single data stream that contains both audio and video than reading data from two files, one containing compressed video and the other compressed audio, as for the latter option the reading device has to jump regularly between physical hard disc positions in both files.

Therefore MPEG decided to address combining audio and video into a single system stream. A simple example of combining an audio stream and a video stream into a system stream is presented in Figure 5.1.

In Figure 5.1, the audio and video streams are depicted as byte streams. In the system stream there are packages with audio data and packages with video data. To form these packages, at certain times data chunks are taken from the audio and video streams. The values of $t_{\text{video-chunk}}(i)$ and $t_{\text{audio-chunk}}(j)$ indicate the duration of video chunk i in the video stream and of audio chunk j in the audio stream, respectively. In Figure 5.1, the system stream has a constant bitrate, slightly higher than the sum of the video bitrate and the audio bitrate, while the duration of each chunk in the video or audio stream is about the same. As the audio bitrate is substantially lower than the video bitrate, audio chunks contain fewer bytes than video chunks, and consequently, the audio packages are substantially smaller than the video packages.[1]

[1] The audio packages could be designed to be about the same size as video packages, but in that case there may be many more video packages than audio packages.

Fundamentals and Evolution of MPEG-2 Systems: Paving the MPEG Road, First Edition. Jan van der Meer.
© 2014 John Wiley & Sons, Ltd. Published 2014 by John Wiley & Sons, Ltd.

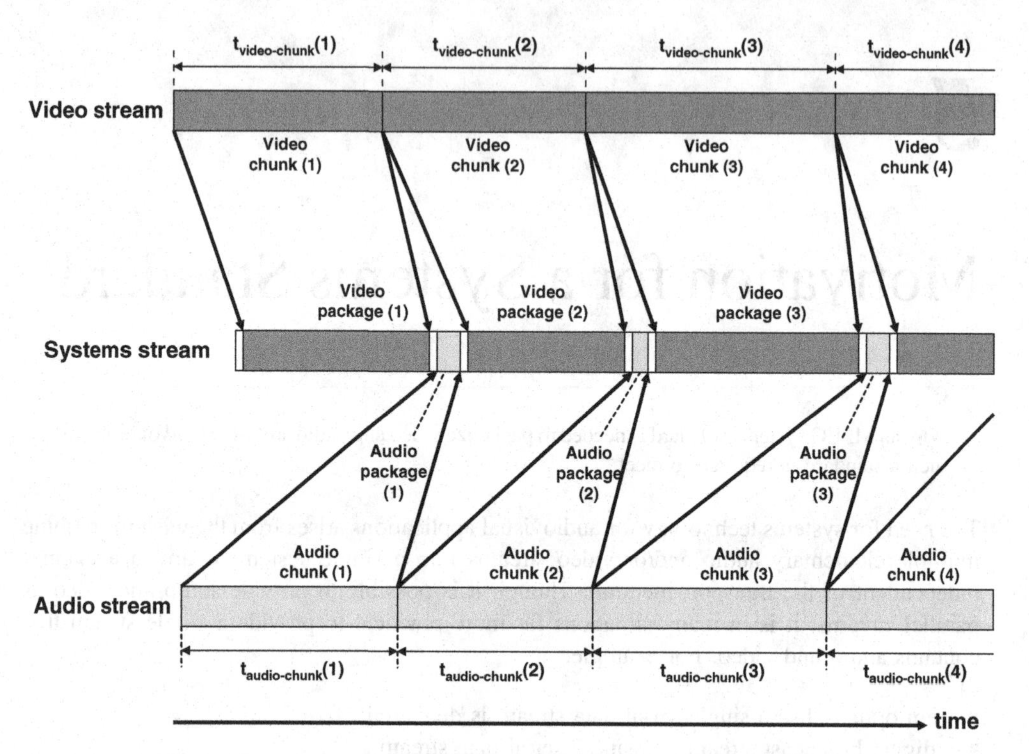

Figure 5.1 Example of combining audio and video into a system stream

To carry multiple elementary streams in a system stream, a format needs to be defined to package the various elementary stream data so that the decoder can conveniently retrieve and decode the elementary stream data, as desired. For example, in Figure 5.1, each audio and video package is preceded by a header that indicates whether audio or video data is carried in the package.

When standardizing a systems format to combine audio and video, it is particularly important to define a format that can meet all requirements of the various delivery systems in an application independent manner. This allows production and management of content irrespective of the delivery network to the consumer, thereby offering a market as broad as possible for equipment needed to produce and manage content. When successful, the systems format will find industry-wide adoption and become the basis for an infrastructure to produce, store, exchange and transport audiovisual content that may exist for many years.

There are many requirements that a systems format to combine audio and video has to meet. For example:

- The format must define how to packetize audio, video and other elementary streams, as well as how to signal the type of content that is carried.
- It must be possible for decoders to conveniently retrieve and decode audio and video from a systems stream.
- It must be possible to synchronize audio and video.

- Random access to audio and video must be possible.
- The format must be robust against errors, either due to data loss or data errors, in particular in error prone environments.
- The format must allow for a high quality of service.
- It must be possible to apply conditional access to audio and video, to disallow access to the content unless certain conditions are met; for example, access to the content may only be possible if a required payment is made.
- It must be possible to carry non-stream data, for example representing data for informational services, such as teletext and closed captioning in TV broadcast.

To meet the above requirements has been the challenge to the MPEG Systems group, but the main challenge was to design a format so that the above requirements are met in a manner independent of the delivery system.

6

Principles Underlying the MPEG-2 Systems Design

The important issues to solve when designing an MPEG system that enables a high Quality of Service. How to achieve a constant end-to-end delay, given the delays due to video coding, audio coding and packetization of video and audio. How buffers are required to multiplex video and audio into a single system stream. The delay caused by a multiplex and demultiplex operation and the video and audio data flow through the buffers. How to synchronize video and audio at players and how to model the decoding process, so that the required buffer resources for players can be specified accurately. How compliancy of streams and players to MPEG-2 Systems can be specified unambiguously. The important timing issues in the design of MPEG-2 systems. Use of a system clock at encoder and decoder and how to convey its value in a stream. Frequency and tolerance of the system clock in MPEG-1 and in MPEG-2 systems. Regeneration of the system clock in a player. How to ensure a high Quality of Service and the importance of transport layer independency.

When designing MPEG-1 and MPEG-2 Systems, the obvious driving concern for the MPEG System experts has been 'how to make MPEG video and audio work' for all target applications. To reach this goal, it has been crucial to achieve a common understanding on a number of important considerations that ultimately govern the design of MPEG Systems. This chapter discusses these design considerations.

6.1 Building an End-to-End System

6.1.1 Constant End-to-End Delay

One of the working assumptions when designing MPEG Systems has been a constant end-to-end delay. The end-to-end delay in an MPEG Coding System is depicted in Figure 6.1 for an audio frame A_1 and a video picture P_1. With a constant end-to-end delay, for each picture and each audio frame the value of the end-to-end delay is the same.[1] This means that, if audio and video are synchronized at the input of the MPEG coding system, then audio and video are also synchronized at its output.

[1] In MPEG, few rules are applied without an exception; even here for a low-delay picture, a constant end-to-end delay does not apply when picture skipping is used.

Fundamentals and Evolution of MPEG-2 Systems: Paving the MPEG Road, First Edition. Jan van der Meer.
© 2014 John Wiley & Sons, Ltd. Published 2014 by John Wiley & Sons, Ltd.

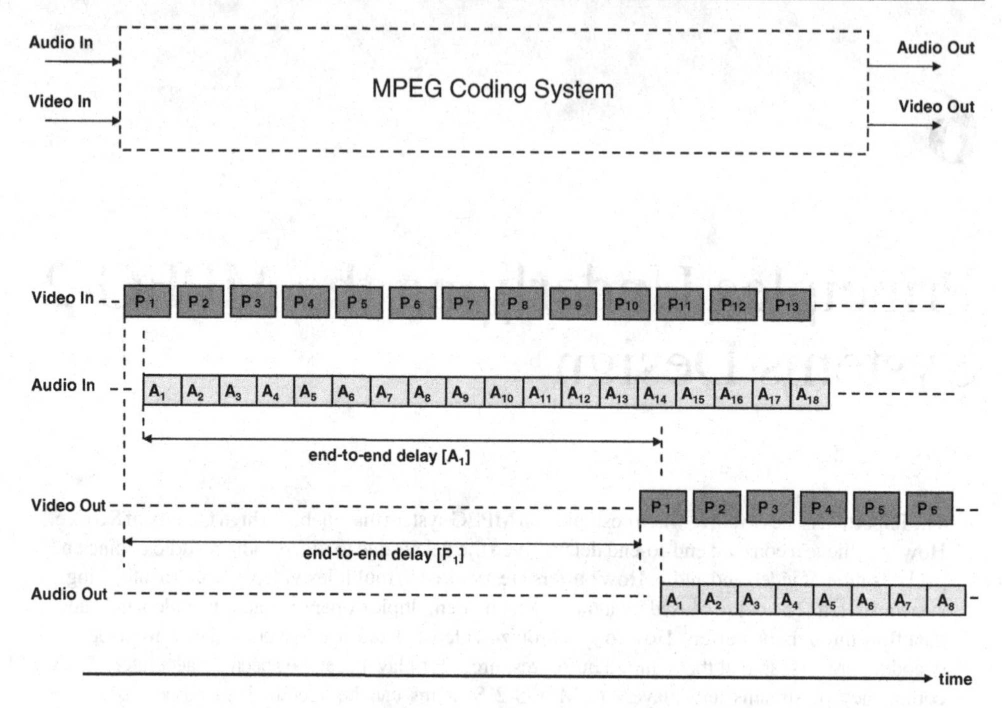

Figure 6.1 Constant end-to-end delay in an MPEG coding system

In an MPEG coding system various delays may occur, depending on the applied processing. Though MPEG systems addresses only delays at the decoder, knowledge of the delays that apply in an MPEG Coding System is useful in understanding MPEG coding systems. Therefore in Sections 6.1.2 and 6.1.3, an analysis is made of the delay typically caused by two important functions in an MPEG Coding System: Video Coding and Audio Coding, while Section 6.1.4 discusses compensation for delay differences that may occur.

For simplicity reasons, the multiplex and demultiplex operation required to packetize the video and audio streams into a system stream at the encoder and to de-packetize video and audio from the system stream at the decoder is not yet taken into account in Sections 6.1.2 and 6.1.3. The multiplex and demultiplex operation can be modelled independently of the video and audio coding, as discussed in Section 6.2. In practice though, the packetization of video and audio may not be fully independent of the video and audio coding, for example to minimize delay or to optimize the implementation of the encoder. Such a decision is fully at the discretion of the encoder; MPEG only requires that the encoder produces a system stream that complies to the applicable MPEG system standard.

6.1.2 Video Coding Delay

Figure 6.2 depicts the processing of video in an MPEG video encoder and decoder. Depending on the coding strategy,[2] the video encoder determines which pictures in the input video are coded as I-, P- or B- picture and applies the associated picture re-ordering and compression.

[2] To determine the most effective encoding strategy, the video encoder may analyse the characteristics of the input video prior to determining the coded picture types.

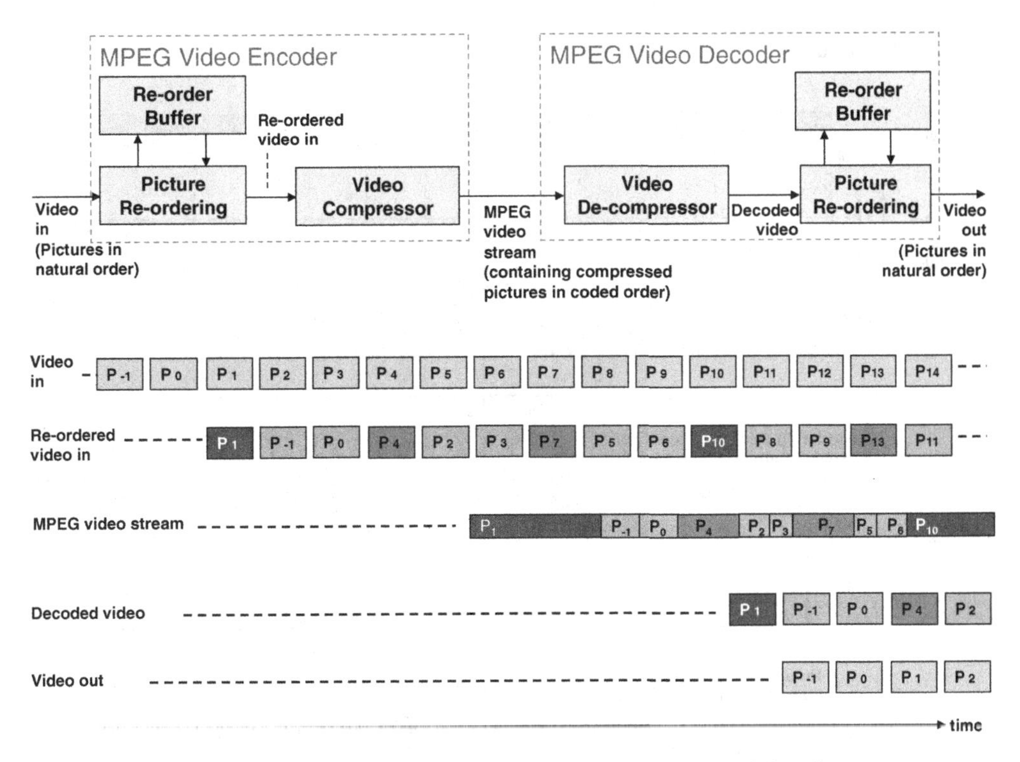

Figure 6.2 Video processing for MPEG video encoding and decoding

The video decoder receives the resulting MPEG video stream and decompresses the coded pictures, upon which the decoded pictures are re-ordered back into natural order prior to producing the output video.

When discussing the principles of MPEG video coding in Chapter 3, it is shown in Figures 3.9 and 3.11 that video encoding and decoding cause a delay, due to the re-ordering of the pictures on one hand and to the video compression on the other. The total video coding delay is equal to the sum of the re-ordering delay and the video compression delay. Both delays can be further analysed separately from each other.

6.1.2.1 Video Re-Ordering Delay

To analyse the delay due to re-ordering, the video compression delay can be ignored by assuming absence of the video compression and decompression. The value of the re-ordering delay depends on the maximum number of consecutive B-pictures that the video encoder may apply. Encoders may use a fixed schedule of two B-pictures between two reference pictures (I- or P-pictures), but encoders may also optimize encoding efficiency by scheduling a variable number of consecutive B-pictures between two reference pictures. For example, in the case of video with little motion, it may be efficient to schedule four consecutive B-pictures. In the hypothetical case depicted in Figure 6.3, the encoder is configured so that the maximum number of consecutive B-pictures is at most equal to four: the encoder can switch between using one, two, three (not used in Figure 6.3) or four consecutive B-pictures.

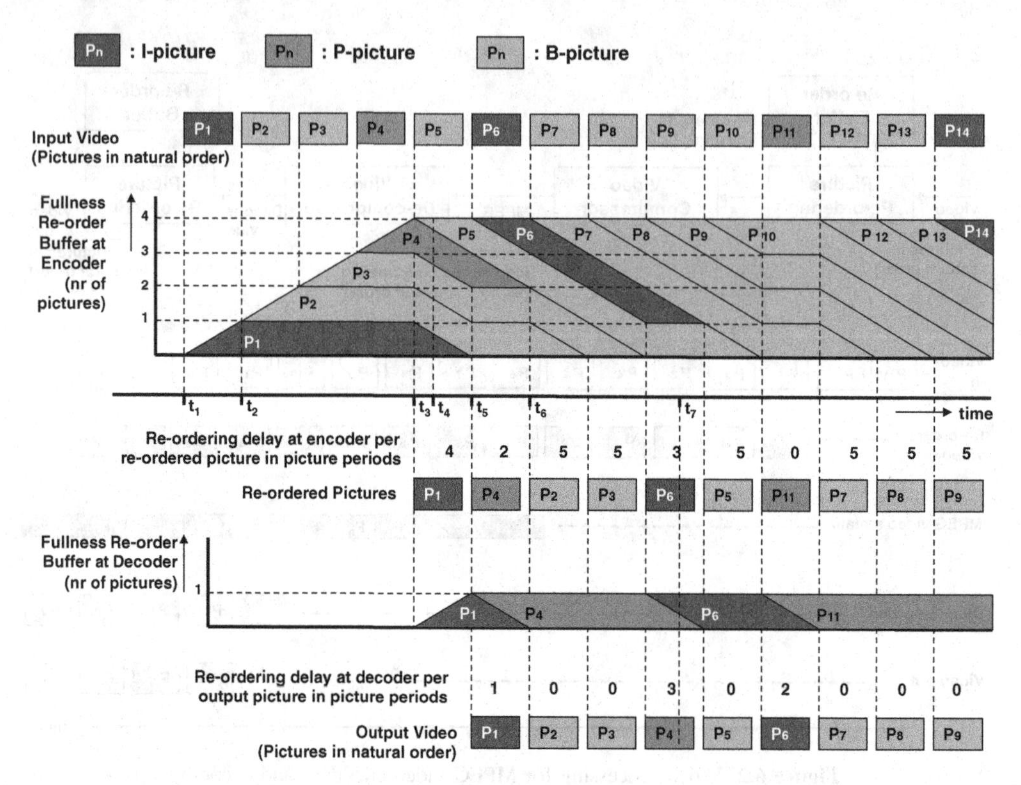

Figure 6.3 Re-ordering delay without video compression and decompression

Figure 6.3 shows the re-ordering delay between the input video and the re-ordered pictures at the encoder and the re-ordering delay between the re-ordered pictures and the output video at the decoder. Also the storage of the pictures in the re-order buffers at the decoder and encoder is shown. Due to the maximum of four consecutive B-pictures, the encoder causes a re-ordering delay of at most five picture periods. This is easily seen in Figure 6.3, where the coding of picture P_7 uses P_{11}; hence the coding of P_7 needs to be postponed until P_{11} is coded, which results in a delay of five picture periods. The coding of picture P_{11} is preceded by coding of subsequently P_1, P_4, P_2, P_3, P_6 and P_5; Figure 6.3 shows that the re-ordering delay caused by the encoder for these pictures is equal to 4, 2, 5, 5, 3 and 5 picture periods, respectively.

At the decoder, B-pictures are not delayed, but immediately presented in the video output instead. Only I- and P-pictures are delayed at the decoder; their presentation in the output video is delayed until the preceding (in presentation order) B-pictures are presented. An I- or P-picture that is preceded by k B-pictures is delayed by (k + 1) picture periods. Hence P_1, P_4, P_8 and P_{11} in Figure 6.3 are delayed at the decoder by 1, 3, 2 and 5 picture periods, respectively. Note that the delay at the encoder for these pictures is respectively equal to 4, 2, 3 and 0 picture periods, resulting in a total re-ordering delay for each picture equal to five picture periods.

In the general case, if n is the maximum number of consecutive B-pictures that an encoder is configured to apply, then the value of the total re-ordering delay is equal to $(n + 1)$ picture periods. The B-pictures undergo no re-ordering delay at the decoder, but only at the encoder, while the I- and P-pictures may undergo a re-ordering delay both at the encoder and at the decoder. If the number of preceding B-pictures in presentation order is equal to k, then the re-ordering delay for an I- or P-picture at the encoder is equal to $(n - k)$ and at the decoder $(k + 1)$ picture periods. The total re-ordering delay is always the same for each picture. MPEG leaves the applied value of n as an encoder feature and therefore the total re-ordering delay is encoder dependent.

Figure 6.3 also shows storage of the various pictures in the re-order buffers at the encoder and the decoder. Between time t_1 and t_2, the data of picture P_1 is received and stored in the re-order buffer at the encoder, followed by pictures P_2, P_3 and P_4. At time t_3, when there are four pictures in this re-order buffer, picture P_5 starts entering, but at the same time the data from picture P_1 starts leaving, so that the content of the re-order buffer remains equal to the data from four pictures. For example, at time t_4, the contents of the buffer is equal to (about) two-thirds of picture P_1, picture P_2, picture P_3, picture P_4 and (about) one-third of picture P_5. Note that picture P_{11}, the one that follows four consecutive B-pictures (P_7, P_8, P_9 and P_{10}), is the only picture that is not delayed at the encoder and therefore not stored in the re-order picture buffer at the encoder.

At the decoder, the received pictures are re-ordered back into natural order. The B-pictures are not delayed, but immediately presented at the video output. Each I- and P- picture is stored in the re-order buffer at the decoder until the arrival of the next reference picture. In Figure 6.3, between time t_3 and t_5, the data of picture P_1 is received and stored in this re-order buffer. Between time t_5 and t_6, data from picture P_4 is received and stored in the buffer, but at the same time data from picture P_1 leaves the buffer. The same happens for each reference picture: at the time the next reference picture is stored in the re-order buffer at the decoder, the previous one leaves it; for example, at time t_7, the content of the buffer is roughly equal to one-third of picture P_4 and two-thirds of picture P_6. The result is that, when the video decompression is ignored, the re-order buffer never contains more than the data of one picture.

In practice, obviously the picture re-ordering is jointly implemented with the video decompression in the same video decoder, using a single picture memory. For the same order of picture types as in Figure 6.3, the flow of the decoded reference pictures trough the picture memory is demonstrated in Figure 6.4.[3] Upon decoding, each reference picture is stored in the picture memory, while each decoded B-picture is immediately presented in the output video. At the instance in time the next reference picture starts entering the picture memory, the presentation of the previous reference picture commences in the output video; at the same instance, the reference picture that is in the picture memory longest starts getting removed from the picture memory. This procedure ensures that each B-picture coded between both reference pictures can be predicted from those two reference pictures. Hence, as can be seen in Figure 6.4, for the prediction of B-pictures, a preceding reference picture must remain longer in the picture memory than required for re-ordering. Figure 6.4 shows that, after startup, there are two reference pictures in the picture memory.

In summary, the re-ordering delay depends on the maximum number of consecutive B-pictures that an encoder is configured to apply. An encoder using continuously two consecutive B-pictures between an I- or P-picture, causes a total re-ordering delay of three picture periods. It should be noted that decoders do not need any knowledge of the maximum number of consecutive B-pictures that an encoder may apply: a decoder simply delays the presentation of

[3] Figure 6.4 shows simplified versions of Figures 3.17 and 3.18.

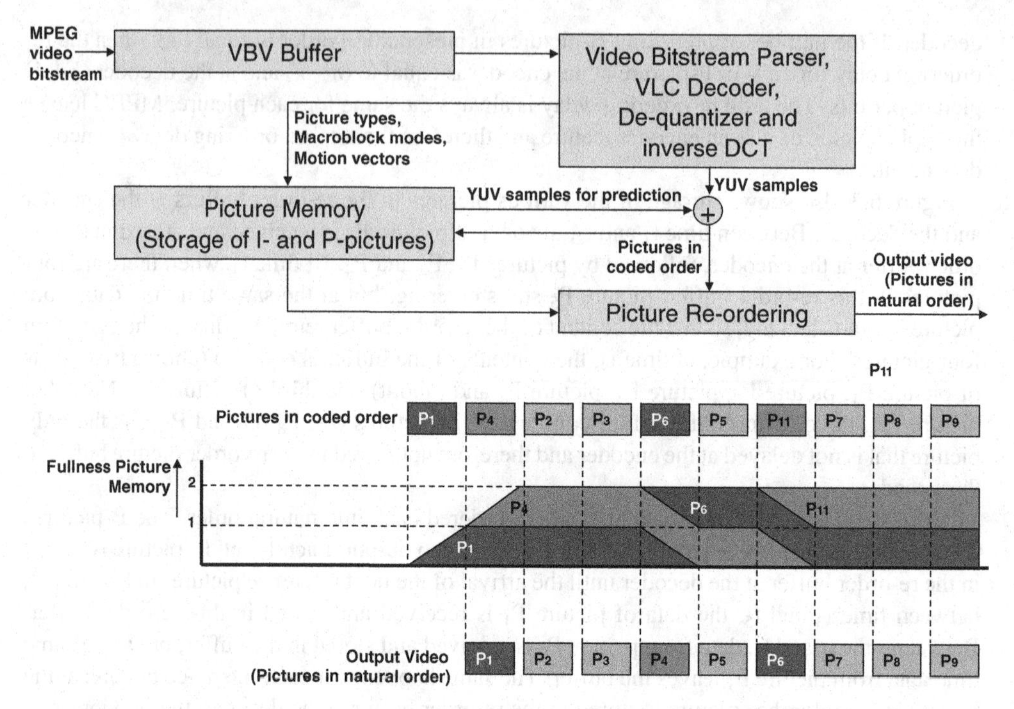

Figure 6.4 Re-ordering in a decoder using a single picture memory

each I- or P-picture until the next reference picture arrives, independent of the number of intermediate B-pictures.

6.1.2.2 Video Compression Delay

The video compression delay caused at encoding and decoding is depicted in Figure 6.5. For each picture P_i, this delay is equal to the delay between the (re-ordered) picture P_i at the input of the video compressor and the decoded (but not yet re-ordered) picture P_i at the output of the video decompressor. As the end-to-end delay is constant, the total delay caused by video compression and decompression must be the same for each picture.

While making reference to the beginning[4] of the compressed data of picture P_i in the coded MPEG video stream, two components are distinguished: the compression delay caused by the encoder and the delay caused by the decoder, in Figure 6.5 referred to as the video compression delay and the video decompression delay. Note that these delays represent the delay caused by buffering prior to and following the transport of the picture data in the video stream at the encoder and the decoder, respectively.

While the sum of the video compression delay and the video decompression delay is the same for each picture, both delays vary depending on the size of the compressed pictures. For example, when the video bitrate is constant, the transport of a large coded picture in the MPEG

[4] The beginning of the compressed picture data is specified accurately by the MPEG systems standard.

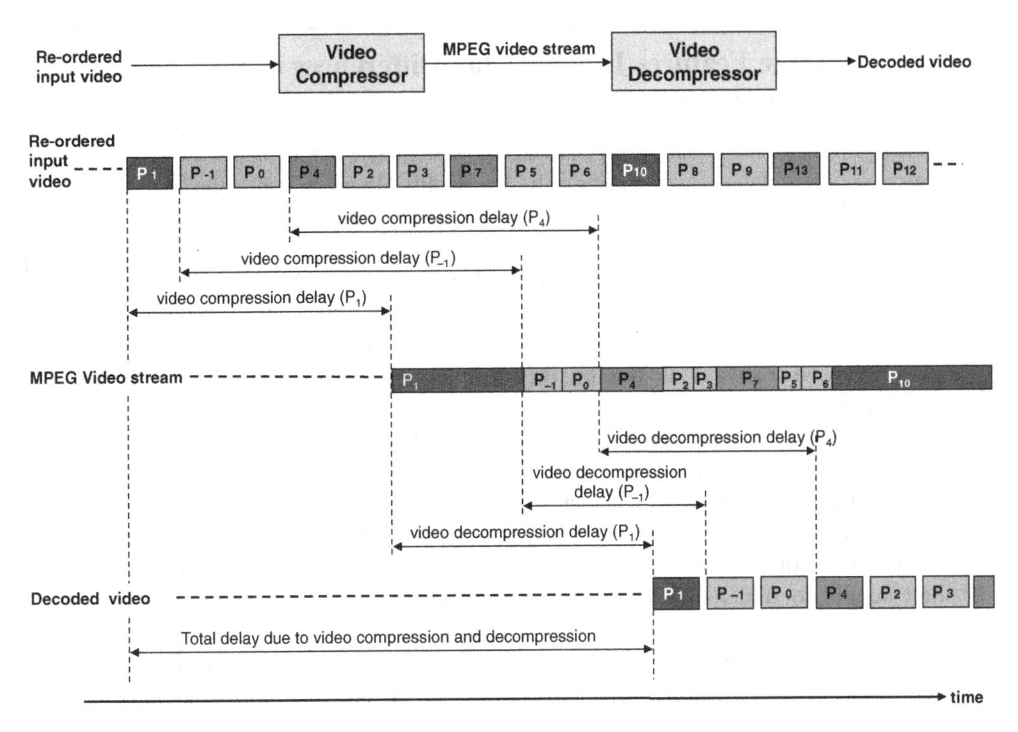

Figure 6.5 Delay due to video compression at the encoder and the decoder

video stream takes more time than the transport of a small coded picture. As a consequence, the decompression delay for a coded picture immediately following a relatively large picture will be smaller than the decompression delay for that large picture. For example, in Figure 6.5, picture P_{-1} follows the large I-picture P_1, due to which video decompression delay (P_{-1}) is smaller than video decompression delay (P_1). Likewise, the coded data for pictures P_{-1} and P_0 is smaller than average, due to which video decompression delay (P_4) is larger than video decompression delay (P_{-1}).[5]

The video decompression delay is tightly coupled to the fullness of the video input buffer at the video decoder. For example, in Figure 6.5 just prior to the instance in time picture P_1 is decoded, the coded data of P_1, P_{-1}, P_0 and about half of P_4 has entered the input buffer at bitrate R_v during a period equal to the video decompression delay for P_1. Furthermore, at the instance in time picture P_4 is decoded, pictures P_1, P_{-1} and P_0 are already decoded and therefore just prior to decoding P_4, the input buffer contains the coded data of P_4, P_2, P_3, P_7, P_5 and about half of P_6; the coded data of these pictures entered the input buffer at bitrate R_v during a period equal to the video decompression delay for P_4. Hence, if the video bitrate R_v is constant, the fullness of the input buffer at the video decoder just prior to the instance in time that coded picture P_k starts being decoded, is equal to R_v times [video decompression delay (P_k)].

[5] Note that also video decoding delay (P_0) is larger than video decoding delay (P_{-1}).

Note 6.1 Some Features Due to Delay Differences in Live Broadcast of Sport Events

Often significant end-to-end delays are introduced in television broadcast services. In many cases these delays are not noticed by the public. However, when the same event is broadcast live over different channels, the delay differences may easily become apparent. Some interesting experiences are described below.

During a nice summer evening in Holland, neighbours were watching an exciting soccer match on television in their gardens, but via different providers; when goals were scored, or just not, they could hear each other's loud reactions, but one family was not amused: they were always late because their neighbours received the pictures more than a second earlier.

Important sports games are often broadcast live by both television and radio; some people prefer to listen to the more enthusiastic radio reporters, instead of to the more minimalistic commentary on television. Radio broadcast is usually barely delayed, hence when the television broadcast undergoes a delay of several seconds, scored goals are announced on the radio several seconds before they can be watched on television. Some people not understanding this feature concluded the radio reporters were capable of predicting the future.

A much larger delay for radio broadcast happens also. When travelling abroad, people may wish to watch a live sports game on television, but without listening to the commentary in a language they do not understand. Instead they can to listen to the radio broadcast in their own language streamed over the Internet. But unfortunately the radio commentary arrives very late; as a result they may listen to what happened on the television screen half a minute ago: an annoying experience.

6.1.2.3 Total Video Coding Delay

The total video coding delay is found from adding the total re-ordering delay, the total video compression delay and any other delays caused by video pre-processing or other operations deemed necessary by the encoder. The maximum number of consecutive B-pictures and the delay applied for video compression are important parameters in the design of efficient video encoders, the choice of which is left by MPEG to the discretion of encoder manufacturers. Typically, a re-ordering plus video compression delay of about 0.7 s is sufficient for an encoder to compress video input signals efficiently with enough flexibility to allocate bits to code the input video. However, in practice often much larger values of the total video coding delay are found, up to several seconds, due to video pre-processing, format conversions, transcoding or other time-consuming operations. Delays of several seconds are significant and may impact user perception, especially in real-time broadcast of sport events (see Note 6.1)

6.1.3 Audio Coding Delay

The delay due to audio coding differs significantly from the delay caused by video coding. The actual delay depends on the applied audio coding standard and on the implementation in the encoder and decoder. For MPEG-1 audio, the theoretical delay is in the range between 20 and

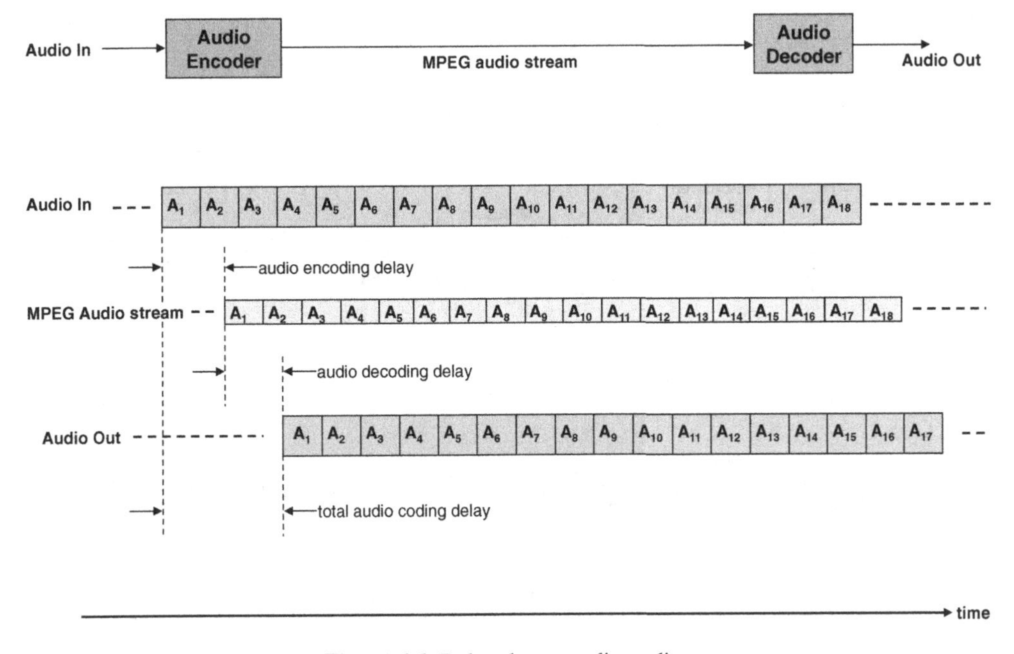

Figure 6.6 Delay due to audio coding

60 ms. More advanced standards apply more sophisticated compression algorithms that usually cause more delay, but in practice, the delay due to audio coding remains significantly below the delay caused by video coding. For an example see Figure 6.6, where the audio coding delay is slightly more than the duration of three audio frames, split in more or less equal delays caused by the audio encoder and audio decoder.

6.1.4 Delay Compensation

Because usually the audio coding delay differs substantially from the delay caused by video coding, there is typically a need to apply an additional delay to audio to achieve the same delay for video pictures and audio frames in an MPEG coding system. There are many ways to implement the compensating delay for audio; it may be applied prior to encoding the audio or in the coded domain on the MPEG audio stream. MPEG does not specify how and where to implement the compensating delay for audio. However, any compensating delay must be implemented at the encoder, not at the decoder. The required delay compensation can only be calculated at the encoder where the actual encoding delays of the applied video and audio encoders are known.

Figure 6.7 presents an example of an encoding system, without multiplexing, where the compensating delay is implemented prior to audio encoding, while the associated signal processing is depicted in Figure 6.8. The value of the compensating delay is equal to [(video coding delay) – (audio coding delay)], which ensures that the time offset between P_1 and A_1 at the output signals is the same as at the input signals of the MPEG coding system.

Within an MPEG coding system, the value of the end-to-end delay is required to be the same for each so-called elementary stream, a generic term for a coded video, coded audio or other

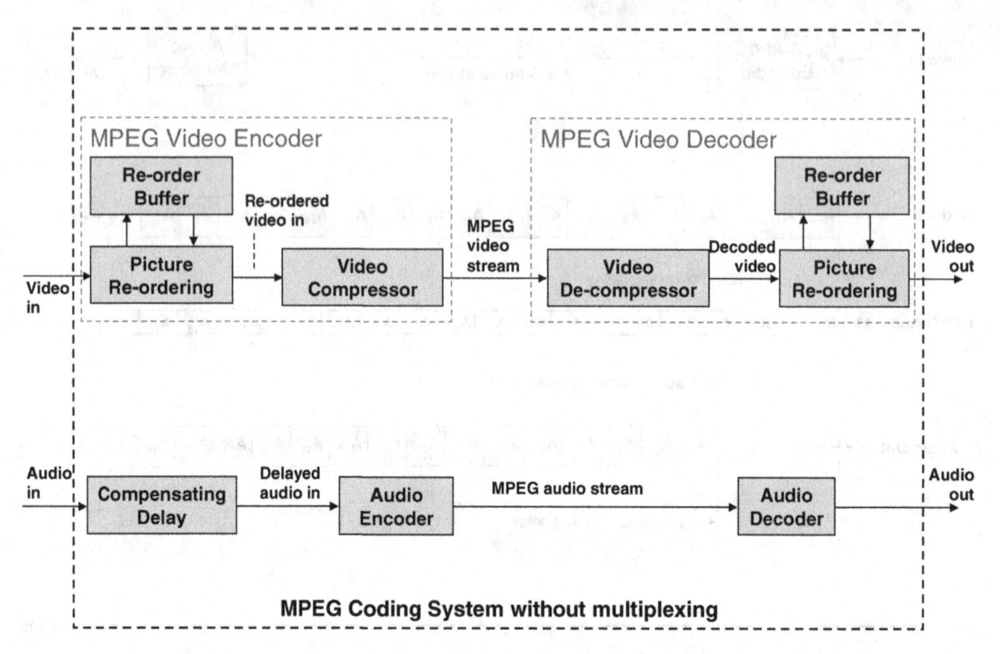

Figure 6.7 Audio delay compensation in MPEG Coding System (no multiplexing)

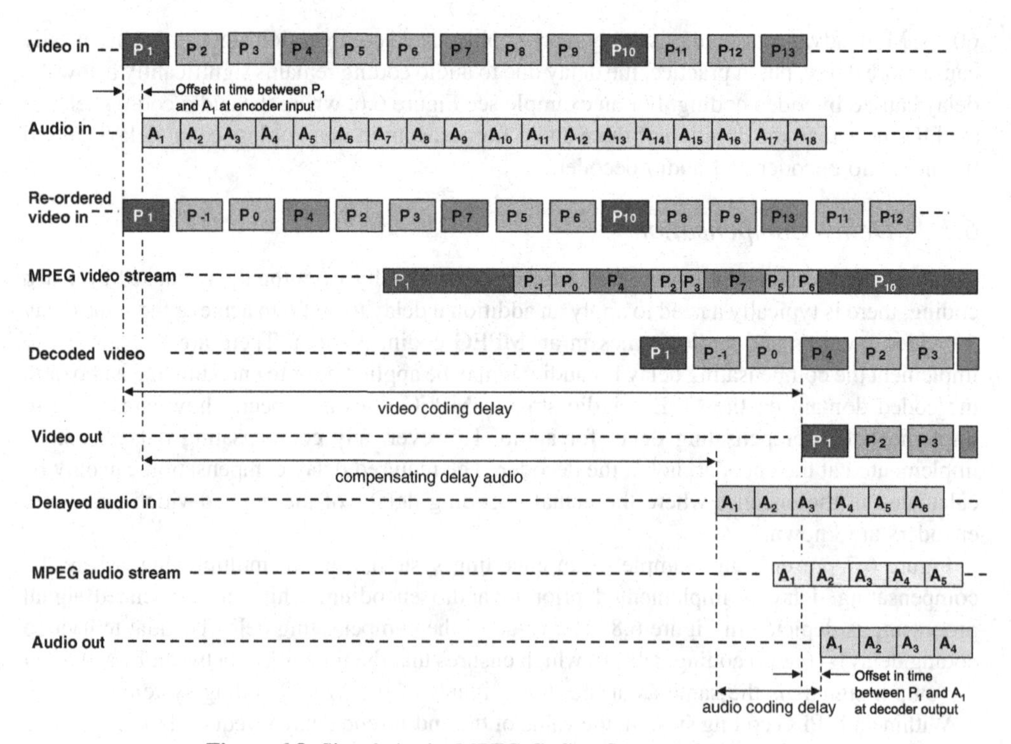

Figure 6.8 Signals in the MPEG Coding System of Figure 6.7

coded stream. To achieve a constant end-to-end delay for each elementary stream, delay compensation may be required for multiple associated elementary streams.

6.2 The Multiplex and Demultiplex Operation

The multiplex and demultiplex operation describes the packetizing and de-packetizing of coded video or coded audio streams into and from a single MPEG system stream. In an MPEG system stream also other coded streams can be carried. To describe any coded stream, MPEG systems introduces 'elementary stream' as a generic term to denote a coded video, coded audio or other coded stream.

To describe the basics of multiplexing and demultiplexing elementary streams, a simple extension of the MPEG coding system of Figure 6.7 is introduced in Figure 6.9. At the MPEG mux, the MPEG video and audio streams are multiplexed into an MPEG system stream, while at the MPEG demux, the system stream is demultiplexed to reconstruct the carried MPEG video and audio streams. For simplicity reasons it is assumed that there are no other streams involved in the multiplex, and that the MPEG demux exactly reproduces at its output the MPEG video and audio streams as received at the input of the MPEG mux.

In the course of this book it will be shown that multiplex and demultiplex operations often differ in several aspects from the approach in Figure 6.9. For example, system decoders may use a single buffer for demultiplexing and elementary stream input buffering prior to decoding, without exactly reconstructing the byte delivery timing of the video and audio streams at the

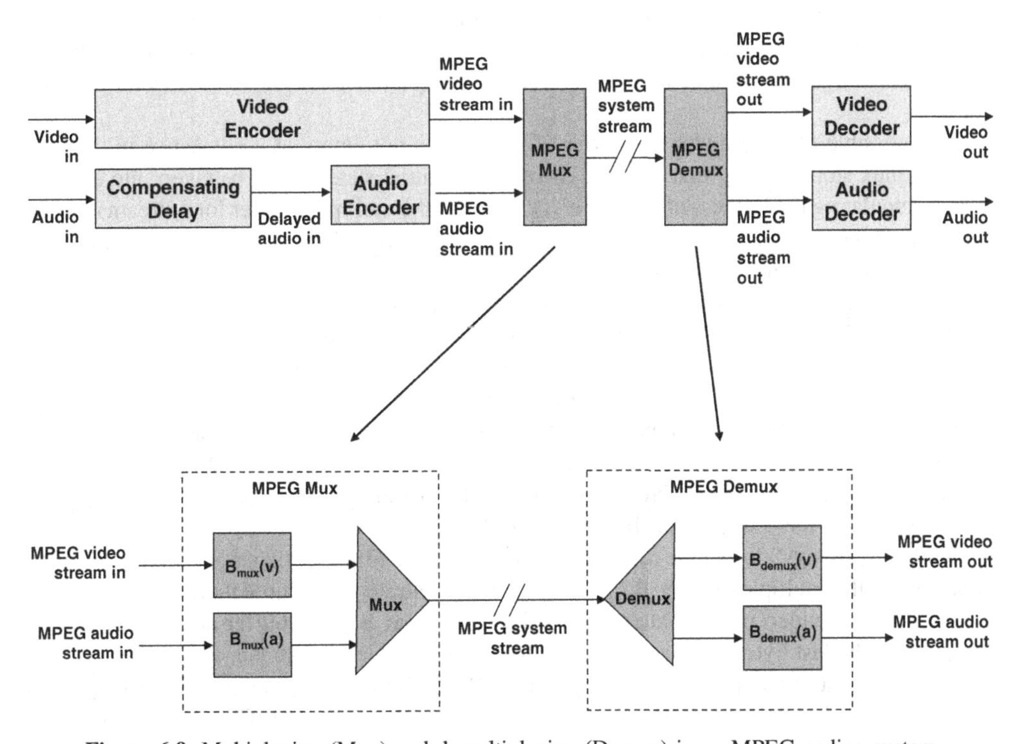

Figure 6.9 Multiplexing (Mux) and demultiplexing (Demux) in an MPEG coding system

output of the demultiplex buffer. Nevertheless, Figure 6.9 provides a good basis to discuss the use of buffers, the delay associated to the multiplex and demultiplex operation, the trajectory of video and audio data through the buffers, and the need for flexibility at multiplexers to position the video and audio packets in time.

Input to the MPEG Mux in Figure 6.9 are an MPEG video and an MPEG audio stream. The MPEG mux switches between serving the video and the audio stream: when serving a stream, the MPEG mux puts a packet with data from that stream in the system stream. The resulting MPEG system stream carries a sequence of packets with data from either the video or the audio stream. The system stream has a bitrate slightly higher than the sum of the video and audio bitrates, to accommodate room for a header preceding each packet with video and audio data. Below, the bitrate of the MPEG system stream is referred to as the mux bitrate.

The MPEG mux continuously receives the MPEG video and audio stream data, also during periods a stream is not served. To prevent losing data when a stream is not served, stream data must be stored in multiplex buffers $B_{mux}(v)$ and $B_{mux}(a)$ prior to packetization. At the MPEG demux, the inverse operation takes place. In Figure 6.9, the MPEG demux receives the MPEG system stream and outputs the MPEG video and audio streams. The MPEG demux puts the video and audio data received from the system stream in the demultiplex buffers $B_{demux}(v)$ and $B_{demux}(a)$, used for smoothing the bursty video and audio data from the system stream, so as to reproduce the video and audio streams at the same rate as received at the input of the MPEG mux.

The required size of the multiplex buffers $B_{mux}(v)$ and $B_{mux}(a)$ is determined by the scheduling of the periods during which the MPEG mux serves and does not serve the stream. When a video or audio stream is not served, its multiplex buffer fullness increases at a rate equal to the video or audio bitrate. When a video or audio stream is served, its multiplex buffer fullness decreases at a rate equal to the bitrate of the system stream minus the video or audio bitrate.

It is possible that the multiplex buffer of a stream is not empty at the instance in time the MPEG mux stops serving that stream. Therefore its initial state should be taken into account when calculating the fullness of the buffer. Obviously, the multiplex buffer for a stream must be larger than the accumulated amount of stream data entering it during 'not served' periods and leaving it during 'served' periods.

A simple example of the use of multiplex and demultiplex buffers for video is shown in Figure 6.10. The video bitrate is constant and equal to 4.0 Mb/s or 0.5 MB/s, while the mux bitrate is equal to 6.0 Mb/s or 0.75 MB/s. The MPEG mux creates video packets containing 9000 bytes each. The duration of such packet in the MPEG video stream is 9000/500 000 = 18 ms, while the duration of the transport of that packet in the MPEG system stream is equal to 9000/750 000 = 12 ms. Figure 6.10 shows a (rather hypothetical) multiplex strategy, whereby the end of each video packet in the incoming MPEG video stream is aligned with the end of the transport of that packet in the MPEG system stream.

With this strategy, during the first 6 ms of a packet in the MPEG video input stream the video bytes enter the multiplex buffer $B_{mux}(v)$ at the video bitrate, while no data leaves the buffer. After 6 ms, the video packet data starts getting read from $B_{mux}(v)$ buffer at the mux bitrate. Right after the last byte of the video packet from the MPEG video stream enters $B_{mux}(v)$, this byte is removed to complete the video packet in the MPEG system stream, so that $B_{mux}(v)$ is empty again when the first byte of the next video packet in the MPEG video stream enters $B_{mux}(v)$.

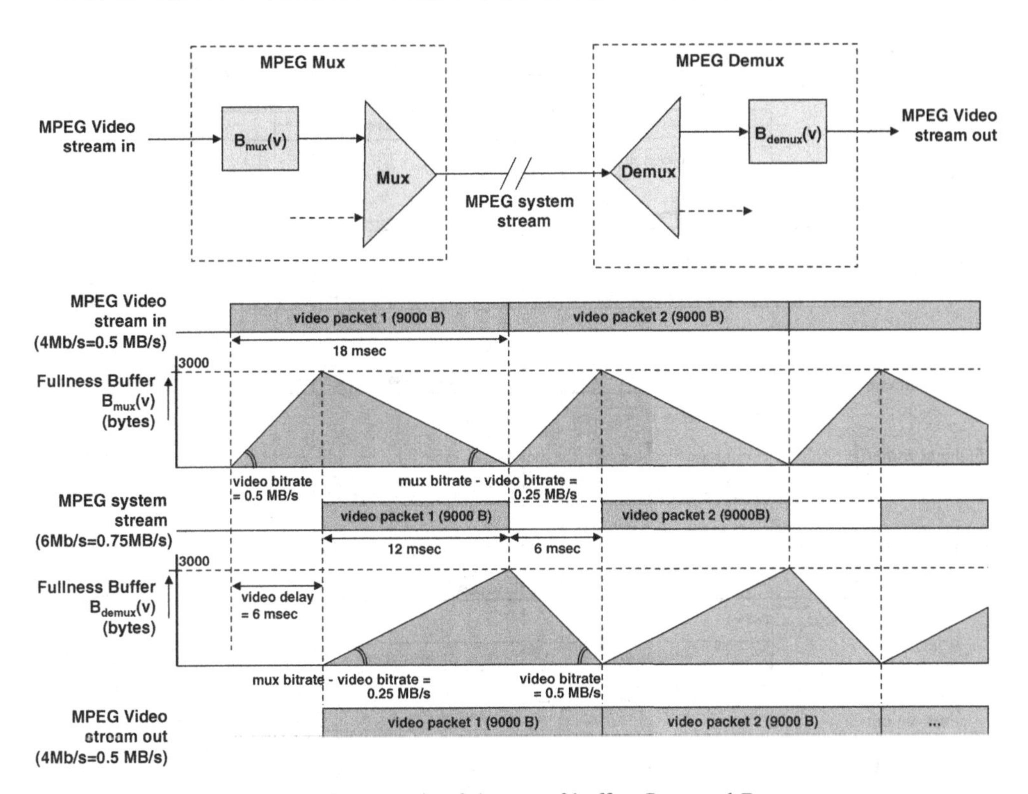

Figure 6.10 Example of the use of buffers B_{mux} and B_{demux}

During the first 6 ms of the video packet from the MPEG video stream, the fullness of $B_{mux}(v)$ increases at the video bitrate of 0.5 MB/s. The fullness reached after 6 ms is equal to $(0.006 \times 500\ 000) = 3000$ bytes. After 6 ms, the video data remain entering the buffer at 0.5 MB/s, but at the same time video data are removed from the buffer at a rate equal to 0.75 MB/s. As a result, the buffer fullness decreases by a rate equal to 0.25 MB/s; after 12 ms, the buffer will exactly be empty.

At the MPEG Demux, the $B_{demux}(v)$ buffer starts getting filled at the mux bitrate as soon as the first byte of the video packet arrives from the MPEG system stream. However, at the same time, the bytes of the video packet start being removed at the video bitrate from the buffer to produce the MPEG video stream at the output of the MPEG demux. The result is that the buffer fullness increases at a rate equal to the mux bitrate minus the video bitrate until the end of the video packet in the MPEG system stream. During this period, the buffer fullness increases to $0.012 \times 250\ 000 = 3000$ bytes. During the next 6 ms, no more video data enter the buffer, so that the contents of the buffer decreases at a rate equal to the video bitrate; during this period $0.006 \times 500\ 000 = 3000$ bytes leave the buffer. As a result, the $B_{demux}(v)$ buffer is empty again at the start of the arrival of the next video packet in the MPEG system stream.

Figure 6.11 shows the B_{mux} and B_{demux} buffers with input and output signals in the case of a multiplex strategy whereby video and audio packets are created that all have the same duration D in the MPEG video and audio stream, respectively. The mux bitrate is equal to the sum of the video bitrate and the audio bitrate; overhead due to system headers is neglected. The video and

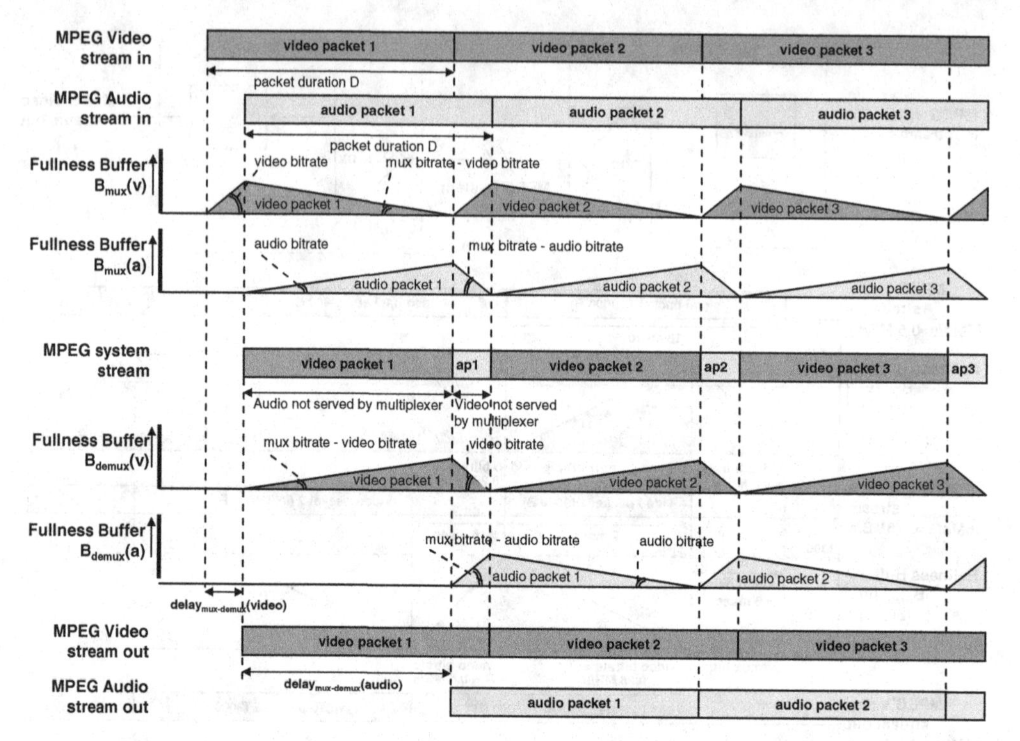

Figure 6.11 A simple strategy to multiplex an MPEG video and audio stream

audio packets in the incoming MPEG video and audio streams are positioned so that for each video and audio packet the end of the packet in the MPEG video or audio stream is aligned with the end of the transport of that packet in the MPEG system stream. In Figure 6.11, the audio packets in the MPEG system stream are denoted as ap1, ap2 and so on.

At the MPEG mux, the fullness of buffers $B_{mux}(v)$ and $B_{mux}(a)$ increases at the video or audio bitrate during the periods that no data leaves the buffer; during periods that data does leave the buffer to create a packet in the MPEG system stream, the fullness decreases at a rate equal to the mux bitrate minus the video or audio bitrate. At the MPEG demux, the fullness of the $B_{demux}(v)$ and $B_{demux}(a)$ buffers decreases at the video or audio bitrate during the periods that no data from the system stream enters the buffer. During periods that packet data from the system stream does enter the buffer, the fullness increases at a rate equal to the mux bitrate minus the video or audio bitrate.

As in Figure 6.10, as soon as the first byte from the video or audio packet in the MPEG system stream arrives in the $B_{demux}(v)$ or $B_{demux}(a)$ buffer, the first byte of the video or audio packet is removed from the buffer to produce the MPEG video or audio stream at the output. Thereby alignment is achieved between the beginning of the transport of the packets in the MPEG system stream and the beginning of the corresponding packets in the outgoing MPEG video and audio streams.

Figure 6.11 shows that the multiplex-demultiplex operation causes a delay of the video and audio streams, but that this delay is not the same for video and audio. With the simple multiplex strategy used in Figure 6.11, the video and audio streams are delayed by the period the

MPEG mux does not serve the stream; see $\text{delay}_{\text{mux-demux}}$(audio) and $\text{delay}_{\text{mux-demux}}$(video) in Figure 6.11. Due to the difference in bitrate, audio is served for a shorter period than video. As a consequence, the 'not served' period for audio stream is longer than for video, and therefore the delay caused for audio is larger than for video.

To achieve the same multiplex delay for video and audio, an additional delayof video is required. In practice, the delay compensation can be implemented by reducing the compensating delay for audio discussed in Figure 6.7. However, it is also possible to implement this delay in the MPEG mux, as shown in Figure 6.12.

In Figure 6.12, exactly the same MPEG video and audio streams are provided at the input as in Figure 6.11, but the MPEG mux imposes a delay for putting the video data in the MPEG system stream. The processing of audio in Figure 6.12 is exactly the same as in Figure 6.11. For video, placing the video packets in the system stream is delayed by storing the video data longer in the $B_{mux}(v)$ buffer, which will obviously increase the fullness of $B_{mux}(v)$.

The additional delay causes the packetization to be no longer on the boundaries of the original video packets in the MPEG video stream. For example, due to the delay the first video packet in the MPEG system stream does not carry the entire video packet 1, but only the first fragment vp 1a thereof. The remaining fragment vp 1b with a duration in the video stream equal to the value of the additional delay is placed in the next video packet in the system stream, together with the first fragment vp 2a of video packet 2.

Figure 6.12 shows the behaviour of the $B_{mux}(v)$ and $B_{demux}(v)$ buffers and the flow of the video packet fragments through these buffers. Note the special processing in the MPEG demux

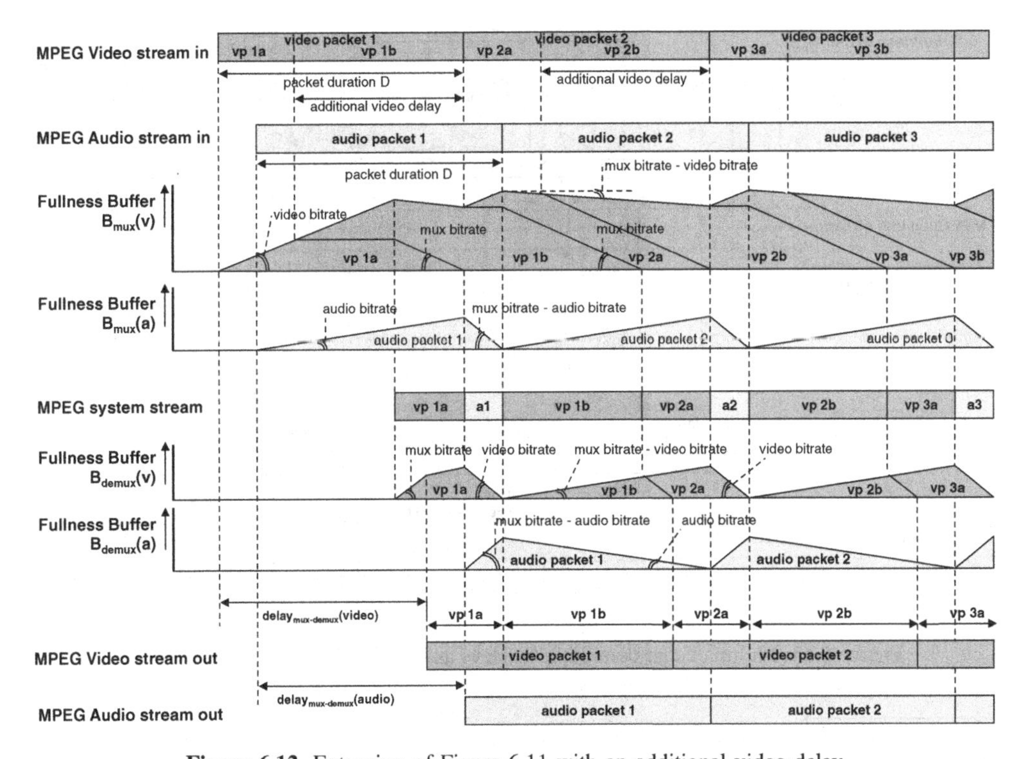

Figure 6.12 Extension of Figure 6.11 with an additional video delay

of the first video packet in the MPEG system stream carrying only the single fragment vp 1a. Figure 6.12 assumes that at the output of $B_{demux}(v)$ a continuous MPEG video stream at a constant bitrate is provided; therefore the first byte of vp 1a is not put in the MPEG video stream immediately upon arrival, but after a short delay.[6] Due to the applied delay of the MPEG video data at the MPEG mux, $delay_{mux\text{-}demux}$(video) equals $delay_{mux\text{-}demux}$(audio) in Figure 6.12. As a result, the MPEG demux produces the MPEG video and audio streams at its output exactly so that the original mutual offset in time at the input of the MPEG mux between the video pictures and audio frames is reproduced.

When an MPEG demux reconstructs at its output an elementary stream with exactly the same bitrate at which the stream was received at the input of the MPEG mux, then the delay caused by the MPEG mux and demux is (by definition) the same for each byte in that stream. For each byte in the elementary stream, this delay is equal to the time elapsed between the occurrences of the byte in the input elementary stream at the MPEG mux and at the output stream of the MPEG demux. However, the delay caused by the MPEG mux typically differs from the delay caused by the MPEG demux. This is shown in Figure 6.13 by means of the trajectories of bytes from an input stream through B_{mux}, the system stream, and B_{demux} into the output stream.

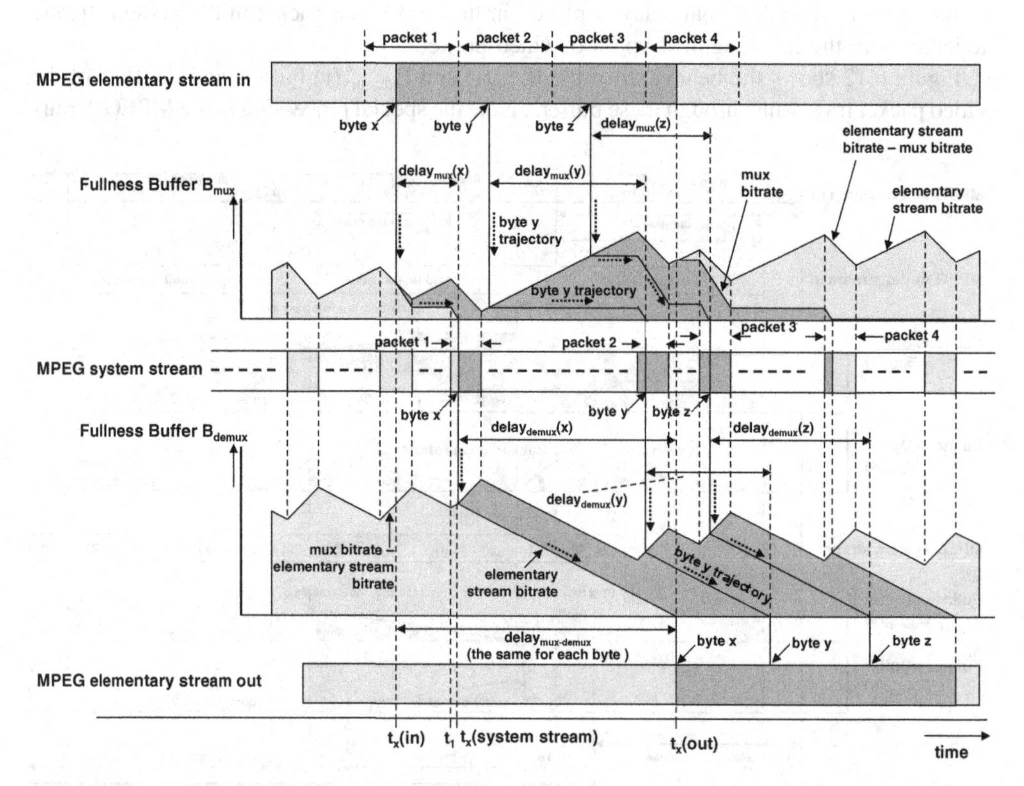

Figure 6.13 Multiplex and demultiplex delays and byte trajectories through buffers

[6] This is a startup complication for fragment vp 1a that does not occur for subsequent fragments vp 2a, vp 3a and following fragments.

Figure 6.13 shows an elementary stream at the input of the MPEG mux with a number of highlighted bytes, starting with byte x and including bytes y and z. For transport in the MPEG system stream, packets are formed in the elementary stream. Packets 1, 2, 3 and 4 are formed to carry the highlighted bytes; non-highlighted bytes precede the highlighted bytes in packet 1 and follow the highlighted bytes in packet 4. At the time a byte enters the MPEG mux, it is put in B_{mux}. At time t_1, the first byte of packet 1 is removed from B_{mux} and inserted in the MPEG system stream; byte x enters B_{mux} at time t_x(in). The bytes of packet 1 are removed from B_{mux} at the mux bitrate; at time t_x(system stream), byte x leaves B_{mux} and is inserted in the system stream.

The MPEG system stream is transported to the MPEG demux; upon arrival, each byte of the MPEG system stream enters B_{demux}. In Figure 6.13, the transport of the system stream is assumed to be instantaneous, and therefore at the same instance in time t_x(system stream), byte x leaves B_{mux} and enters B_{demux} via the system stream. To form the elementary stream, bytes in B_{demux} are removed at the elementary stream bitrate; at time t_x(out), byte x is removed and put in the elementary stream. The delay caused by the MPEG mux is $delay_{mux}(x) = [t_x(\text{system stream}) - t_x(\text{in})]$ and by the MPEG demux $delay_{demux}(x) = [t_x(\text{out}) - t_x(\text{system stream})]$, as depicted in Figure 6.13.

Figure 6.13 also presents $delay_{mux}(y)$, $delay_{demux}(y)$, $delay_{mux}(z)$ and $delay_{demux}(z)$ as the delays caused by the MPEG mux and MPEG demux for bytes y and z. Because the value of the total delay of the mux-demux operation is equal to $delay_{mux\text{-}demux}$ for each byte, the delays caused by the MPEG mux and by the MPEG demux vary per byte in a complementary manner. If the delay for any byte i at the MPEG mux is equal to $delay_{mux}(i)$, then the delay for byte i at the MPEG demux is $delay_{demux}(i) = [delay_{mux\text{-}demux} - delay_{mux}(i)]$. Hence, the value of each delay may vary between zero and $delay_{mux\text{-}demux}$.

Practical multiplexers need flexibility to position the video and audio packets in time, for example to create room for other streams in the multiplex. By means of a simple example, Figure 6.14 shows consequences of a more flexible positioning of audio packets with the multiplex strategy of Figure 6.11. The video and audio packets from Figure 6.11 are denoted in Figure 6.14 as vp1, vp2 and so on, and as ap1, ap2 and so on. Compared to the original multiplex strategy, audio packets can be positioned earlier or later in the system stream by exchanging position with the immediately preceding or following video packet. Figure 6.14 presents three system streams. MPEG system stream 1 is the original system stream from Figure 6.11. MPEG system stream 2 is a modified version of MPEG system stream 1, in which ap3 is positioned early in the stream, prior to vp3, which causes late positioning of vp3. Furthermore ap6 is positioned late, after vp7, which causes early positioning of vp7. Early and late positioning both cause delay:

- Early positioning causes delay at the multiplexer, because the producing of the system stream has to be postponed until the early packet arrives at the input. For example, the MPEG system stream 2 in Figure 6.13 cannot be produced because ap3 in the MPEG audio input stream is not yet fully available at the early position of ap3 in the system stream. Therefore the delayed MPEG system stream 2 is also depicted in Figure 6.14, with a delay equal to the early displacement of ap3. Note that vp7 also undergoes an early displacement, but that the displacement of vp7 is smaller than for ap3.
- Late positioning causes delay at the demultiplexer, because the creation of the MPEG video or audio stream has to wait for the delivery of a late packet. Figure 6.14 shows the MPEG

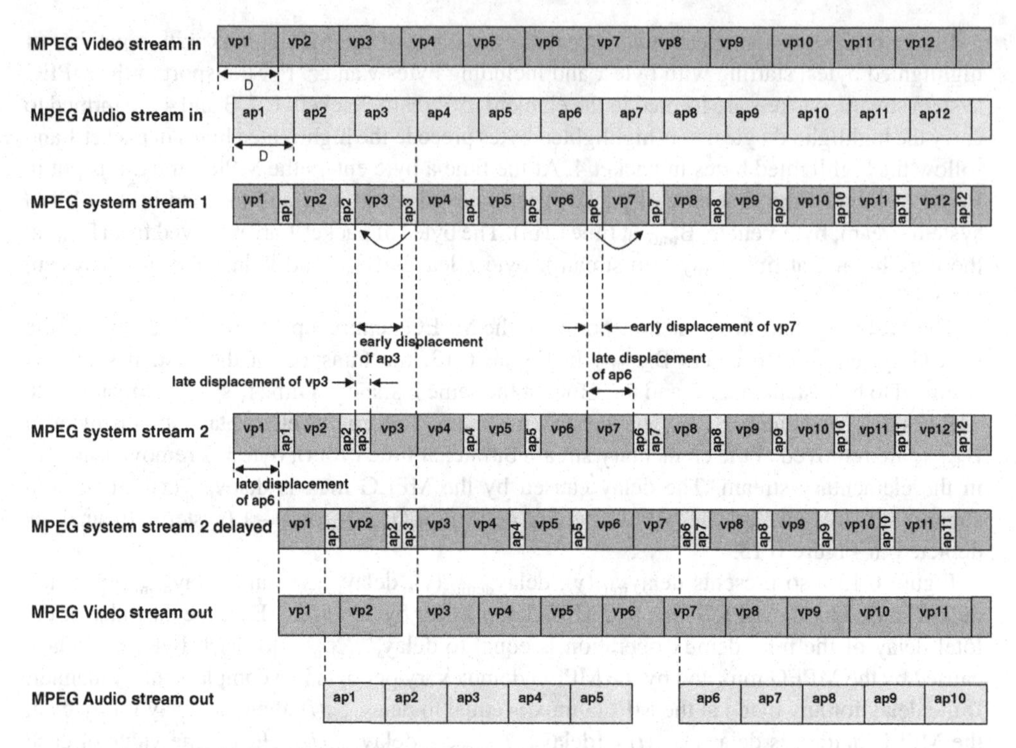

Figure 6.14 Introducing some flexibility in the multiplex strategy of Figure 6.11

video and audio output streams as produced by the MPEG demux of the delayed MPEG system stream 2. Packets vp1, ap1, vp2 and ap2 are positioned as in MPEG system stream 1. Therefore the MPEG video and audio output streams start initially the same as in Figure 6.11. However, vp3 in the system stream is delivered late, due to which vp3 and all following video packets in the MPEG video output stream are delayed by the late displacement of vp3. Likewise, in the MPEG audio output stream, ap6 and all following audio packets are delayed by the late displacement of ap6.

In other words, compared to the evenly spaced positioning, the early and late positioning of packets cause a total multiplex and demultiplex delay equal to the maximum early displacement plus the maximum late displacement. When the delays differ for audio and video, a compensating delay is needed for the stream that is delayed least.

Figure 6.15 depicts the behaviour of the multiplex and demultiplex buffers for the delayed MPEG system stream 2 from Figure 6.14. As the MPEG mux has knowledge of the applied displacements, the MPEG mux delays the creation of the system stream with the maximum value of the applied early displacement so as to prevent a system stream with a discontinuity. Obviously, the delay causes an increase of the fullness of the multiplex buffers $B_{mux}(v)$ and $B_{mux}(a)$. The flow of the video and audio packets through $B_{mux}(v)$ and $B_{mux}(a)$ is shown. When the MPEG mux finishes putting the early packets ap3 and vp7 in the system stream, a minimum in the fullness of $B_{mux}(a)$ and $B_{mux}(v)$ is caused. On the other hand, when the MPEG mux starts

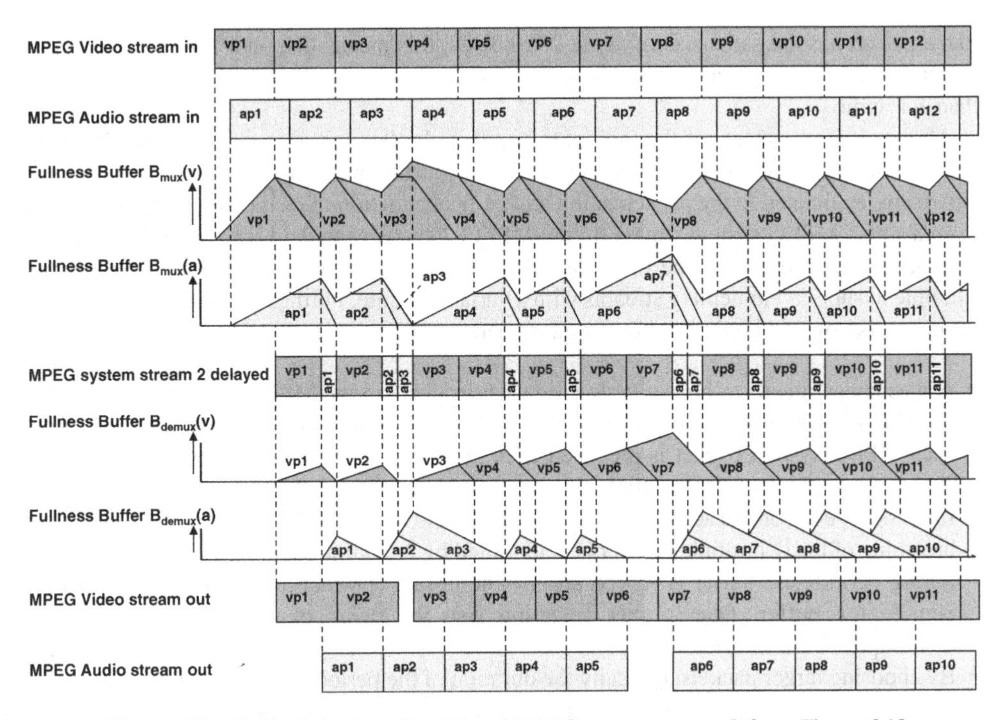

Figure 6.15 Buffer behaviour for delayed MPEG system stream 2 from Figure 6.13

putting the late packets ap6 and vp3 in the system stream, there is a maximum in the fullness of the multiplex buffer.

At the MPEG demux, Figure 6.15 shows that, after putting vp2 and ap5 in the MPEG video and audio output streams, packets vp3 and ap6 in the system stream appear to be delivered late, resulting in empty demultiplex buffers until the delivery of vp3 and ap6 on one hand and a discontinuity in the MPEG video and audio output streams on the other. When a late packet starts to arrive from the system stream, the demultiplex buffer fullness reaches a minimum, while a maximum is reached at the end of the arrival of an early packet.

As discussed in Figure 6.13, when the MPEG demux exactly reproduces the audio or video stream received at the input of the MPEG mux, then the delay caused by the mux-demux operation is constant for each byte in that stream: for any byte i the total delay $[\text{delay}_{\text{mux}}(i) + \text{delay}_{\text{demux}}(i)] = \text{delay}_{\text{mux-demux}}$. In such a case, for a constant bitrate elementary stream, the sum of the fullness of B_{mux} and the fullness of B_{demux} is also constant; if the fullness of one is minimum, the fullness of the other is maximum, as can be seen in Figure 6.15.

In MPEG, flexibility is required for the construction of audio and video packets, as the requirements may vary considerably per application. Some applications require fixed packet sizes, or packetization on audio frame or video picture boundary. To avoid unnecessary restrictions of the application scope of the MPEG system standard, the MPEG systems group decided for a flexible standard, with neither strict requirements on the packet size, nor on alignment at audio frame or video picture boundary; packets may contain fractions of an audio frame or video picture. In practice, obviously, applications specify certain constraints for the construction of audio and video packets. Video CD for example requires that each audio and video packet has the same size of about 2000 bytes; this means that on a Video CD there are typically many more video than audio packets, to cope with the much higher bitrate for video than for audio.

Above, various aspects of the multiplex and demultiplex operation are discussed by using simple examples, suitable to explain the important principles with respect to buffer behaviour and delays associated to the packetization and de-packetization of elementary streams into and from a single MPEG system stream. In the MPEG-2 system standard, various constraints and parameters are defined for the multiplexing and demultiplexing of elementary streams, for example to support elementary streams with a variable bitrate. Further details will be discussed in the course of this book, but here some remarks can be made already:

- A mechanism needs to be available to identify packets in the MPEG system stream and to signal its content, for example whether the packet carries audio or video data; for this purpose each MPEG system packet is preceded by a header.
- The buffers required for multiplexing and demultiplexing are 'breathing', that is their fullness depends on the actual bitrates at the input and output of the buffers. For example, the breathing of the demultiplex buffer is determined by the bitrate of the MPEG system stream during periods the buffer is served and by the bitrate at which data is removed from the demultiplex buffer. Due to this breathing also the delay of coded data through the demultiplex buffer varies.
- By choosing larger packets, typically the duration of the periods an elementary stream is not served by the system stream is increased. This leads to larger multiplex and demultiplex buffers, as well as to a larger total delay by the multiplex–demultiplex operation.
- By choosing smaller packets, the overhead due to headers preceding the packets increases.

6.3 Delivery Schedule of MPEG System Streams

Transport of an MPEG system stream causes a delay; typically, this delay is relatively small, but not necessarily constant. When the transport causes a variable delay, time jitter is produced in the delivery of the system stream. The caused delay may differ significantly per network packet, such as in ATM networks, where the delay may a priori even be unknown. A more subtle jitter is caused when, prior to transport, each packet of an MPEG system stream is interleaved with channel specific data for error correction purposes. The interleave of MPEG system packets and this data causes a (slight) change in the delivery time of the bytes in the packets of the MPEG system stream. This is depicted in Figure 6.16, where an MPEG system stream is shown that consists of a concatenation of packets of fixed size. In between two MPEG system packets, the error correction data is inserted. In Figure 6.16, prior to packet 1, the error correction data D1 is inserted. The D1 data can be used to correct errors that may occur during the transport of packet 1; similarly, D2 applies to packet 2. In this example, the size of the error correction data for each packet is assumed to be fixed.

In Figure 6.16, the insertion of the error correction data is made possible by somewhat squeezing each MPEG system packet. The packets are squeezed, so that the delivery time of the first byte of each packet is shifted by the time taken by the transport of the error correction data, while the delivery time of the last byte of each packet remains the same. For example, in Figure 6.16, room is made for the delivery of D2 between t_1 and t_2 by postponing the delivery of the

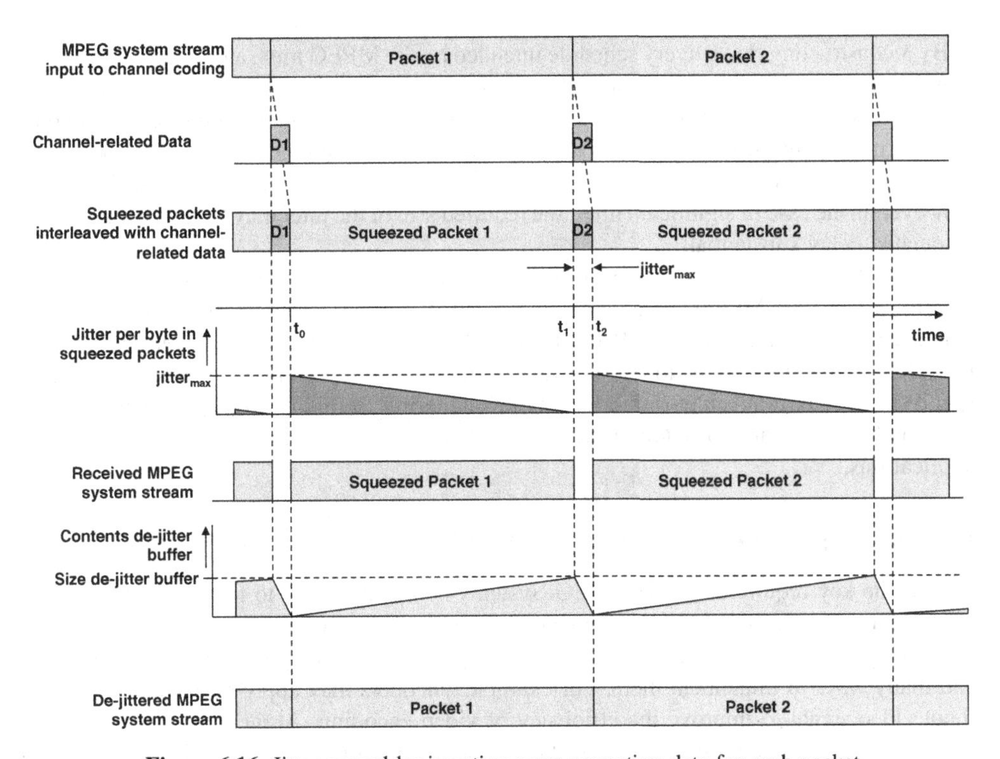

Figure 6.16 Jitter caused by inserting error correction data for each packet

first byte of packet 2 from t_1 to t_2. After the transport, the error correction data is removed and may be used to correct errors, if any. As a result, the received system stream with the squeezed packets is bursty. Compared to the original MPEG system stream, the transport layer introduced a jitter in the delivery of the bytes of each squeezed packet, ranging from $\text{jitter}_{max} = (t_2 - t_1)$ for the first byte to zero for the last byte of the packet.

In the simple case of Figure 6.16, the caused jitter can be removed by using a de-jitter buffer in which the squeezed packets enter, while the buffer is emptied at a rate equal to the bitrate of the original system stream, so that the original MPEG system stream is reproduced at the output of the buffer, but with a constant delay, without any jitter. For each byte in the system stream, the value of the delay is the equal to $(t_2 - t_1)$.

When jitter is introduced in the transport of an MPEG system stream, a receiver decoding such stream must be capable of handling such jitter. However, the jitter characteristics strongly depend on the transport channel used. Therefore it was decided that transport jitter is out of scope for the MPEG-1 and MPEG-2 system standards and should be handled by the transport layers. Thereby the delivery schedule of an MPEG-1 or MPEG-2 system stream is the schedule at which the MPEG mux produces the system stream, prior to any transport or storage. MPEG systems requires the MPEG mux to accurately specify the timing of the MPEG system stream produced, by means of parameters contained in the system stream. This ensures that the delivery schedule of an MPEG-2 system stream, as produced by the MPEG mux, can always be reconstructed accurately from information contained in the stream, independent of how an MPEG-2 system stream is delivered and where it originated from.

By reconstructing the delivery schedule intended by the MPEG mux, any jitter caused by the transport can be removed conveniently. The removal of jitter requires the use of a buffer to store the jittered stream, so that at its output a stream without any jitter can be produced. The amount of buffering depends on the amount of jitter. In the case shown in Figure 6.16, a relatively small buffer is needed, with a required size of $(t_2 - t_1)$ times the bitrate of the system stream. However, in the case of significant jitter, the required size of the jitter may become large and the associated delay substantial.

In conclusion, the delivery schedule of an MPEG-1 or MPEG-2 system stream is determined by the MPEG mux, and the handling of jitter is left to the transport layer. One exception is made however. Part 9 of the MPEG-2 standard defines a generic 'real time interface for low jitter applications' for the low jitter transport commonly used for broadcast, specifying a maximum jitter value for streams adhering to this interface (see Chapter 15). Thereby MPEG-2 part 9 provides a standardized method for handling jitter in most broadcast applications.

6.4 Synchronization of Audio and Video

One of the key requirements for MPEG Systems is the capability to play back audiovisual content in a synchronized manner, independent of any pre-processing at encoding or post-processing at decoding. There are many options available for such pre- and post-processing, and many ways to implement them. For example, encoders may apply pre-processing of the input video signal to improve the efficiency of video encoding. At the decoder, video post-processing tools can be used to improve the picture quality of the decoded video by reducing the visibility of coding artefacts. If a decoder produces an output video signal at another frame rate than the picture rate of the coded video, then the post-processing will include frame rate conversion.

An example of an MPEG coding system is provided in Figure 6.17, where an audio/video source, such as a video camera, a film scanner or a video recorder provides the video and audio input signals that are assumed to be in (perfect) synchronization. For reasons that are at the discretion of the encoder, the video is pre-processed in Figure 6.17; this usually causes a delay of the video input signal of one or more picture periods. The additional video delay due to the pre-processing needs to be taken into account in the compensating audio delay at the encoder. The video and audio encoder compress the video and audio into the MPEG video and audio streams, that are multiplexed into the MPEG system stream.

The MPEG system stream is transported in a medium specific manner, for example over an optical disk such as a Video CD, a DVD or a Blu-ray disc, or over a TV broadcast channel. In case jitter is caused, the jitter must be removed at the medium specific transport layer, as discussed in Section 6.3, so as to ensure that at the input of the MPEG demux, the MPEG system stream is delivered at the schedule intended by the MPEG mux. After demultiplexing the system stream, the MPEG video and audio streams are available for decoding, presuming that during transport no unrecoverable errors occur. The resulting video and audio output streams are provided to a display and loudspeaker, for example in a TV set or a monitor. The MPEG coding system achieves audio and video synchronization if the mutual timing relationship between the video pictures and the audio frames delivered by the audio/video source at the encoder is reproduced at the output of the decoder, irrespective of the processing applied in the coding system.

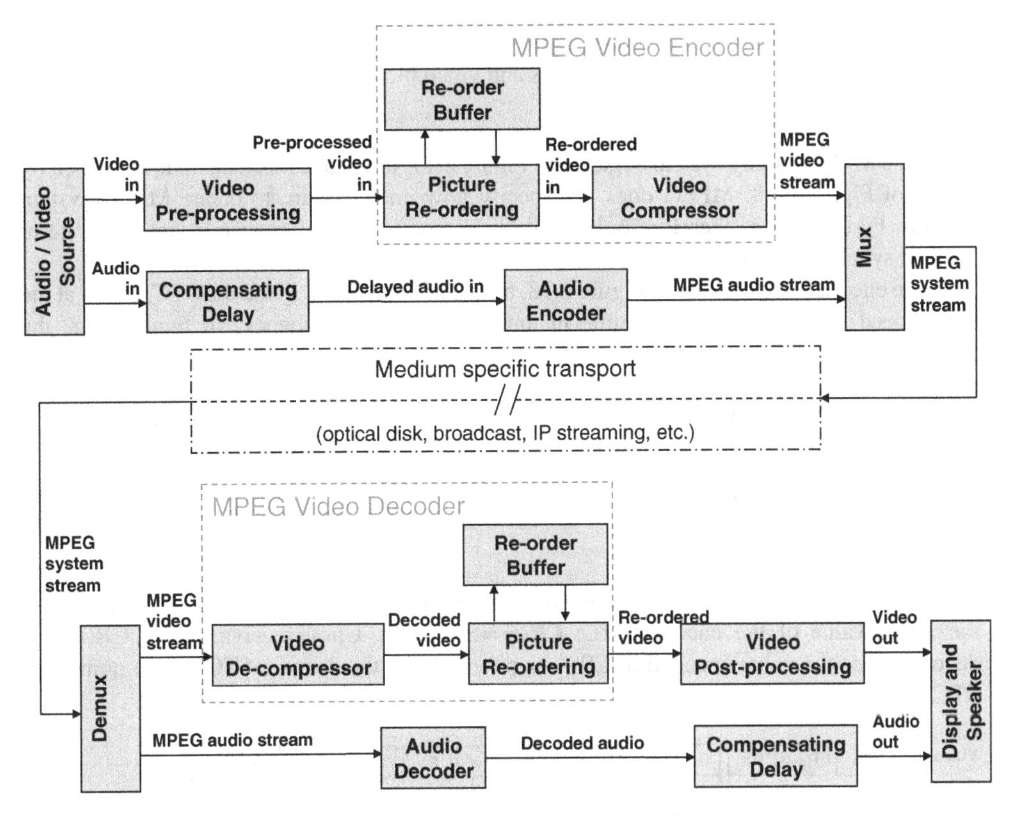

Figure 6.17 Example of a typical MPEG coding system

MPEG systems uses a relatively simple and straightforward mechanism to recover the original mutual timing relationship between audio and video by introducing a timing model as follows.

- The encoder and decoder share a common system time clock, the STC. Each STC value represents the time in units of the STC frequency. The STC is used to accurately determine the original positions in time of audio frames and video pictures at the input of the MPEG coding system and to re-position the decoded audio frames and video pictures at the output of the decoder. Usually, the audio sampling frequency and the picture period are locked to the STC, which means that the STC has an exactly specified ratio to the audio sampling rate as well as to the video picture rate.
- To allow receivers to correctly construct the STC, the MPEG system stream is required to carry samples of the STC, here referred to as clock references (CRs). Each CR provides the value of the STC at the instance in time the CR is inserted in the MPEG system stream.
- For the positioning in time of audio frames and video pictures at the output of the decoder, presentation time stamps are used. Each presentation time stamp (PTS) provides the value of the STC at the 'beginning' of an audio frame or video picture. The MPEG system stream is required to convey the PTS values to the receiver.

- It is the responsibility of the encoder to determine PTS values that reflect the original mutual timing relationship between audio frames and video pictures from the audio/video source at the encoder.

For the MPEG coding system depicted in Figure 6.17, the above mechanism is explained by means of Figure 6.18. MPEG does not specify how encoders are to create MPEG system streams; Figure 6.18 is therefore only a simple example of how an encoder may operate to achieve synchronization.

At the encoder and decoder in Figure 6.18, both STCs, the STC_{enc} and the STC_{dec}, run at the same speed and share the same time in units of the STC frequency. In Figure 6.18, the following parameters are used:

- i indicates the index of a Clock Reference (CR) in the MPEG system stream.
- $STC_{enc}(i)$ indicates the value of the STC at the encoder at the instance in time $CR(i)$ is transported in the MPEG system stream by the encoder.
- $STC_{dec}(i)$ indicates the value of the STC at the decoder at the instance in time the decoder receives $CR(i)$.

The same STC time is shared between the encoder and decoder by conveying in each $CR(i)$ the actual value of the encoder STC: $CR(i) = STC_{enc}(i)$. Upon receiving each $CR(i)$, the decoder takes the value encoded in $CR(i)$ as reference for the decoder STC; hence nominally $STC_{dec}(i) = CR(i) = STC_{enc}(i)$.

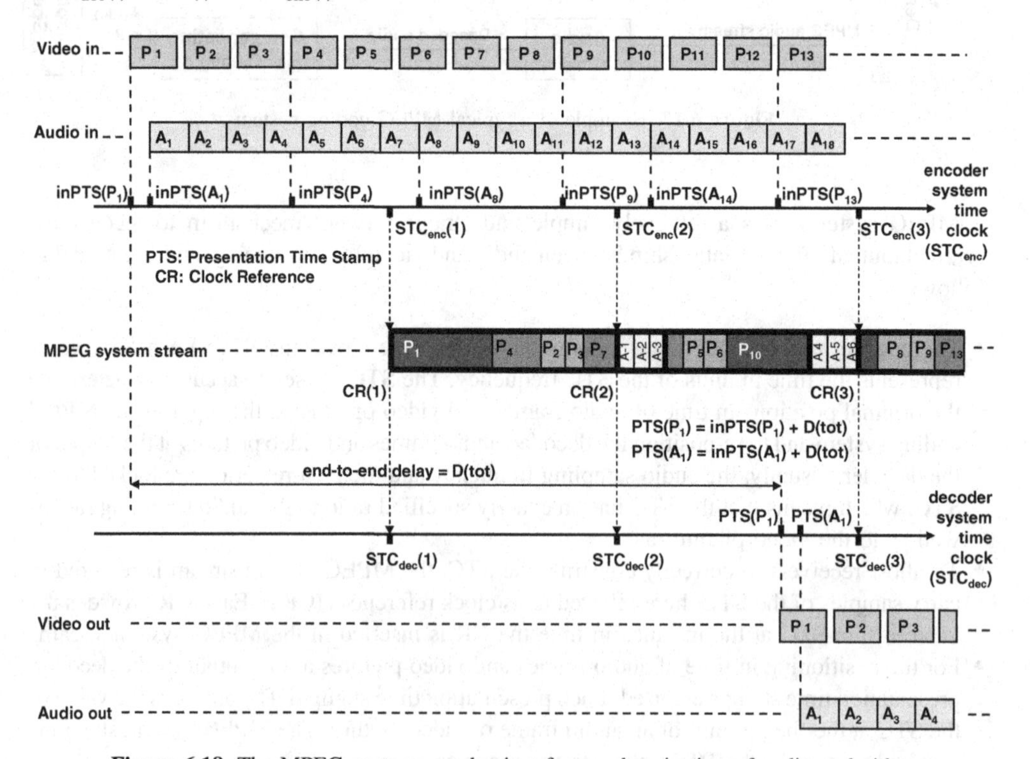

Figure 6.18 The MPEG systems mechanism for synchronization of audio and video

The encoder STC is used to determine the PTS values of audio frames and video pictures contained in the audio and video input signals, here referred to as inPTS. For example, inPTS (P_1) and inPTS(A_1) indicate the values of the encoder STC at the beginning of video picture P_1 and audio frame A_1 in the audio and video input signals, respectively. Likewise, inPTS(P_4) indicates the value of the encoder STC at the beginning of video picture P_4 in the video input signal, while inPTS(A_8) indicates the value of the encoder STC at the beginning of audio frame A_8 in the audio input signal.

The delay caused by the MPEG coding system between the audio and video at its input and the audio and video at its output is equal to the value D(tot) of the end-to-end delay. As the encoder and decoder in Figure 6.18 share the same time, the intended PTS for an audio frame or video picture is found by adding D(tot) to the inPTS value. For example, PTS(P_1) = inPTS (P_1) + D(tot) and PTS(A_1) = inPTS(A_1) + D(tot).

As an alternative to adding D(tot) to the inPTS values, the encoder may also use inPTS as PTS value [e.g. PTS(P_1) = inPTS(P_1)], and compensate for D(tot) by means of CR(i) values [e.g. CR(1) = $STC_{enc}(1)$ – D(tot)]. This achieves a constant offset in time between the STC in encoder and decoder, whereby the STC time in the decoder is D(tot) earlier than in the encoder: $STC_{dec}(i) = CR(i) = STC_{enc}(i) - D(tot)$.

The above method requires that the encoder has knowledge of the end-to-end delay value D(tot). The total end-to-end delay is the sum of the delay caused at encoding and the delay caused at decoding. The encoder obviously has knowledge of the first, but not of the latter, as the delay caused at decoding can vary widely. This is one of the reasons why MPEG systems defines a reference model for system decoding, the so-called System Target Decoder, STD. The STD is a hypothetical decoder in which the decoding process is specified in mathematical terms. In reality a physical decoder will operate in a manner that differs from the STD, but in practice such decoder can compensate for the way its design differs from that of the STD. The ideas behind the STD will be discussed in the next section; what matters here is that the CR and PTS values in an MPEG system stream do not refer to a physical decoder with unknown properties, but instead to the mathematical decoding process in the STD that is specified by MPEG and exactly known to the encoder.

The synchronization method must be capable of handling presentation discontinuities in MPEG video and audio. Such discontinuities may occur due to editing of coded audio and video streams at the encoder.[7] When editing MPEG video and audio streams, it needs to be taken into account that the offset between audio frames and video pictures varies. For example, MPEG-1 coded video with a frame rate of 29.97 Hz has a nominal picture period of 33.367 ms, while an MPEG-1 Layer 2 audio frame has a nominal frame period of 26.1224 ms in the case of a 44.1 kHz sampling frequency, which results in a continuously changing time offset between video pictures and audio frames. To ensure accurate synchronization of audio and video after the edit, it is required to maintain the mutual timing relationship between audio and video after the edit point. Due to the varying offset, the first decoded video picture and audio frame after the edit point can typically not be positioned immediately after the last decoded picture and frame before the edit point. Usually, after the edit point, either the audio frame or the video picture will start later than expected from the nominal frame or picture period. In Figure 6.19

[7] To ensure a good decoder performance after such edit, special measures in video coding are required, in particular with respect to predictive coding and VBV buffer usage for the pictures immediately preceding and following the edit point.

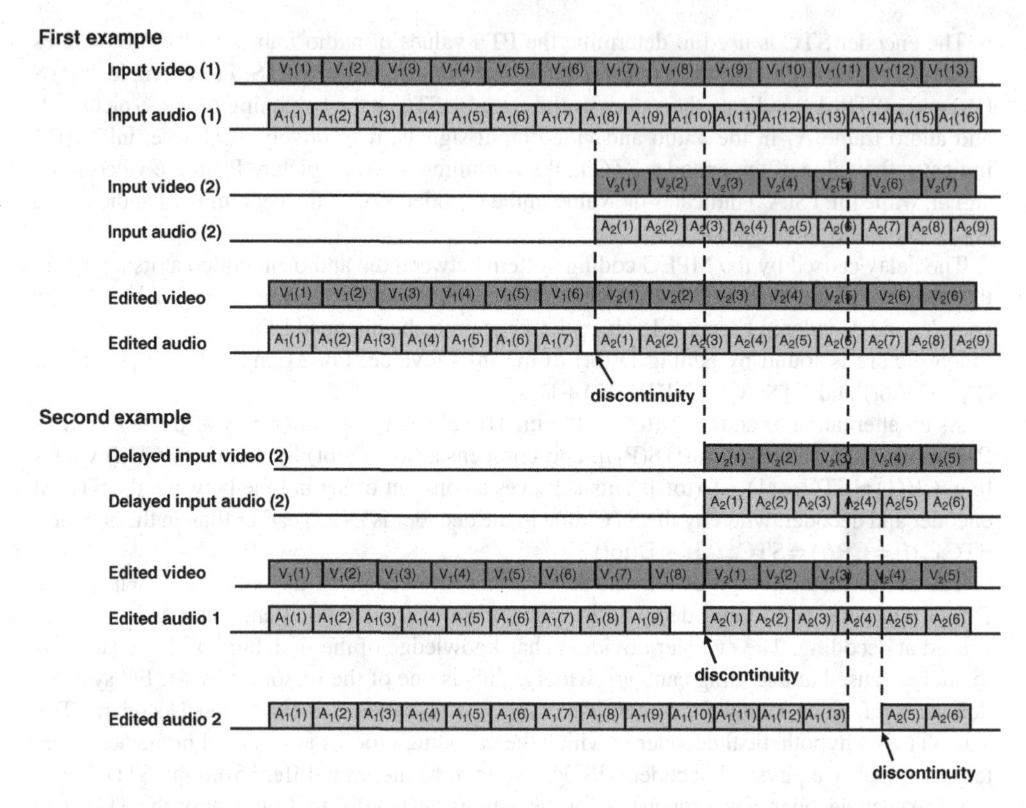

Figure 6.19 Examples of presentation discontinuities due to editing

two examples are given; at the encoder an edit is made within the input audio (1) and video (1) to switch to input audio (2) and video (2). In both examples, the caused discontinuity is in audio. To ensure correct presentation after a discontinuity, a PTS must be present for the first audio frame or video picture after a discontinuity.

In the first example, the edit point in the input video is after picture $V_1(6)$. In the edited video, picture $V_1(6)$ immediately follows picture $V_2(1)$ from input video 2. Audio frame $A_2(1)$ starts at the same time as picture $V_2(1)$ and therefore the edit point in input audio (2) is after audio frame $A_1(7)$, because the time between the end of $A_1(7)$ and the beginning of $A_2(1)$ is less than the audio frame period. The resulting 'time gap' between $A_1(7)$ and $A_2(1)$ causes a discontinuity at the start of $A_2(1)$. In the second example, input audio and video (2) are delayed for an edit after picture $V_1(8)$. In the edited video, now picture $V_1(8)$ is immediately followed by $V_2(1)$. In the input audio, the edit point is now after audio frame $A_1(9)$; the resulting 'time gap' between $A_1(9)$ and $A_2(1)$, larger than in the first example, causes the discontinuity at the start of $A_2(1)$, as depicted in the edited audio 1. Sometimes, a time offset between the editing positions of audio and video is needed; for example, there may be a need to postpone editing the audio, because a speaker needs to complete a sentence. This is shown in edited audio 2, in which the audio edit point is postponed until after $A_1(13)$.

6.5 MPEG-2 System Streams and the STD Model

For the MPEG-2 target applications, such as broadcast and optical media, the MPEG system standard needs to ensure that a number of very important requirements are met. Some of those requirements need to ensure that decoders can be designed of high quality. Examples of such requirements are:

- exact synchronization between audio and video;
- minimum delay of audio and video;
- minimum buffer resources;
- no buffer overflow;
- no buffer underflow, unless explicitly allowed;
- a good random access.

Furthermore, for target applications it is essential to ensure that each valid MPEG-2 system stream can be properly decoded by a compliant decoder. To achieve this, the following two requirements are very important:

- possibility to unambiguously determine whether an MPEG-2 system stream complies with the MPEG-2 system standard;
- possibility to unambiguously determine whether an MPEG-2 system decoder complies with the MPEG-2 system standard.

How to achieve the above requirements has been discussed between MPEG system experts at length; there are several potential solutions. As usual, it takes time to reach common understanding of the problems and to develop consensus on the most desirable approach. However, the end result was a unique and very powerful concept, in which the System Target Decoder model, in short STD model, forms the core of MPEG systems. The STD model was born during MPEG-1 system development and developed further for MPEG-2 systems. The STD model is the brilliant solution of the MPEG system experts for resolving crucial system issues. The STD not only provides a framework that allows accurate A/V synchronization combined with an efficient and robust buffer management, but offers also a very important tool for determining compliancy of encoders and decoders and for ensuring interoperability between an encoder and a decoder. The main characteristics of the STD are summarized below.

- The STD is a hypothetical model, operating with mathematical accuracy, for the decoding of MPEG-2 system streams carrying one or more MPEG video or audio streams.
- To ensure that the STD is capable of decoding any MPEG audio and video stream contained in any MPEG-2 system stream, the STD characteristics with a dependency on MPEG video, audio and system streams are described by parameters, such as buffer sizes and bitrates.
- In each MPEG-2 system stream, the applicable values of the STD parameters are either provided by encoded fields in the MPEG-2 system stream or provided otherwise. As a result, each MPEG video, audio and system stream provides its own STD characterization.
- The STD is designed so that an actual decoder can compensate for ways in which its design differs from that of the STD.

- The STD serves as a reference decoder in various ways:
  - The MPEG-2 system standard uses the STD to accurately specify the timing of byte arrival and decoding events, as well as other semantics and associated constraints of MPEG-2 system streams.
  - One of the conditions[8] for an MPEG-2 system stream to be 'valid', that is to be fully compliant with the MPEG-2 system standard, is to meet each requirement and constraint for decoding by the STD. Therefore an MPEG-2 system encoder must use the STD to ensure that a produced MPEG-2 system stream meets each appropriate requirement for decoding by the STD.
  - A physical MPEG-2 system decoder designed to decode MPEG-2 system streams within certain application constraints, must properly decode each such stream that does not violate any requirement or constraint for decoding by the STD. In other words, the decoder designer must ensure, possibly within certain application constraints, that each MPEG-2 system stream that plays correctly on the STD, also plays properly on the designed physical decoder.

The STD model is an hypothetical model for the decoding of elementary streams contained in an MPEG-2 system stream. Elementary stream is a generic term for a coded video, coded audio or other coded stream. Initially, MPEG-2 systems only defined the STD model for the decoding of MPEG audio and MPEG video streams, but when MPEG extended its scope, the MPEG-2 systems STD was extended for the decoding of other MPEG defined media, such as MPEG-7 metadata, as will be discussed in subsequent chapters of this book. The use of the STD is not limited to MPEG: it is allowed to define private, in terms of MPEG, extensions of the STD. For example, various application standardization bodies defined STD extensions for the decoding of application specific media, such as subtitling.

For decoding in the STD, each contained elementary stream has its own 'branch' in the STD where, prior to decoding, the elementary stream data is stored in one or more input buffers. In MPEG-2 systems multiple versions of the STD model are defined that are closely related but that differ in the use of input buffers. In Figure 6.20 the general concept of the STD model is presented with support for the decoding of MPEG video, audio and systems data. Though Figure 6.20 does not yet address any detail on the use of input buffers, it should be noted that the sizes of STD input buffers are provided by parameters in the MPEG-2 system stream. Also data rates at the input and removal schedules at the output of these buffers are exactly specified.

At the input of the STD, the MPEG-2 system stream is provided at a specified bitrate. The MPEG-2 system stream carries samples of the System Time Clock to determine the STC time in the STD in an unambiguous and exact manner. Along the STC time line, the arrival time of each byte in the MPEG-2 system stream is exactly specified. Immediately upon arrival, the MPEG-2 system stream is parsed so as to ensure that the elementary stream data is passed to its decoding branch. Each elementary stream has exactly one decoding branch and in one decoding branch no more than one elementary stream is decoded. System control and timing data, such as decoding and presentation time stamps, are passed to the systems decoder, to ensure that each video picture and audio frame is decoded and presented at the specified decoding and presentation time. The decoding branch for MPEG video contains an explicit

[8] Another condition is that the syntax of the stream is fully correct.

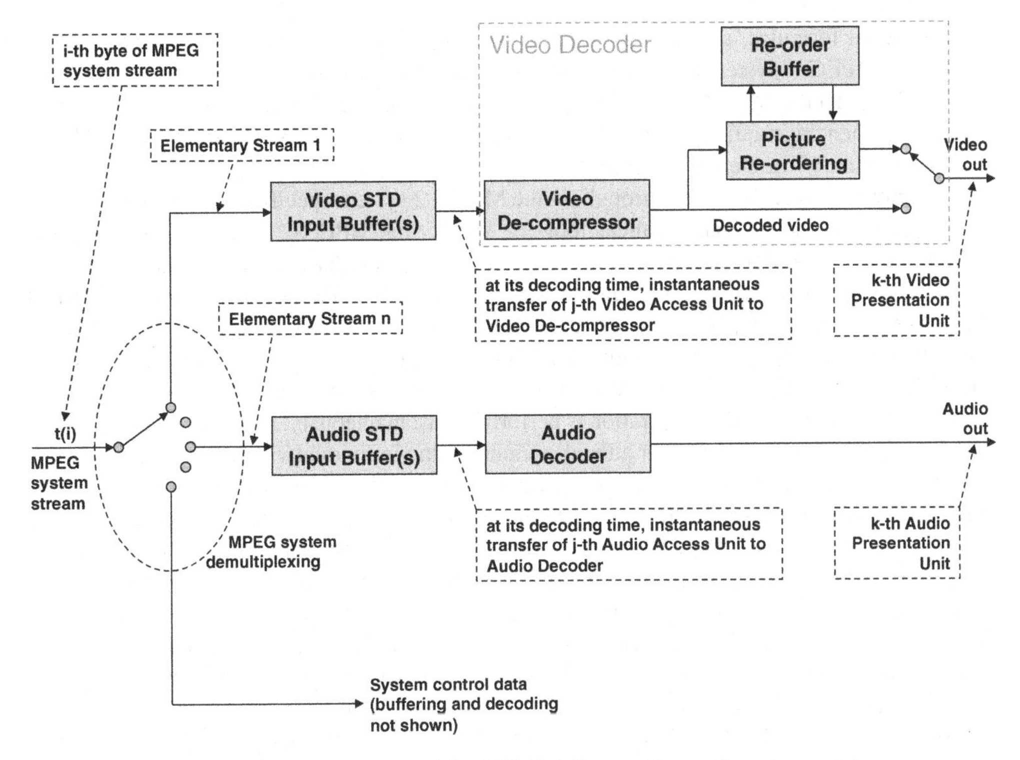

Figure 6.20 The concept of the MPEG-2 System Target Decoder model

option to switch off picture re-ordering for video streams that do not use B-pictures. Picture re-ordering causes a delay of one picture period for each I and each P picture, but if no B-pictures are used in the video stream, the re-ordering can be switched off to prevent this delay.

The STD model is based on instantaneous decoding, comparable to the VBV model in MPEG video; at the specified decoding time for a video picture or audio frame, the compressed data representing the video picture or audio frame is instantaneously decoded and removed from the input buffer. To mathematically describe requirements and constraints for STD decoding, a number of parameters is defined; some examples are provided below:

- i – the index of a byte in the MPEG-2 system stream;
- $t(i)$ – the time of arrival of the ith byte of the MPEG system stream at the input of the STD;
- n – the index of an elementary stream;
- j – the index in decoding order of an access unit in an elementary stream; an access unit is a generic term for the 'coded representation of a presentation unit', in the case of video a coded video picture and in the case of audio a coded audio frame;
- k – the index in presentation order of a presentation unit.

By means of above parameters, each coded video picture or audio frame is identified uniquely as access unit (j) of elementary stream n, and each decoded video picture or audio frame as presentation unit (k) of elementary stream n. By using STD parameters in mathematical formulas, a generic formalism is available to specify constraints and requirements for MPEG-2 system compliancy with mathematical accuracy. For example, when the delay of any data through the STD input buffers for a specific elementary stream should not exceed 1 s, then this requirement is formally expressed as $td_n(j) - t(i) \leq 1$ s for all j, and all bytes i in access unit (j) in elementary stream n.

The above formalism is used throughout the MPEG-2 system specification. Hence, the STD model is in essence a mathematical tool for specifying MPEG-2 system constraints and requirements. On the other hand, the concept of the STD model is close enough to physical decoder designs to serve as a reference for implementations. To ensure that each compliant MPEG-2 system stream playing properly on the STD also plays on a physical decoder, decoder manufacturers are required to compensate for ways in which their actual decoder design differs from that of the STD, as discussed in the next two paragraphs.

In the STD, each decoding operation is performed instantaneously. In a real decoder system, the decoding of a video picture or an audio frame by the individual video and audio decoders will take time and the caused delays must be taken into account in the design of the decoder system. Assume for example, that the decoding of each video picture takes exactly one picture presentation period 1/P, where P is the video frame rate. During that period, all coded data of the decoded picture is removed from the input buffer, but details of the removal schedule are unknown, as these depend on the amount of data spend on coding each macroblock within the picture. Within the input buffer, each possible removal schedule is allowed by assuming that all coded picture data is removed when decoding the last macroblock of the picture. With this assumption, the video decoding process in the real decoder is the same as in the STD, but delayed by exactly 1/P. During that delay compressed video data arrives at the input buffer of the video decoder. If this data arrives at a maximum bitrate R, then this amount of coded video data is at most R/P. Hence, to compensate for the caused delay, the size of the input buffer of the video decoder must be larger than in the STD model by R/P. The video presentation is likewise delayed with respect to the STD. Since the video is delayed, the audio presentation should be delayed by a similar amount to provide correct synchronization.

For decoding in the STD, each MPEG-2 system stream is produced with an intended delivery schedule of mathematical accuracy. In practice however, system streams are transported across data channels that lack such accuracy, causing deviations of the intended delivery schedule. To compensate for any transport specific deviation, each MPEG-2 system stream is required to carry all information needed to reconstruct the intended delivery schedule. Hence, prior to system decoding the impact of the transport channel can be eliminated, independent of the applied transport. Typically, additional buffering is needed to compensate for channel characteristics, as discussed in Sections 6.3 and 6.4.

MPEG-2 system streams are self-contained; each system stream contains all timing and other information needed for delivery to and decoding in the STD and for verifying whether all applicable constraints and requirements are met. The transport channel does not have any impact on this information and hence after transport, independent of the applicable transport characteristics, it is always possible to verify whether the received system stream is compliant.

The role of the STD in MPEG-2 system compliancy verification is depicted in Figure 6.21. A system stream is MPEG-2 compliant if it plays correctly on the STD, that is if each requirement and constraint for decoding by the STD is met by the system stream. A decoder complies with MPEG-2 systems if it can properly decode each compliant MPEG-2 system stream.

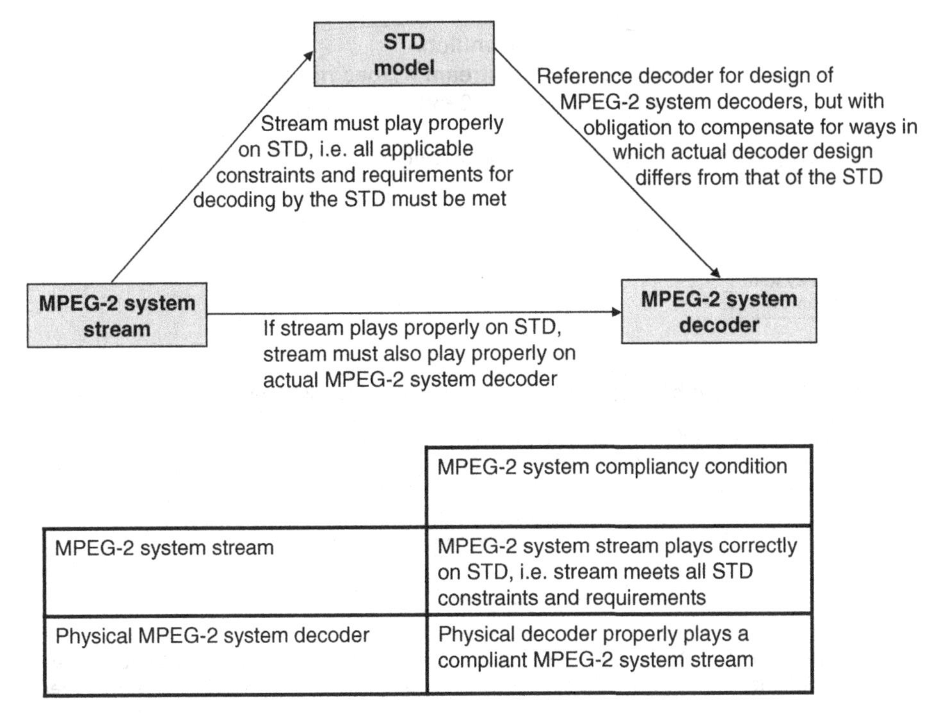

	MPEG-2 system compliancy condition
MPEG-2 system stream	MPEG-2 system stream plays correctly on STD, i.e. stream meets all STD constraints and requirements
Physical MPEG-2 system decoder	Physical decoder properly plays a compliant MPEG-2 system stream

Figure 6.21 The role of the STD in MPEG-2 system compliancy verification

The use of a mathematical STD model for determining compliancy of MPEG-2 system streams and for use as a reference for MPEG-2 system decoder design is an extremely powerful concept. It allows the development of reliable and high quality encoders and decoders as required for MPEG to take-off in the industry. It also allows the building of accurate, and thereby unambiguous, compliancy tools to verify whether a stream produced by an encoder meets all applicable STD requirements and constraints. Such tools are very valuable 'referees' when an actual decoder fails to play a certain MPEG-2 system stream. If a stream is compliant indeed, it must play properly on an actual decoder with the required properties; if not, the decoder does not comply with MPEG systems (see also Figure 6.22).

In Figure 6.22 a conflict is depicted where an actual MPEG-2 system decoder Y fails to properly play a certain MPEG-2 system stream X. In such conflict it is very important, both technically and from a business perspective, to determine whether the encoder produced an invalid stream X or whether the manufacturer of decoder Y made a mistake. This case may even become more complicated, if other decoders, as opposed to decoder Y, do play stream X properly. To resolve this issue it is sufficient to verify whether stream X is MPEG compliant.

This verification must include the contained elementary streams, as an error in the MPEG audio or video data may also cause the decoder to fail. Moreover, the verification should take the decoder capabilities into account; decoders are designed to support some particular elementary stream features and not every possible one.[9] A decoder is only required to be

[9] Audio and video coding standards usually have a wide range of capabilities to avoid unnecessarily constraining of potential applications in future.

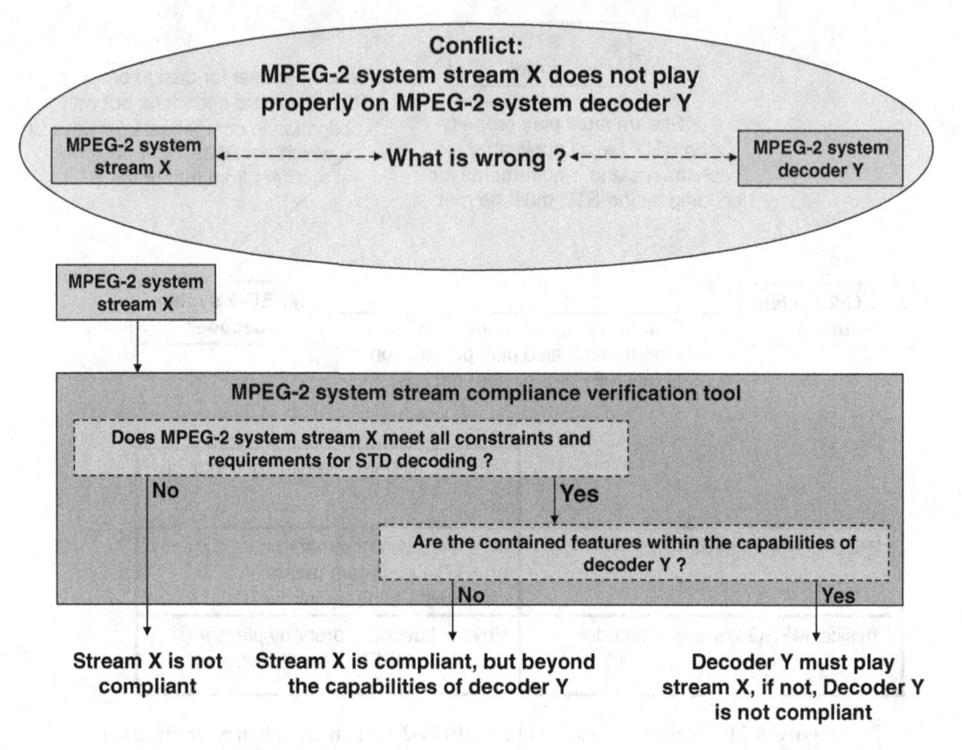

Figure 6.22 MPEG-2 systems compliancy verification for a conflict

MPEG compliant within its own capabilities, and therefore a stream verifier needs to check whether the elementary stream features contained in the stream are supported by the decoder, such as certain picture sizes and sampling rates. In Figure 6.22, if stream X is MPEG compliant and the contained features are within the capabilities of decoder Y, then decoder Y must be capable to play stream X properly. If not, then decoder Y fails to be MPEG compliant. If stream X does not meet all STD requirements and constraints, stream X is not compliant, even it plays properly on some decoders that are designed with larger margins than decoder Y.

6.6 Timing Issues

During the design of MPEG-1 and MPEG-2 systems, timing issues have been subject of much debate. Examples of such debated issues are:

- the frequency and tolerance of the System Time Clock, STC, and locking of the audio and video sampling frequencies to the STC frequency;
- coding of STC samples included as Clock References, CRs, in MPEG system streams:
  - how many bits per STC sample;
  - how often to include an STC sample in the MPEG system stream;
- accuracy of decoding and presentation time stamps and the number of bits to code them.

As a result of these debates important system choices were made when MPEG-1 systems was developed. These choices were discussed again when some additional MPEG-2 systems requirements were identified, related to the regeneration of the STC and sampling frequencies in system decoders. This section discusses first the MPEG-1 choices for the frequency, tolerance and coding of STC samples and time stamps, followed by a review of methods to regenerate the STC in system decoders. Finally, the problems of the MPEG-1 choices that led to changing the frequency and tolerance of the STC for MPEG-2 systems are discussed.

6.6.1 Frequency and Tolerance of the STC in MPEG-1 Systems

The STC samples in the MPEG-2 system stream specify the value of the STC at the time the STC sample enters the STD, so as to allow reconstruction of the STC frequency in receivers. In many, if not all, MPEG-2 system applications the video and audio sampling clocks are locked to the frequency of the STC.[10] This means that there is a fixed rational relationship between the STC frequency and the video and the audio sampling frequencies. One advantage of such locking is that the clock regeneration circuitry in receivers can be simplified; one of the clocks can serve as a master time base from which the other clocks are derived based on the fixed rational relationship between the clocks.

Another advantage of above mentioned locking is derived from the fact that video and audio are typically periodic signals with a fixed video picture or audio frame rate. If locked to the audio and video sampling clocks, the STC progresses with a fixed value during each video picture and each audio frame. For example, if the video picture rate is equal to 25 Hz, and if the STC frequency is 90 kHz, then for each picture the STC clock progresses 90 000/25 = 3600 ticks, and for a picture rate of 29.97[11] with 90 000/29.97 = 3003 ticks. Hence, if the STC clock is locked to the audio or video sampling clock, and if the time stamp is known for one picture or audio frame, then the time stamp for the next picture or audio frame can be calculated from the nominal picture or audio frame period. In which case, as long as the signal remains periodic, there is no need to encode the time stamps for each video picture and audio frame.

The STC frequency should be suitable both in 525-line/60 Hz and in 625-line/50 Hz systems. Already in the 1980s, a digital video interface standard was specified for use in both the 525-line/60 Hz and in 625-line/50 Hz television environments. This specification uses a sampling frequency for luminance of 13.5 MHz and for chrominance of 6.75 MHz. For some more background see Note 6.2.

In selecting the STC frequency, Rec 601 was an important guideline; it was agreed that the STC frequency should have an 'easy' rational relationship to 13.5 MHz. In MPEG systems the choice for the STC frequency was discussed for the first time during the development of MPEG-1 systems. At that time it was decided to use an STC frequency of 90 kHz, corresponding to 13.5 Mhz divided by 150. This choice was made mainly because 90 kHz provides enough accuracy for the time stamps specifying the decoding and presentation times of video pictures and audio frames. To synchronize audio and video, an error of several ms is permissible, far more than the 90 kHz period of about 11.1 μs.

---

[10] To some extent this is optional, though; MPEG-1 systems allows but does not require this locking, while MPEG-2 systems requires locking in some but not all cases. See also Sections 12.1 and 13.4.

[11] The video picture rate of 29.97 Hz is not fully accurate; the exact value of this frame rate is 30 000/1001 (= 29.970 029 970 029 970 etc).

Note 6.2 The Use of the 13.5 MHz Sampling Frequency for Analogue Video

Around the early 1980s, in the ITU-R, then called the CCIR-R, a specification was developed for a single component (Y, U and V) based digital video standard, suitable for usage both in 525-line/60 Hz and in 625-line/50 Hz environments. This resulted in ITU-R Recommendation 601, in which 13.5 MHz is specified as the sampling frequency for luminance Y and 6.75 MHz as the sampling frequency for the chrominance signals U and V. With these frequencies each horizontal line in both environments contains (about) 720 Y pixels and 360 U and V pixels.

To explain a bit more, some analogue TV background is needed. The analogue TV signal formats 525-line/60 Hz and 625-line/50 Hz were invented to control the electron beam of a Cathode-Ray Tube (CRT). This electron beam is deflected horizontally and vertically to illuminate the fluorescent screen of the CRT display so as to draw the video pictures, line by line.

To draw one horizontal video line, the intensity of the electron beam is varied corresponding to the brightness of the video picture to be displayed, while the horizontal deflection ensures that the beam travels across the target line. After writing a line, the beam must return to the horizontal and vertical starting position of the next line; this takes time, and to allow for that, the horizontal blanking interval was defined. A video line has a total duration of about 64 μs, with roughly 52 μs of active video and 12 μs for the horizontal blanking interval. Upon reaching the bottom of the screen, the beam needs to return to the first line at the top of the screen and for this purpose also a vertical blanking interval, VBI, is needed. In a 625-line video system, 49 lines are used for the VBI and in a 525-line video system 45 lines. Below are some figures:

	60 Hz/525-line video	50 Hz/625-line video
Total video line duration	63.555 . . . μs	64 μs
Number of 13.5 MHz cycles per line	858	864
Number of luminance pixels per line	720	720
Number of active video lines per picture	480	576
Number of VBI lines per picture	45	49

The 720 luminance pixels per line carry active video with a duration of 720/13.5 = 53.333 . . . μs, slightly more than the maximum active video period per line in the various video systems. As a consequence, some of the leading and trailing pixels carry no video data, resulting in black pixels. For coding the video, often only 704 of the 720 pixels are used to prevent small vertical black bars on the right and left hand side of the picture.

To convey STC samples and time stamps in system streams, in particular the number of bits used for the fields carrying these parameters is important, as this number determines the periodicity of the STC. For example, if the number of bits is equal to 16, then with an STC frequency of 90 kHz, the duration of one STC period is equal to $2^{16}/90\ 000 = 65\ 536/90\ 000 = 0.728$ s. Hence, an STC counter would reach the same value twice within 1 s. The STC however must be suitable to position events in time uniquely during normal use, for example during the playback of an audio or video stream. Which means that the STC period must be larger than the duration of a stream. Of course such duration can be extremely long. While MPEG does not impose a limit on the duration of a stream, the MPEG systems group decided that in practice the uniqueness of time stamps would be good enough with an STC period larger than 24 h. Therefore it was decided to encode STC samples and time stamps with 33 bits,[12] resulting in an STC period of $2^{33}/90\ 000 \approx 95\ 444$ s ≈ 26.5 h.

The MPEG-1 system specification requires the 90 kHz STC frequency to be in the range between 90 000 − 4.5 Hz and 90 000 + 4.5 Hz; this accuracy of ($4.5/90\ 000 = 50 \times 10^{-6}$) 50 ppm (parts per million) allows the use in STC clock circuitry of cost-effective standard crystals, as opposed to more expensive precision crystals. Next to the STC accuracy, MPEG systems also specifies the STC slew rate: the maximum rate of change of the STC frequency per second. For the 90 kHz STC in MPEG-1 the slew rate is 250×10^{-6} Hz/s, corresponding roughly to 2.778 ppm/s.

The accuracy and slew rate of the STC are important parameters for the design of clock circuitry needed to reconstruct the STC in a receiver or player, but also for the delivery of MPEG system streams. For example, an MPEG system stream may be provided by an optical disk player, such as a DVD or CD Video player, consumer devices for which the use of expensive precision crystals is less suitable.

6.6.2 Regeneration of the STC in System Decoders

There are many ways to design the timing control in a device for the playback of MPEG-2 system streams; such designs are beyond the scope of this book. However, to understand the timing problems caused by the MPEG-1 choices for MPEG-2 systems, it is important to discuss a few basics of designing clock circuitry. One of these basics in the choice of the master clock, that is the clock that drives the timing of the device. For this purpose two methods should be distinguished for the delivery of an MPEG-2 system stream to the receiver: 'push' and 'pull'.

With 'push' delivery, the MPEG-2 system stream is pushed to the device with a timing determined by the delivery device, for example a broadcast encoder or an optical disk. With this method, the STC clock at the decoder must slave to the STC clock at the encoder by using the STC samples in the received MPEG-2 system stream.

Figure 6.23 depicts the regeneration of the decoder STC in the case of a 'push' delivery. The received MPEG system stream drives the regeneration of the decoder STC. For this purpose usually a Phase Locked Loop (PLL) is used to control the decoder STC. The PLL contains an oscillator that produces the STC frequency that runs the STC counter. At the output of this counter the current decoder STC values are available, representing the decoder 'time' in terms of the STC, as used for processing the time stamps for video and audio decoding and presentation. When a playback starts, the first received STC sample is loaded in the STC

[12] Even though 33 is a weird number in a byte orientated environment.

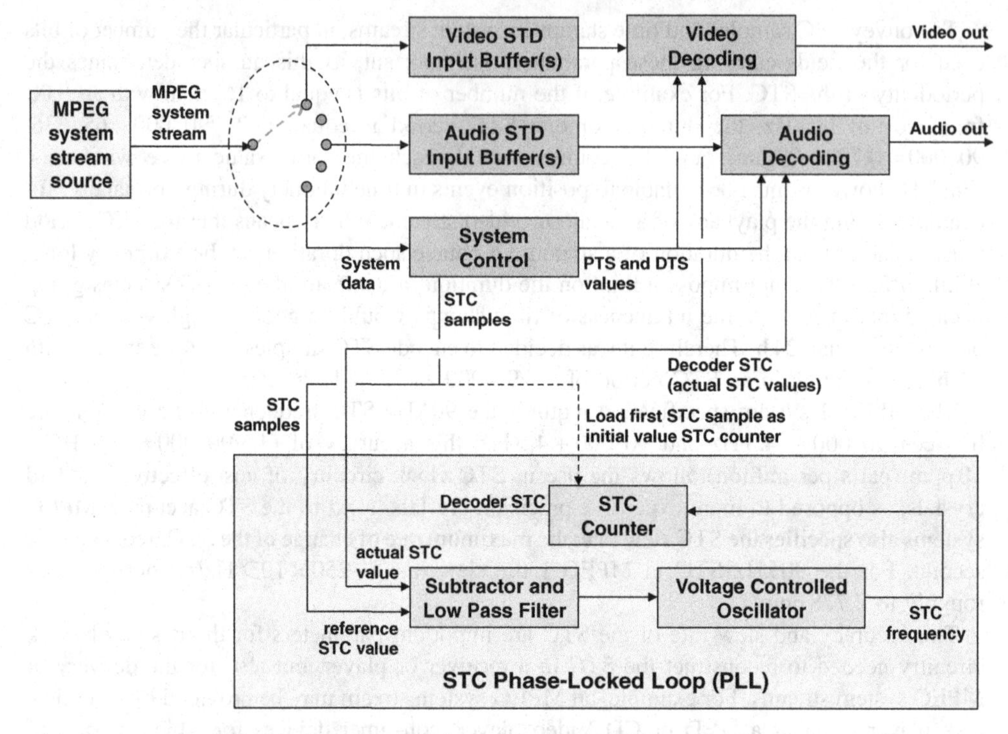

Figure 6.23 Regeneration of the STC for a 'Push' delivery

counter as the initial value of the decoder STC. When the next STC sample arrives, a subtractor within the PLL compares the value of the received STC sample with the actual value of the decoder STC; depending on the calculated difference an error signal is produced. This error signal drives a Voltage Controlled Oscillator (VCO) that operates at the nominal STC frequency. The STC samples are reference values for the decoder STC. If an STC sample arrives before the decoder STC reaches the value of this sample, then the STC frequency is slightly increased to ensure that the STC starts running somewhat faster. Similarly, the STC frequency is decreased if the STC sample arrives after the decoder STC reaches the sample value. In perfect operation the difference between both values is zero for each STC sample. To ensure that the VCO responds in an appropriate manner on identified differences, a low-pass filter is applied on the output of the subtractor.

With 'pull' delivery, the timing of the decoding of the MPEG-2 system stream is controlled by the receiving device; with this method, it must be ensured that the MPEG-2 system stream is delivered in a timely manner. For example, a stream from an optical disk may be delivered by a player at a delivery rate controlled by the receiving device. Another example is the receiving device playing an MPEG-2 system stream that is already available to the receiver on a hard disk. A more recent example is progressive download, whereby an MPEG-2 system stream is downloaded to the player; as long as the download rate is higher than the bitrate at which the stream is played back, it is ensured that the player is not running out of data during playback.

In case of 'pull' delivery, the receiver designates one of the clocks in the receiver as master clock. From the designated master clock, the decoder STC can be derived by applying the

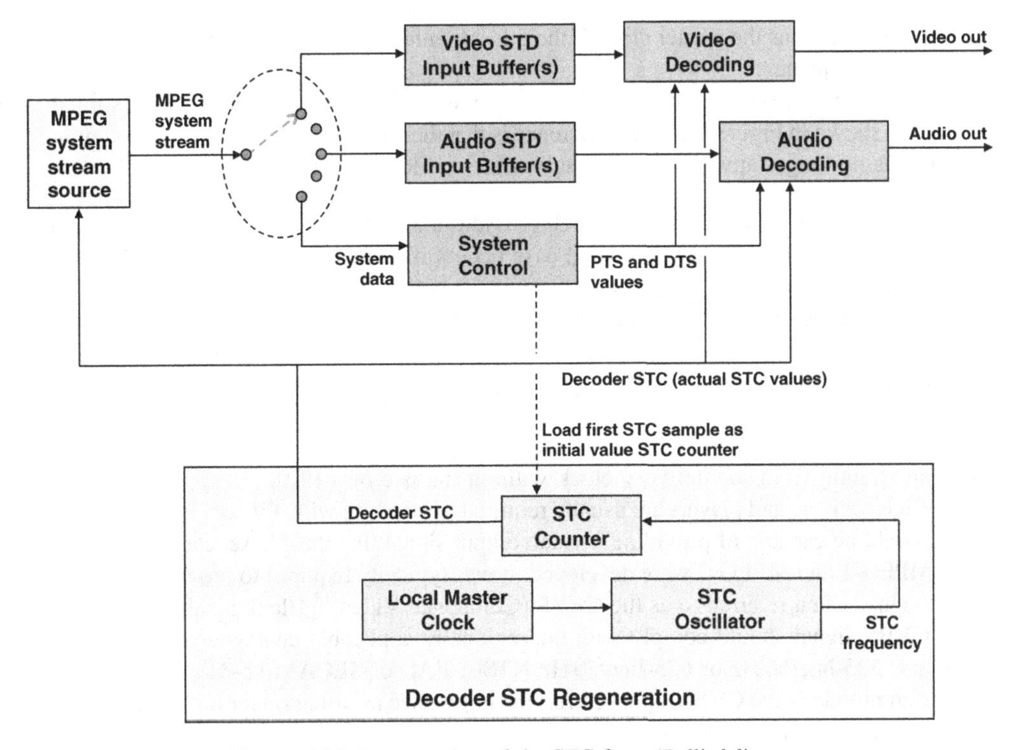

Figure 6.24 Regeneration of the STC for a 'Pull' delivery

required mathematical multiplication or division. Figure 6.24 shows the decoder STC driving the delivery of the MPEG system stream to the receiver. The receiving device has its own 'free-running' autonomous master clock, and uses this clock to drive the decoder STC for controlling the timing of the playback of the stream. In all cases the initial value of the decoder STC at the start of a playback is set by loading the first received STC sample in the STC counter. The receiver may choose to apply the delivery schedule to the STD according to the STC samples in the stream. However, because the decoding does not depend on the timing of stream delivery, the receiver may also choose any other delivery schedule that is considered appropriate by the receiver, as long as the system stream is decoded properly. While other options are available too, one option to implement such stream delivery is a bursty delivery by the source device to a buffer preceding the STD, followed by stream delivery from that buffer based on read STC samples. Another delivery option is to keep its input buffers sufficiently full to ensure that decoding can take place without interruption due to buffer underflow.

In case of 'pull' delivery, the receiver is completely free in choosing its master clock. Typical options include a separate STC oscillator and the audio or video sampling frequency, as briefly discussed below.

- The video or audio sampling frequency is the master clock. If the sampling frequencies are locked to the STC, then there is a fixed and exact relationship between the audio or video sampling clock and the STC frequency, which allows to derive the STC frequency straight from the audio or video sampling frequency. For example, the video sampling frequency may be selected as the master clock if there is a (mathematically) convenient relationship to the STC frequency.

- A separate STC oscillator is used. In this case the STC regeneration can be quite similar to the one shown in Figure 6.23; the difference is that the VCO is autonomously 'free-running'. There is no longer any need for the subtractor and low-pass filter.

When the MPEG system stream source provides a delivery clock synchronously with the delivered data to the receiver, a special case occurs if such delivery clock is locked to the STC frequency at encoding. Such locking requirement is beyond the scope of MPEG, but may be imposed by an application standardization body or other entity in charge of application requirements. If the delivery clock is not locked to the STC, 'push' delivery applies and the solution provided in Figure 6.23 can be used. However, if the delivery clock is locked to the STC, the receiver design can be simplified. Due to the fixed and exact relationship between the frequencies of the STC and of the delivery clock, the STC can be derived straight from the delivery clock without the use of a PLL.

MPEG receivers and players are usually required to interface with TV sets. For this purpose, they should be capable of providing a video output signal that the TV set can handle. At the time MPEG-1 and MPEG-2 were developed, it was typically required to provide a composite video signal, often referred to as the CVBS (Composite Video,[13] Blanking and Sync) signal. This CVBS signal should comply with the regionally applicable analogue colour television standard: 525-line/60 Hz or 625-line/50 Hz NTSC, PAL or SECAM [1–3].

The amplitude of the CVBS signal carries the luminance information during the active video period of a video line, while the chrominance information is modulated on a so-called colour sub-carrier. NTSC and PAL use amplitude modulation (AM) on a single sub-carrier, while SECAM applies frequency modulation (FM) on two sub-carriers. In the analogue TV standards, the frequency of the colour sub-carrier is locked to the horizontal line frequency. The frequency of the colour sub-carriers in NTSC, PAL and SECAM is around 4 MHz. More detailed information on the applied frequencies and the locking to the horizontal line frequencies is provided in Note 6.3. The applied amplitude modulation of the chrominance information in NTSC and PAL requires a very accurate frequency of the colour sub-carrier; the tolerance is only ±10 Hz, less than 3 ppm. In NTSC and PAL, also the slew rate is constrained stringently; often the recommendation is a maximum short term jitter of 1 ns per horizontal line time, with a maximum long-term drift of 0.1 Hz/s. As also known from radio broadcast, FM is less susceptible to interference and other disturbances than AM, and therefore SECAM tolerates a less accurate frequency of the colour sub-carrier than NTSC and PAL: in SECAM, the tolerance is ±2 kHz. Due to required locking between the frequency of the colour sub-carrier and the horizontal line frequency, the accuracy constraints in ppm for the colour subcarrier(s) also apply to the line frequency and the frame period.

It should be noted though that television receivers are usually more tolerant than the NTSC, PAL and SECAM standards. For colour reproduction it is important that the frequency of the sub-carrier is stable, but the actual value of this frequency and its locking to the horizontal line frequency are less critical. This is usually exploited in VHS and other analogue consumer video tape recorders, where the process of reading video from the magnetic tape typically produces a

[13] In this context composite video means a single analogue video signal carrying both luminance and chrominance information.

Note 6.3 The Colour Sub-carriers in NTSC, PAL and SECAM

In NTSC and PAL, for both chrominance signals, a single sub-carrier is used with a frequency of about 3.6 and 4.4 MHz, respectively. SECAM applies vertical sub-sampling of chrominance; two colour sub-carriers are used, one for each chrominance signal, transmitted on alternate lines; the frequencies of the colour sub-carries are (about) 4.25 and 4.41 MHz.

In either case the frequency of the colour sub-carrier is locked to the horizontal line frequency. The following applies:

- In 525-line/60 Hz NTSC, the frequency of the colour sub-carrier is 455/2 times the line frequency = $(455/2) \times 525 \times (30/1.001) \approx 3.579\ 545$ MHz.
- In 625-line/50 Hz PAL, the frequency of the colour sub-carrier is equal to 283.75 times the line frequency + 25 Hz = $283.75 \times 625 \times 25 + 25$ Hz = 4.433 618 75 MHz.
- In 625-line/50 Hz SECAM, the frequency of the two colour sub-carriers is equal to 272 and 282 times the horizontal line frequency:
  - $272 \times 625 \times 25 = 4.25$ MHz.
  - $282 \times 625 \times 25 = 4.406\ 25$ MHz.

rather unstable video output signal. When such video is played back and converted into an NTSC or PAL CVBS signal, the frequency of the colour sub-carrier is not locked to the horizontal line frequency, but instead made free-running, to avoid an unstable colour sub-carrier. The resulting CVBS signal is obviously not NTSC or PAL compliant[14] and may cause problems in professional environments; however, it is well handled by virtually any television receiver.

6.6.3 Frequency and Tolerance of the STC in MPEG-2 Systems

When MPEG-1 systems was designed, requirements for professional receivers were not addressed, and the 3 ppm timing accuracy required by NTSC and PAL was considered a minor issue. When generating a CVBS output, the MPEG-1 player can use a free-running colour sub-carrier, the same approach as commonly used in VHS recorders, and therefore the tolerance and slew rate defined for MPEG-1 were considered appropriate. However, MPEG-2 systems also finds application in more professional products, such as professional broadcast devices that receive digital TV services at a cable head-end for distribution as analogue TV broadcast services across a cable network to consumers (see Figure 6.25). Such devices decode incoming digital TV services and produce an analogue NTSC, PAL or SECAM 525-line/60 Hz or 625-line/50 Hz TV signal that may be required to meet the accuracy constraint on the colour sub-carrier in accordance with the applicable standard for the analogue TV output signal.

The generation of NTSC, PAL or SECAM compliant video in an MPEG-2 receiver requires specific attention. In particular issues associated with the NTSC and PAL requirement for a 3 ppm accurate colour sub-carrier should be addressed; see below.

[14] Also other adjustments may have been made, for example to the number of lines per video picture in certain playback modes.

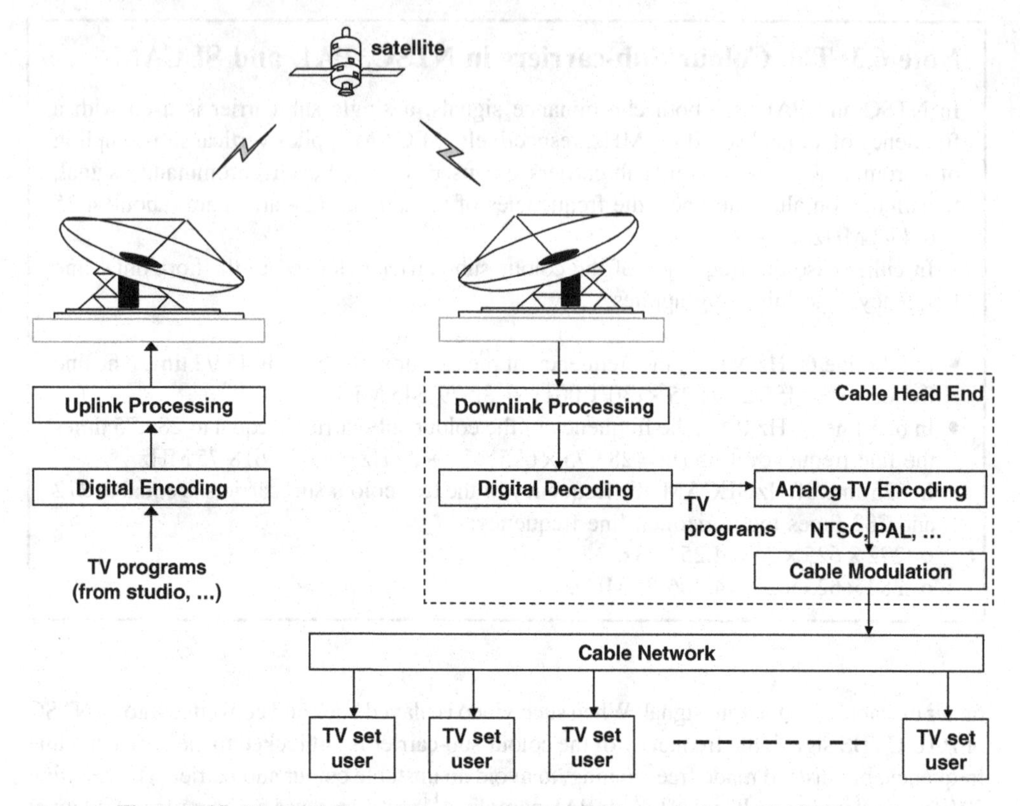

Figure 6.25 Digital delivery TV programs via satellite to analogue Cable Network

- In case of 'pull' delivery, the receiving device is in control of stream delivery as well as clock generation, and so the device could generate a colour sub-carrier within 3 ppm accuracy; however, most likely the costs of generating a clock with the required accuracy and stability will only be spent in professional applications and not in consumer devices.
- In case of 'push' delivery, the clocks in the receiver are generated from STC samples in the received MPEG system stream. A fully NTSC or PAL compliant CVBS signal can only be generated if the applied STC frequency at the encoder has a tolerance less than 3 ppm. Thereby the decoder capability to construct a fully NTSC or PAL compliant CVBS signal would depend on the accuracy of the applied encoder STC. Not a favourable characteristic.
- If the receiver uses a VCO with some tens of ppm accuracy, then the receiver would be able to lock to a less than 3 ppm accurate encoder STC. However, at startup, the initial VCO frequency may be beyond the target 3 ppm from the encoder STC; in such case the start of decoding may be postponed until the locked state is reached. In practice, also jitter in stream delivery may occur, which may cause the VCO to produce an STC frequency beyond the 3 ppm accuracy requirement, unless a very slow VCO is used (see also next bullet).
- If a 3 ppm accurate clock is used, the VCO will have a very slow behaviour, resulting in a significant period during which the MPEG system stream data arrives at the input of the receiver at a different rate than it is extracted and decoded. In other words, during some period, the access units are decoded at different times than presumed in the STD model. As a

result, the input buffers become more or less full than in the STD, depending on the trajectory of the recovered STC frequency with respect to the encoder STC frequency. If the recovered STC runs slower than the encoder STC, buffer overflow may occur. To prevent this, additional input buffering is needed. If the recovered STC runs faster than the encoder STC, buffer underflow may occur. This can be prevented by delaying all decoding operations, which also causes a need for additional input buffering. The receiver needs to prevent both buffer underflow and overflow and therefore both measures described above are to be implemented. While these measures to compensate for VCO behaviour are always needed, they are usually relatively marginal, but become more significant with a slow VCO.

The original intention of the MPEG system experts was to define a single accuracy constraint for the encoder STC frequency, also for MPEG-2 systems. However, given the considerations listed above, this did not prove feasible. For MPEG-2 systems the MPEG-1 accuracy of 50 ppm was considered too loose, and a compromise was found in a minimum STC frequency accuracy of 30 ppm. To address the problem that this 30 ppm tolerance is still too large to achieve the accuracy of the colour sub-carrier required in NTSC and PAL, the MPEG-2 systems specification enables signalling of the applied STC accuracy. For example, broadcast encoders in professional applications may follow a tight tolerance, the value of which can be signalled in the MPEG-2 system stream, so as to facilitate a fully NTSC and PAL compliant operation of receivers, recorders and play back equipment in such applications.

When an NTSC and PAL compliant colour sub-carrier must be derived from a receiver STC locked to the encoder STC, the use of an STC frequency as low as 90 kHz is undesirable. When the VCO compares the value of the local STC to the value of the received STC reference, the minimum phase difference that can be determined with a 90 kHz clock is equal to 11.1 μs. By using a (much) higher STC frequency, this value can be decreased significantly, thereby avoiding VCO design complications associated to a 90 kHz STC. For this reason, it was decided for MPEG-2 systems to use an STC frequency of 27 MHz, equal to 300×90 kHz, so that phase differences of multiples of 0.037 μs can be determined. Note that with a 27 MHz STC, the 30 ppm accuracy corresponds to a tolerance of ± 810 Hz. The time stamp resolution however was left at 90 kHz, the same as in MPEG-1 systems. Also the slew rate (in ppm/s) was left as in MPEG-1, which results in a slew rate of the 27 MHz STC of 75×10^{-3} Hz/s, equal to 300 times 250×10^{-6} Hz/s, the slew rate of the 90 kHz STC in MPEG-1 systems.

6.7 Quality of Service Issues

Enabling a high Quality of Service, QoS, was an important design criterion for MPEG-2 systems. Quality of Service is a general term and in many respects the QoS is an application issue beyond the scope of MPEG. However, MPEG is in a position to enforce some minimum conditions to achieve the high QoS level deemed necessary for MPEG-1 and MPEG-2 applications; most of them are related to the use of audio and video coding in MPEG-2 systems.

For the Quality of Service of MPEG-2 system based audiovisual systems obviously high quality audio and video is important. To produce audio and video output signals of high quality at a given bitrate, a player needs to apply MPEG audio and video decoders that perform well, without introducing additional artefacts. For this purpose, requirements for audio and video

Note 6.4 MPEG Audio and Video Measures to Ensure High Quality Output

To make sure that an MPEG audio decoder produces high quality audio, MPEG defined accuracy constraints for reconstructed audio samples. For this purpose, MPEG defined a number of coded audio streams with associated decoded audio samples. These audio streams and decoded audio samples serve as a reference for the design of audio decoding devices. When decoding the reference audio stream, each compliant MPEG audio decoder must produce output audio samples that approximate the provided reference samples within a certain accuracy range.

For MPEG video decoders a slightly different approach has been defined. The video decoding process is specified unambiguously, except the inverse transform, in the case of MPEG-2 video the IDCT. As only the IDCT may yield different results amongst different implementations, only the accuracy of the IDCT is specified. In this context it should be noted that verification of the accuracy of the Y, U and V samples at the output of an MPEG video decoder is usually less practical, due to up-sampling in actual decoders to produce a suitable video output signal. Such sub-sampling is needed if the display format differs from the format of the coded picture; for example the chrominance signals U and V are usually sub-sampled in vertical direction, while the decoded video may also have a different temporal resolution than the displayed video. Therefore the values of the reconstructed Y, U and V samples of the decoded pictures are usually lost by processing not defined by MPEG.

In case of insufficient IDCT accuracy, decoded pictures will contain errors. When such errors occur in reference pictures used for predictive coding, most likely serious error propagation is caused to a series of pictures. MPEG defined IDCT accuracy constraints so that no such errors will occur in compliant video decoders. In addition to specifying IDCT constraints, MPEG defined a number of coded video streams for testing whether an actual video decoder decodes the video stream 'properly', specifically in some worst case situations.

decoders have been defined (see Note 6.4). Meeting these requirements is of course essential, but when used in an MPEG-2 system environment, a number of additional requirements apply:

- Good random access; it should be possible to start the playback of an MPEG-2 system stream at random positions in the system stream.
- Accurate synchronization between audio and video.
- Reasonable delay caused by the MPEG-2 systems decoder.
- No buffer overflow and underflow.
- Good error resiliency.

Good random access: Many MPEG-2 applications require a good random access to an MPEG-2 system stream. An example is a broadcast application with multiple MPEG-2 system streams carrying various broadcast programs; the user may switch from one program to another in a more or less instantaneous manner. As a rule of thumb, upon switching to another program, audio and video decoding should resume within 1 s. Another example is a player of an MPEG-2 system stream from an optical medium, where the user has the feature to play the stream fast forward or fast reverse. Usually fast forward and fast reverse is implemented by jumping from one random access point to another, either in forward or in backward direction. Therefore in the stream random access points should be defined in a sufficiently frequent manner. Finally, if the decoding is interrupted due to an error or loss of data in the coded system stream, then it should not take a long time to find an access point for the decoding to resume, once the system stream is correctly received again.

The desire to achieve a good random access to audio and video bitstreams has usually little impact on audio encoding. However, for MPEG video encoders applying predictive coding, this desire imposes a number of requirements, such as regularly coding an I-picture or an equivalent measure. To access a video stream randomly, requires a player to parse the stream. Of course this is possible, even in reverse direction, but in many cases the required processing effort can be simplified significantly by signalling random access points at system level. Therefore MPEG-2 systems supports signal information to aid random access at certain points in the system stream. However, the use of this signalling is not mandated, as during the development of the MPEG-2 system standard it was claimed that some applications do not need any other access point beyond the one at the beginning of the stream. Therefore it was concluded that mandating the use of random access signalling was better left to applications and their standardization bodies.

Accurate synchronization: In MPEG-2 systems, presentation time stamps signal the instances in time to present video pictures and audio frames in the STD. An actual decoder needs to compensate for ways its design differs from that of the STD. In the STD video pictures and audio frames are presented immediately at 'PTS time', but in practice presentation inaccuracies occur, for example due to mapping of the video to the output video frame rate. Moreover there may be an additional delay between the 'PTS time' and the actual presentation time, for example due to video post-processing. As the additional delay usually differs for audio and video, the delay difference needs compensation, but if not applied correctly, the user may perceive audio and video not synchronized. To ensure a sufficient QoS level in this respect each MPEG-2 system decoder is required to synchronize audio frames and video pictures within ±40 ms. If the audio and the video decoder present audio and video at the output within accuracy ranges $(-d_{early,a}, +d_{late,a})$ and $(-d_{early,v}, +d_{late,v})$ respectively, then the above requirement means in practice that the worst case synchronization errors caused by video late and audio early and by video early and audio late are both smaller than or equal to 40 ms, or in formula: $(d_{early,a} + d_{late,v}) \leq 40$ ms and $(d_{early,v} + d_{late,a}) \leq 40$ ms. See Figure 6.26. In any case, when the user perceives a synchronization error, it cannot be caused by a compliant MPEG-2 system decoder.

Reasonable delay: One of the conditions for a high QoS is that upon delivering an MPEG system stream to the decoder it does not take too long until the decoded video and audio are presented to the user. Therefore the MPEG-2 systems standard specifies that any byte in the audio and video bitstream that enters the STD must be decoded in the STD within 1 s, which is considered a 'reasonable' maximum delay value. Two exceptions are defined. One is for still picture streams and another for MPEG-4 streams. For those streams other constraints apply; for still picture data the maximum STD delay is 60 s; applications using still picture data are foreseen to use a low bitrate to broadcast still pictures for 'slide-show' presentations, for which a decoder delay of at most 60 s was considered sufficient. For MPEG-4 streams, the maximum decoder delay was increased to 10 s, because a maximum delay of 1 s was considered too low to achieve the maximum coding efficiency under specific circumstances. If needed, application standardization bodies can specify for their domain a maximum delay smaller than 10 s.

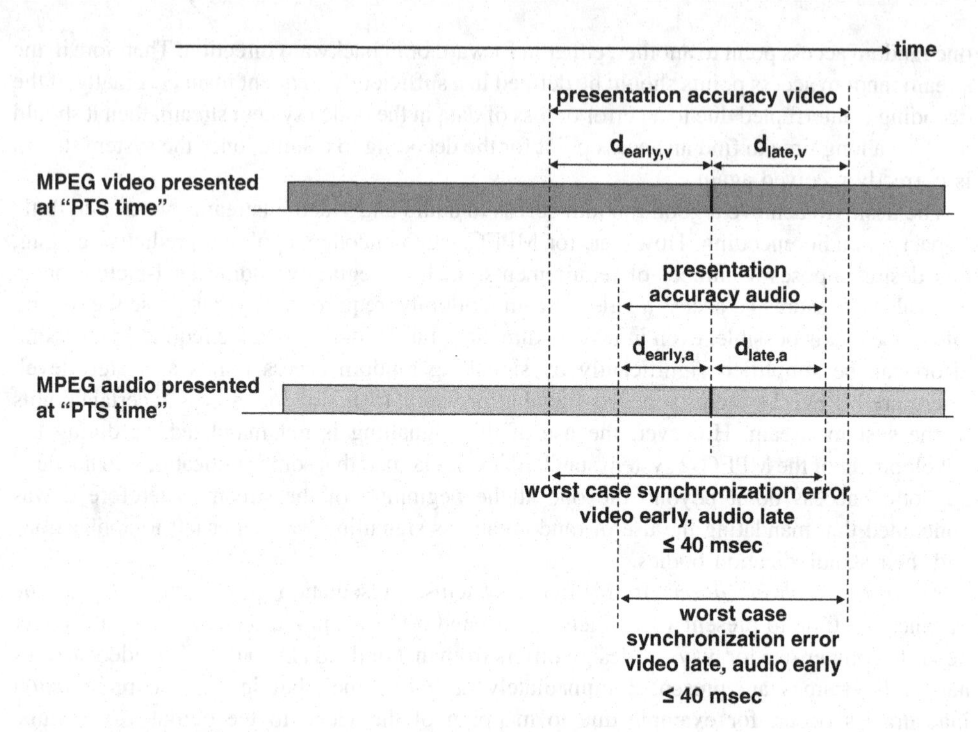

Figure 6.26 Tolerances in synchronization of audio and video

Buffer overflow/underflow: An important condition for a high QoS is that overflow of the input buffers does not occur at any time. Buffer overflow causes the loss of coded data in the STD, due to which the decoding process in the STD will be corrupted. As a result of such corruption, for some period no correctly decoded audio or video can be presented.[15] In particular for video the duration of the interruption may be considerable, due to the use of predictive coding. Also underflow of the input buffers must not occur, unless explicitly allowed. Underflow occurs for example when all coded data of a picture or audio frame is not present in the input buffer at decoding time. In such case the decoding must be postponed, which is likely to interrupt the presentation process, and possibly also the decoding process. However, in cases where underflow is explicitly allowed, its processing is taken into account in the decoding process.

Good error resiliency: During transport of MPEG system streams, data loss and errors may occur. For a high QoS, the decoding impact of lost or erroneous system stream data should be minimized. The MPEG audio and video standards already contain means to minimize the impact of data loss and error. Moreover, manufacturers often use privately defined error concealment tools[16] to their audio and video decoders, as a competitive means to ensure the best possible audio and video output in case of data loss or data error. However, also at system level error resiliency should be taken into account.

[15] Of course a physical decoder may tolerate some buffer overflow by designing buffer margins.

[16] In case of audio, the concealment may successfully hide small errors conceptually, but for video the errors and their propagation will often be so serious that erroneous behaviour will be perceived.

When system stream data loss occurs, the delivery of the stream is interrupted for some time, for example due to poor transport conditions. In case of data errors, the amount of received data is correct but the value of the data is changed during transport. In both cases in receivers the identification of the system packets may get corrupted. If this occurs, it is important upon receiving correct system data again, to rapidly resume identification of the system packets. A system packet is identified by recognizing synchronization words in the packet header; the speed of identification can therefore be increased by using small system packets. However, when smaller system packets are used, the packet frequency increases and consequently also the system overhead due to packet headers.

If a system packet contains only a few errors, the use of error correction technology on a per system packet basis is possible; the use of smaller system packets improves the feasibility of such error correction techniques, as will be discussed in subsequent chapters of this book. The use of error correction technology is transport medium specific, but the MPEG system is designed so that a convenient mapping between the system packets and error correction technology is possible.

6.8 Transport Layer Independence

The objective of MPEG is to provide audiovisual standards that are interoperable amongst applications, so as to create the best possible conditions for the success of the MPEG standards on the market. To achieve such interoperability, the MPEG Systems group designed system formats that can be used in a generic manner for all target transport media, as far as practically possible. One of the major features of MPEG-2 systems in this respect is that each MPEG-2 system stream is self-contained, that is that the stream contains all information needed for its decoding. If some information essential for the decoding of the stream would not be present in the stream, then it would be needed to convey the missing information out of band or by some other means, possibly media-specific or non-standardized.

To transport the systems format on a specific medium, typically a set of constraints specific for that medium need to be applied, as well as the mapping of the stream to that medium. These constraints are applied by the systems encoder, so as to meet requirements on the bitrate of the medium and on other medium specific parameters, such as package sizes for convenient mapping to the medium, Also constraints on audio and video encoders may be needed, such as the used audio sampling rate and a list of permitted picture sizes.

Which constraints apply and how to map the systems format to a transport medium is not specified by MPEG, but by medium specific application specifications, such as the Video CD, DVD, Blu-ray Disc, ARIB, ATSC and DVB specifications. Often the mapping to the transport medium is very straightforward. For example, for usage on Compact Disc, the system constraints are chosen so that the system packages fit easily with the sector format of CDROM. For broadcast applications, MPEG-2 systems defined a single format with a packet size that can be broadcast conveniently over terrestrial, satellite or cable networks using error protection on a per packet basis, but in addition the same format can also be transported over ATM networks by splitting each such packet into four packet fragments of equal size, each of which can be carried in an ATM cell, preceded by a one byte header signalling amongst others which fragment is carried in the ATM cell; see Section 9.4.3.

Figure 6.27 shows the transport medium independence of MPEG-2 systems. The medium specific constraining of the MPEG audio, video and systems encoding is shown, as needed to

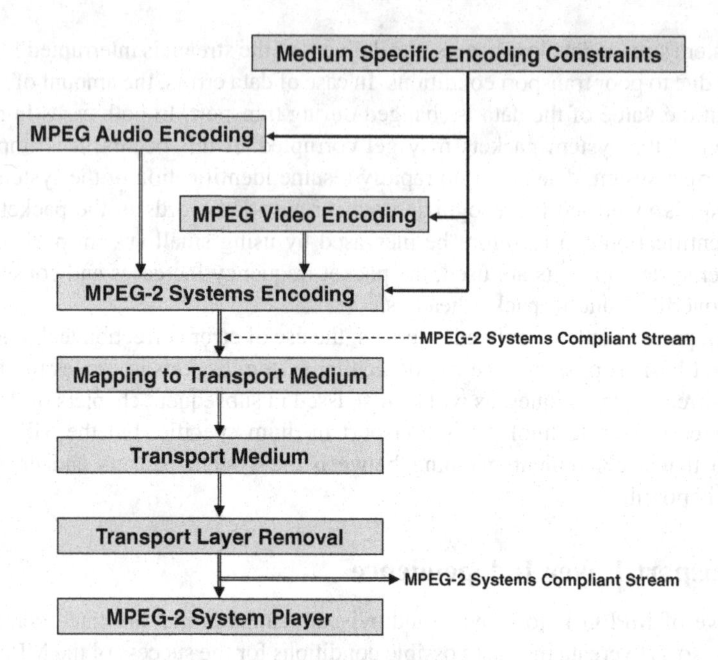

Figure 6.27 Transport medium independency in MPEG-2 systems

conveniently map the system packages to the transport medium and as required by the application specification for that medium. At playback, after removal of the medium specific transport layer, an MPEG-2 system stream is produced. Though several encoding constraints may have been applied, this MPEG-2 system stream is fully compliant, so that it can be decoded not only by a decoder specifically designed for the specific transport medium, but by any compliant MPEG-2 player, presuming that the player has the decoding capabilities specified by the medium application specification.[17]

References

1. ITU (2005) The NTSC, PAL and SECAM standards are published by ITU-R as Rec. ITU-R BT.470-7, referencing ITU-R BT.1700 and Rec. and Rec. ITU-R BT.1701 http://www.itu.int/rec/R-REC-BT.470-7-200502-I/en; http://www.itu.int/rec/R-REC-BT.1700-0-200502-I/en; http://www.itu.int/rec/R-REC-BT.1701-1-200508-I/en.
2. Background on the NTSC, PAL and SECAM standards, as well as on the 4: 2: 0 and other YUV formats is found in: Keith, J. (2007) *Video Demystified, a Handbook for the Digital Engineer*, 5th edn, Newnes/, ISBN: 978-0-7506-8395-1.
3. Wikipedia (2013) Further background on the NTSC, PAL and SECAM standards and their usage is found on Wikipedia http://en.wikipedia.org/wiki/NTSC; http://en.wikipedia.org/wiki/PAL; http://en.wikipedia.org/wiki/SECAM.

[17] For example, one may obviously not expect an MPEG video decoder designed for at most standard resolution to be capable of decoding MPEG video coded at high definition resolution, as the latter will require more processing power and more memory, most likely beyond the decoder's capabilities.

7

MPEG-1 Systems: Laying the MPEG-2 Foundation

The important issues to solve when building an MPEG system that works fine. How to achieve a constant end-to-end delay, given the delays due to video coding, audio coding and packetization of video and audio. How to synchronize video and audio at players and how to model the decoding process, so as accurately specify required buffer resources for players. Use of the decoding model to accurately specify compliance of streams and players.

7.1 Driving Forces

At the time the MPEG-1 effort started, developments were ongoing in the consumer electronics industry to define a system for the playback of audiovisual content from Compact Disc; see also Chapter 2 of this book. The aim of MPEG was to offer an open standard suitable not only for Compact Disc, but also for other media, such as Hard Disc and telecommunication networks, so as to enable technology convergence between consumer, computer and telecom applications and to improve the conditions for a mass market on multimedia to evolve.

The timely availability of the MPEG-1 standard was an important condition for its success. Failure to get timely results would likely cause the establishment of multiple incompatible standards. Video compression is the most critical element in achieving the above, and therefore MPEG agreed a tight methodology to specify the MPEG video standard. For the work to be carried out, three phases were defined [1–3] (see also Note 7.1). The aim was to define the MPEG-1 video coding scheme as the best possible compromise between coding efficiency on the one hand and costs of decoder implementation on VLSI silicon on the other.

When a company develops a product containing new technology, there is often a preference for using own technology and, if appropriate, to establish a de facto standard based thereon. Establishing a de facto standard is risky, as it is expensive and there is no guarantee on success. On the other hand, using an open standard such as defined by MPEG, can also be an acceptable outcome, in particular when because of standardization, the application scope can be widened, with the promising prospect of enlarging the potential market and improving the economies of scale. Intellectual property issues play an important role in these choices (see Note 7.2).

Fundamentals and Evolution of MPEG-2 Systems: Paving the MPEG Road, First Edition. Jan van der Meer.
© 2014 John Wiley & Sons, Ltd. Published 2014 by John Wiley & Sons, Ltd.

Note 7.1 The Three Phases for Standardizing MPEG-1 Video

For MPEG-1 video standardization, MPEG used a methodology consisting of three phases, each with their own tight schedule:

1. Requirements phase: during this phase, MPEG-1 video requirements were specified and documented in a Call for Proposals for a video compression system; in addition, a number of test sequences were agreed. Interested parties were invited to respond to this CfP by submitting the specification of a video compression algorithm, video files containing test sequences compressed by this algorithm and an indication of associated implementation costs.
2. Competitive phase: during this phase, the 14 received CfP responses were analysed; subjective tests were performed on the coded test sequences, so as to identify promising techniques in video coding and to trade-off coded picture quality and implementation costs.
3. Convergence phase; during this phase, a collaborative effort was made to integrate promising ideas and techniques into a single solution. Techniques with considerable merit were abandoned in favour of slightly better or slightly simpler ones. Improvements were considered and if appropriate agreed. At the end of the convergence phase, upon addressing and resolving all inaccuracies, the final MPEG-1 video standard was produced.

It should be noted that the above phases are generally applied for MPEG compression standards.

Note 7.2 Intellectual Property Rights and Standards

Inventions can be protected by Intellectual Property Rights, IPR, based on one or more granted patents. The IPR specify the availability of patented technology to others. It is possible to keep the use of patented technology exclusive to the inventor and possibly some others, for example to protect a market position. On the other hand, it is also possible to allow usage of the covered technology to third parties, for which usually a license fee is due.

If a de facto standard is established by a private party, then that private party controls in principle the licensing conditions for that standard. However, also third party patents may be infringed, which may or may not be available for licensing. If available, an arrangement may be useful with the third party patent holders as to the due license fee and its sharing amongst patent holders. If a third party patent that is infringed is not made available for use in the de facto standard, it may be necessary to remove the related technology from the de facto standard.

For an MPEG standard, a very general and practical principle applies for holders of patents applicable to the MPEG standard. The MPEG standard is given the benefit of using the invention covered by the patent, and in exchange, the patent holders get the right to ask a 'reasonable' fee of the users of the standard. On MPEG standards usually several parties own IPR; therefore normally a patent pool is established, so that a one-stop license for the use of an MPEG standard can be obtained. Patent pools and other licensing arrangements are beyond the scope of MPEG and ISO/IEC, and are managed in separate negotiations.

Chad Fogg, e-mail message 2 December 1994, Top 10 romantic pick-up lines overheard at MPEG meeting:
Nr 9: 'Oh, I've heard of the patent pool, but have you heard of my patent Jacuzzi?'

For any MPEG standard, it is essential not to apply patented technology that is not available to all parties: any interested party should be permitted to use an MPEG standard. Therefore MPEG participants are required to submit a so-called IPR statement with a commitment that any patent applicable to the future standard will be licensed under 'Reasonable And Non-Discriminatory' (RAND) terms and conditions. MPEG leaves any judgement on what 'reasonable' is to the market. When a standard is used in various application areas, complications may occur if a license fee that is considered 'reasonable' in one application area, is considered unreasonable in another application area or market. Such complications may hinder market adoption of the MPEG standard, but are beyond the scope of MPEG and ISO/IEC, and need to be resolved in the market.

Source of text in italics: Chad Fogg. Reproduced with permission from Chad Fogg.

Already in the early days of MPEG-1, the need was recognized for a system standard that specifies how to carry compressed video and audio in a single stream. To specify such a standard, a System sub-group was established within MPEG. This was an important first step, but a far bigger initial challenge was to get the 'right' system experts to participate; for some background on this challenge see Note 7.3. For the systems sub-group to be successful, MPEG needed to prove its added value, in particular to the consumer electronics industry. The MPEG

Note 7.3 Conditions for MPEG Participation

Experts on video and audio coding usually have a history in research; their participation in activities that focus on 'state of the art' technology is usually part of their task in research. Therefore video and audio coding experts and their research organizations are typically eager to participate in activities such as undertaken by MPEG.

System experts need in depth knowledge of the issues relevant for transport of coded video and audio. Though such knowledge is essential to successfully apply video and audio compression in practice, it is not an enabling key technology by itself. Transport knowledge is more application orientated and within a company usually not sufficiently available in research. Instead, transport of coded audio and video is often addressed during architectural studies in product pre-development groups. For such groups participation in MPEG is far less obvious than for research; in MPEG systems they will only participate if the company is convinced that an MPEG system standard is – or may become – essential for the product to be successful in the market.

Note 7.4 The Need for a Wide Scope in Standardization

When MPEG-1 Systems was designed, most input was provided by system experts from the consumer electronics industry, the telecommunication industry and the computer industry. The most active experts were involved with projects running in the company that employed them, and when producing a proposal, they obviously tried to solve the problems as identified and understood within the scope of their own project. However, the problems needed to be resolved within a wider scope, and therefore, to agree on a more general system solution, the participating experts needed not only to understand their own problems, but also the problems of the other participants.

For example, when one of the first proposals for the MPEG-1 systems standard was specifically produced with the Compact Disc application in mind, the proposer was given a 'painful brainwash' by the other participants, so as to make fully clear which problems were not adequately addressed by this too narrow minded proposal. But the good news was that once a common understanding of all problems was achieved, the basics for what became the final MPEG-1 Systems standard could be agreed rapidly.

view that the use of common audiovisual coding technology across applications is important for market take-off was rapidly acknowledged, but for the participation in the MPEG-1 systems group it was essential that the collaborative research activity in MPEG on video compression yielded superior results in a timely manner. Once this became obvious, the industry provided the MPEG-1 system activity with real focus and momentum.

The challenge for the MPEG-1 systems sub-group was to design a generic format for the system stream so that the stream can be transported efficiently across all target transport media. It would be short-sighted to optimize a transport format specifically for a certain transport medium, without considering its use on other target media. As long as the requirements for the various transport media do not differ too much, an efficient generic system format can be specified, but it requires the involved experts to understand the specific issues associated with all target media. Initially, participants are usually focussed on their own application, with a too narrow scope for MPEG systems, in which case sometimes an interesting learning process was undergone (see Note 7.4).

The main application driving MPEG-1 system was playback of audiovisual content from Compact Disc, with important feedback from the computer industry that playback from Hard Disc should also be supported. The telecommunication industry defined a transport protocol, Rec. H.221, for audiovisual services across ISDN networks (see also Note 2.1), but when it became clear that this protocol was not suitable for use by MPEG, the telecommunication industry participated in the system discussions as follower, verifying that use of the evolving MPEG-1 system transport format across telecommunication networks was potentially feasible.

7.2 Objectives and Requirements

The MPEG-1 target media, Compact Disc and Hard Disc, are digital storage media with a low probability of data errors; typically neither storing data nor reading data introduces errors. Nevertheless measures are taken to ensure a good error performance. For example, on any Compact Disc, random errors and burst errors can be corrected up to some level. For remaining errors on an audio CD, a player can apply error concealment by replacing erroneous audio samples by samples interpolated from error-free ones. To address errors remaining after the first error correction layers, for a CD-ROM two modes can be used, one with additional error correction codes, and one without such codes. The first is useful for very error sensitive (computer) data, but comes at the expense of about 12% of the available data capacity. Also Hard Disc drives apply error correction and may apply stronger levels of error correction in future, when deemed necessary, for example when higher data densities are used.

Because of the error robustness of the MPEG-1 target media, MPEG decided to design the MPEG-1 system stream format without paying special attention to error protection and correction. Thereby the objective of the design of the MPEG-1 system standard became to specify a generic and flexible transport format, operating in a relatively error-free environment.

At the start of the MPEG-1 system discussions not all transport format requirements may have been fully understood, but during the design and the ongoing discussions, the following requirements did evolve:

- The MPEG-1 system stream format must define packetization and signalling of one or more MPEG-1 video streams, MPEG-1 audio streams, private streams, and a padding stream,[1] while leaving headroom for adding support for other elementary streams in future.
- It must be possible for a decoder to conveniently retrieve and decode an elementary stream from an MPEG-1 system stream.
- It must be possible to carry System Time Clock samples in an MPEG-1 system stream, so as to enable convenient reconstruction of the STC at the decoder.
- An MPEG-1 system stream must allow sufficiently accurate synchronization of video and audio to achieve lip synchronization; for tolerances see Section 6.7 and Figure 6.26.
- The MPEG-1 system format must allow random access to the MPEG-1 audio and video streams in a convenient manner.
- MPEG-1 system streams must be robust against errors, even if the likelihood of an error is low.
- The MPEG-1 system format must allow for a high quality of service; in particular, buffer overflow and buffer underflow shall not occur.[2]
- It must be conveniently possible for MPEG-1 system encoders to construct a compliant MPEG-1 system stream.
- The MPEG-1 system overhead must be low.

The MPEG-1 systems layer is constructed in two layers: the outermost layer is the system layer, and the innermost is the compression layer. The system layer provides the functions necessary for the decoding of one or more compressed data streams carried in an MPEG-1 system stream. The interface between the system layer and the compression layer is defined in a very transparent manner. Though the system layer carries information on the contained compressed data, the compressed data itself is carried unaltered by the system layer.

[1] A brief discussion on the need for padding and the purpose of a padding stream is found in Section 7.3.

[2] As opposed to MPEG-2, there is in MPEG-1 no support for low-delay video, and therefore the buffer underflow discussed in Figure 3.14 in Section 3.2.2 is not permitted in MPEG-1 systems.

7.3 Structure of MPEG-1 System Streams

An MPEG-1 system stream [2] carries one or more elementary streams, multiplexed together. Elementary stream data is stored in so-called packets. A packet consists of a packet header followed by packet data consisting of a variable number of contiguous bytes from one elementary stream. The packet header identifies by means of a stream-ID the elementary stream of which data is carried in the packet and may also carry time stamp information for the decoding and presentation of that elementary stream.

The packets are organized in packs; a pack commences with a pack header and carries zero or more packets. The pack header contains timing and bitrate information of the MPEG-1 system stream. The pack may also carry a system header containing a summary of the system parameters defined in the system stream. A decoder may use the system header to determine whether it is capable of decoding the entire stream. The system header must be included in the first pack of an MPEG-1 system stream and may be repeated unaltered in other packs. If a system header is contained in a pack, then it must precede any packet in the pack.

An example of an MPEG-1 system stream and its construction is given in Figure 7.1, where contiguous video and audio data chunks are defined in an MPEG-1 video and audio stream for transport in an MPEG-1 system stream. Each audio and video packet starts with a packet header, followed by the contained audio and video data chunk. In Figure 7.1 the bitrate of the MPEG-1 system stream is assumed constant and the duration of the audio and video data chunks in the MPEG-1 audio and video streams is about the same. Because the audio bitrate is much lower than the video bitrate, an audio packet is significantly smaller than a video packet.

In Figure 7.1, the MPEG-1 system stream commences with a pack header, followed by the system header, the first video packet, the first audio packet, the second video packet, the second

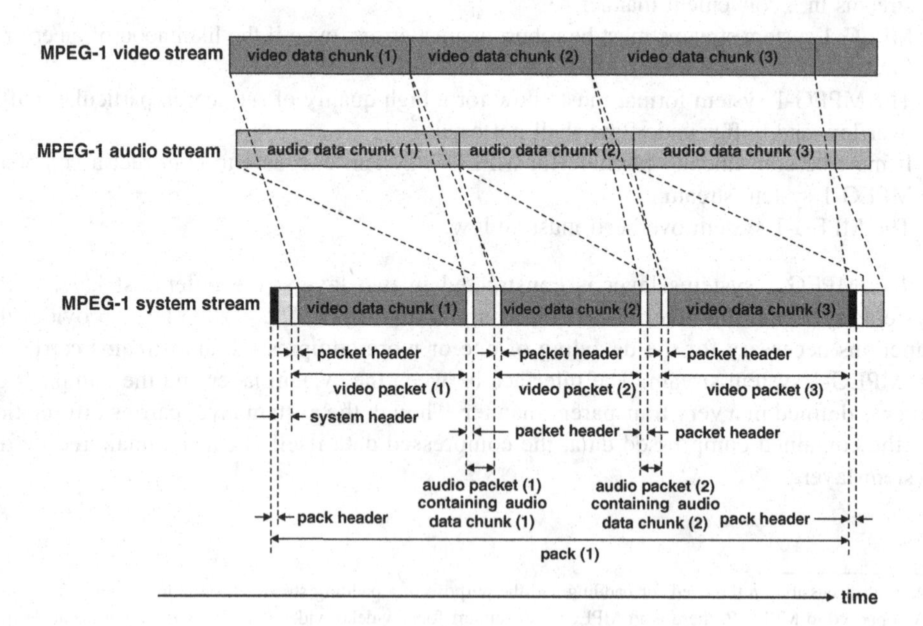

Figure 7.1 Example of constructing an MPEG-1 system stream

audio packet and the third video packet. This completes the first pack; after the third video packet a pack header follows, indicating the start of the next pack. After this pack header the third audio packet follows; the option to repeat the system header immediately after the pack header is not used.

Applications are free to organize packets in packs, but there is a constraint on the maximum pack duration. The pack header carries a sample of the System Time Clock and at least every 0.7 s an STC sample is required to be provided. Therefore the duration of a pack cannot exceed 0.7 s. Obviously, applications may specify additional constraints, such as including a pack header prior to each audio or video packet.

Upon receiving an MPEG-1 system stream, a system decoder can parse and decode the system stream and the contained audio and video streams as soon as the structure of the system stream is identified. When starting to play an MPEG-1 system stream from the beginning, the decoder knows that the stream starts with a pack header followed by a system header and zero or more packets. However, when accessing the stream at a random position, it is unknown whether the first byte of the accessed system stream is a video, audio or system byte. In such case obviously the decoding of the system stream has to wait until its structure is recognized by the system decoder.

The system stream structure is known as soon as pack, packet and system headers are identified. To recognize those headers, MPEG considered the use of synchronization words, preferably unique bit patterns that are not found in the system stream except for synchronization words. The use of unique synchronization words prevents the need to verify whether or not a found bit pattern represents a synchronization word. Such verification requires additional processing, will delay access to the stream and is therefore preferably prevented.

A system stream carries multiplexed video and audio streams. Therefore a bit pattern can only be unique in a system stream, if it is also unique in each contained audio and video stream. Selecting system synchronization words has been subject to debate between the MPEG system group on one hand and the MPEG video and the MPEG audio groups on the other, but no bit patterns were identified that were unique in both video and audio. Moreover, a system stream may also carry private data streams, and to expect such streams to prevent the use of certain synchronization words is not practical.

Because the use of unique synchronization words in MPEG-1 system streams is not feasible, a verification mechanism is needed to check whether a found bit pattern indeed represents a synchronization word. To rapidly identify the system stream structure, it is desirable that the probability of a successful verification is high. This is achieved by using an extended version of video synchronization words, based on start codes. These start codes are unique in video, and also the system layer is designed so that start codes are unique. In audio streams it is unlikely, but possible, to find start codes. Also in private streams start codes may be found. However, in many practical situations, when a bit pattern corresponding to a system synchronization word is found in an MPEG-1 system stream, then its verification will most likely be successful.

Each start code consists of a start code prefix, followed by a start code value. The start code prefix consists of three consecutive bytes with the hexadecimal values 00, 00 and 01, respectively, hence in terms of bits 23 times a '0' followed by a single '1'. The start code value is a byte that identifies the type of start code. For video, the start code values 0 up to 184 are used, as listed in Table 3.2 in Chapter 3 of this book. The MPEG video and MPEG system groups agreed to make the remaining values 185 to 255 available for use by MPEG systems.

Table 7.1 Usage of start code values in MPEG-1 systems

Start code type	Start code value (decimal)	Start code value (hexadecimal)
Video start codes (for details see Table 3.2)	0 through 184	00 through B8
End of MPEG-1 system stream	185	B9
Pack header	186	BA
System header	187	BB
Stream-ID values for packet header usage (see Table 7.2)	188 through 255	BC through FF

In MPEG-1 systems each pack header, system header and packet header begins with a start code, followed by the start code value signalling the type of header. One start code value indicates a pack header, another a system header, while the end of an MPEG-1 system stream is also indicated by a start code value (see Table 7.1). An end of MPEG-1 system stream indication did prove useful for timing purposes and when multiple MPEG-1 system streams are concatenated in one data stream. For example, each system stream may represent a video clip, each with its own timing. How to play such concatenated streams is an application issue; for example, the streams can be played back as a series of video clips or the next video clip can be played upon request of the user; however, all of this is beyond the scope of MPEG.

For a packet header multiple start code values are needed to uniquely identify each elementary stream in the multiplex. The start code value in the packet header is therefore called the stream-ID of the contained elementary stream. An MPEG-1 system stream may contain multiple video and audio streams; 16 start code values are assigned to signal an MPEG-1 video stream and 32 to signal an MPEG-1 audio stream. Furthermore, one start code value is assigned to signal a padding stream and two values to signal a private stream, one with and one without the option to encode time stamps in the packet header. The other values are reserved for future use. In total 68 start code values are assigned as stream-ID for packet header usage, so that an MPEG-1 system stream can carry at most 68 different elementary streams at the same time. See Table 7.2 for the actual assignments of stream-ID values to type of streams.

Padding may be useful to achieve a constant total rate of the system stream, when packs and/or packets need to be aligned with a physical transport mechanism, or as 'an emergency measure' to prevent buffer overflow in the STD. However, a padding packet cannot be used if the required number of stuffing bytes is smaller that the minimum length size of a padding

Table 7.2 Usage of stream-ID values for the type of streams in MPEG-1 systems

Type of stream	Stream-ID (decimal)	Stream-ID (hexadecimal)
Reserved for future use	188	BC
Private stream 1	189	BD
Padding stream	190	BE
Private stream 2	191	BF
MPEG-1 audio stream 0 through 31	192 through 223	C0 through DF
MPEG-1 video stream 0 through 15	224 through 239	E0 through EF
Reserved for future use	240 through 255	F0 through FF

Table 7.3 Usage of stream-ID values for type of streams in MPEG-1 systems

Type of stream	Stream-ID (binary)	Stream-ID (hexadecimal)
MPEG-1 audio stream number xxxxx (0–31)	110x xxxx	C0 through DF
MPEG-1 video stream number xxxx (0–15)	1111 xxxx	E0 through EF

The notation x means that the values 0 and 1 are both permitted; the stream number is given by the values taken by the xs.

packet. Therefore the packet header also provides a padding capability by means of optionally adding one or more stuffing bytes (see Figure 7.3). This can be used for purposes similar to that of the padding stream, and is well suited to providing word (16 bit) or long word (32 bit) alignment in applications where eight-bit alignment is not sufficient.

In the range of system start code values, the audio and video stream-ID values are chosen so that the least significant bits of the stream-ID field, an unsigned integer, indicate the number of the audio and video stream (see Table 7.3).

The design choice for carriage of at most 16 video and 32 audio streams was mainly driven by the usual desire in standardization to specify the standard so that the standard itself does not constrain potential applications. The capability to carry multiple video and audio streams in one MPEG-1 system stream is very useful indeed. Applications may wish to include audio in multiple languages with the same video or may wish to include along with the main video a small sized video with the director's commentary or with signing for the bad of hearing. Also in the context of broadcast applications, carriage of a large number of video and audio streams in one system stream may prove very relevant.[3]

Figure 7.2 presents an MPEG-1 system stream that is followed by another one; both streams could for example represent a video clip. Only the global structure is depicted. The first system stream starts with a pack header, immediately followed by the system header; the first pack further contains an interleave of three video packets and two audio packets. In subsequent packs, the same system header may be repeated. The first MPEG-1 system stream is concluded by an 'end of MPEG-1 system stream' start code. Figure 7.2 shows that a time gap exists between the conclusion of the first MPEG-1 system stream and the beginning of the next one. This time gap may vary from zero to any positive value. System decoders should be aware that after the conclusion of an MPEG-1 system stream another one may (or may not) commence.

To conveniently verify a system start code found in the system stream, a simple mechanism is made available by means of the number of bytes contained in pack headers, system headers and packets. Following each pack header, system header and packet, there is either another pack header, system header, packet or a start code indicating the end of the MPEG-1 system stream. Hence, upon finding a system start code, it should be verified whether after the number of bytes contained in the found pack header, system header or packet, another system start code is present in the system stream. If no system start code is present at that position, then there is either an error in the system stream or the found start code is an emulated version, for example in an audio stream or a private stream.

[3] However, at the time MPEG-1 systems was developed, the suitability of MPEG-1 for broadcast was not considered yet.

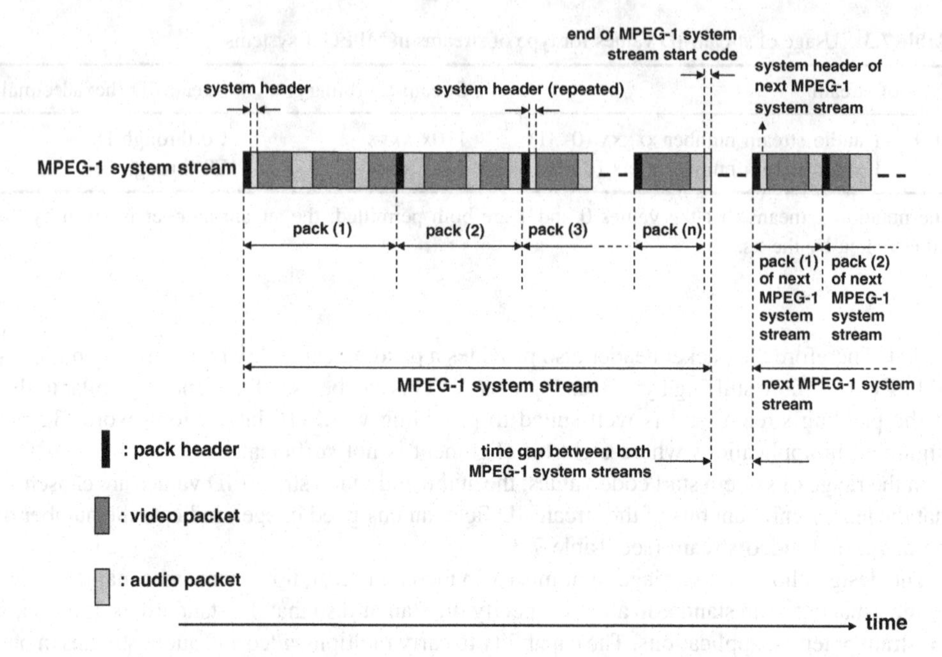

Figure 7.2 An MPEG-1 system stream followed by another one

Upon verifying one or more system start codes, the decoder may conclude that the system stream parser is 'in sync' with the system stream structure. The actual number of system start codes to be verified for achieving 'in sync' status is at the decoder's discretion. But as soon as this status is achieved, the system stream parser will pass the audio and video data contained in the packets to the appropriate elementary stream buffers and decoders.

The pack header has a fixed length in number of bytes and therefore the system parser knows how many bytes are contained in the pack header: after the pack start code always eight bytes follow. A system header and a packet contain a variable number of bytes and therefore after the system header start code and the packet start code a 16 bit field follows, specifying the number of immediately following bytes in the system header or packet (see also Figure 7.3). Typically, the system start code verification is a continuously running process, only requiring the reading and processing of start codes and length information. The system header length and packet length fields are 16 bit integers that can take a maximum value equal to $2^{16} - 1 = 65\,535$. As a consequence, the total number of bytes in each system header and in each packet, including the start code and length field, is at most 65 541 bytes.

The pack header contains the STC sample, in MPEG-1 called the System Clock Reference, SCR, and the value of the bitrate of the MPEG-1 system stream during the arrival of the pack, called the mux-rate. The packet header optionally carries stuffing bytes and fields specifying the size of the STD input buffer for the decoding of the contained elementary stream in the STD, and time stamps for the first coded video picture or audio frame that begins in the packet. In an audio packet header only a PTS may be present, while in a video packet header both a DTS and a PTS may be present, but only if the value of the DTS differs from the PTS value. Figure 7.3 presents an example of the MPEG-1 system stream data structure. Further details are discussed in the following Sections 7.4 and 7.5.

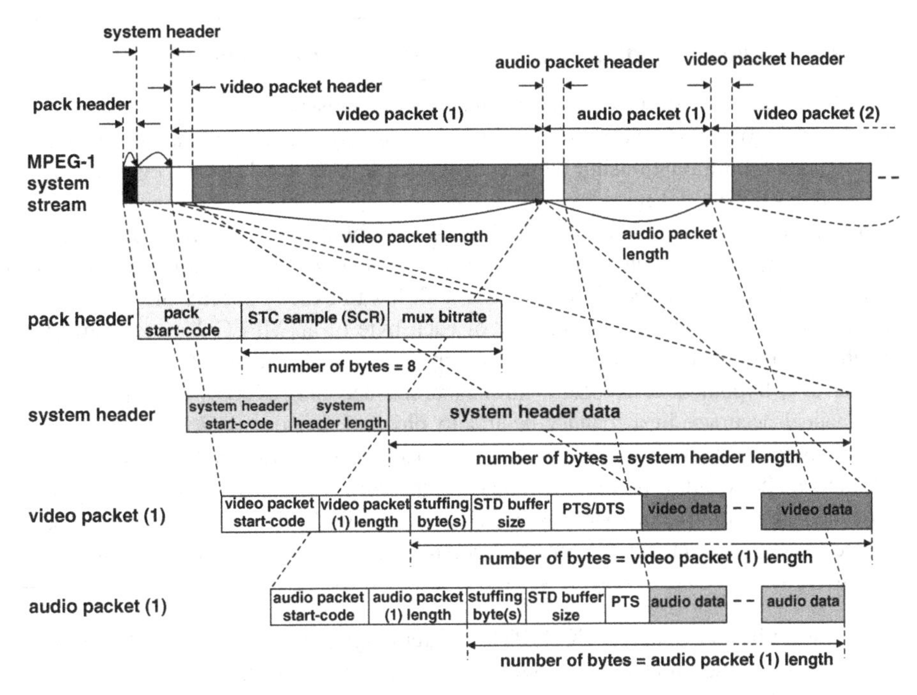

Figure 7.3 Structure of an MPEG-1 system stream in more detail

7.4 The MPEG-1 System Target Decoder

During MPEG-1 system standardization, the concept of a System Target Decoder was 'born' as a hypothetical model for the decoding of elementary streams transported in an MPEG system stream. The STD model operates with mathematical accuracy, and serves as a reference for compliancy of both system streams and system decoders, as discussed in Section 6.5 of this book. The STD does impose requirements on the construction and playback of MPEG-1 system streams, but not on the architecture and other implementation aspects of decoders and encoders. For some background see Note 7.5.

The MPEG-1 System Target Decoder, depicted in Figure 7.4, defines two decoding branches, one for MPEG-1 video and the other for MPEG-1 audio elementary streams. There may be multiple audio and video streams in a system stream and each elementary stream has its own decoding branch.

At its input, an MPEG-1 system stream enters the STD; byte i in the stream arrives at time $t(i)$. At the input of the STD the MPEG-1 system stream is demultiplexed; upon arrival, each packet data byte from elementary stream n is passed instantaneously to the input buffer B_n for stream n. For each elementary stream, only a single input buffer is defined in the MPEG-1 STD. The size BS_n of B_n is specified by the STD buffer size field in the packet header of stream n. Stream n is decoded by decoder D_n. In the case of a video stream utilizing B-pictures, decoded pictures are re-ordered back into natural order by means of re-order buffer O_n. For video streams without B-pictures, the re-ordering can be switched off. Bytes present in the pack,

Note 7.5 Development of the MPEG-1 System Target Decoder

The MPEG systems group discussed various system issues on packetization, buffer control, synchronization and timing, but the biggest challenge was to define an unambiguous timing relationship between a system stream and decoding events, comparable to the VBV model in video.

An important step in this approach has been an input document from Juan Pineda, working for Apple Computer Inc. at that time, proposing a mathematical formalism to specify the decoding process. This formalism allows to exactly specify the time of each decoding event and the time of delivery of each byte of an MPEG-1 system stream.

With this formalism, the MPEG system group could specify the MPEG-1 System Target Decoder as a hypothetical decoder with an internal decoding process that operates with mathematical accuracy in a manner neutral to physical decoder implementations. Upon adoption by the systems group, the video group decided to use the same formalism for the VBV model. The decoding process in the STD extents the VBV model of video and introduces instantaneous decoding of audio frames. The STD enabled the MPEG systems experts to define very precise requirements for the construction and decoding of MPEG-1 system streams.

system or packet headers, such as SCR, PTS, packet length and STD input buffer size are not delivered to any of the input buffers, but are made available to control the system.

For the processing of stream *n* in the STD, access units and presentation units are distinguished, whereby an access unit is the coded representation of a presentation unit. Hence, decoder D_n decodes successive access units $A_n(j)$ in stream *n* and produces presentation

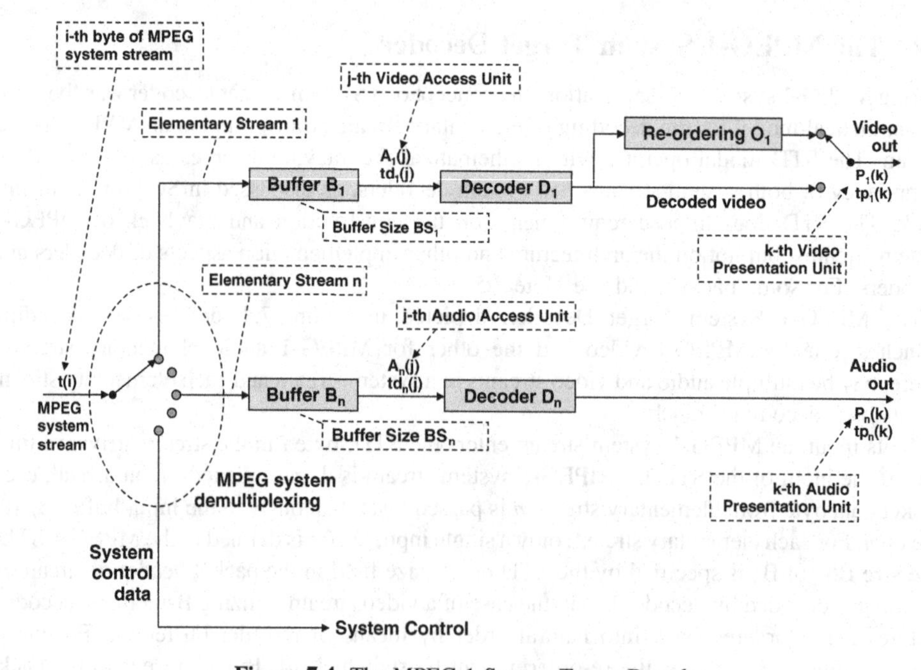

Figure 7.4 The MPEG-1 System Target Decoder

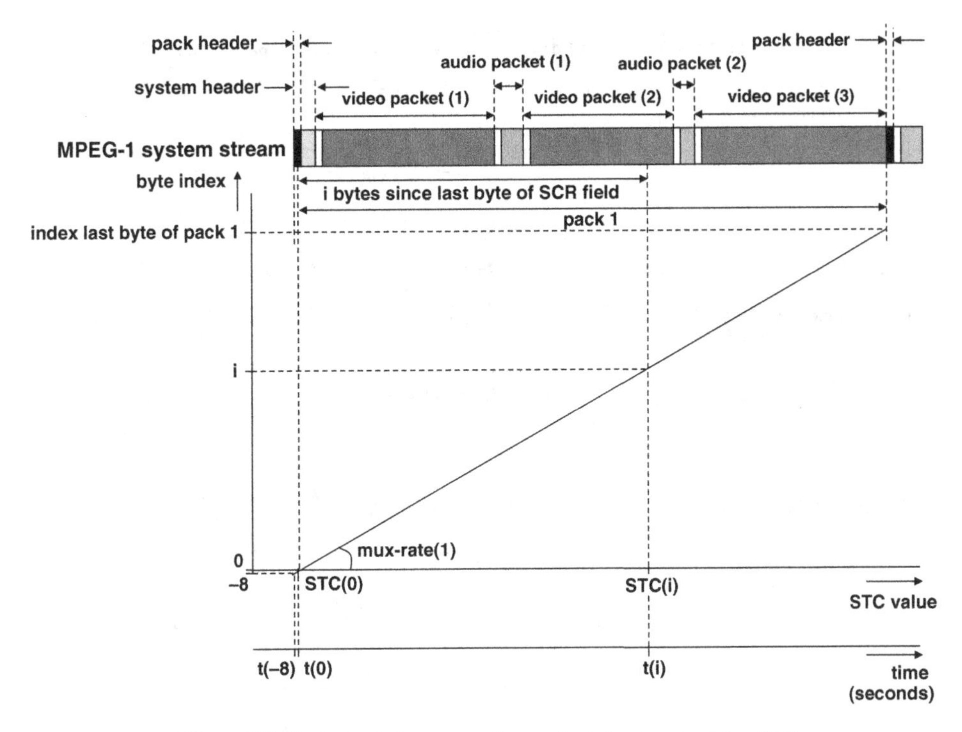

Figure 7.5 Arrival of the bytes of a pack at the input of the STD

units $P_n(k)$, that is decoded pictures or audio frames at its output. The parameters j and k are indices of A_n and P_n in coded order and presentation order, respectively. Access unit $A_n(j)$ is decoded instantaneously at its decoding time $td_n(j)$, while presentation unit $P_n(k)$ is presented at its presentation time $tp_n(k)$.

In each pack header the SCR field is encoded in units of the System Time Clock frequency, specifying the intended arrival time at the input of the STD of the final byte of the SCR field. All bytes in a pack arrive at the input of the STD at a bitrate equal to the mux-rate, as encoded in the pack header. The value of the mux-rate is measured in 50 bytes/s, rounded upwards, and is used to define the arrival time at the input of the STD of each byte contained in the pack (see Figure 7.5).

Figure 7.5 depicts the delivery of the bytes of a pack in an MPEG-1 system stream at the input of the STD. Horizontally, the time is depicted, both in values of the System Time Clock STC and in seconds; byte i in the stream arrives at time $STC(i)$ in values of the STC and $t(i)$ in seconds. Obviously, $STC(i) = t(i) \times (\text{STC frequency})$. Vertically, the index of bytes in the MPEG-1 system stream is presented. The index of the final byte of the SCR field in the header of pack 1 has index zero. In the pack header, eight bytes precede the final byte of the SCR field, hence the first byte of pack 1 has index minus eight.

In Figure 7.5, the final byte of the SCR field in the header of pack 1 enters the STD at the time encoded in that SCR field. Therefore:

$$STC(0) = SCR_{\text{pack1}}, \text{ while } t(0) = SCR_{\text{pack1}}/(\text{STC frequency}).$$

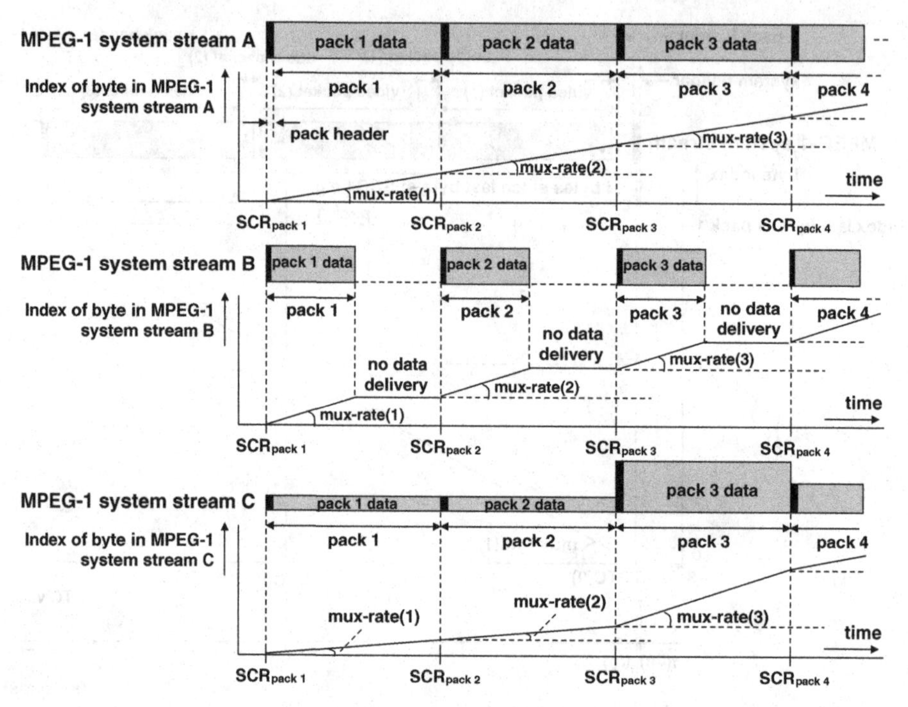

Figure 7.6 Constant, bursty and variable bitrate MPEG-1 system stream examples

The mux-rate value encoded in the header of pack 1 indicates that (50 × mux-rate) bytes enter the STD each second; hence i bytes enter in [i/(mux-rate × 50)] seconds. Therefore:

$$t(i) = [\mathrm{SCR_{pack1}}/(\mathrm{STC\ frequency})] + [i(\text{mux-rate} \times 50)], \text{ while}$$

$$\mathrm{STC}(i) = \mathrm{SCR_{pack1}} + [i/(\text{mux-rate} \times 50)] \times (\mathrm{STC\ frequency}).$$

The above formulae are also valid for the eight bytes in pack 1 preceding the final byte of the SCR field. For example, the values t(−8) and STC(−8) for the arrival time of the first byte of pack 1 are found by using $i = -8$ in above formulae.

The MPEG-1 system specification supports a constant bitrate, bursty delivery and a variable bitrate of the system stream; Figure 7.6 presents three examples thereof. These examples demonstrate the available flexibility for constructing MPEG-1 system streams. All MPEG-1 system streams have in common that the bitrate during the delivery of each pack is constant. Between the delivery of two packs, there may be some period during which no data is delivered at the input of the STD and at each next pack the delivery rate may change.

Stream A in Figure 7.6 has a constant bitrate. During constant bitrate operation, for each byte in the MPEG-1 system stream the following linear equation applies for its arrival time t(i) at the input of the STD:

$$t(i) = C_1 \times i + C_2,$$

where:

i is the index of a byte in the MPEG-1 system stream,

t(i) is the arrival time in seconds at the input of the STD of the ith byte in the MPEG-1 system stream,

C_1 and C_2 are real-valued constants valid for the entire stream.

The mux-rate parameter is an approximation of C_1 in units of 50 bytes/s, due to which inaccuracies occur when calculating byte arrival t(i) based on the encoded SCR and mux-rate values. However, in the case of a constant bitrate, each encoded SCR value must be derived from the equation:

$$SCR(i) = [C_1 \times i + C_2] \times [\text{STC frequency}],$$

where:

i is the index of the final byte of the encoded SCR field,
SCR(i) is the encoded arrival time at the input of the STD in units of the STC frequency of the final byte of the encoded SCR field,
C_1 and C_2 are real-valued constants valid for the entire stream.

Note however that the SCR field is encoded with a limited number of bits, due to which the associated modulus operation should be taken into account.

The value of the mux-rate measures C_1 in 50 bytes/s, rounded upwards, due to which the bytes in the pack may be delivered at the input of the STD slightly faster than based on the real value of C_1. As a consequence, after delivery of the last byte of a pack from a system stream with a constant bitrate, there may be a small time interval during which no bytes are delivered to the input of the STD.

Stream B in Figure 7.6 has a bursty bitrate; the value encoded in the mux-rate field is relatively high; the bytes of the packs are delivered at the STD faster than required for constant bitrate operation. Upon delivery of all bytes of the pack to the STD, there is a time interval without data delivery until the data of the next pack enters the STD. A bursty delivery can be made to duplicate the actual delivery schedule of the transport channel, for example when in the transport channel a burst of payload data is interleaved with a burst of other data, such as transport specific data.

Stream C is an example of an MPEG-1 system stream with a variable bitrate. In this example, the mux-rate and SCR values are encoded so that the bytes of subsequent packs enter the STD according to a continuous delivery schedule of which the rate may change from pack to pack. During pack 1 and pack 2, the mux-rate value is about the same, but during pack 3 the mux-rate is increased by about a factor four. Hence, pack 3 contains about four time more bytes than pack 1 and pack 2, as indicated vertically. For pack 4, the mux-rate is decreased to about twice the value for pack 1 and pack 2.

The STD input buffer B_n for each carried elementary stream n has size BS_n, specified by an encoded STD buffer size field in the header of a packet carrying stream n. The STD buffer size must be provided in the header of the first packet in the MPEG-1 system stream that carries data of elementary stream n and again whenever its value changes. A changed value of the STD buffer size takes effect immediately upon arrival of the changed value at the input of the STD.

At its decoding time $td_n(j)$, all data of access unit $A_n(j)$ are instantaneously removed from buffer B_n, and decoded into presentation unit $P_n(k)$. In case of a video stream using B-pictures, presentation units that represent an I-picture or P-picture are delayed in re-order buffer O_n before being presented at the output of the STD. They are delayed until the next P-picture or

Note 7.6 Example Precise Definition of Video Access Unit Data Removal from Bn

In an MPEG-1 video stream, a picture start code signals the beginning of coded picture data. Upon decoding this picture, not only the coded data of this picture is removed from the STD input buffer B_n, including the picture start code, but also any sequence start code and group of pictures start code that may immediately precede the picture start code.

The end of the coded data of a picture is signalled by a start code, indicating either the start of a next picture, group of pictures, or sequence, or the end of the sequence. In MPEG-1 video, each start code may be preceded by one or more stuffing bits to achieve byte alignment and by one or more stuffing bytes. In case a start code signalling the end of coded picture data is preceded by such stuffing data, then this stuffing data is considered trailing coded picture data. However, for the very first picture of a video sequence there is no previous picture, and therefore, if any stuffing data precedes the very first picture of the sequence, this stuffing data is removed at the same time as the coded data of that very first picture.

I-picture is decoded; while being stored in O_n, the subsequent B-pictures are decoded and presented. For an example of re-ordering see Figure 6.3.

To specify the decoding process in the STD in an unambiguous manner, a very precise specification of the removal of access unit data from B_n is required. For all bytes entering B_n, the removal needs to be defined, including any stuffing data that may have been applied in the audio or video streams. An example of the very precise definition in MPEG-1 of the removal of video access unit data from B_n is discussed in Note 7.6.

The instantaneous video decoding applied in the STD is comparable to the VBV model in MPEG-1 video; however, in the STD the scope is wider: not only decoding events and input buffer control are addressed, but also presentation to achieve synchronization between audio and video. For a video stream, the decoding time stamp accurately specifies when the bytes of a video access unit are removed from the STD input buffer, while the presentation time stamp accurately specifies when the decoded video picture starts to be displayed.

In the MPEG-1 audio standard no hypothetical decoder is used and also no other reference is defined as to how an audio decoder should perform. For single audio-only applications, such as radio broadcast, there is also no real need for that: usually there is only some tens of msec difference in delay between different audio decoder implementations, which is no major problem for audio broadcast applications. However, for synchronizing audio and video, and for mixing multiple (decoded) audio streams in a player, a precise definition of the presentation of audio frames is needed, and for this purpose the MPEG systems experts decided to also apply an instantaneous decoding model for MPEG-1 audio.

At audio elementary stream level, the introduction of instantaneous decoding requires the use of an input buffer. An audio access unit can only be decoded instantaneously if all bytes of the audio AU are available to the decoder. For that purpose, the audio AU data must be stored in the input buffer; once the last byte of the coded audio AU is stored in the input buffer, instantaneous decoding of the audio AU can take place.

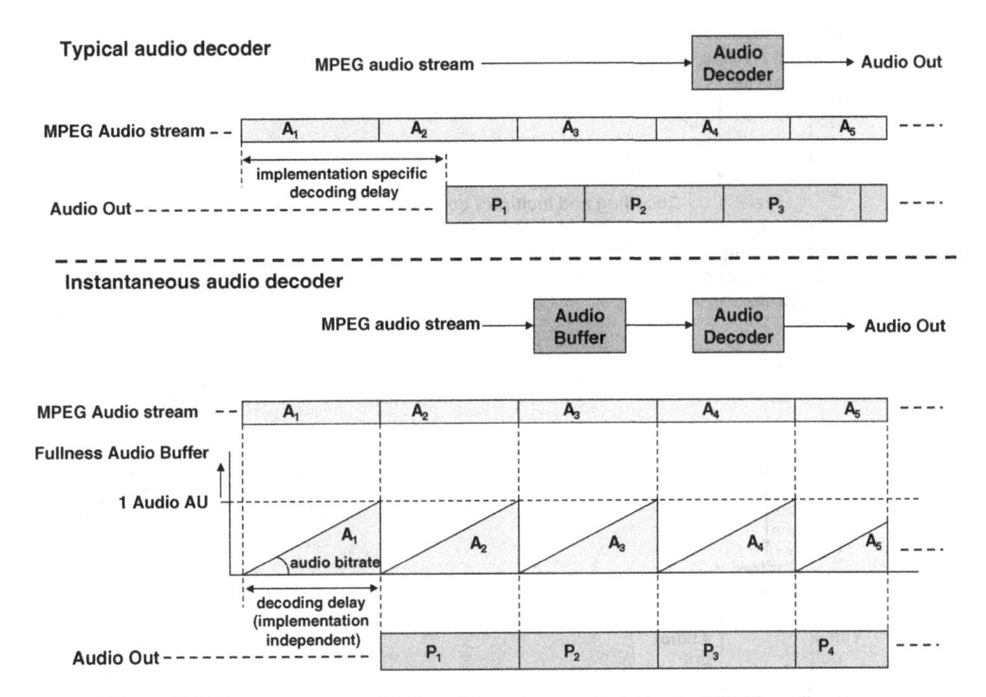

Figure 7.7 Instantaneous audio decoding of a constant bitrate MPEG audio stream

Figure 7.7 presents an example of the different performances of a typical audio decoder and of the instantaneous audio decoder when receiving a (non-multiplexed) MPEG audio stream. The MPEG-1 audio stream in Figure 7.7 has a constant bitrate, due to which each audio access unit A_1, A_2, A_3 and so on has the same size. In both decoders, each audio access unit A_n is decoded into a presentation unit P_n.

The typical audio decoder produces its audio output after a certain delay due to the processing internal to the decoder; the caused decoding delay is implementation specific. In the instantaneous audio decoder, each audio access unit A_n enters the audio buffer. Initially the audio buffer is empty; when the A_n data starts entering the audio buffer, its fullness increases at the audio bitrate. As soon as the last byte of A_n enters the buffer, all A_n data is removed from the audio buffer and decoded instantaneously into presentation unit P_n, upon which the first audio sample of P_n is presented at the decoder output, followed by the other audio samples of P_n at the audio sampling rate. This repeats for each following audio access unit; in this example the required size of the input buffer is equal to the size of one audio access unit, which is the minimum size required by instantaneous decoding. Obviously, the instant in time of decoding could be postponed, which would increase the required size of the audio buffer.

Hence, in the STD for both video and audio, the following applies. At its presentation time $tp_n(k)$, presentation unit $P_n(k)$ is presented at the output of the STD. Due to the instantaneous decoding of the access units, for presentation units that are not re-ordered, $td_n(j)$ is equal to $tp_n(k)$, where k is the index of the presentation unit reconstructed from access unit with index j. Though in practice various delays may occur between decoding and presentation of audio and video, the STD models these delays for video and audio as zero. In the STD a video picture is

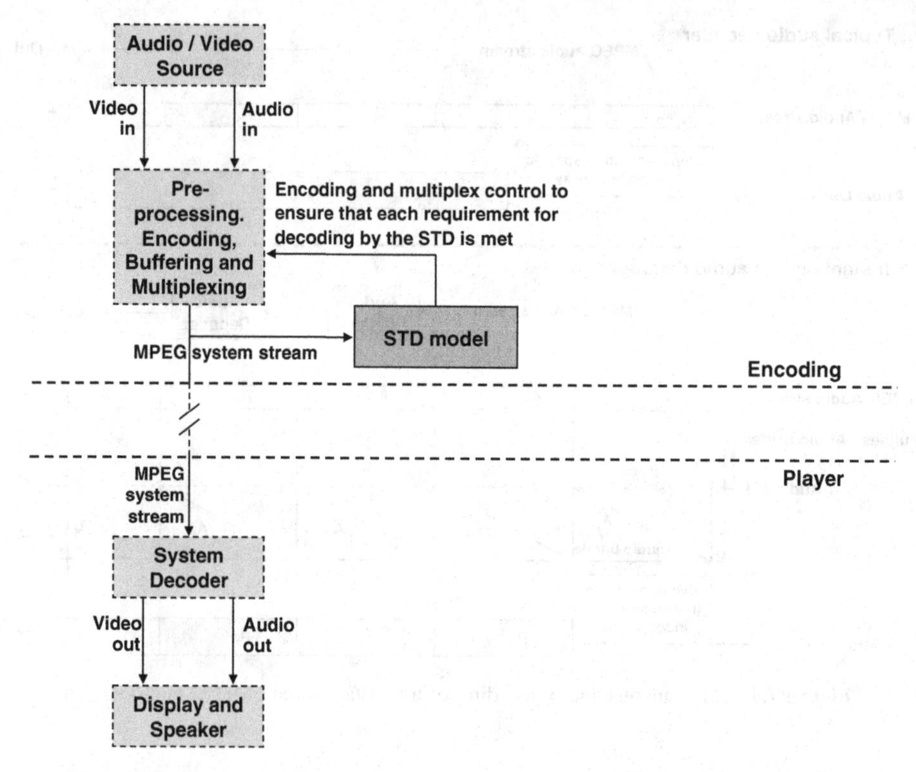

Figure 7.8 Use of the STD model to control encoding and multiplexing at encoder

displayed instantaneously at its presentation time $tp_n(k)$, while the presentation of an audio frame starts at time $tp_n(k)$ when the first audio sample of the decoded audio frame is presented instantaneously. Subsequent samples of the audio frame are presented in sequence at the audio sampling rate.

The implementation and architecture of encoders and decoders is not prescribed by the hypothetical decoder defined by the STD model. However, encoders have to construct MPEG-1 system streams based on the performance of the STD, ensuring that each requirement for decoding by the STD is met. For example, the multiplexer at the encoding must ensure that the input buffers in the STD do not overflow at any time (see Figure 7.8). Furthermore, decoders have to ensure the successful decoding of each system stream that meets the requirements of decoding by the STD. A physical decoder will obviously not decode video and audio instantaneously, but must compensate for ways in which its design differs from that of the STD, as discussed in Section 6.5.

In the STD the input buffer is needed for elementary stream decoding, as well as to accommodate for demultiplexing. Therefore the size of the STD input buffer must be larger than the size of the input buffers of the elementary streams in the VBV based video decoder and the instantaneous audio decoder. Elementary stream decoding and demultiplexing can be considered being mutually independent operations; therefore each could have its own input buffer in the STD. However, the MPEG systems group decided to use a single buffer, shared for elementary stream usage and demultiplexing, so as to offer maximum flexibility in utilizing the available buffer space. For example, the video VBV buffer is almost never full, which can be exploited by a system encoder to exploit free VBV buffer space, by assigning more buffering to demultiplexing, as long as the utilized buffering capacity in total does not exceed the value BS_n.

Examples of the video buffering capacity available over time for the multiplexer at the encoder are given in Figures 7.9 and 7.10. In both examples the same MPEG-1 system stream is produced carrying the same video data; in both examples also the size BS_n of buffer B_n is the same. In Figure 7.9, the multiplexer operates without knowledge and control of the VBV buffer fullness, while in Figure 7.10 the VBV fullness is taken into account by the multiplexer.

For MPEG-1 video streams, the additional buffer room assigned for (de)multiplexing is equal to (BS_n – VBV buffer size); the first is specified in the MPEG-1 system stream, the latter in the MPEG-1 video stream. In the example of Figure 7.9, the multiplexer has no knowledge of the actual usage of the VBV buffer by the MPEG-1 video stream and therefore the multiplexer 'plays safe' by using a demultiplex buffer with a size equal to (BS_n – VBV buffer size). In Figure 7.9, during the delivery of the video data in the system stream at the mux-rate, the video data leaves the demultiplex buffer at the video bitrate, due to which the fullness of the demultiplex buffer increases at a rate equal to (mux-rate – video bitrate). When no video data enters the demultiplex buffer, its fullness decreases by the video bitrate. Figure 7.9 also shows the elementary video stream, reconstructed from the decoding time stamps and the VBV

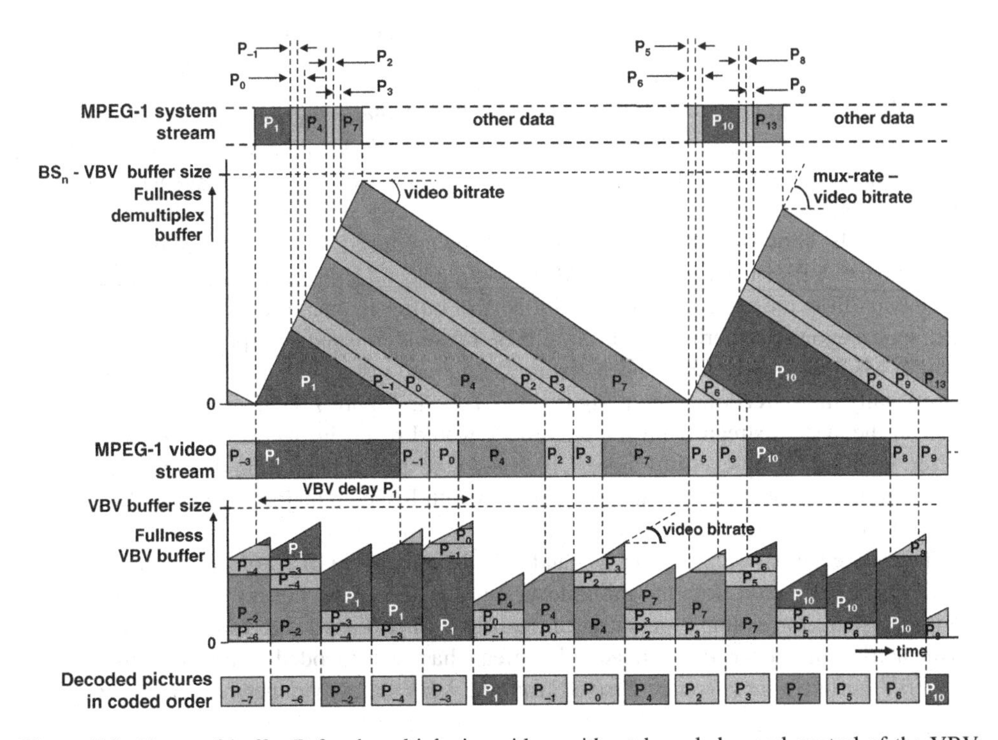

Figure 7.9 Usage of buffer B for demultiplexing video, without knowledge and control of the VBV buffer fullness

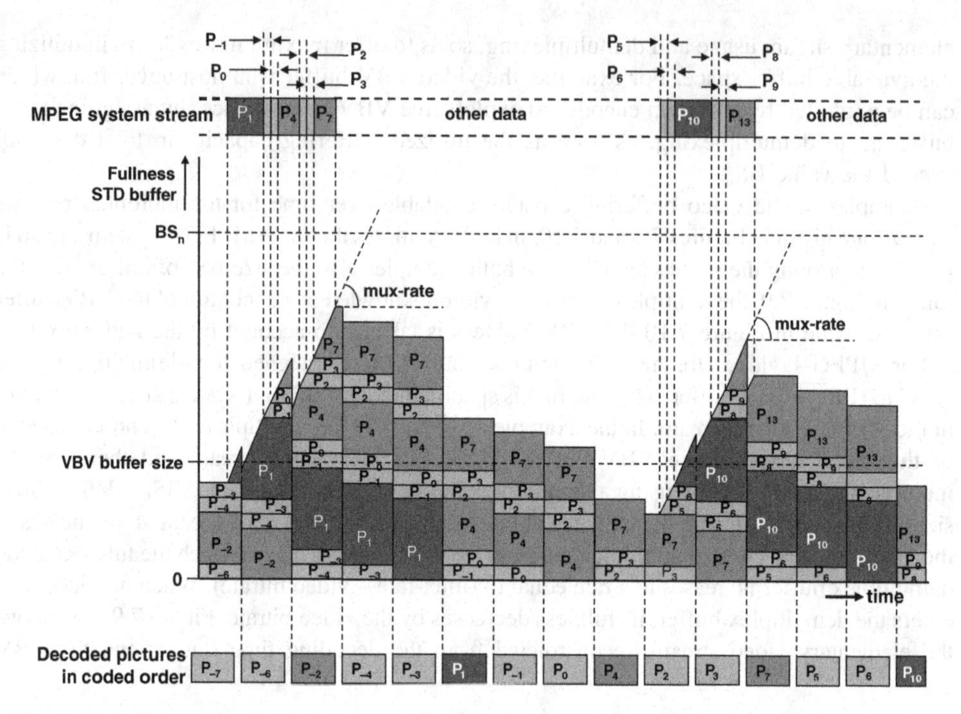

Figure 7.10 Usage of buffer B for demultiplexing video, with the VBV fullness taken into account

delays.[4] As an example, the VBV delay of picture P_1 is depicted; the reconstruction is discussed in more detail in Figure 7.11. Without knowledge of the VBV fullness, the multiplexer has to ensure that the demultiplex buffer never overflows or underflows, so as to prevent overflow and underflow of BS_n.

Figure 7.10 shows an MPEG-1 system stream produced by multiplexer operating with knowledge of VBV buffer usage at the MPEG-1 video decoder. During the delivery period of a video burst from the system stream, the fullness of buffer B_n increases at a rate equal to mux-rate, while beyond that period the buffer fullness remains constant, except at decoding time of a picture, when the coded data of such picture is removed instantaneously. The multiplexer has significantly more freedom than in Figure 7.9. While in Figure 7.9 extension of the first video burst in the system stream with the data of P_5 would result in overflow of the demultiplex buffer, with the multiplexer of Figure 7.10, due to the available room in the VBV buffer, that burst can be extended not only with P_5, but also with P_6 and a major part of P_{10} data without overflowing B_n.

The difference between Figures 7.9 and 7.10 is knowledge of VBV buffer fullness at the video decoder. In case both the multiplexer and the video encoder are an integral part of a coding system, for example operating in real-time, such knowledge can be made available easily at the multiplexer. If the coded video stream has been encoded separately, knowledge of the VBV buffer fullness can be derived from the VBV delay parameter in the MPEG video

[4] Such reconstruction is obviously only possible if the value of the VBV delay is encoded in the video stream. Note that the VBV delay parameter may be coded with a value that is not meaningful.

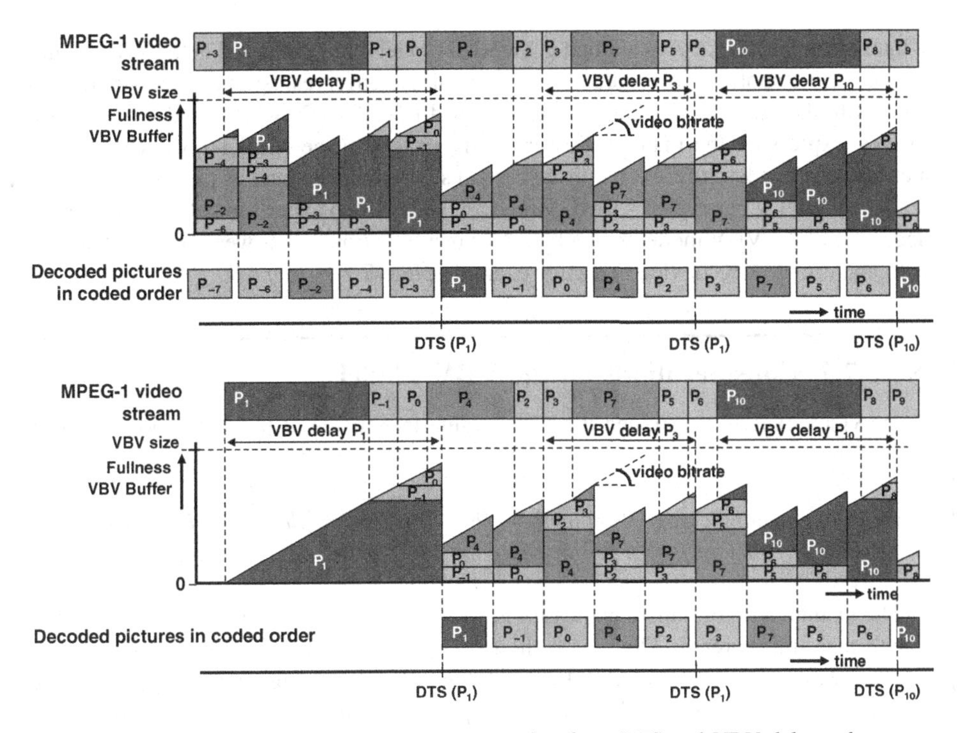

Figure 7.11 MPEG-1 video stream reconstruction from DTS and VBV delay values

stream, if encoded, and the decoding time stamps. An example is given in Figure 7.11. At the top, the ongoing decoding of an MPEG-1 video stream is depicted; for P_1, P_3 and P_{10} the VBV delays are indicated, as well as the decoding time stamps. For each picture j, the coded data in the video stream starts getting delivered at the instant in time found from [DTS(P_j) – VBV delay (P_j)]. At the bottom of Figure 7.11, the same video stream is given, but now the decoding starts at picture P_1; it is shown that the value of the VBV delay applies independently of the picture at which the decoding of the stream commences.

The reconstruction of a video elementary stream from the DTSs and VBV delays can prove useful when it is to be determined whether compliancy problems due to underflow or overflow of buffer B_n in the STD are caused by the video encoder or by the system encoder. If the reconstructed video stream does not violate VBV buffer constraints, then the compliancy problem is caused by the system encoder. It should be noted however that analysing whether a compliancy problem is caused by a video encoder or a system encoder may only be relevant in very specific circumstances, in which different entities are responsible for video and system encoding. In practice, a single entity is usually responsible for both.[5]

[5] At the time MPEG-1 was developed however, video and audio encoding were usually performed by software in non-real time, whereby the system encoder multiplexed coded data from independently produced files containing the compressed video and audio data.

The mathematical accuracy, as applied in the STD model for systems and in the VBV model for video, is very important to ensure compliancy of system streams and video streams. Both models with the associated accuracy are also used in MPEG-2 systems and MPEG-2 video. However, expressing the mathematic formalism in text in the MPEG-2 video standard did sometimes result in lengthy text requiring careful study to understand its meaning, in particular when the vbv-delay parameter is used. Moreover, some video experts may have questioned the added value of the VBV model in practice. As a result, various jokes were produced during the development of MPEG-2 video (see for example Note 7.7).

Note 7.7 Considerations on the VBV Model

Some video experts questioned the added value in practice of the VBV model with its formalism.

Chad Fogg, e-mail message 9 September 1994, Top 10 Complaints of MPEG Delegates: Nr 9: Live in constant fear that some day the VBV model might actually make sense.

The mathematic formalism expressed in words did sometimes result in text with more length than transparency. Obviously choice (c) for question 12) below is correct.

Chad Fogg, e-mail message 3 August 1995, MPEG Quiz number 3:
12) VBV overflow and underflow can be prevented in a MPEG-2 video stream by:

(a) *adjusting rate control.*
(b) *setting vbv-delay = 0xFFFF.*
(c) *as per Annex C of ISO/IEC 13 818-2, insuring that:*

*[vbv-delay*R(n)]/90 000 ≤ vbv-buffer-size, where R*(n) *is the piecewise constant rate computed as the number of bits after the final bit of the nth picture start code before and including the final bit of the (n + 1)th picture start code, divided by the time interval between the examination of the VBV buffer of picture n and n + 1, minus the decoding time between picture n and n + 1, where the time interval is a multiple of the inverse of the frame rate, T, which can be computed as: (a) for progressive sequences with low-delay = 1, 1*T when the nth picture is a B-picture with repeat-first-field = 0; is 2*T when the nth picture is a B-picture with repeat-first-field = 1 and top-field-first = 0; is 3*T when repeat-first-field = 1 and top-first-field = 1; is T when the nth picture is a P- or I-picture and if the previous P-picture or I-picture has repeat-first-field = 0; is 2*T if the nth picture is a P or I picture and if the previous P or I picture has repeat-first-field = 1 and top-first-field = 0; is 3*T if the n-th picture is a P-picture or I-picture and if the previous I- or P-picture has repeat-first-field = 1 and top-field-first = 1, where T is the inverse of the frame rate, unless the previous I- or P-picture does not exist, then ....*

... deleted 24 more lines of text ....

... *when the* nth *picture is a field-structure coded P-frame or I-frame, where T is the inverse of two times the frame rate.*

Source of text in italics: Chad Fogg. Reproduced with permission from Chad Fogg.

7.5 The MPEG-1 System Stream

7.5.1 Data Structure and Design Considerations

Each MPEG-1 system stream consists of a concatenation of one or more packs; to conclude the MPEG-1 system stream, the last pack is followed by an 'End of MPEG-1 system stream' start code. A pack may carry a system header and zero or more packets. To recognize pack headers, system headers and packet headers, start codes are used (see Tables 7.1 and 7.2). One of the important design considerations of the MPEG-1 system standard has been to ensure that start codes are unique in the MPEG-1 system layer, and therefore measures are taken to prevent start code emulation in the system layer.

A start code consists of a three-byte start code prefix, followed by a one-byte start code identifier. The three-byte prefix represents a bit string of 23 × '0' followed by a single '1'. Hence, to prevent start code emulation, the system layer must be designed so that a bit string of more than 22 × '0' can never occur in the system layer. For this purpose, so-called marker bits are used. Each marker bit has the value '1' and is placed in the system data structures at appropriate positions between parameters; moreover, parameters with more than 22 bits are split in multiple parts, separated by a marker bit.

The data structures of the pack header and the system header are presented in Figure 7.12. The pack header carries the 33-bit of the STC sample, called the SCR, System Clock Reference, and the mux-rate parameter. The mux-rate is coded as a 22-bit unsigned integer;

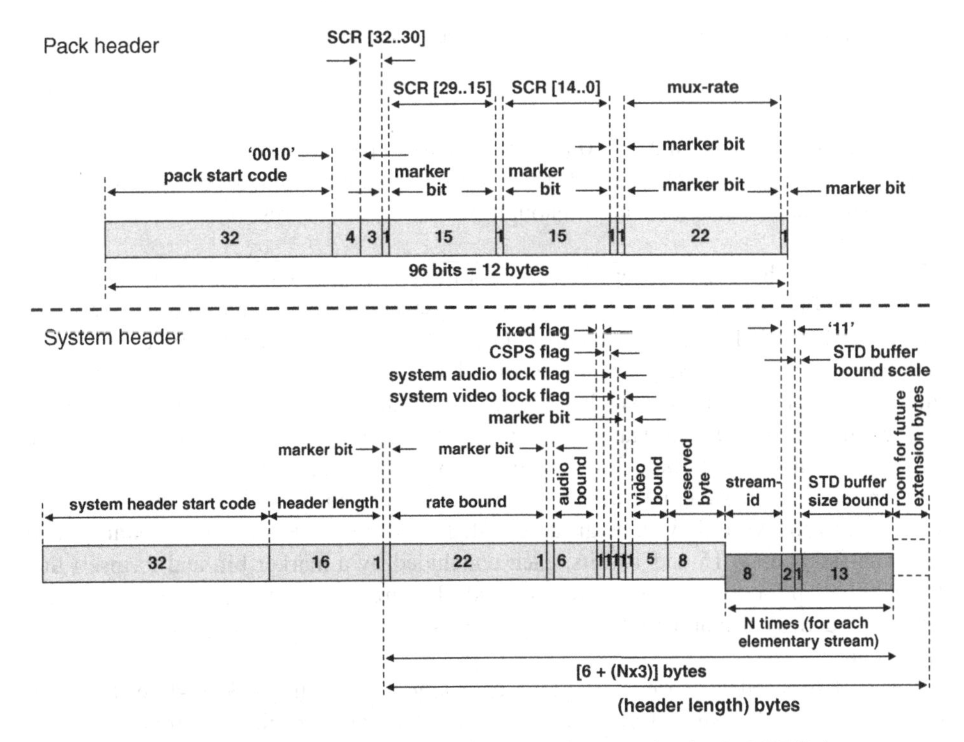

Figure 7.12 Structure of the pack header and system header in MPEG-1

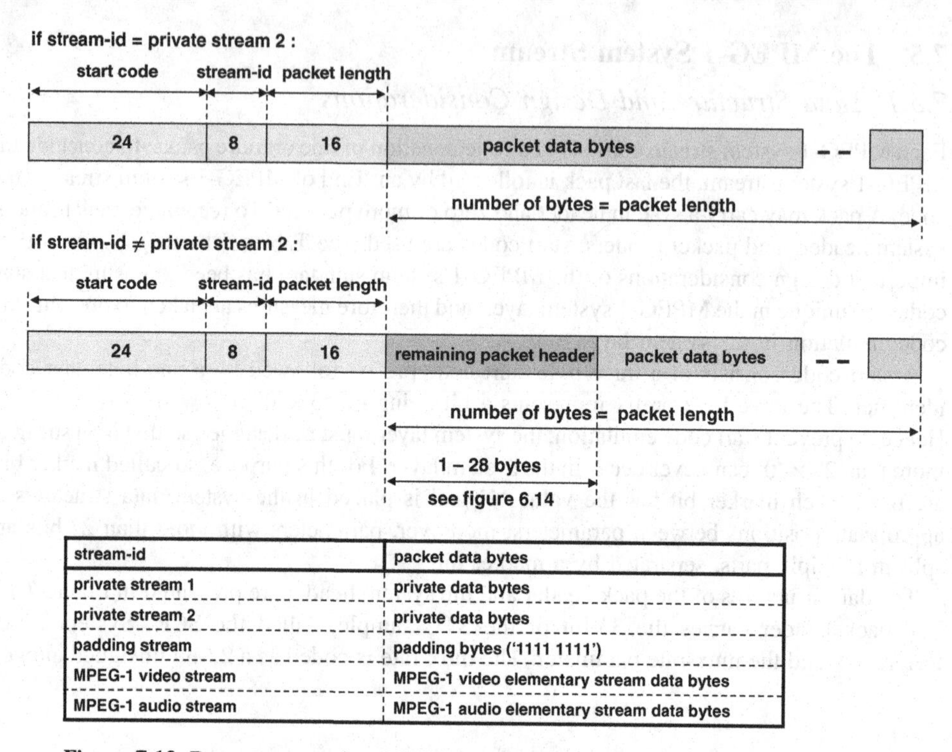

stream-id	packet data bytes
private stream 1	private data bytes
private stream 2	private data bytes
padding stream	padding bytes ('1111 1111')
MPEG-1 video stream	MPEG-1 video elementary stream data bytes
MPEG-1 audio stream	MPEG-1 audio elementary stream data bytes

Figure 7.13 Data structure of an MPEG-1 packet and its header; see also Figure 7.14

its value is measured in units of 50 bytes/s, rounded upwards; the value zero is forbidden. Hence, the maximum bitrate of an MPEG-1 system stream is equal to $(2^{22} - 1) \times 50$ bytes/s, more than 1600 Mb/s and high enough to cover any foreseeable MPEG-1 system application.

The fields in the pack header are defined to be consistent with the fields in the packet header (see Figures 7.12 and 7.13), so as to ease the design of parsers. This applies in particular to the SCR and time stamp fields. In the packet header, the time stamp fields, if present, are immediately followed by elementary stream data, and therefore the time stamp fields in the packet header are concluded by a marker bit to prevent start code emulation in the time stamp field and the following elementary stream data. Start code emulation is also to be avoided in the 33 bits of the SCR and time stamp fields, and therefore each such field is split into subfields separated by a marker bit. For parsing convenience, the MPEG systems group decided to split each SCR and time stamp field in three sub-fields of 3, 15 and 15 bits, each concluded by a marker bit, and to insert the bit string '0010' prior to the first SCR sub-field. The mux-rate field of 22 bits is preceded and concluded by a marker bit, not to prevent start code emulation, but to achieve byte alignment.

The system header carries important characteristics of the MPEG-1 system stream by summarizing bounds of the system parameters defined in the stream, so that a system decoder can verify whether its capabilities are suitable for decoding the entire system stream. The system header provides subsequently the following information (see Figure 7.12):

- Rate-bound: indicates an upper bound for the mux-rate value coded in any pack in the system stream.
- Audio-bound: indicates an upper bound for the maximum number of MPEG-1 audio streams simultaneously active in the STD, hence the maximum number of branches in the STD simultaneously decoding audio,[6] or, in yet other words, the maximum number of MPEG-1 audio decoders required to decode the system stream. Note that the audio-bound field is coded with six bits to allow the inclusive range from 0 to 32.
- Fixed-flag: indicates that the mux-rate parameter in each pack header in the system stream is encoded with the same value, thereby signalling either constant bitrate operation or bursty bitrate operation with a constant mux-rate; for the related discussion of constant and other bitrate operation see Figure 7.6.
- CSPS flag: indicates whether the system stream meets each requirement specified for a CSPS (Constrained System Parameter Stream), as discussed in Section 7.5.2.
- System-audio-lock-flag: indicates whether for each audio frame in each carried audio stream, the audio sampling rate in the STD is locked to the STC frequency, that is that the division (STC frequency)/(audio sampling rate in the STD) yields the same result for each nominal audio sampling frequency.
- System-video-lock-flag: indicates whether for each video picture in each carried video stream, the picture rate in the STD is locked to the STC frequency, that is that the division (STC frequency)/(picture rate in the STD) yields the same result for each nominal video picture rate.
- Video-bound: indicates an upper bound for the maximum number of MPEG-1 video streams simultaneously active in the STD, hence the maximum number of branches in the STD simultaneously decoding video or, in yet other words, the maximum number of video decoders required to decode the system stream.[7] Note that the video-bound field is coded with five bits to allow the inclusive range from 0 to 16.
- Stream-ID, STD-buffer-bound-scale and STD-buffer-size-bound: indicates for each carried elementary stream in the system stream an upper bound for the STD buffer size, as follows:
  - The stream-ID field indicates to which elementary stream(s) the following STD-buffer-size-scale and STD-buffer-size-bound refer, either by providing the unique stream-ID value of an elementary stream, or by providing the wildcard values B8 and B9, signalling all audio streams and all video streams carried in the system stream, respectively.
  - The STD-buffer-bound-scale indicates whether STD-buffer-size-bound is encoded in units of 128 or 1024 bytes.
  - The STD-buffer-size-bound specifies a value greater than or equal to the maximum size of BS_n encoded in any packet carrying stream n.

The three bytes with the stream-ID, the STD-buffer-bound-scale and the STD-buffer-size-bound may be present multiple times; for each elementary stream present in the system stream,

[6] During MPEG-1 development it was assumed that applications require decoding of each elementary stream present in a system stream, as opposed to the model later introduced in MPEG-2 systems, where applications can select which elementary stream(s) to decode. For example in broadcast applications, in one MPEG-2 system stream one video stream can be present next to multiple associated audio streams, each in a different language.

[7] See the previous footnote on audio-bound.

the STD-buffer-size-bound must be specified exactly once by this mechanism. The first (most significant) bit of the stream-ID field is a '1' for each valid stream-ID value, which is used in the system header syntax as a flag signalling the presence of the stream-ID, the STD-buffer-bound-scale and the STD-buffer-size-bound fields.

MPEG left room in the system header for extension in future by allowing trailing bytes to follow the final three byte combination of stream-ID and STD buffer bound fields, but the first of these following bytes is required to have the first (msb) bit set to the value '0', so as to indicate that this byte is not a stream-ID field. The number of trailing extension bytes is equal to the remaining number of bytes from the number indicated by the header length. In this context it may be interesting to note that the system header is followed by either a packet or a pack header, both with a first byte that has its first (msb) bit set to '0'.

The MPEG-1 packet header begins with a 24-bit start code prefix, followed by the eight-bit stream-ID and the 16-bit packet length fields (see Figure 7.13). The packet length field signals the number of bytes in the packet immediately following the packet length field. If the stream-ID signals private stream 2, there is no more header data and private data bytes follow the packet length field; consequently there is no possibility to encode time stamps and other fields in the packet header for a private stream 2. For other stream-ID values,[8] there is a remaining packet header with a length between 1 and 28 bytes, as discussed below (see Figure 7.14).

The remaining packet header presented in Figure 7.14 starts with optional stuffing bytes; any number between zero and sixteen stuffing bytes may be present. Each stuffing byte is a fixed eight-bit value equal to '1111 1111'. Parsers may interpret the first '1' as a stuffing byte flag that indicates the presence of a stuffing byte. After the optional stuffing byte fields, optional STD buffer data follows consisting of 16 bits. The presence of the STD buffer data is indicated by the first two bits being '01'. Parsers may interpret the second bit, the '1', as an STD buffer data flag that indicates the presence of the STD buffer data. The other 14 bits of the STD buffer data consist of the one-bit STD buffer scale and the 13-bit STD buffer size field. The STD buffer size field is an unsigned integer that measures the STD buffer size in units of 128 bytes or 1024 bytes. The STD buffer scale signals which value is used:

```
if STD buffer scale = '0', then BSn = STD buffer size x 128 bytes;
if STD buffer scale = '1', then: BSn = STD buffer size x 1024 bytes.
```

For audio streams the STD buffer scale must have the value '0', and for video streams the value '1'. For other streams the most appropriate value may be used.

After the optional stuffing and STD buffer data, the time stamp information follows. Due to the typically periodic character of video and audio, there is no need to encode a time stamp for each access unit, and therefore the time stamps may be absent. If a time stamp is encoded, then often the DTS value is equal to the PTS value; in that case only the PTS is encoded. If the DTS

[8] Padding streams are not associated with decoding and presentation and have therefore no input buffer in the STD, while also time stamps are meaningless. An informative section of the MPEG-1 system standard states that a padding stream should have no STD buffer and time stamps, but nevertheless MPEG-1 systems provides the option to encode an STD buffer size and time stamps in the packet header of a padding stream. The option to encode these fields for a padding stream is meaningless, and therefore it was decided for MPEG-2 systems to normatively treat a padding stream the same as a private data stream 2, by disallowing any further packet header data after the packet length field.

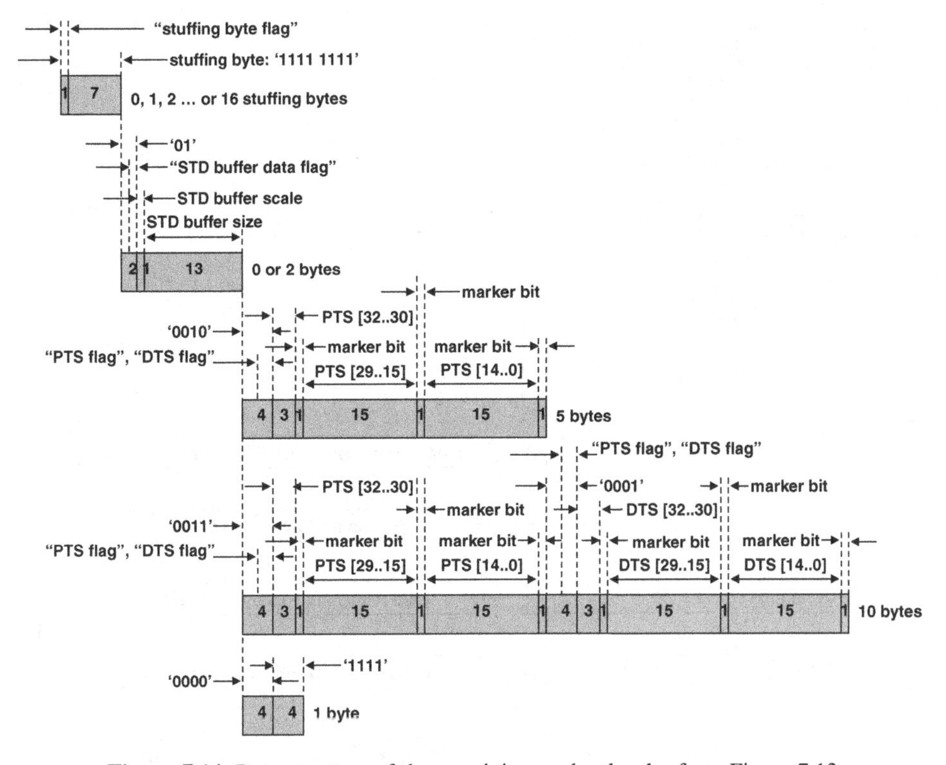

Figure 7.14 Data structure of the remaining packer header from Figure 7.13

value differs from the PTS value (only for I- and P-pictures in MPEG-1 video using B-pictures), then both the PTS and the DTS are encoded.[9] Hence, there are three options for the time stamp information:

- Only the PTS is encoded; the DTS value equals the PTS value.
- Both the PTS and DTS are encoded (only allowed when the DTS and PTS values differ).
- The PTS and the DTS are both absent.

The encoding of a PTS consumes five bytes, and of a PTS and a DTS 10 bytes. Hence, for the time stamp coding in the packet header either five or 10 bytes are needed. Which option is applied in the packet header is indicated by the first four bits of the field. The first two of these bits are '00',[10] while the following two bits signal the presence of a PTS or DTS field. A parser may therefore interpret these bits as PTS and DTS flags. The value '00' of these bits signals

[9] In MPEG-1 video, the frame rate is constant, and therefore the difference between a DTS and PTS value for an I- or P-picture is equal to the number of immediately preceding B-pictures times the nominal picture period; see Figure 6.3 in Chapter 6 of this book. Hence the DTS and PTS values for an I- or P-picture are redundant for MPEG-1 video, and as a consequence there is no fundamental need to provide the DTS and the PTS. However, the MPEG systems group decided to encode both in the packet header, to prevent giving benefit to DTS or PTS based decoder architectures.

[10] Expressing that the following field neither represents a stuffing byte, nor STD buffer data.

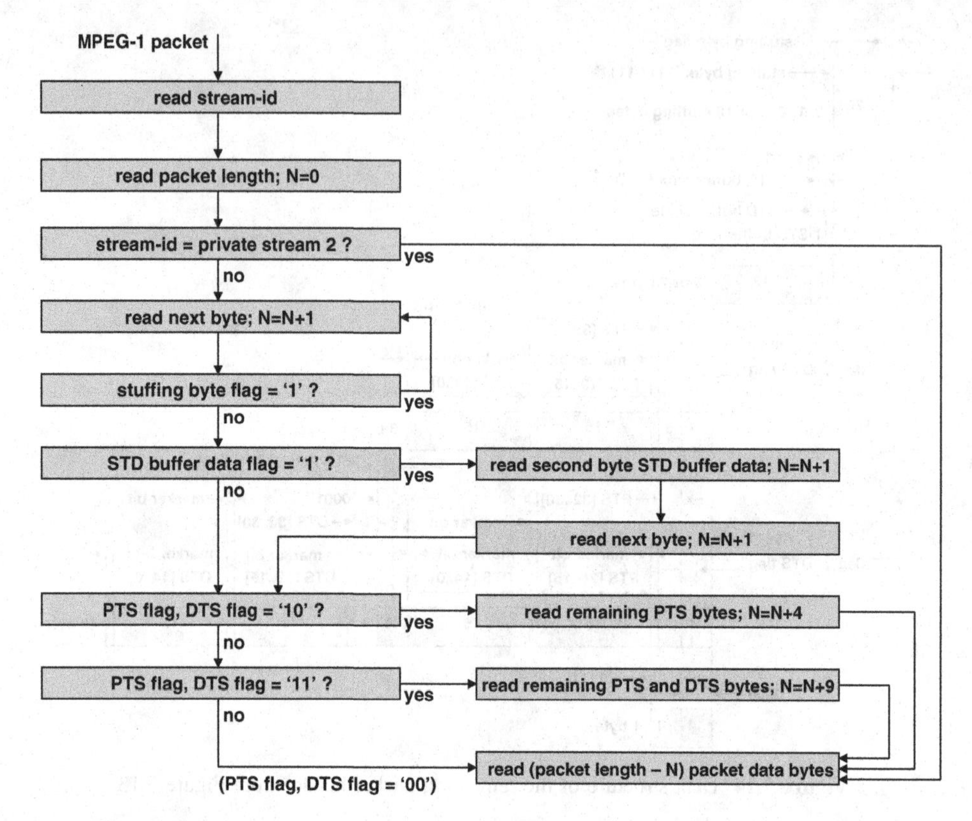

Figure 7.15 Example of a global flow diagram of an MPEG-1 packet header parser

absence of PTS and DTS, '10' that only a PTS is present, '11' that both a PTS and a DTS are present; if a DTS is encoded, for consistency reasons, the first four bits preceding the DTS data are coded with the value '0001', indicating that only a DTS remains.

Figure 7.15 depicts an example of a global flow diagram of an MPEG-1 packet header parser using the stuffing byte flag, the STD buffer data flag and the PTS and DTS flags. The parser starts with reading the stream-ID value and the packet length field, upon which the initial value of an integer parameter N is set to zero. By means of this parameter, the number of bytes in the packet header that follow the stream-ID field is determined. For each byte of the header read after the stream-ID field, the value of N is increased by one. After completing the parsing of the packet header, the value of N varies from N = 0 to at most N = 28; N = 0 is only found in packets with a stream-ID indicating a private stream 2; for other stream-IDs, the value N = 1 is found when stuffing bytes, STD buffer data and PTS/DTS data are absent. The value N = 28 corresponds to the presence of 16 stuffing bytes, two bytes STD buffer data and 10 bytes PTS and DTS data. Once all packet header bytes are processed, (packet length – N) packet data bytes remain; these bytes are passed to the input buffer of the appropriate decoder. However, in the case of a padding stream the packet data bytes are ignored.

If the stream-ID in the packet header signals a format that is unknown to the decoder, the decoder should ignore the remainder of that packet, including contained packet data bytes. In other words, the presence of such packets should not effect the decoder. This ensures compatibility when creating new formats: a decoder remains capable of decoding 'known streams' from an MPEG-1 system stream when a new format is introduced. For example, when a new subtitle format is defined and used for a movie, than this movie can also be played on an existing decoder that does not understand the subtitling format, though such decoder will obviously only present audio and video, without the subtitles.

7.5.2 Constrained System Parameter Streams

As many other standards, MPEG-1 systems is very flexible, potentially allowing a huge application range. However, initial applications benefit from a compliancy point for the industry to focus on. In this respect, the feasibility of a highly integrated MPEG-1 system decoder with a memory space of 4 Mbit (4×2^{20} bit) was an important requirement for the target application, Video CD. To ensure the feasibility of such system decoder, capable of decoding one MPEG-1 video stream and one MPEG-1 audio stream carried in an MPEG-1 system stream, various trade-offs were discussed and agreed in the MPEG video and system groups, including bounds to the size of the STD input buffers. Another concern was the packet rate of an MPEG-1 system stream; at the time MPEG-1 was developed, the computer industry wanted to ensure the feasibility of software parsing of an MPEG-1 system stream[11] and therefore the packet rate should not be too high.

For both these reasons, the Constrained System Parameter Stream, CSPS, concept was specified. An MPEG-1 system stream is a CSPS if it conforms to constraints on the STD input buffer sizes for video and audio streams and on the packet rate. A Constrained System Parameter Stream is identified by the CSPS flag set to '1' in the system header[12] (see Figure 7.12)

For each MPEG-1 video and MPEG-1 audio stream carried in a CSPS, the STD buffer size must be bounded to a value considered reasonable and meaningful for the target application on Compact Disc. For this purpose, the required multiplex buffering, in addition to the elementary stream buffering, can be analysed by modelling each STD input buffer as a multiplex buffer B_{MUX} followed by an elementary stream buffer B_{VBV} or B_{AAU}. At the output of B_{MUX}, a video or audio stream is provided at a bitrate equal to the elementary stream bitrate R_v or R_a (see Figure 7.16). The size B_{VBV} is specified in the video stream and the size of B_{AAU} is equal to the size of one audio access unit.

The analysis of the required amount of multiplex buffering for a CSPS is based on the multiplex method used for Compact Disc, where each MPEG-1 system packet is transported in a sector; on CD-ROM, each sector has the same size and the sector rate is constant and equal to 75 sectors/s; for a brief discussion of this format see Section 7.6.1. Furthermore, a typical target case is considered with a system bitrate of 1.4 Mb/s, a video bitrate of 1.2 Mb/s and an audio bitrate of 0.2 Mb/s. In this case, for every seven sectors on average there are six sectors with a video packet and one sector with an audio packet.

The size of B_{MUX} for an elementary stream depends on the bitrate R_m of the system stream, the elementary stream bitrate R_v or R_a and the maximum amount of time that the multiplex is

[11] At the time MPEG-1 was developed, computers were capable of parsing the MPEG-1 system stream by software, while hardware accelerators were commonly used for video and audio decoding.

[12] If the CSPS flag in the system header is not set to '1', it is not specified whether the MPEG-1 system stream meets the constraints for CSPS streams.

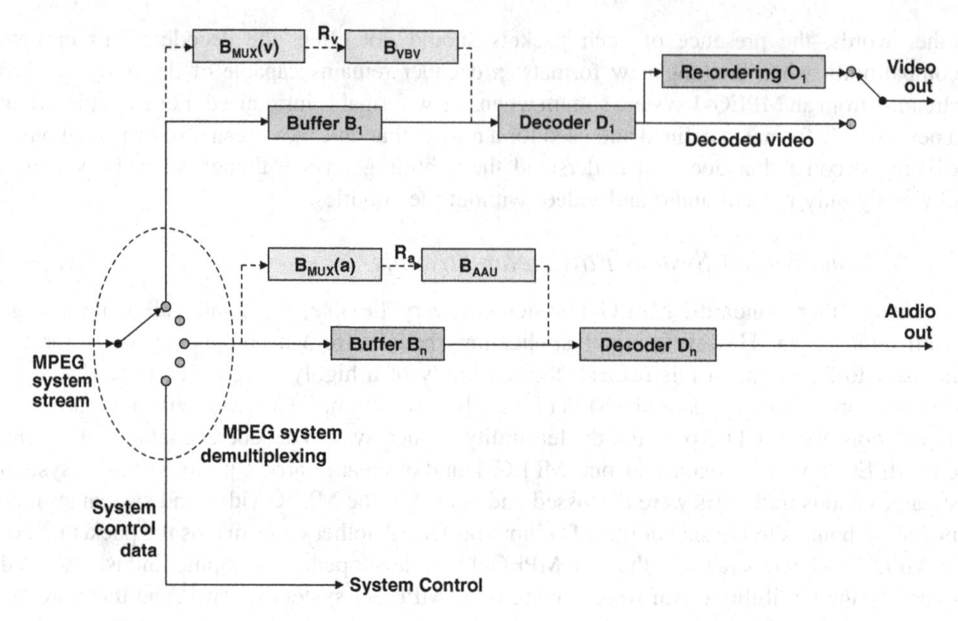

Figure 7.16 Modelling of the STD input buffers to analyse multiplex buffering

not serving that elementary stream. During the delivery of an audio packet, the fullness of $B_{MUX}(a)$ increases at a rate equal to $(R_m - R_a) = 1.2$ Mb/s; when the delivery of the audio packet ends, its fullness is increased by (1 200 000/8)/75 = 2000 bytes. During the delivery of a video packet, no audio data enters $B_{MUX}(a)$, due to which its fullness decreases at a rate equal to $R_a = 0.2$ Mb/s; during every three successive video packets, the fullness of $B_{MUX}(a)$ decreases by 3 × (200 000/8)/75 = 1000 bytes. Likewise, during the delivery of an audio packet, the fullness of $B_{MUX}(v)$ decreases at the video bitrate $R_v = 1.2$ Mb/s with 2000 bytes, while during the delivery of every three successive video packets the fullness of $B_{MUX}(v)$ increases at a rate equal to $(R_m - R_v) = 0.2$ Mb/s with 1000 bytes.

Figure 7.17 depicts three multiplex strategies. For each strategy, the buffer fullness curve is given for both $B_{MUX}(v)$ and $B_{MUX}(a)$. With the multiplex strategy applied in MPEG-1 system stream 1 in Figure 7.17, the size of B_{MUX} is minimized by distributing packets of the same stream as regularly as possible; each audio packet is followed by six video packets. During the delivery of an audio packet, the fullness of $B_{MUX}(a)$ increases from empty state to 2000 bytes and during the following six video packets these 2000 bytes are removed again from $B_{MUX}(a)$. Furthermore, during the delivery of the six successive video packets, $B_{MUX}(v)$ increases from empty to 2000 bytes, while during the delivery of the audio packet, these 2000 bytes are removed again from $B_{MUX}(v)$. In conclusion, the multiplex buffer for Compact Disc application requires a size of at least 2000 bytes, both for video and for audio.

It is reasonable to require a multiplexer to be careful with interspersing relatively low data rate audio packets fairly evenly within the higher data rate video packets, but also flexibility in multiplexing is important. Therefore Figure 7.17 also considers multiplex buffering requirements for MPEG-1 system stream 2, in which two successive audio packets are followed by 12 successive video packets. For this stream, the multiplex buffers for video and audio require a

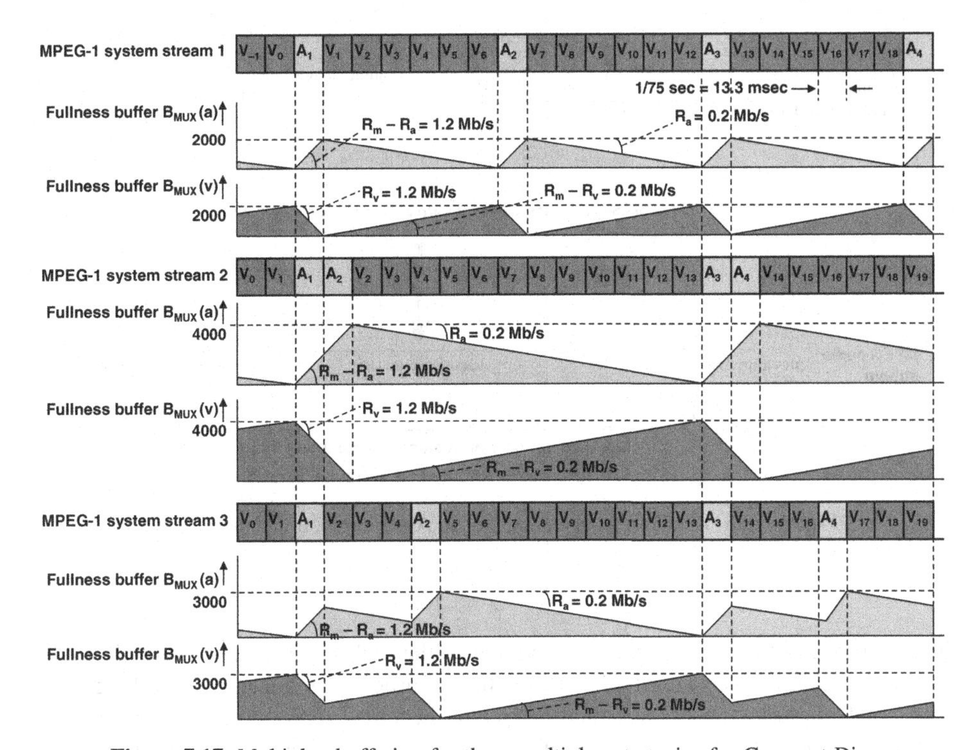

Figure 7.17 Multiplex buffering for three multiplex strategies for Compact Disc

size of at least 4000 bytes each. MPEG-1 system stream 3 offers a compromise whereby audio packet A_2 is moved forward by three instead of six video packets. This reduces the required minimum size of $B_{MUX}(v)$ and $B_{MUX}(a)$ to 3000 bytes.

For an MPEG-1 audio stream carried in a CSPS, a multiplexing room of about 3000 bytes in the STD input buffer seems reasonable. However, the STD input buffer also needs to accommodate B_{AAU}, storing one audio access unit. With a constant audio bitrate, the size of an MPEG-1 audio access unit is found from the audio bitrate multiplied by the audio access unit duration. The AAU duration is equal to the number of samples per audio frame divided by the sampling frequency (see also Figure 7.18).

In Layer I the highest bitrate is 448 kb/s, in Layer II 384 kb/s and in Layer III 320 kb/s. However, in Layer II and III there are three time more samples per audio frame than in Layer I (1152 vs 384). Therefore the maximum AAU size is found in Layer II with an audio bitrate of 384 kb/s at the lowest sampling frequency of 32 kHz. Hence, maximum AAU size = [(AAU duration) × (audio bitrate)]/8 = [(1152/32 000) × (384 000)]/8) = 1728 bytes. In the more practical cases of a sampling frequency of 44.1 kHz and an audio bitrate of 192 or 256 kb/s, the AAU size is equal to 627 bytes or 836 bytes, respectively.[13] It was decided therefore to bound the size of the STD input buffer for MPEG-1 audio streams carried in a CSPS to 4096 bytes, which provides a headroom of more than 3000 bytes for multiplexing MPEG-1 audio at

[13] The actual values are about 626.94 and 835.92, which means most frames contain 627 or 836 bytes, while a few contain only 626 or 835 bytes.

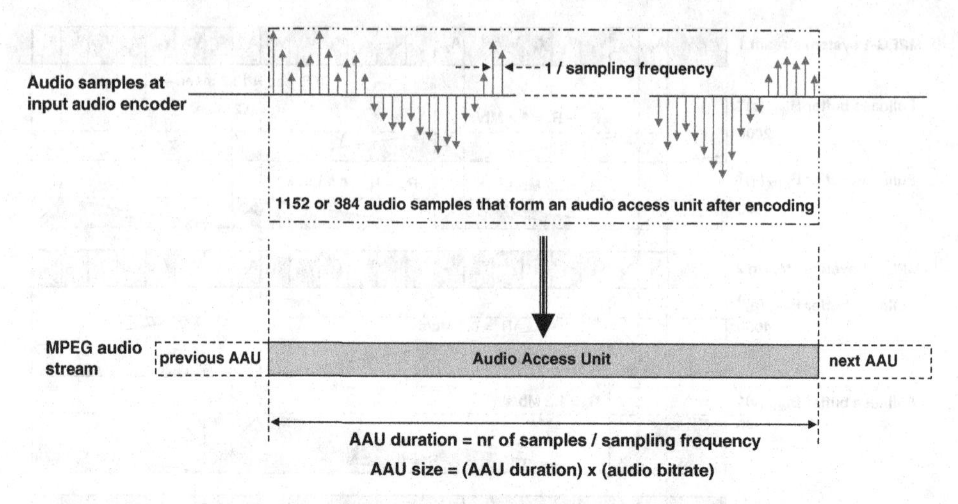

Figure 7.18 Relation between AAU size, AAU duration and audio bitrate

bitrates up to 256 kb/s. At higher bitrates however, the multiplexing flexibility will decrease. For the maximum AAU size of 1728 bytes, the headroom is reduced to 2368 bytes, which means that multiplexing of such audio streams becomes impossible with the presumed multiplex strategy for Compact Disc, if the applied packets contain each more than 2368 bytes.

For video, the CSPS concept is tied to the similar concept of a constrained parameter video stream that has been defined in MPEG-1 video; see Note 7.8. For decoding a constrained parameter video stream in the highly integrated decoder mentioned above, an STD input buffer size of 46 KB is sufficient; in addition to the VBV buffer size bound, this leaves 6 KB available as headroom to accommodate for multiplexing. This headroom for video is about twice as much as for audio. This is because video may be multiplexed with multiple (relatively low bitrate) audio or subtitling streams. In which case there may be more packets not serving video than assumed in the multiplex examples for Compact Disc depicted in Figure 7.17. It should be noted though that the multiplex flexibility can be increased by exploiting the available room in the STD input buffer when the video stream uses a smaller VBV buffer or does not fully utilize the available size of the VBV buffer.

For video however, the scope of CSPS expands beyond constrained parameter video streams and the target application. This is driven by two considerations. Firstly, when coding video at a picture resolution allowed within constrained parameter video streams, but using a higher bitrate than 1.856 Mb/s, then the needed amount of VBV buffering is not larger than that required at 1.856 Mb/s. This is because the ratio of largest versus smallest coded picture usually decreases with higher bitrates. Secondly, an STD buffer bound is also useful in case the MPEG-1 video standard would find application at higher resolution than possible within constrained parameter video streams.

Based on the above considerations, the following STD buffer bounds were decided for carriage of any MPEG-1 video stream within a CSPS. If the picture size of the video stream is allowed within constrained parameter video streams, then the STD buffer is bounded to 46 KB.

Note 7.8 The Concept of a Constrained Parameter Stream in MPEG-1 Video

A constrained parameter MPEG-1 video stream is identified by the constrained parameter flag in the sequence header. If this flag is set to '1', then the video stream is required to meet each of the following requirements, mainly derived from Compact Disc applications:

- horizontal picture size ≤ 768 pixels;
- vertical picture size ≤ 576 lines;
- number of macroblocks per picture ≤ 396;
- number of macroblocks per second $\leq 396 \times 25$;
- picture rate ≤ 30 pictures/s;
- some motion vector constraints;
- VBV buffer size $\leq 40\,960$ bytes $= 40$ KB;
- video bitrate $\leq 1.856 \times 10^6$ bits/s $= 1.856$ Mb/s.

The above constraints are specified for flexible support of picture sizes and picture rates, but it is not possible to apply the maximum value of each constraint at the same time. For example, the maximum number of macroblocks per picture and per second does allow 352 pixels by 288 lines at 25 Hz, but if a horizontal picture size of 704 pixels is to be used for 25 Hz pictures, then the vertical picture size needs to be reduced to 144 lines to meet the constraints on the number of macroblocks per picture and per second.

If the picture size of the MPEG-1 video stream is larger than allowed within constrained parameter video streams, then the memory requirements will be greater than can be accommodated within the 4Mbit memory of the target IC anyway, and so a larger STD buffer was no major problem. Moreover, a VBV buffer larger than 40 KB will be needed for efficient coding. Therefore, the MPEG system and video experts decided a very reasonable and simple approach with linear scaling of the STD buffer size bound with the bitrate from 1.856 Mb/s onwards. See Figure 7.19.

Expressed in formulas, the following must apply for an MPEG-1 video stream carried in a CSPS. If the picture size of the video stream is allowed within a constrained parameter video stream, then:

$$BS_n \leq 47\,104 \text{ bytes} = 46 \text{ KB}.$$

If the video resolution is larger than allowed within a constrained parameter video stream, then if R_{vmax} is the maximum bitrate applied in the video stream:

$$BS_n \leq \max[46, 46 \times (R_{v\,max}/1.856 \times 10^6)]\text{KB}.$$

For each MPEG-1 audio stream carried in a CSPS the following applies: $BS_n \leq 4096$ bytes $= 4$ KB.

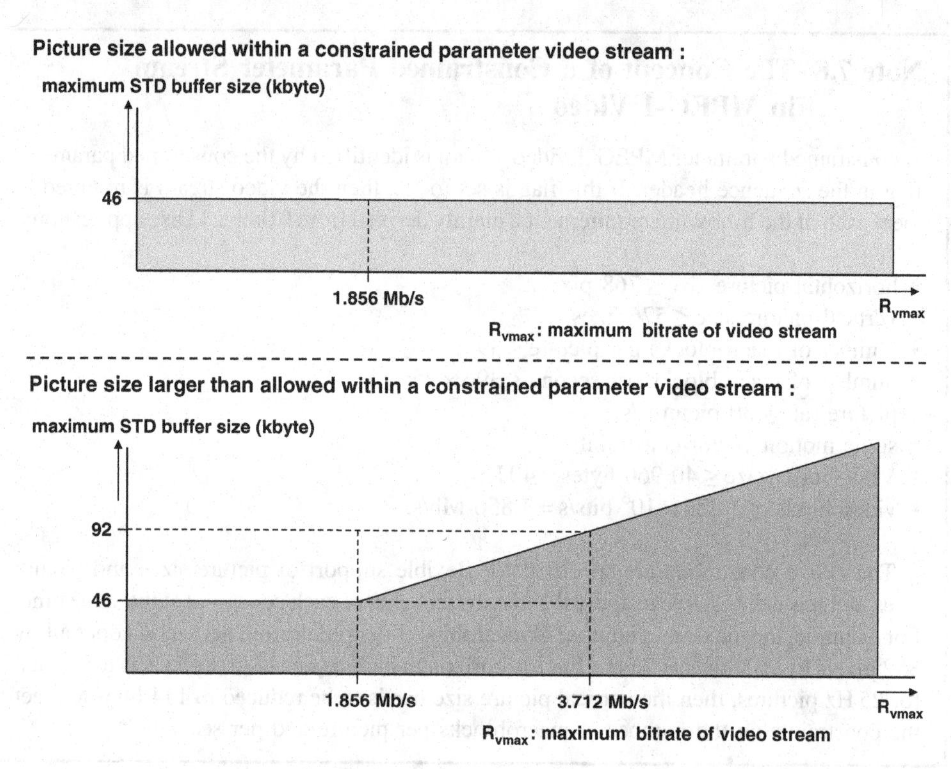

Figure 7.19 Allowed STD buffer sizes for an MPEG-1 video stream in a CSPS

To ensure the feasibility of software parsing in computers at the time MPEG-1 was developed, in a CSPS the packet rate at the input of the STD is constrained. The packet rate is not a stream average, but measured in each interval between two adjacent packs or between the last pack and the end of MPEG-1 system stream start code. For system bitrates up to 5 Mb/s, the packet rate in a CSPS is bounded to 300 per second, four times as high as the packet rate for the Compact Disc application. For higher system bitrates, the use of this constraint would mean that only large packets are allowed, which is less desirable. For example, with a larger packet size, the required STD buffer size increases, while the coded data in the packet covers a longer (presentation) time window; the latter may create problems in the case of low bitrate streams. For system bitrates higher than 5 Mb/s, the CSPS packet rate is therefore bounded by a linear relation to the value of the maximum mux-rate, that is the rate bound coded in the system header. The packet rate bound in a CSPS in relation to the maximum mux-rate is depicted in Figure 7.20.

7.5.3 Compliancy Requirements of MPEG-1 System Streams

Each MPEG-1 system stream needs to comply to all syntactical and semantical constraints specified in the MPEG-1 system standard. In the context of this book, a few semantical constraints are particularly relevant:

- An MPEG-1 system streams must be constructed so that the STD input buffer of an elementary stream neither overflows nor underflows at any point in time during the delivery

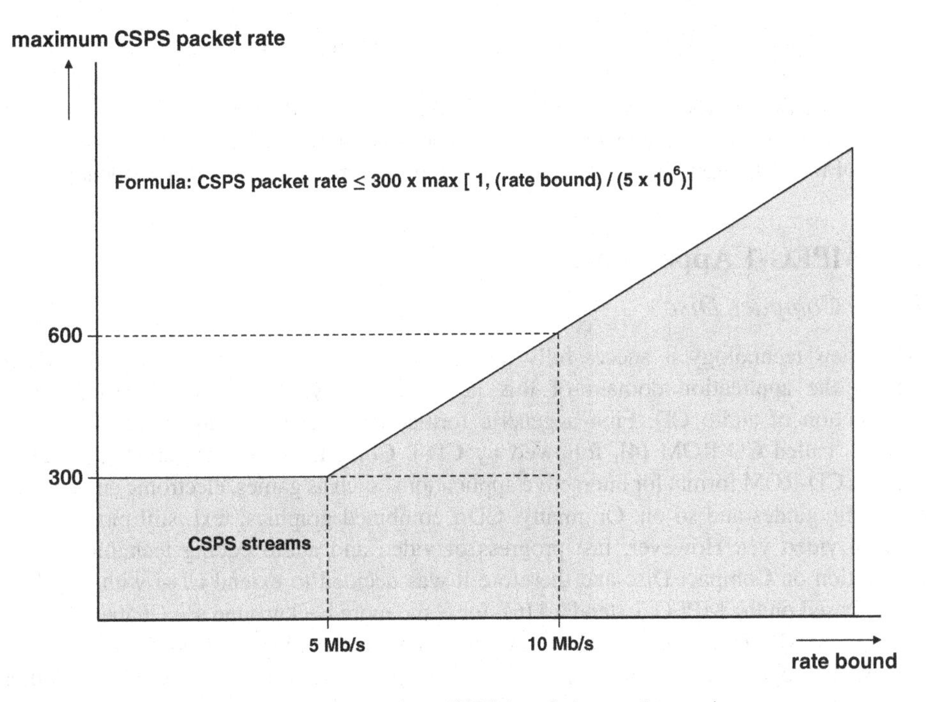

Figure 7.20 Permitted packet rates in a CSPS in relation to the maximum mux rate

and decoding of that stream at the STD. Underflow occurs if the data of an access unit is not completely present at its decoding time, due to which that access unit cannot be decoded at its decoding time. Overflow occurs if the amount of access unit data in the STD buffer is more than the size of the STD input buffer, so that the decoding of one or more access units in the STD will fail, due to access unit data loss.

- The delay caused by STD input buffering must not exceed one second. This constraint is driven by Quality of Service considerations, in particular for very low bitrate video streams, possibly coding small pictures, where the STD buffer is large enough to contain video access units covering a decoding time of more than 1 s. By ensuring that each access unit delivered to the STD is decoded within 1 s, long delays are prevented.
- The time interval between the delivery of two successive SCR values at the input of the STD must not exceed 0.7 s. This constraint is important for the design of clock circuitry regenerating the STC in a system decoder.
- An encoded PTS value in the packet header applies to the first access unit that commences in that packet; no PTS value must be present if no access unit commences in the packet.
- A DTS must be encoded in the packet header, only if the following two conditions are met: (a) a PTS is encoded in that packet header and (b) the decoding time in the STD differs from the presentation time. When these two conditions are not met, the encoding of a DTS is disallowed.
- To ensure an acceptable decoder performance at decoding startup and random access, a PTS must be encoded for the first access unit of an elementary stream, while the maximum time difference between two successively coded PTS values must not exceed 0.7 s. Furthermore, a PTS must be encoded for the first access unit following a discontinuity in an elementary

stream. Such discontinuity may be caused by editing, as discussed in Section 6.4 and Figure 6.19.

- To ensure that an informed decision can be taken by a decoder prior to starting the decoding of a system stream, a system header must be included in the first pack and information on the size of the STD input buffer must be encoded in the first packet of each elementary stream.

7.6 MPEG-1 Applications

7.6.1 Compact Disc

When new technology is successfully introduced to the market, efforts are often made to expand the application domain of this new technology. This also happened after the introduction of audio CD. First a generic format for carriage of digital data on CD was defined, called CD-ROM [4], followed by CD-i, Compact Disc Interactive [5]. CD-i is a specific CD-ROM format for interactive applications, such as games, electronic encyclopaedia, electronic guides and so on. Originally, CD-i combined graphics, text, still pictures, but no moving video yet. However, fast progress of video and audio coding technology enabled application on Compact Disc and therefore it was decided to extend CD-i with Full Motion Video based on the MPEG-1 standard [6]; for some more background see Chapter 2. Also the potential of MPEG-1 usage on CD for Karaoke and distribution of movies and music video clips was recognized, for which many CD-i features were not needed. Therefore a dedicated format for this purpose was defined: Video CD [7].

Both CD-i and Video CD did not achieve the same overwhelming success in the market as audio CD, partly because at the time of development, the picture quality achieved by the encoders was not always high enough. Furthermore, no mechanism was available to prevent making perfect copies, due to which the focus of the industry moved to MPEG-2 and DVD. Nevertheless, the improved performance of video encoders made Video CD a very viable and cost-effective alternative for DVD in some markets. On those markets, mostly in Far East countries, Video CD became a major success. In the following paragraphs the Video CD format is briefly discussed.

Video CD is built on the CD-ROM format; CD-ROM is a generic format for storage of digital data; the data are stored in so-called sectors. Each sector contains 2352 bytes; there are 75 sectors/s at nominal playback speed. The raw bitrate of a CD-ROM is the same as the bitrate of an Audio CD, but CD-ROM applies additional error protection and correction, as briefly discussed in Section 7.2. A Mode 1 CD-ROM has three layers of error correction data, resulting in a net storage capacity of 2048 data bytes/sector, while a Mode 2 CD-ROM has less error correction data, with a net capacity of 2336 bytes. For Video CD, the Mode 2 CD-ROM format is used; however, for signalling and other purposes 12 more bytes are needed, so that a Video CD sector contains 2324 bytes, resulting in a bitrate of $75 \times 2324 \times 8 = 1.3944$ Mb/s of Video CD.

The mapping between Video CD sectors and the MPEG-1 system structure with packs and packets is straightforward. Each Video CD sector carries exactly one pack with one or more packets. However, there is a small complication due to the fact that the header of each sector is required to signal the type of data that is carried in the sector. As a consequence, on Video CD, a pack carries either video, audio or still picture data, but not a mix thereof. Hence, the flexibility of MPEG-1 systems with mixing of packets with various content types in one pack, is constrained. A pack may contain a system header, and one or more packets; next to the content type signalled in the sector header, also padding packets are allowed. A pack within a Video CD sector may be concluded by an 'end of MPEG-1 system stream' start code.

Also application specific constraints were defined for Video CD, similar as applied later at a much wider scale for many MPEG-2 applications. On Video CD, the video is coded using MPEG-1 video, either as moving video or as still pictures for a slide show. For moving video two resolutions are permitted: 352 pixels by 240 lines at 29.97 Hz or 352 pixels by 288 lines at 25 Hz. The still pictures have the same spatial resolution as moving video or use an increased spatial resolution of 704 pixels by 480 lines or 704 pixels by 576 lines, respectively. The audio is coded using MPEG-1 audio, Layer II at an audio sampling frequency of 44.1 kHz. At bitrates in the range between 64 and 192 kb/s mono audio is supported, while stereo, intensity stereo and dual channel audio are supported with bitrates between 128 and 384 kb/s.

The Video CD format allows to play a single piece of content, such as a movie, from a Video CD, but it is also possible to play a series of content items, such as Karaoke songs, music clips, or chapters of a video course. The user may select the items to play, but a Video CD may also include one or more playlists, whereby selected items are played one by one, with a selectable waiting time in between. The items may have their own time line; hence, upon finishing the playback of an item and starting to play the next one, a discontinuity in the STC may occur. How the player handles such discontinuity depends on the implementation; there may be some minimum delay, but the transition between two items may also be seamless. While the impact of STC discontinuities is not discussed further here, it should be noted that this subject is addressed in more detail when local program insertion in broadcast programs is discussed in Section 11.3.

7.6.2 *Computers*

The MPEG-1 standard was successfully prepared for playback on computers, but usage on computers did not drive MPEG-1. Playback of Video CD is often supported on computers, though initially there were complications due to the free-running clocks for audio sampling and video presentation, which meant that lip-sync could be achieved only by skipping or repeating the presentation of an audio frame or a video picture. When computers became more powerful, more advanced tools became available to improve the synchronization performance, such as audio re-sampling. Due to the rapid increase of the processing power, PCs even became an attractive tool for the encoding of Video CDs. However, in general, application on computers is more driven by MPEG-2 than by MPEG-1.

7.7 Conclusions on MPEG-1

The MPEG-1 specification forms the very important first step of MPEG on its ladder to success. MPEG-1 did not only prove the feasibility of video and audio compression for many applications, including MPEG-2, but also developed various system approaches essential for the success of MPEG-2 in the market, such as transport layer independency and the STD.

In particular the invention of the STD concept resolved many fundamental system issues. The System Target Decoder allows to determine unambiguously whether or not streams and decoders are compliant, and serves at the same time as a reference for the design of minimal costs decoders and encoders. By specifying minimum buffering requirements and disallowing the buffer under- and overflow that would disturb proper decoding, the perfect performance of compliant minimum costs systems is guaranteed.

While the availability of excess processing power and memory in recent players, including PCs, can cover up a lot of problems, it remains essential to accurately define minimum buffer resources. First of all, minimum buffer resources remain important for the design of minimal costs systems, particularly in the low costs market. But also from quality of service perspective the definition of minimum buffer resources is important. In a perfectly performing MPEG compliant system, the absence of buffer under- and overflow and the operation with acceptable decoder delays can only be guaranteed if buffer resources are limited in the accurately defined environment of the STD. Unlimited buffering easily leads to delays that are unacceptably long.

References

1. Le Gall, D.J. (1991) MPEG: A video compression standard for multimedia applications. *Communications of the ACM*, **34** (4), 46–58.
2. Le Gall, D.J. (1992) The MPEG video compression algorithm. *Signal Processing: Image Communication*, **4** (2), 129–140.
3. MacInnis, A.G. (1992) The MPEG systems coding specification. *Signal Processing: Image Communication*, **4** (2), 153–159.
4. ISO (1995) The CD-ROM standard is available from ISO as ISO/IEC 10149:1995 at http://standards.iso.org/ittf/PubliclyAvailableStandards/index.html Information on the CD-ROM standard is found at http://en.wikipedia.org/wiki/CD-ROM.
5. Information on the CD-I standard is found at http://en.wikipedia.org/wiki/CD-i.
6. van der Meer, J. (1992) The Full Motion System for CD-I. *IEEE Transactions on Consumer Electronics*, **38** (4), 910–920.
7. Wikipedia (2013) Information on the Video CD standard is found at www.videohelp.com/vcd http://en.wikipedia.org/wiki/Video_CD.

Part Two

The MPEG-2 Systems Standard

8

The Development of MPEG-2 Systems

Why broadcast, telecommunication and optical media drove the development of MPEG-2 systems. The requirements and objectives for the initial MPEG-2 systems development. The establishment of a future-proof MPEG-2 systems based infrastructure for the production, distribution and delivery of audiovisual content. The need for support of new content formats within this infrastructure, in particular the extension of the transport capabilities within MPEG-2 system streams. The evolutionary role of the MPEG-2 systems standard on the standardization efforts undertaken by MPEG after MPEG-2 and the produced amendments to MPEG-2 systems.

8.1 Driving Forces

When the momentum of the MPEG-1 activity steadily increased, the awareness of the MPEG activity on the market place improved significantly. In particular, applications not addressed by MPEG-1 considered the suitability of MPEG technology for exploitation in their business models. Until that time, in the TV broadcast area the main focus was on introducing HDTV and wide-screen services based on analogue broadcast technology. For example, in Japan the MUSE [1] system was developed and in Europe the HDMAC [2] and PAL-plus [3] systems. While some of these systems found usage at the market place, none of them was a major success.

Already at the end of the 1980s, a digital TV system [4] was demonstrated in the United States, but the focus was solely on HDTV, and a very challenging business model was not identified yet. This changed dramatically when the market realized that MPEG technology could be used to exploit the existing TV broadcast channels in a far more efficient manner. Across satellite, cable and terrestrial networks, a broadcast channel used for a single analogue TV service, such as BBC-1, can be replaced by a digital channel with a net bitrate in the range of 20–40 Mb/s. The actual bitrate depends on channel characteristics such as the channel bandwidth and on the applied channel coding technology. If the net channel bitrate is equal to 25 Mb/s, and if the video and audio of an analogue TV are compressed to a total average bitrate of 5 Mb/s, then each such channel will be capable of carrying five digitally compressed TV

Fundamentals and Evolution of MPEG-2 Systems: Paving the MPEG Road, First Edition. Jan van der Meer.
© 2014 John Wiley & Sons, Ltd. Published 2014 by John Wiley & Sons, Ltd.

Note 8.1 The Shifting Focus of MPEG Experts in Early MPEG-2 Developments

In 1991 there was a Call for Proposals for MPEG-2 video, and during 1992 the actual development of the tools for the MPEG-2 video standard took place. At that time, the MPEG-1 systems standard was finished, and though it was obvious that also an MPEG-2 systems standard would be needed, there was very little interest to work on such technology in the first half of 1992. One of the MPEG meetings mid-1992 in a wonderful exotic place was attended by over 100 video experts, but only by very few systems experts. However, this changed drastically when the MPEG-2 video standard was proven feasible. In the second half of 1992, the number of system experts attending MPEG meetings started to increase rapidly. By the end of 1993, there were over 100 experts attending the MPEG systems meetings, while the number of experts attending the MPEG video meetings, where the MPEG-2 video standard was finalized, decreased to a few tens.

services. In other words, by replacing analogue TV by digital TV at the same SD resolution, across the same channels, many more digitally encoded TV services can be introduced. This promising prospect provided a potentially very viable business case for MPEG compression technology and caused thereby a lot of interest from the TV broadcast community for the ongoing developments in MPEG.

For many applications, video compression is the most critical enabler: if insufficient picture quality is achieved at the target bitrate, the application may not be feasible. As a result, at the time the MPEG-2 activity started, the focus of many MPEG experts was mainly on MPEG-2 video compression, so as to determine its feasibility. But once the MPEG-2 video feasibility was proven, the focus moved to MPEG-2 Systems for solutions how to combine (multiple) audio and video streams in a single system stream (see Note 8.1).

Broadcast was not the only area paying substantial attention to MPEG developments; also the optical disc industry was eagerly following what happened in MPEG. At the Compact Disc bitrate it would be possible to distribute movies at 'VHS' quality, but at that time obviously not at the SD video resolution of 720 pixels by 480/576 lines @ 60/50 Hz. For this purpose a higher density optical disc format would be needed, and upcoming MPEG technology provided an opportunity for defining such a format, which ultimately became the DVD standard [5].

Meanwhile the telecommunication industry was investing heavily in networks for the home, using technologies such as ATM and ADSL. The objective was to deliver video content to consumers, for example as a Video on Demand (VoD) service,[1] for which the consumer pays for viewing the content at any time. Though the introduction of these networks was technically successful, the availability of content was often a problem, initially resulting in various commercial failures. Nevertheless, next to the broadcast and optical media industry the telecommunication industry was an important driver of MPEG-2 developments.

[1] Another (competing) option was Near-VoD, using high-bandwidth distribution mechanisms, typically via cable or satellite, whereby multiple copies of a programme were broadcast at short time intervals (such as 10–20 min) providing convenience for viewers, who can watch the programme without needing to tune in at a scheduled point in time.

The MPEG-2 systems requirements imposed by the optical disc industry on one hand and the broadcast and telecommunication industry on the other did have a lot of commonality, but differed in some important respects. For the new optical disc format the basics of the MPEG-1 system standard were very suitable, though the need for a few extensions was considered essential. However, for the broadcast industry and, to a lesser extent, the telecommunication industry the MPEG-1 assumption of operation in an error-free environment was not realistic, given the error-prone networks exploited by these industries. Furthermore, the use of MPEG-2 systems in a broadcast system, with content distribution in high volume to many users at the same time, required solutions not addressed by MPEG-1 systems. As a result, in particular the broadcast industry drove MPEG-2 systems towards a standard that is not only suitable for use in a relatively error-free environment, but also in environments where errors are likely, such as storage or transmission in lossy or noisy media.

When the MPEG-2 system specification was published in 1996, the work on MPEG-2 systems did not stop. Usually, when the development of a standard is finished, the standardization effort is done. This also applies to many MPEG standards; the need for a corrigendum or an amendment may evolve, but after publication of the standard and possibly some after-care period, it is up to the market to deploy (or ignore) the standard, and the experts involved in the development of the standard move to their next challenges. However, MPEG-2 systems is a transport standard for digital audio and video; based on MPEG-2 systems, the industry made huge investments to create an extensive infrastructure for the production, distribution and delivery of audiovisual content. Within this infrastructure, it is possible to define support for new generation video and audio formats without any impact on the delivery of existing formats. This future-proof infrastructure, ensuring maximum service compatibility at the market place, is expected to exist for many years, but requires regular updates.

While the development of the MPEG-2 systems and the video and audio standards was highly market driven, this was not necessarily the case for later MPEG efforts. When the work on MPEG-2 was finished and the standard published, many MPEG-2 experts left MPEG to assist in developing MPEG-2 based products and markets. But at the same time, by the major success of the MPEG-1 and MPEG-2 standards, many other experts from the audio and video research community got attracted to MPEG, eager to get involved in MPEG's success story by working on the next generation MPEG standards. This introduced a difficult dilemma: should the MPEG group continue its standardization effort by defining new work items, even if the market need for new standards was not obvious yet, or should its standardization effort be put at low speed, thereby risking MPEG to lose its momentum?

In this context it is important to realize that the superior compression results achieved for coding of video and audio were obtained by cooperation between virtually all major research institutes active in the audiovisual coding domain. In other words, MPEG proved that cross-industry cooperation can serve as a very successful research accelerator to achieve technically superior results in a relatively short period and in an efficient manner. Moreover, the fact that an MPEG standard is defined by an industry-wide effort improves the opportunities of these standards at the market place. The huge number of new participants when MPEG-2 was finished indicated that the industry at large clearly recognized the benefits of participating in MPEG.

Since 1996, the MPEG momentum has indeed been exploited in developing new standards, but which standardization effort to undertake has not always been obvious. As a result, some of the developed MPEG standards were more technology than market driven, but also new formats of potential interest to MPEG-2 systems based applications evolved. For such formats, amendments to MPEG-2 systems were produced. These amendments specify the usage of the new formats in an MPEG-2 system environment, in particular carriage of the new format in an MPEG-2 system stream and its decoding by an MPEG-2 system decoder.

The produced amendments caused an evolution of the MPEG-2 systems standard that was driven by technology and associated market expectations for optical media and broadcast applications. For some new formats, the market potential for the target applications did prove overestimated, while some other amendments were applied very successfully at the market place, such as the amendment specifying transport of AVC video over MPEG-2 systems. The evolution of MPEG-2 systems and the role of other MPEG standards therein is discussed in detail in Section 8.3.

8.2 Objectives and Requirements

MPEG-2 systems is designed to be a generic format for combining elementary video and audio streams, as well as other data, into a system stream, while meeting the requirements for its target applications. By designing the format so that MPEG-2 system streams can be transported conveniently across any broadcast and telecommunication channel, the MPEG-2 system format becomes a very suitable basis for an infrastructure not only for delivering, but also for producing, storing and distributing audiovisual content. Moreover, by leaving room for other formats, this infrastructure can be updated easily when new formats evolve.

For MPEG-2 systems, requirements very comparable to the ones for MPEG-1 apply, but also many requirements not addressed by MPEG-1 systems, in particular to deal with broadcast of multiple programs in an error-prone environment. To express the requirements, the notion of a program was introduced, whereby a program is defined as a collection of program elements, such as elementary video and audio streams, subtitling streams and any other data streams or data carousels. A program element does not need to have a time base, but those that do, share a common time base, and are intended for synchronized presentation. Note that this definition of a program is typically similar to an analogue TV channel (sometimes called a service), such as BBC-1, without signalling events within such service, such as a news program or talk show.

Prior to or during the initial design of MPEG-2 systems, the following requirements were identified:

1. The MPEG-2 system stream format must be suitable for transport, storage and retrieval at error prone and at relatively error free media and networks.
2. MPEG-2 system streams must be robust against data loss and errors, both in error prone, and in relatively error free environments.
3. The MPEG-2 system overhead must be low.
4. It must be conveniently possible for MPEG-2 system encoders to construct a compliant MPEG-2 system stream.
5. As far as practical, the MPEG-2 system format should be compatible with the MPEG-1 system format.
6. In broadcast applications, it must be possible to simultaneously transport multiple programs within one MPEG-2 system stream.

7. The MPEG-2 system stream format must define packetization and signalling of one or more MPEG-1 or MPEG-2 video streams, MPEG-1 or MPEG-2 audio streams and private streams.
8. The MPEG-2 system stream format must leave headroom for adding support for other elementary stream formats in future.
9. The MPEG-2 system stream format must allow the use of padding.
10. It must be possible for a decoder to simultaneously retrieve and decode the elementary streams from the same program in an MPEG-2 system stream.
11. It must be possible to determine the decoder capabilities required for the decoding of an elementary stream prior to starting its decoding.
12. The MPEG-2 system format must support trick mode operation.
13. While programs may share the same time base, a single time base (STC) must be associated to each program in an MPEG-2 system stream.
14. To enable convenient reconstruction of the STC of a program at the decoder, it must be possible to regularly carry STC samples for each program in an MPEG-2 system stream.
15. To achieve lip synchronization in decoders, an MPEG-2 system stream must allow for accurate synchronization of video, audio and other elementary streams within the same program; for synchronization tolerances see Section 6.7 and Figure 6.26.
16. The MPEG-2 system format must allow for convenient random access to the MPEG-1 and MPEG-2 audio and video streams, as well as to other elementary streams.
17. It must be possible to prevent overflow and underflow of buffers in decoders, unless explicitly allowed.
18. The delay caused by a decoder must be 'reasonable'.
19. Copy protection and copy control of content carried in an MPEG-2 system stream must be possible.
20. The MPEG-2 systems format must allow Conditional Access (CA) to content of a program, based on altering data of program elements by means of scrambling and descrambling and controlled by one or more CA systems, so as to ensure that the content can be accessed only if authorized.
21. It must be possible for CA systems to alter program element data, without altering important system data, such as the headers of packets carrying the program element data.
22. The MPEG-2 systems format must allow for re-multiplexing of programs in broadcast networks.
23. For MPEG-2 system streams in broadcast networks, it must be possible to temporarily replace audio and video streams in a program, for example to insert locally targeted weather forecasts or advertisements in a global broadcast service.
24. The MPEG-2 systems format must allow convenient mapping to the transport layers for broadcast networks, in particular telecommunication (ATM), cable, satellite and terrestrial networks.

On the last requirement, some notes should be made. The transport layer, including any channel coding schemes for broadcast networks are beyond the scope of MPEG. At the time MPEG-2 systems was defined, ATM was known to use ATM cells with a payload of 48 bytes. However, the channel coding schemes for digital TV broadcast across satellite, cable and terrestrial networks were largely undefined yet. Hence, the challenge for the MPEG systems group was to define an MPEG-2 system stream format so that the stream can be transported conveniently over ATM, while at the same time the format can serve as basis for the channel coding designs for satellite, cable and terrestrial networks. By an optimized interface between the systems format and the channel coding, convenient transport of MPEG-2 system streams across these networks can be ensured. Which should result in a rather seamless transition from one network to another, with minimum processing at the network transition. This is for example important when a satellite is used to transport programs carried in an MPEG-2 system stream to a cable head-end for distribution to subscribers of the cable network. For a further discussion of this subject see Sections 9.1 and 9.4.3.

8.3 The Evolution of MPEG-2 Systems

Since the development of the MPEG-2 standard, the MPEG group expanded its scope significantly, resulting in a large number of produced MPEG standards. MPEG develops its standards around work items; typically each work item results in a suite of MPEG standards, organized in various parts with the same assigned ISO/IEC number, such as ISO/IEC 13818 for MPEG-2 [6]. Table 8.1 lists the work items addressed by MPEG between 1988 and 2013.

The MPEG standards bookshelf requires considerable room. By October 2013, all suites of MPEG standards contained 140 parts, many of which are thick documents of hundreds of pages. Tables 8.2 and 8.3 present an overview of all standards produced until October 2013 by MPEG under the work items listed in Table 8.1.

For many MPEG standards, transport over MPEG-2 systems is irrelevant; there may be various reasons for this. For example, some MPEG standards address streaming or download over IP, which may provide a complementary delivery method for content to MPEG-2 system based applications, but without any impact on MPEG-2 systems. Also, MPEG standards may

Table 8.1 MPEG work items as of 2013

Topic	Standard	Area covered by standard
MPEG-1	ISO/IEC 11172	Coding of moving pictures and associated audio at up to about 1.5 Mb/s
MPEG-2	ISO/IEC 13818	Generic coding of moving pictures and associated audio
MPEG-4	ISO/IEC 14496	Coding of audio-visual objects
MPEG-7	ISO/IEC 15938	Multimedia content description interface
MPEG-21	ISO/IEC 21000	Multimedia framework
MPEG-A	ISO/IEC 23000	Multimedia application formats
MPEG-B	ISO/IEC 23001	MPEG systems technologies
MPEG-C	ISO/IEC 23002	MPEG video technologies
MPEG-D	ISO/IEC 23003	MPEG audio technologies
MPEG-E	ISO/IEC 23004	MPEG multimedia middleware
MPEG-V	ISO/IEC 23005	Media context and control
MPEG-M	ISO/IEC 23006	Multimedia service platform technologies
MPEG-U	ISO/IEC 23007	MPEG rich media user interface
MPEG-H	ISO/IEC 23008	High efficiency coding and media delivery in heterogeneous environments
MPEG-DASH	ISO/IEC 23009	Dynamic adaptive streaming over HTTP
EXPLORATION		A suite of exploration activities likely to become standards or parts of standards

Table 8.2 Produced MPEG-1, -2, -4, -7 and MPEG-DASH standards, as of 2013

MPEG-1 (ISO/IEC 11172)		MPEG-2 (ISO/IEC 13818)	
Part 1	Systems	Part 1	Systems
Part 2	Video	Part 2	Video
Part 3	Audio	Part 3	Audio
Part 4	Compliance	Part 4	Conformance testing
Part 5	Software simulation	Part 5	Software Simulation
		Part 6	DSM-CC
MPEG-4 (ISO 14496)		Part 7	Advanced Audio Coding (AAC)
Part 1	Systems	*Part 8*	*10 Bit video – discontinued*
Part 2	Visual	Part 9	Real-time Interface for systems decoders
Part 3	Audio		
Part 4	Conformance testing	Part 10	Conformance testing for DSM-CC
Part 5	Reference software	Part 11	IPMP on MPEG-2 systems
Part 6	Delivery multimedia integration framework (DMIF)	**MPEG-7 (ISO/IEC 15938)**	
Part 7	Optimized software for MPEG-4 tools	Part 1	Systems
Part 8	MPEG-4 on IP framework	Part 2	Description definition language
Part 9	Reference hardware description	Part 3	Visual
Part 10	Advanced video coding (AVC)	Part 4	Audio
Part 11	Scene description and application engine	Part 5	Multimedia description schemes
Part 12	ISO base media file format	Part 6	Reference software
Part 13	IPMP extensions	Part 7	Conformance
Part 14	MP4 file format	Part 8	Extraction and use of MPEG-7 descriptions
Part 15	AVC file format		
Part 16	Animation framework extension (AFX)	Part 9	Profiles
Part 17	Streaming text format	Part 10	Schema definition
Part 18	Font compression and streaming	Part 11	Profile schemas
Part 19	Synthesized texture stream	Part 12	Query format
Part 20	Lightweight application scene representation	Part 13	Compact descriptors for visual search
Part 21	MPEG-J extensions for rendering	**MPEG-DASH (ISO/IEC 23009)**	
Part 22	Open font format	Part 1	Media presentation description and segment formats
Part 23	Symbolic music representation		
Part 24	Audio–system interaction	Part 2	Reference software and conformance
Part 25	3D Graphics compression model	Part 3	Implementation guidelines
Part 26	Audio conformance	Part 4	Format independent segment encryption and Authentication
Part 27	3D Graphics conformance		
Part 28	Composite font		
Part 29	Web video coding		
Part 30	Timed text and associated images in ISO base media file format		
Part 31	Timed text and other visual overlays in ISO base media file format		

address issues beyond the scope of MPEG-2 system based applications or may suffer lacking of interest. However, within MPEG also new standards evolve that are important to MPEG-2 system based applications and for which carriage over MPEG-2 systems is to be defined. In particular for new video and audio formats evolving over time, as discussed in Sections 3.2.3

Table 8.3 Produced MPEG-21 and MPEG-A, B, C, . . . standards, as of 2013

MPEG-21 (ISO/IEC 21000)	
Part 1	Vision, technologies and strategy
Part 2	Digital item declaration
Part 3	Digital item identification
Part 4	IPMP components
Part 5	Rights expression language
Part 6	Rights data dictionary
Part 7	Digital item adaptation
Part 8	Reference software
Part 9	File format
Part 10	Digital item processing
Part 11	Evaluation methods for persistent association
Part 12	Test bed for MPEG-21 resource delivery
Part 13	*Void* (published elsewhere)
Part 14	Conformance
Part 15	Event reporting
Part 16	Binary format
Part 17	Fragment identification
Part 18	Digital item streaming
Part 19	Media value chain ontology
Part 20	Contract expression language
Part 21	Media contract ontology

MPEG-C (ISO/IEC 23002)	
Part 1	Accuracy specification for implementation of integer-output iDCT
Part 2	Fixed-point 8 × 8 DCT/IDCT
Part 3	Auxiliary video data representation
Part 4	Media tool library
Part 5	Reconfigurable media coding conformance and reference software

MPEG-V (ISO/IEC 23005)	
Part 1	Architecture
Part 2	Control information
Part 3	Sensory information
Part 4	Virtual world object characteristics
Part 5	Data formats for interaction
Part 6	Common types and tools
Part 7	Reference software and conformance

MPEG-U (ISO/IEC 23007)	
Part 1	Widgets
Part 2	Advanced user interaction interface
Part 3	Reference software and conformance

MPEG-A (ISO/IEC 23000)	
Part 1	Purpose for multimedia application formats
Part 2	Music player application format
Part 3	Photo player application format
Part 4	Musical slide show application format
Part 5	Media streaming application format
Part 6	Professional archival application format
Part 7	Open access application format
Part 8	Portable video application format
Part 9	Digital multimedia broadcasting application format
Part 10	Surveillance application format
Part 11	Stereoscopic video application format
Part 12	Interactive music application format
Part 13	Augmented reality application format
Part 14	Augmented reality reference model
Part 15	Multimedia preservation application format

MPEG-B (ISO/IEC 23001)	
Part 1	Binary MPEG format for XML
Part 2	Fragment request unit
Part 3	XML representation of IPMP-X messages
Part 4	Codec configuration representation
Part 5	Bitstream syntax description language
Part 6	*Void* (became MPEG-DASH)
Part 7	Common encryption for ISO base media file format files
Part 8	Coding independent media description code points
Part 9	Common encryption for MPEG-2 transport stream

MPEG-D (ISO/IEC 23003)	
Part 1	MPEG surround
Part 2	Spatial audio object coding
Part 3	Unified speech and audio coding

MPEG-E (ISO/IEC 23004)	
Part 1	Architecture
Part 2	Multimedia API
Part 3	Component model
Part 4	Resource and quality management
Part 5	Component download framework
Part 6	Fault management framework
Part 7	Integrity management framework
Part 8	Reference software and conformance

Table 8.3 (*Continued*)

	MPEG-H (ISO/IEC 23008)		MPEG-M (ISO/IEC 23006)
Part 1	Media transport	Part 1	Architecture
Part 2	High efficiency video coding	Part 2	MPEG extensible middleware API
Part 3	3D Audio	Part 3	Reference software and conformance
		Part 4	Elementary services
		Part 5	Service aggregation

and 3.3.2, an almost continuous standardization effort is needed to extend MPEG-2 systems with support for the new formats defined within the scope of MPEG.[2]

The extensions are defined so that the delivery of existing formats is not effected, to ensure maximum compatibility at the market place. Typically, for each extension an MPEG-2 system amendment is produced. As a result, the MPEG-2 systems is a 'living' standard that grows almost every year. Produced MPEG-2 systems amendments are consolidated regularly, typically about every 5 years, into a new edition of MPEG-2 systems. The growth of the MPEG-2 system standard does not significantly contribute to the size of the MPEG bookshelf however; on average, the MPEG-2 systems specification grows each year about five pages.

Since the publication of the first edition of MPEG-2 systems, roughly every year an amendment to MPEG-2 systems is produced. Some of these amendments address minor extensions, for example to address a new profile in MPEG video or audio, but in particular the amendments specifying transport over MPEG-2 systems of new coding formats have caused an ongoing evolution of transport capabilities within MPEG-2 system streams.

The main developments within this evolution are depicted in Figure 8.1. The first edition of the MPEG-2 system standard specifies transport of MPEG-1 and MPEG-2 video and audio over MPEG-2 system streams. In 2000, the transport of MPEG-4 streams over MPEG-2 systems was specified, and in 2003 transport of metadata, such as MPEG-7 streams, but leaving explicitly room for application standardization bodies to specify other metadata formats (see also Section 4.1). In 2004, transport of AVC video was specified (see Section 3.2.3), while in 2007 transport of MPEG-4 timed text and MPEG-4 lossless audio (see Sections 4.2 and 4.3). In 2008 the amendment on transport of auxiliary video, more in particular of depth information associated to 2D pictures, was published (see Section 4.5.2). In 2009 two amendment were published, one on transport of SVC (see Section 3.2.3) and the other on transport of multiview video (discussed in Sections 4.4 and 4.5.3).

Figure 8.2 shows the complete list of MPEG-2 system amendments published prior to October 2013. Also the consolidation of these amendments into the second, third and fourth revision of the MPEG-2 system specification, published in 2000, 2007 and 2013 is shown. The amendments are numbered with respect to the latest edition of the MPEG-2 systems standard, and the year of publication is indicated both of the amendment, and of the amended edition of MPEG-2 systems. For example, ISO/IEC 13818-1:2000/Amd.3:2004 indicates that amendment 3, published in 2004, applies to the MPEG-2 systems edition published in 2000.

[2] The use of non-MPEG standards within MPEG-2 systems is beyond the scope of MPEG but may be defined by application standardization bodies such as DVD Forum, ATSC and DVB.

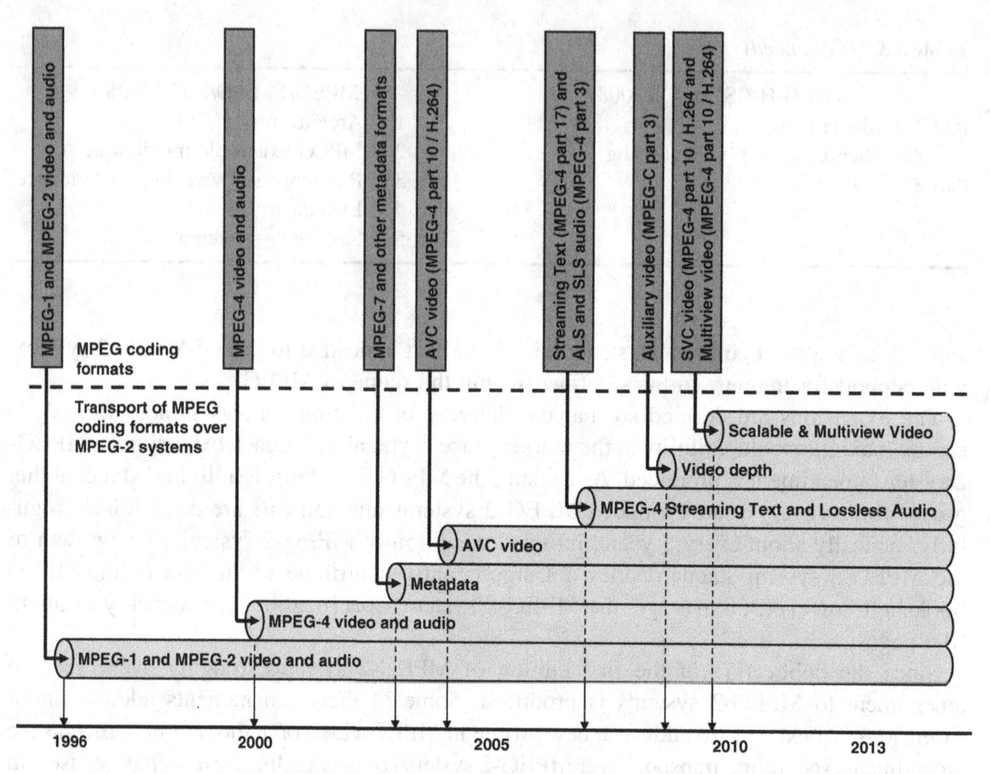

Figure 8.1 Evolution of transport capabilities within MPEG-2 system streams

Below the produced MPEG-2 system amendments are summarized.

- Amendments to the first edition of MPEG-2 Systems (ISO/IEC 13818-1:1996):
  - *ISO/IEC 13818-1:1996/Amd.1:1997 – A Registration Procedure for Copyright Identifier.* To copyright audiovisual works, such as films, documentaries, television programs, sports events and advertising clips, a unique identification of each work is needed. For this purpose, the first MPEG-2 systems edition defines a 32-bit copyright identifier field and a variable length field with additional copyright information. The copyright identifier signals the syntax and semantics of the additional copyright information. To allow unique signalling, the identifier must be unique, and to achieve this, registration of copyright identifier values is required by a Registration Authority. Amd.1 specifies the registration procedure of copyright identifiers and duties of the Registration Authority (RA).

    For example, the copyright identifier value may signal the use of ISAN (International Standard Audiovisual Number), a voluntary numbering system and metadata schema, that is defined as the audiovisual equivalent of ISBN for books. ISAN is developed by major players of the audiovisual industry under the auspices of the International Standards Organization (ISO). ISAN is registered as ISO 15706-1:2002 and ISO 15706-2:2002, as discussed in Section 4.1. However, the use of ISAN is voluntary and other methods to uniquely identify audiovisual works are possible too.

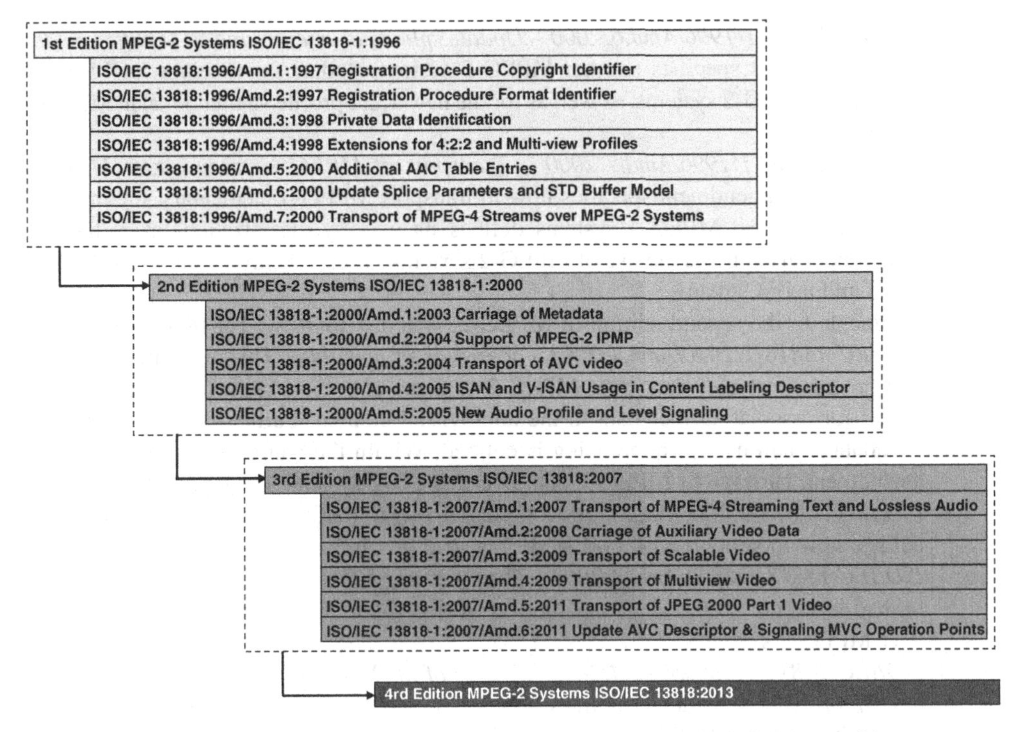

Figure 8.2 Consolidation of produced amendments in new MPEG-2 system editions

- *ISO/IEC 13818-1:1996/Amd.2:1997 – A Registration Procedure for Private Data Formats.* To uniquely and unambiguously identify private data formats, MPEG-2 systems uses a 32 bit format identifier field followed by a variable length field with additional identification information. The syntax and semantics of the additional identification information is signalled by the value of the format identifier. To ensure uniqueness, format identifier values are assigned by a Registration Authority. Amd.2 specifies the registration procedure of format identifiers and the duties of the Registration Authority (RA). After publication of this amendment, the Society of Motion Picture and Television Engineers (SMPTE) has been assigned as the RA for the format identifier.

  For example, the format identifier value may signal an application standardization body, while the additional identification information signals a format specified by that application standardization body.
- *ISO/IEC 13818-1:1996/Amd.3:1998 – Private Data Identification.* This amendment addresses a minor omission in the MPEG-2 systems standard published in 1996.
- *ISO/IEC 13818-1:1996/Amd.4:1998 – Extensions for 4:2:2 and Multi-view Profiles.* This amendment specifies the required MPEG-2 system extensions associated with new 4:2:2 and multiview profiles specified for MPEG-2 video.
- *ISO/IEC 13818-1:1996/Amd.5:2000 – Additional AAC table entries.* For transport of MPEG-2 AAC over MPEG-2 systems, the required additional table entries are specified in this amendment.

  - *ISO/IEC 13818-1:1996/Amd.6:2000 – Update splice parameters and STD buffer model.* This amendment specifies another MPEG-2 systems update for transport of MPEG-2 AAC over MPEG-2 systems, as well as for the new 4:2:2 profile @ high level in MPEG-2 video.
  - *ISO/IEC 13818-1:1996/Amd.7:2000 – Transport of MPEG-4 streams over MPEG-2 systems.* This amendment specifies how to transport MPEG-4 elementary streams over MPEG-2 systems. MPEG-4 systems defines its own synchronization layer, but this amendment also allows the use of MPEG-2 systems based synchronization for MPEG-4 video and audio streams.
- Amendments to the second edition of MPEG-2 Systems (ISO/IEC 13818-1:2000):
  - *ISO/IEC 13818-1:2000/Amd.1:2003 – Carriage of metadata.* This amendment defines how to transport metadata over MPEG-2 systems and specifies a metadata time-line model for associating metadata to the audiovisual content. Furthermore, a method for metadata signalling and referencing is defined. Within the constraints specified in this amendment, carriage of MPEG-7 data over MPEG-2 systems is supported.
  - *ISO/IEC 13818-1:2000/Amd.2:2004 – Support of MPEG-2 IPMP.* This amendment defines how to support MPEG-2 IPMP, as specified in part 11 of MPEG-2.
  - *ISO/IEC 13818-1:2000/Amd.3:2004 – Transport of AVC video.* This amendment defines how to transport AVC video, specified as MPEG-4 part 10, and also known as H.264, over MPEG-2 systems.
  - *ISO/IEC 13818-1:2000/Amd.4:2005 – Usage of ISAN and V-ISAN in content labelling descriptor.* This amendment defines how to use ISAN and V-ISAN as defined in ISO 15706 for content identification in an MPEG-2 system stream.
  - *ISO/IEC 13818-1:2000/Amd.5:2005 – New audio profile and level signalling.* In response to new profiles defined for MPEG-2 and MPEG-4 AAC, this amendment defines how to signal the new profiles.
- Amendments to the third edition of MPEG-2 Systems (ISO/IEC 13818-1:2007):
  - *ISO/IEC 13818-1:2007/Amd.1:2007 – Transport of streaming text and lossless audio specified in MPEG-4.* This amendment defines how to transport MPEG-4 streaming text and MPEG-4 lossless audio over MPEG-2 systems, which may prove useful for applications requiring closed captioning/subtitling or very high quality audio, respectively.
  - *ISO/IEC 13818-1:2007/Amd.2:2008 – Transport of auxiliary video data, specified as ISO/IEC 23002-3 in MPEG-C.* This amendment specifies how to transport auxiliary video streams, in particular depth information for 2D pictures contained in an associated video stream.
  - *ISO/IEC 13818-1:2007/Amd.3:2009 – Transport of scalable video.* This amendment specifies how to transport SVC video, as defined in MPEG-4 part 10, over MPEG-2 system streams.
  - *ISO/IEC 13818-1:2007/Amd.4:2009 – Transport of multiview video.* This amendment specifies how to transport MVC video, as defined in MPEG-4 part 10, over MPEG-2 system streams.
  - *ISO/IEC 13818-1:2007/Amd.5:2011 – Transport of JPEG 2000 video.* This amendment specifies how to transport JPEG 2000 video, as defined in part 1 of JPEG 2000, over MPEG-2 system streams.

- o *ISO/IEC 13818-1:2007/Amd.6:2011 – Update AVC and MVC signalling*. This amendment updates the AVC descriptor and the signalling of MVC operation points.

All above amendments are included in the fourth edition of the MPEG-2 system standard, ISO/IEC 13818-1:2013. A number of amendments to this fourth edition are prepared, such as an amendment for carriage over MPEG-2 systems of HEVC video, as specified in ISO/IEC 23008-2, also published as H.265.

The support of most content formats in MPEG-2 systems is introduced in Chapters 3 and 4. Details on the decoding of the various content formats supported by MPEG-2 systems are found in Chapter 14.

References

1. Wikipedia (2013) Background information on MUSE is found at: http://en.wikipedia.org/wiki/Multiple_sub-Nyquist_sampling_encoding.
2. Wikipedia (2013) Background information on HDMAC is found at: http://en.wikipedia.org/wiki/HDMAC.
3. Wikipedia (2013) Background information on PAL-plus is found at: http://en.wikipedia.org/wiki/PAL-plus.
4. This digital HDTV system, later called Digicipher, was developed by General Instruments; a paper on Digicipher is found at: Paik, W. (1990) DigiCipher – all digital, channel compatible, HDTV broadcast system. *IEEE Transactions on Broadcasting*, **36** (4), 245–254.
5. See: Taylor, J. (2006) *DVD Demystified*, 3rd edn, Elsevier, ISBN: 978-0-0714-2396-0 and the associated link: http://www.dvddemystified.com/dvdfaq.html Information on the DVD standard and the DVD Forum can be accessed via: http://www.dvdforum.org/forum.shtml Information on DVD is also found at: http://en.wikipedia.org/wiki/DVD.
6. Chiarligione (2013) Actual information on MPEG is found at the MPEG Home page at: http://mpeg.chiariglione.org/ For example, overviews of MPEG standardization efforts and work items are found at: http://mpeg.chiariglione.org/standards; http://mpeg.chiariglione.org/content/terms-reference.

9

Layering in MPEG-2 Systems

Distinction of program streams and transport streams within MPEG-2 systems and PES packets as a common layer to carry content in both. Forming a program stream from PES packets. The structure of a transport stream and the use of transport packets. Navigation information on multiple programs and streams in a transport stream. Carriage of PES packets in transport packets. Consideration on the size of a transport packet. The use of sections for carrying information in a transport stream, and how to convey sections in transport packets.

9.1 Need for Program Streams and Transport Streams

MPEG-2 systems is build on top of MPEG-1 systems and supports the same basic functionality. This includes packetization and multiplexing of multiple video, audio or other elementary streams into a single system stream, synchronization of video and audio, buffer management and timing control. Also, the concept of a System Target Decoder model is used in MPEG-2 systems. As a result, MPEG-2 systems has a high level of commonality with MPEG-1 systems, but to properly address the various requirements on MPEG-2 systems, it did not prove practical to define MPEG-2 systems as a fully compatible extension of MPEG-1 systems.

The main reason why MPEG-2 systems is not fully compatible with MPEG-1 systems is found in the error prone networks used for digital television broadcast, such as satellite, cable and terrestrial. While those networks typically have a high bandwidth allowing to broadcast multiple programs simultaneously, they require error correction. The use of retransmission of erroneous data is not practical in a broadcast environment for several reasons. For example:

- Error bursts may impact a huge number of broadcast receivers.
- The required return channel is often absent in broadcast receivers.
- Retransmission and its associated processing may cause an additional delay that is unacceptable for broadcast applications.

As an alternative method for error correction, broadcast networks typically use Forward Error Correction (FEC), whereby redundant bytes are transmitted that allow a receiver to correct transmission errors by itself, without any data retransmission. By means of forward

Fundamentals and Evolution of MPEG-2 Systems: Paving the MPEG Road, First Edition. Jan van der Meer.
© 2014 John Wiley & Sons, Ltd. Published 2014 by John Wiley & Sons, Ltd.

error correction a certain maximum number of errors can be corrected; but if more errors occur in the data, for example due to a significant error burst, data will be lost. Forward error correction requires the broadcast data to be organized in small transport packets of constant size, in the order of magnitude of 200 bytes. Small transport packets allow cost-effective implementation of forward error correction in transmitters and receivers, and cause a less severe impact of packet loss and un-correctable errors in decoders.

Hence, for error correction purposes in broadcast environments, the use of small MPEG-2 system packets is required. As a consequence, when designing MPEG-2 systems, the MPEG system group needed to handle two conflicting requirements:

- For operation in error prone environments, small MPEG-2 system packets of constant size are required.
- For operation in relatively error free environments, reasonably large MPEG-2 system packets are required to allow system packet parsing in software.

To address both requirements, MPEG decided to define two different system stream formats with maximum commonality, the MPEG-2 Program Stream (PS) format for use on relatively error free media such as optical discs and the MPEG-2 Transport Stream (TS) format for use on error prone media, such as broadcast channels. The commonality between both formats is largely determined by the common use of PES packets, the MPEG-2 equivalent of MPEG-1 packets.

9.2 PES Packets as a Common Layer

The PES packet structure is derived from and very similar to MPEG-1 system packets. Each PES packet starts with a PES header, followed by video, audio or other elementary stream data. A PES header commences with a start code prefix followed by a stream-ID. The same 24-bit start code prefix is used as in MPEG-1 system packets: a string of 23 '0's followed by a single '1', while also the one-byte stream-ID is derived from MPEG-1 systems. The PES header structure is discussed in more detail in Section 13.2.

Conceptually, PES packets are produced prior to forming an MPEG-2 program stream or transport stream.[1] An example is depicted in Figure 9.1; video and audio encoders produce an MPEG video and an MPEG audio stream, respectively. Both streams enter a PES packetizer to produce a series of video and audio PES packets. Each PES packet commences with a PES header, followed by a chunk of contiguous MPEG video or MPEG audio data, respectively, as delivered by the video or audio encoder.

To form a program stream, the PES packets enter the program stream multiplexer, together with any other program stream data. While program streams are designed to carry elementary streams of one program, all sharing the same time base, transport streams are required to be capable of carrying elementary streams of multiple programs, with program specific time bases. Hence, input to the transport stream multiplexer are not only the video and audio PES packets and the other program related data depicted in Figure 9.1, but also PES packets and data from other programs.

[1] A practical multiplexer may combine the forming of PES packets and the program or transport stream in a single operation.

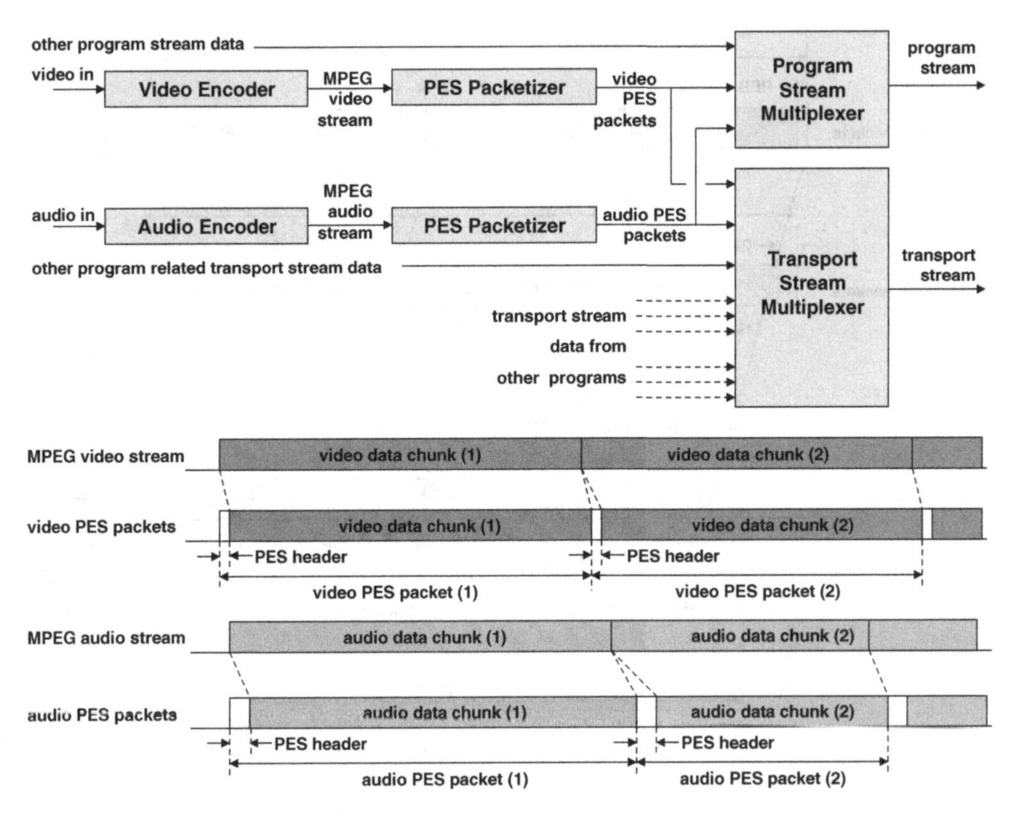

Figure 9.1 Producing PES packets prior to forming a program or transport stream

9.3 Program Streams

The forming of an MPEG-2 program stream is depicted in Figure 9.2. The program stream format is defined for usage by relatively error free media such as optical discs, and its structure is similar to the MPEG-1 system stream format. Each MPEG-2 program stream consists of a concatenation of packs, each containing zero or more PES packets, and is concluded by an MPEG program end code.

Each pack commences with a pack header; other than in MPEG-1, the pack header may include a system header. The first pack header of the program stream must carry a system header that may also be included in other packs. The system header contains a summary of important program stream parameters and may be used by a receiver to determine whether it is capable of decoding the entire program stream. In Figure 9.2, the first pack consists of a pack header, a system header, two video PES packets and one audio PES packet. The second pack header does not carry a system header. In the elementary streams of Figure 9.2, the duration in time of the video and audio data chunks is about the same, and so the video PES packets contain more bytes than the audio PES packets due to the higher bitrate for video.

A minor clarification: in Figure 9.2, it is assumed that the PES headers for the video and the audio stream have about the same size; their transport in the program stream takes

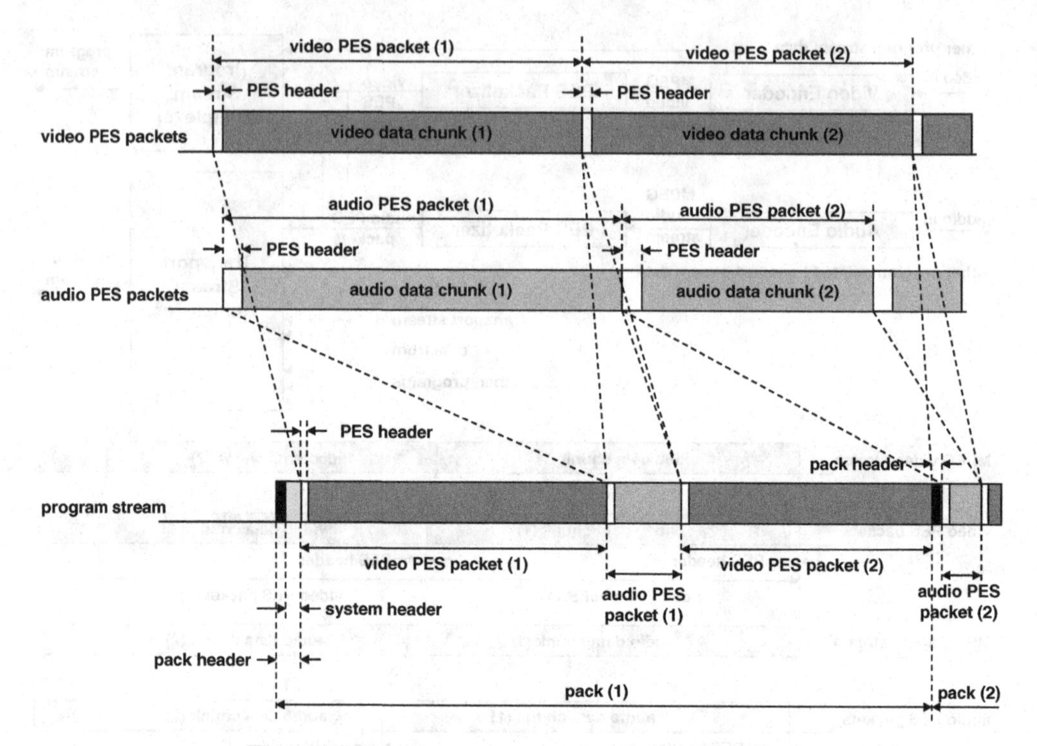

Figure 9.2 Forming a program stream from PES packets

about the same time. However, due to the difference in bitrate, the transport of a PES header takes more time in the stream of audio PES packets than in the stream of video PES packets.

Similar to the MPEG-1 packs discussed in Chapter 6, a sample of the time base of the program, the System Time Clock, STC, is carried in each pack header, along with an encoded value of the program-mux-rate. The STC sample is called the System Clock Reference SCR, the same name as applied in MPEG-1. The SCR specifies the value of the STC when the last byte of the SCR field is intended to arrive at the input of the MPEG-2 System Target Decoder, discussed in Chapter 12. For the pack in which header it is encoded, the program-mux-rate specifies the bitrate at which the bytes of the pack arrive at the input of the decoder. All of the above is very similar to MPEG-1 packs; the main difference is that the STC has a frequency of 27 MHz, instead of 90 kHz, as discussed in Section 6.6. As a result, the byte delivery of the MPEG-2 program stream is very similar to a MPEG-1 system stream. Examples of MPEG-2 program streams with a constant, bursty and variable bitrate are given in Figure 9.3. Note that these examples are very similar to the ones depicted in Figure 7.6, except that the MPEG-1 system parameters are replaced by MPEG-2 system parameters.

As in MPEG-1 systems, each pack header, system header and PES header commences with a start code, followed by a one-byte stream-ID value signalling the type of header. A program end code is a start code followed by the stream-ID value indicated in Table 9.1, without any

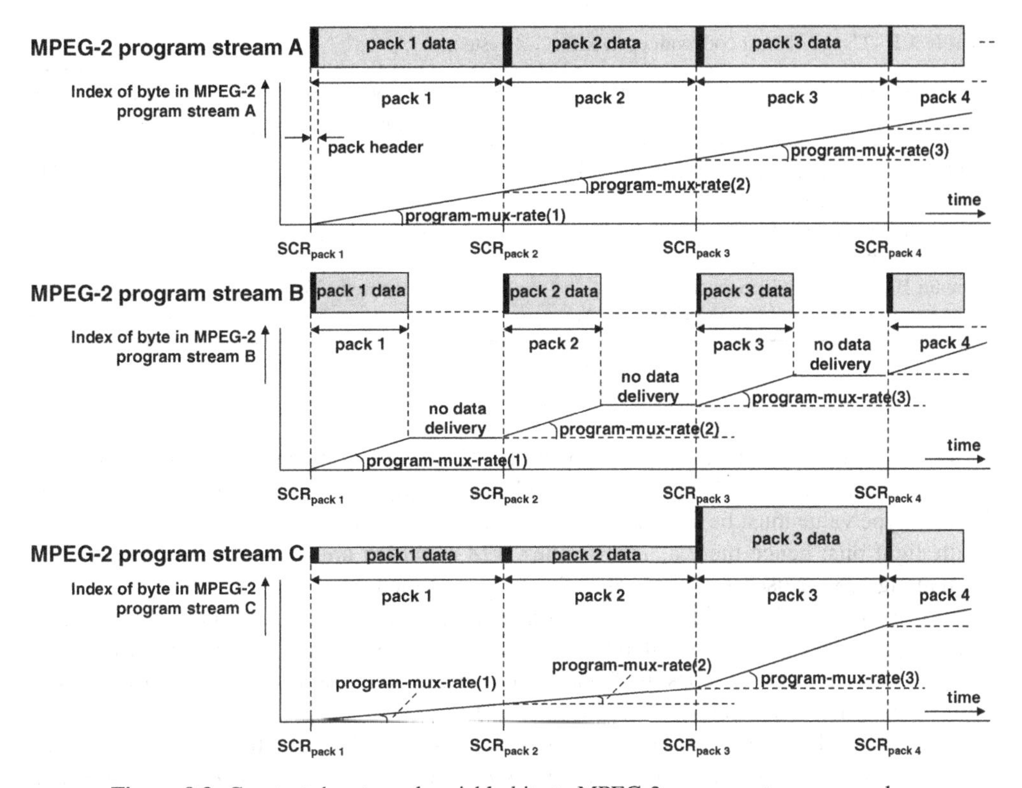

Figure 9.3 Constant, bursty and variable bitrate MPEG-2 program stream examples

further trailing bytes. Program stream decoders have knowledge on the number of bytes contained in a pack header, a system header and a PES packet, and therefore the use of start codes allows parsers in decoders to synchronize to the data structure in a programs stream. By verifying the presence of a start code after the number of contained bytes in the pack header, system header or PES packet, it can be verified whether data loss occurred. See Section 7.3 and Figures 7.2 and 7.3 for a more extensive discussion on this subject and the use of the program end code.

MPEG-2 video is defined as an extension of MPEG-1 video and uses the same start codes and start code types as MPEG-1 video, as listed in Table 3.1. The system start codes made available to MPEG-1 systems (see Table 6.1) remain available to MPEG-2 systems. The MPEG system group decided therefore to copy the use of start code type values in MPEG-2 systems from MPEG-1 systems; however, the MPEG-1 terms are replaced by MPEG-2 terms (see Table 9.1). Furthermore, an important change was made to the semantics of some PES packet stream-IDs due to the fact that, compared to MPEG-1, in MPEG-2 systems many more MPEG defined formats need to be supported. Therefore in MPEG-2 systems the semantics for the 16 MPEG-1 video stream-ID and the 32 MPEG-1 audio stream-ID values were changed into 16 'defined by MPEG' video formats and 32 'defined by MPEG' audio formats. The specification of the actually applied coding format was thereby left to a new parameter, called stream-type, suitable for use throughout the MPEG-2 system specification.

Table 9.1 Usage of start code values in MPEG-2 systems

Start code type	Start code value (decimal)	Start code value (hexadecimal)
Video start codes (for details see Table 3.1)	0 through 184	00 through B8
MPEG program end code	185	B9
Program stream pack header	186	BA
Program stream system header	187	BB
Stream ID values for PES packet header usage (see Section 13.2 and Chapter 14)	188 through 255	BC through FF

For each content format supported by MPEG-2 systems, a stream-type value is assigned. Hence, whenever MPEG-2 systems is extended with support for a new content format, a new stream-type value must be assigned to such new content format. The stream-type is encoded with eight bits; hence there are 256 values, 128 of which are defined by MPEG, and the remaining 128 are user private. Table 9.2 lists the assigned stream-type values for carriage of MPEG-1 and MPEG-2 video and audio and for carriage of private data; other assigned stream-type values are discussed in Chapter 14 for each content format supported by MPEG-2 systems. Stream-type values 0x05 and 0x06 indicate two different methods for carriage of private data: in sections and in PES packets; sections and transport of sections are discussed in Sections 9.4.4 and 9.4.5. Some content formats use multiple transport mechanisms, each signalled by a separate stream-type value, as discussed in Chapter 14. For metadata even five different transport mechanisms are defined (see Sections 14.11). In a program stream, the stream-type is assigned to an elementary stream by means of the Program Stream Map (PSM); the PSM may also contain descriptive information on the elementary streams in the program stream and on the program stream itself. The PSM is discussed in Section 13.4.

Table 9.2 Stream-type assignments for signalling content formats

Stream-type value	Signalled content format
0×00	Reserved
0×01	MPEG-1 video
0×02	MPEG-2 video or MPEG-1 constrained parameter video
0×03	MPEG-1 audio
0×04	MPEG-2 audio
0×05	Private data carried in sections
0×06	Private data carried in PES packets
0×07–0×23	Discussed per content format in Chapter 14; see Table 14.30
0×24–0×7E	Reserved for future use by MPEG-2 systems (as of 2013)
0×7F	Discussed in Section 14.3.5
0×80–0×FF	User private

Program streams are similar to, but not fully compatible with MPEG-1 system streams. To meet the MPEG-2 system requirements, changes were needed in the program stream pack header and in the PES header that could not be implemented in a compatible manner within the data structure of MPEG-1 packs and packet headers. As a consequence, an MPEG-1 system stream is not a valid program stream. However, as the main application focus of both MPEG-1 system streams and MPEG-2 program streams is on optical media, the MPEG system group mandated that, within its video and audio decoding capabilities, each MPEG-2 system compliant program stream decoder must also be capable of decoding MPEG-1 system streams.

Note though that the above requirement may be insufficient to ensure that each MPEG-2 compliant playback device is also capable to playback a MPEG-1 system stream. In practice, an MPEG-1 system stream is usually available on a Video CD, and a program stream on a DVD. Hence, in practice, an MPEG-1 system stream can only be played in an MPEG-2 program stream player, if that player is also capable of playing Video CDs.

9.4 Transport Streams

9.4.1 Transport Packets

A transport stream is a concatenation of transport packets of a constant (small) size, each commencing with a transport packet header and followed by payload bytes (see Figure 9.4). Each transport packet header consists of a fixed and a variable part. The fixed part contains four bytes and commences with a sync byte that has the fixed value 0x47. The fixed part of the

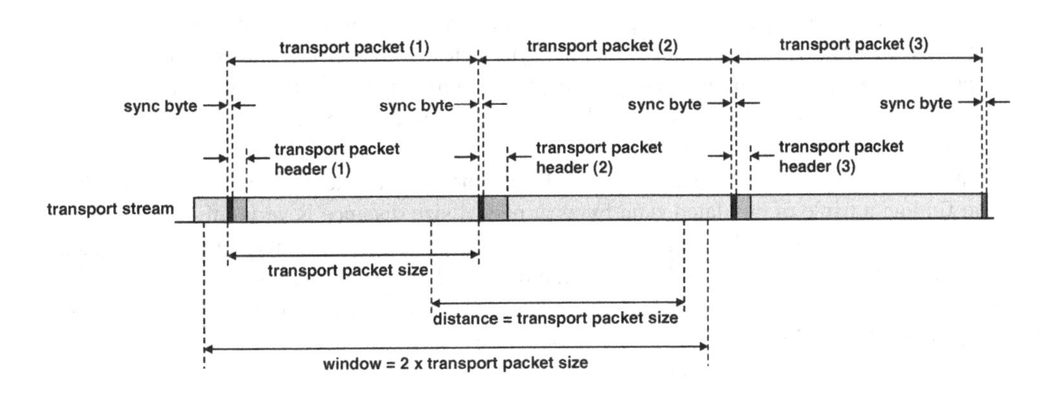

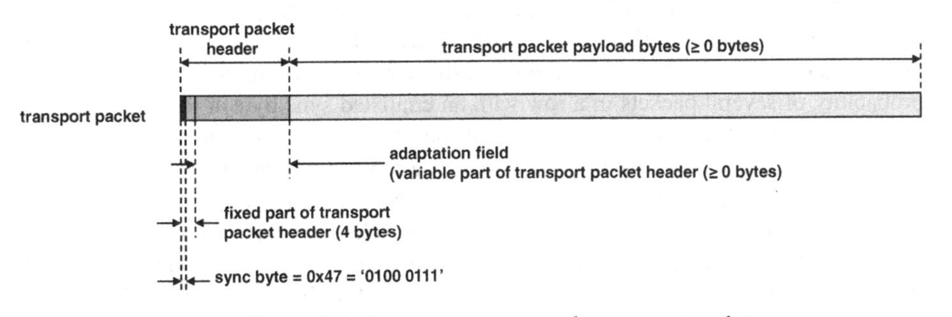

Figure 9.4 A transport stream and a transport packet

header is followed by the variable part, if present. The variable part is called the adaptation field and its presence or absence is signalled in the fixed part. The variable part may take all remaining bytes of the transport packet, in which case there is no payload. In practice, the adaptation field is often absent or takes only a few bytes.

The fixed part of the header carries the PID field; the PID is the packet identifier, used to signal the type of data carried in the transport packet payload. In packets with the same PID value, the payload carries one specific data stream. The PID field is coded with 13 bits, allowing the use of 8192 different PID values. The content carried in a transport stream is organized in programs. When decoding a program, only transport packets with data belonging to that program are processed. Based on the encoded PID values, the decoder selects transport packets for the program to be decoded; all other transport packets may be ignored by the decoder. The use of PID values for carrying elementary streams and other data for a program is discussed in Section 9.4.4.

By means of the sync byte, parsers of transport stream decoders can synchronize to the data structure of transport streams. For example, when accessing a transport stream, the parser may look for two matching bytes, each with the value of the sync byte, at a mutual distance equal to the size of a transport packet. Within any window equal to two times the transport packet size there are two sync bytes present (see Figure 9.4). A parser that is aware of byte alignment, for example by external means, such as a channel decoder, can do this verification byte by byte, but as long as no byte alignment is achieved yet, the sync byte verification must be done bit by bit.

The value of the sync byte is not unique in the transport stream; other bytes in the transport stream may take the same value. When a parser looks within a window of two packet sizes for a couple of matching sync bytes at packet size distance, and only identifies a single couple, which is most likely, then these bytes are the sync bytes, each indicating the start of a transport packet. However, when multiple couples of such matching bytes are found within the window, then sync byte emulation at packet size distance is identified, upon which further verification is needed to find the real sync bytes.

Within the payload, finding a pair of emulated sync bytes at packet size distance is unlikely, and finding a triple of emulated sync bytes at packet size distance is very unlikely. However, the encoding of fields in the packet header may cause a periodic sync byte emulation at packet size distance. Therefore, encoders are recommended to avoid sync byte emulation when encoding fields in the packet header. In particular, in the choice of PID values the periodic emulation of sync bytes at the same position in the packet header should be avoided, both within the PID field and in conjunction with adjacent fields.

Avoiding sync byte emulation is a recommendation, not a mandatory requirement, because in many practical situations an emulated sync byte in one of the packet headers does not pose a problem. A transport stream carries usually many different elementary streams, each contained in transport packets with their own PID value, causing a rather random packet sequence with a low probability of several packets in a row with an emulated sync byte at the same position. However, the recommendation should be taken serious; for example, after re-multiplexing into a transport stream with a much lower bitrate, the above probability may increase significantly.

The value of the sync byte was selected from the perspective of tolerance for transmission errors. This tolerance depends on the bit pattern of the selected value. Various studies have been performed to measure this tolerance; an overview provided by Bylanski and Ingram [1], showed that for use in a sync byte, the bit pattern '0100 0111', corresponding to the hexadecimal value 0x47, exhibits a good tolerance. The MPEG systems group selected therefore this value for the sync byte. In this context it should be noted that the complementary and reverse bit patterns have the same tolerance for transmission errors, but these values were considered to have a slightly higher likelihood of being emulated in the transport packet header.

One of the PID values (see Section 9.4.4) is assigned to signal null packets. Null packets carry no meaningful data and are intended for padding of transport streams. Padding may for example be used when the bitrate of an MPEG-2 system stream is (slightly) higher than required to carry the constituent streams that form the transport stream. See Section 11.6 for a general discussion on padding. Padding packets may be inserted or deleted during transport and therefore delivery of null packets to the decoder cannot be assumed.

9.4.2 Conveying PES Packets in Transport Packets

The PES packets of each elementary stream are carried in transport packets with the same PID value. Because a PES packet is usually much larger than a transport packet, PES packets are usually fragmented prior to carriage in transport packets. Contiguous PES packet fragments are carried unaltered in the payload of transport packets. Hence, the receiver can reconstruct the original PES packets by concatenating the payload data of the transport packets carrying the PES packet data.

The PES header carries important information, such as time stamps, for the decoding of the elementary stream contained in the PES packet. To simplify the access to this information, the MPEG systems group decided to carry PES packets aligned with the payload of transport packets, that is the first byte of each PES header must be the first byte in the payload of a transport packet. As a consequence, the last byte of a PES packet must be the last byte of the payload of a transport packet.

As an example, Figure 9.5 depicts the fragmentation of a video PES packet into eight transport packets. The video PES packet in this example is split into fragments fv1 up to fv8. Note that fv1 also contains the PES header and that fv8 is significantly smaller than the other fragments. The size of fragments fv1 up to fv7 is chosen so as to match the number of available payload bytes after the foreseen transport packet header. Obviously, each fragment in the transport packet is preceded by the packet header. Fragment fv8 is smaller than the available number of payload bytes, and therefore the adaptation field in the transport header is extended by stuffing bytes to ensure that the last byte of fragment fv8 is the last byte of the payload of the transport packet. The example in Figure 9.5 shows the use of stuffing bytes in the adaptation field for achieving alignment between PES packets and transport packets. Note though that in practice the PES packet size is usually optimized for carriage in transport packets, so as to avoid the efficiency loss due to the use of stuffing bytes.

For the video and audio PES packets shown in Figure 9.1, the carriage of the formed transport packets in a transport stream is presented in Figure 9.6. Video PES packet 1 is carried in eight transport packets (as in Figure 9.5), video PES packet 2 in seven transport packets. In the elementary stream, the duration of the video and audio PES packets in Figure 9.6 is about the same; the audio bitrate is much lower than the video bitrate and because the size of each fragment carried in a transport packet is roughly the same, the audio PES packets in Figure 9.6 are split in only two fragments. In the transport stream, the resulting transport packets are interleaved with the other transport packets that constitute to the forming of the transport stream. In the adaptation field of the transport packets carrying fv8, fv15 and fa4, stuffing bytes are included to ensure that the last byte of the fragment is the last byte of the payload of the

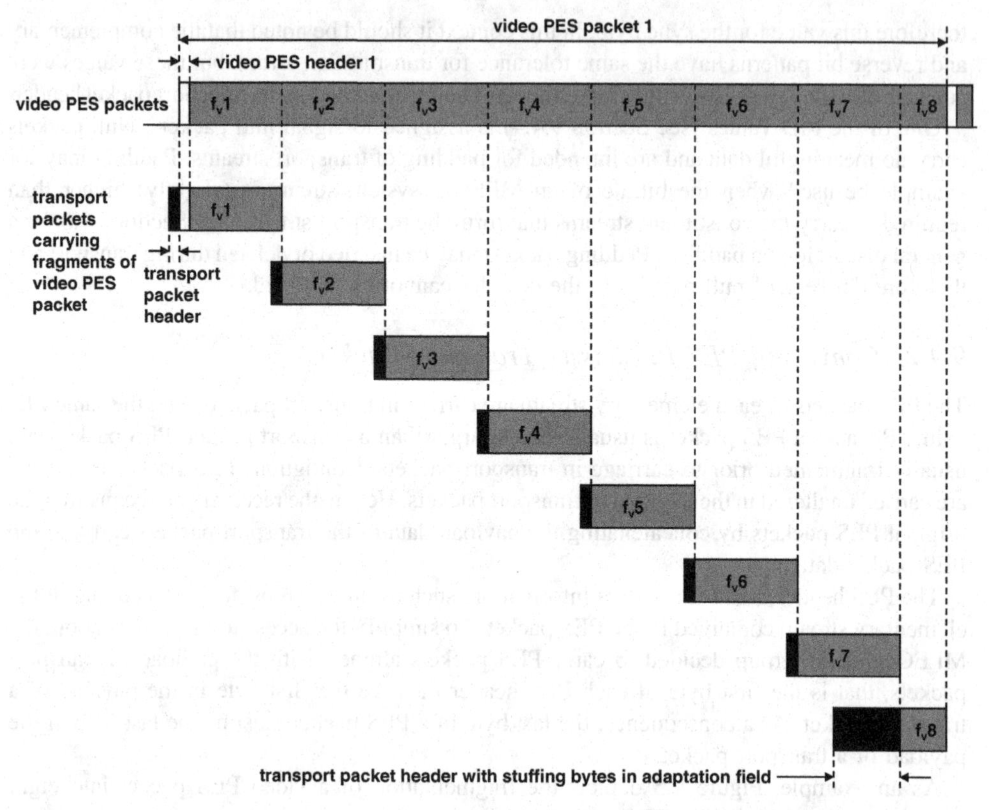

Figure 9.5 Carriage of video PES packet fragments in transport packets

transport packet. Note that the size of audio PES packet 2 is chosen so that no stuffing is required in the header of the transport packet carrying fa2. Which would have made sense for the other PES packets too.

9.4.3 The Size of Transport Packets

The choice of the transport packet size has been subject to much debate in the MPEG systems group. Initially there were proposals for multiple sizes and even a flexible size indicated by a parameter, but soon broad support evolved for defining one fixed transport packet size only. A fixed transport packet size is the only practical solution if multiple networks are involved to convey a transport stream. When each network would apply its own size, then at each network node, a complete re-packetization would be needed, requiring almost a complete system encoder with knowledge of the data carried in the transport packets.

To agree among the MPEG system experts on the need for a fixed size was therefore easy, but to choose its actual value proved far more difficult. The challenge was to find a size suitable for transport across all target networks, in particular networks for digital TV broadcasting and

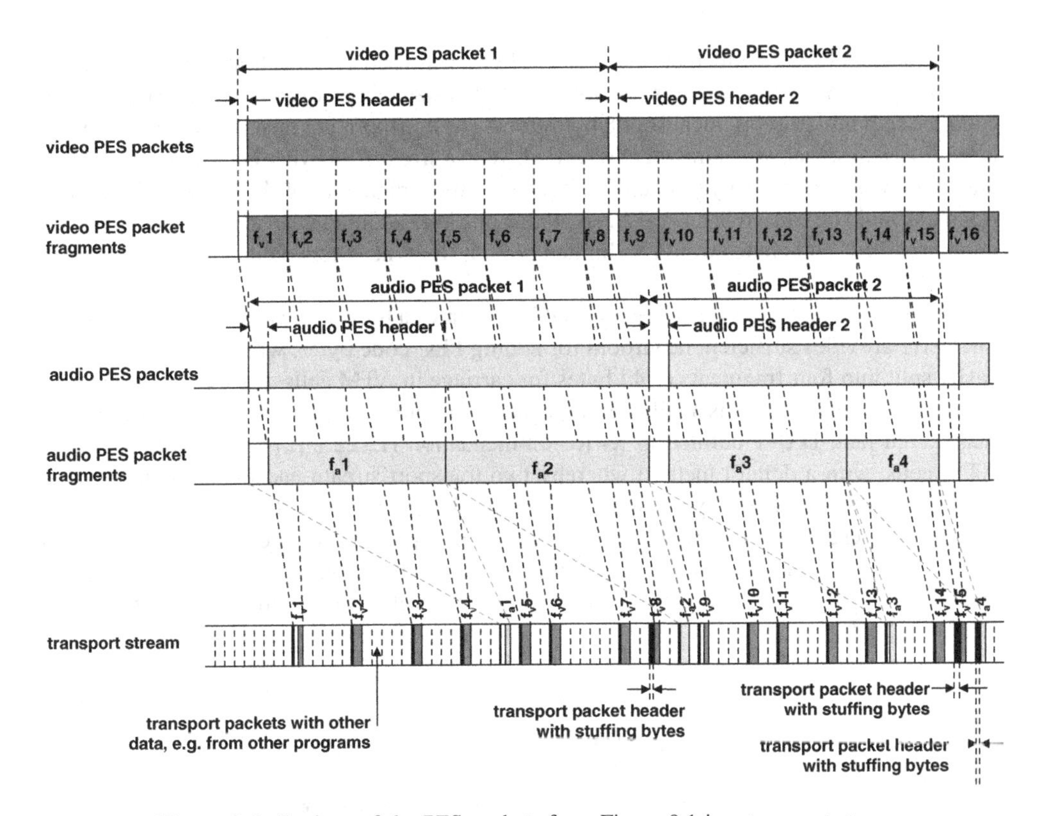

Figure 9.6 Carriage of the PES packets from Figure 9.1 in a transport stream

ATM networks. Though ATM did not find any significant deployment on the market in later years, interfacing with ATM played an important role in defining the size of transport packets. Various issues required a trade-off:

- For a practical and cost-effective implementation of forward error correction in digital TV receivers, the size of a transport packet plus the added FEC code bytes should be at most 255 bytes. How many FEC code bytes are applied depends on the application and/or on the channel; the size of transport packets should leave sufficient flexibility for applications to add FEC code bytes.
- For transport across ATM networks, transport packets are to be fragmented for carriage in ATM cells with a payload of 48 bytes; it should be conveniently possible to fragment a transport stream for carriage in ATM cells, as well as to re-compose a transport stream from ATM cells after such transport.
- Alignment of PES packets with the beginning of the payload of transport packets may require stuffing in the transport packet header. The loss of bandwidth due to such stuffing becomes more severe for larger packet sizes.
- With smaller transport packets, the overhead caused by the packet header and by adding FEC code bytes increases.

Several sizes were considered. A size of 132 bytes was used by some trial systems, but has a relatively poor ATM interfacing performance. With a size of 192 bytes, each transport packet could be split into four fragments of 48 bytes for ATM transport, but then the ATM cell header is to be used to signal which fragment is carried. With a size of 236 bytes, each transport packet, after removing the sync byte, could be split into five fragments of 47 bytes, so that each fragment could be carried in an ATM cell, with a one byte additional header (or trailer), but the number of FEC code bytes would be limited to 20 if the sync byte would be removed prior to channel coding.

However, all above options were rejected in favour of a transport packet size of 188 bytes. Such size provides sufficient headroom for adding FEC code bytes, while the packet can be easily split into four fragments of 47 bytes for carriage in ATM cells. For such carriage, one byte per 47-byte fragment is available for header information, which can be used in a flexible manner. In fact, ITU-T defined in its Recommendation H.222.1 [2] a sub-layer on top of ATM cells, with a default method whereby two transport stream packets are concatenated and followed by an eight-byte trailer for carriage in the payload of eight ATM cells. An example is presented in Figure 9.7; every two transport packets are interleaved with an eight byte ATM sub-layer trailer, upon which the two transport packets and trailer are split into fragments of 48 bytes for carriage in eight ATM cells. For more details see ITU-T Rec H.222.1.

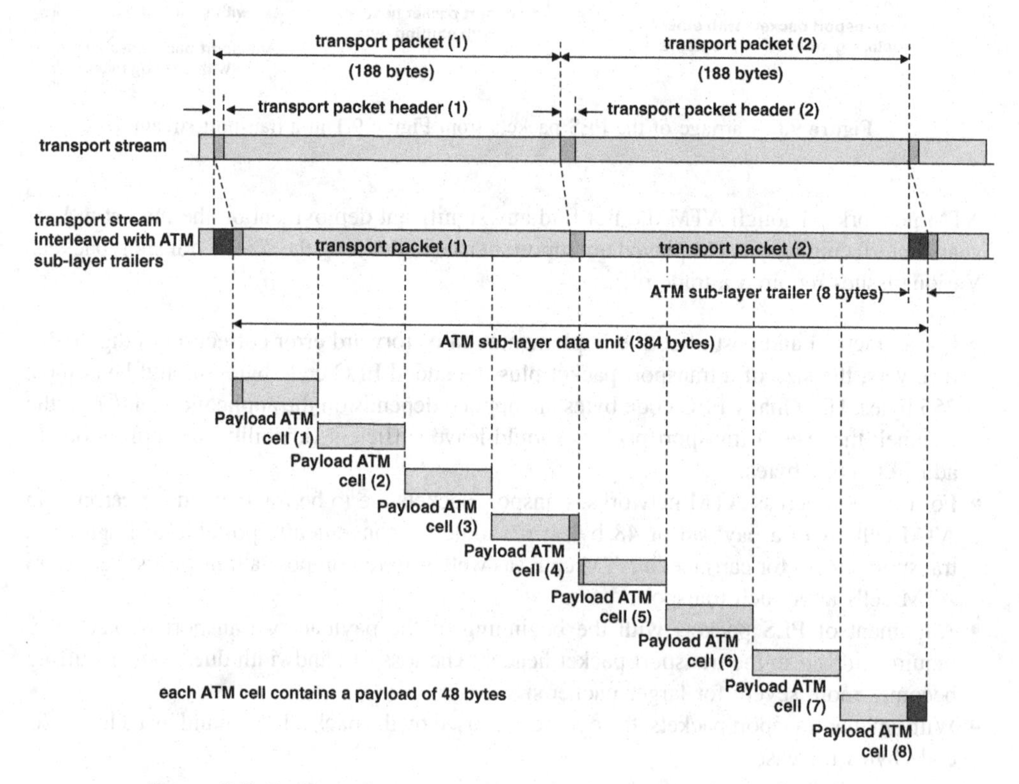

Figure 9.7 Carriage of two transport packets in payload of eight ATM cells

For broadcast across terrestrial, cable and satellite networks, the FEC coding and concatenation of transport packets with FEC code bytes is performed as an integral part of channel coding. After this operation, any supplementary channel coding takes place; for example, byte interleaving may be applied to spread burst errors. After the channel coding, digital modulation [3] is performed, for example QPSK (Quadrature Phase-Shift Keying) [4] or QAM (Quadrature Amplitude Modulation) [5], followed by transmission across the broadcast channel. The receiver demodulates the received input signal and applies the required channel decoding, thereby using the FEC code bytes in an effort to correct errors in the received transport packets, if any.

The number of errors that can be corrected per transport packet depends on the assigned number of FEC code bytes per transport packet. When designing a forward error correction system, the maximum number of errors per transport packet to be corrected is important, but also the overhead of the FEC code bytes and cost-effectiveness of implementations, in particular at receivers. Which FEC coding to use and how many FEC code bytes to concatenate with a transport packet is often channel specific and usually defined by application standardization bodies such as DVB, ATSC and ARIB. However, in any case, when designing receivers, it should be taken into account that transport packets may contain un-correctable errors; for example, under severe weather conditions serious error bursts may occur, or even complete loss of the broadcast signal. MPEG-2 system receivers therefore need to be designed to handle erroneous and lost transport packets.

A conceptual[2] example of the above processing using FEC code bytes is given in Figure 9.8. The broadcast data at the input of the transmitter consists of a series of transport packets. Each packet is concatenated by a number of FEC code bytes to form a broadcast data block. After further channel coding, modulation, broadcasting at the desired frequency, demodulation and some initial channel decoding, the received broadcast data blocks are available after a certain broadcast delay at the input of the FEC decoder. In case of errors, the FEC decoder uses the FEC code bytes in an effort to correct the broadcast (bit) errors in packets. For example, in Figure 9.8, FEC code bytes (1) are used to correct errors in transport packet tp(1). As long as no un-correctable errors occur, the FEC decoder reconstructs at its output the same series of transport packets that was input to the channel encoding.

The above approach with network specific channel coding and a network independent transport stream format with transport packets of 188 bytes in length ensures seamless transport across networks with very minimal processing in the network. At transport stream level no channel adaptations are required. Adding, processing and removing FEC code bytes, as well as splitting and joining ATM cell fragments can be performed without any knowledge of the data carried in the transport packets. Which ensures that an important condition for the widespread use of MPEG technology is met.

9.4.4 *Multiple Programs, PSI, Descriptors and Sections*

Within a transport stream, multiple programs can be carried. A program contains one or more program elements, such as video, audio and private streams, as well as other data related to that program. A program element is not required to have a time base, but those that do, share the

[2] In practice, the FEC encoding and decoding may be more closely integrated in channel coding than suggested in Figure 9.8.

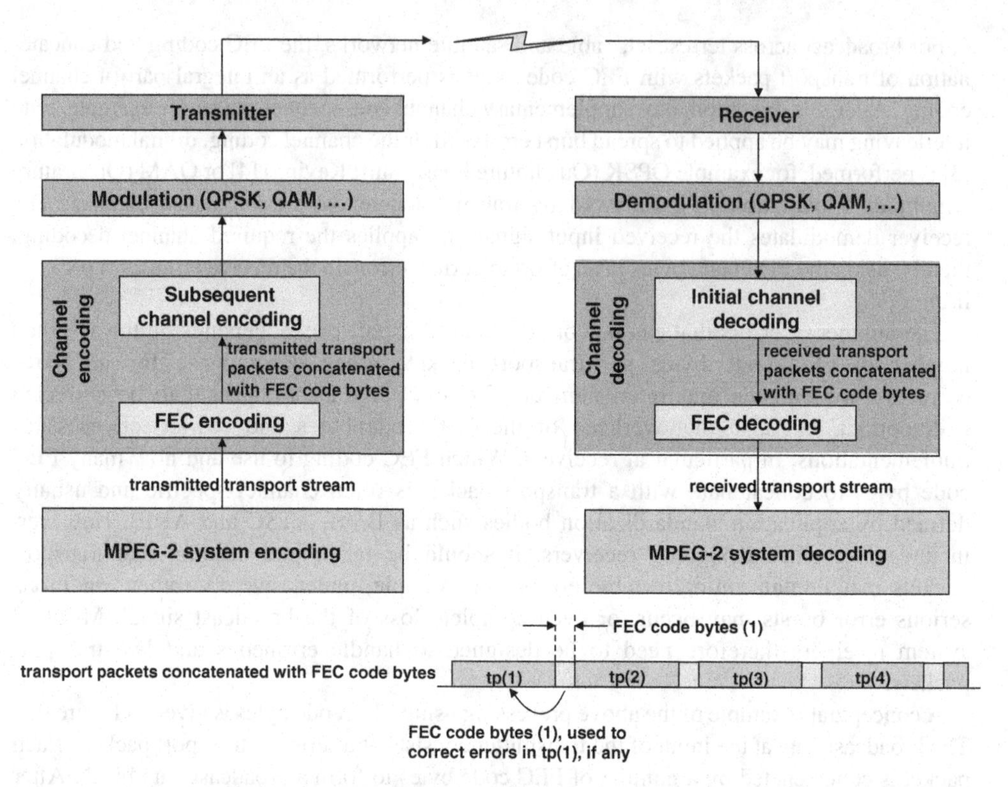

Figure 9.8 Example of adding error-correction code bytes for broadcast

same program time base. Each program has its own time base,[3] the System Time Clock (STC). Note though that the same STC may be shared by multiple programs. Information on the STC is carried by STC samples encoded in the Program Clock Reference (PCR) field in the adaptation field of the header of transport packets. The PCR specifies the value of the STC at the intended time of arrival of the final byte of the PCR field at the input of the MPEG-2 System Target Decoder, discussed in Chapter 11.

The frequency at which a PCR sample is present in the adaptation field is left to encoders, but a PCR must be encoded at least once every 0.1 s. Each transport packet carries 188×8 bits, about 1500 bits. Hence, if a program in a transport stream, including video, audio, other program elements and system overhead, consumes a total bitrate of 3 Mb/s, then the transport stream conveys in one second about 2000 transport packets with data of that program. Using those transport packets for carriage of the PCR, means that out of 200 transport packets with data of that program, at least one must carry a PCR.

By means of the encoded PCR values, the receiver can reconstruct the STC of a program. Figure 9.9 depicts an example of a transport stream carrying transport packets tp_1, tp_2, . . . , tp_{13}

[3] A transport stream may carry programs with only private streams that do not have an MPEG defined time base; such program may not have a time base at all. However, MPEG systems does not specify timing related constraints for programs without an MPEG defined time base, and consequently, a transport stream only offers a transparent data channel for such programs, without imposing requirements for its decoding.

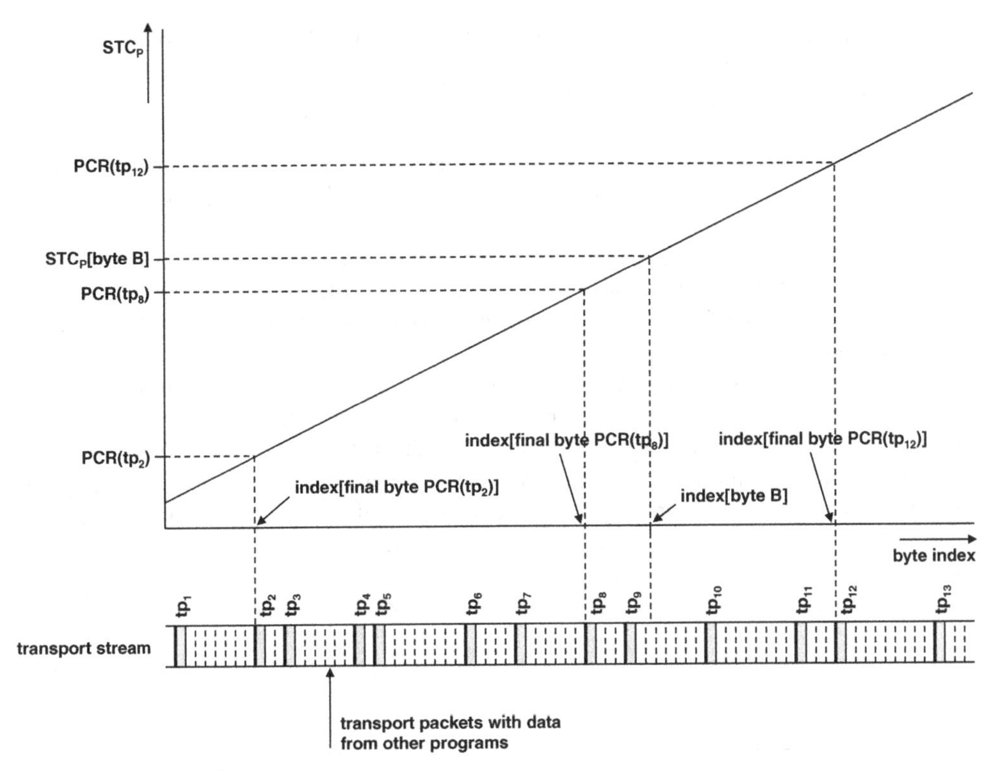

Figure 9.9 Constructing the time base of a program from coded PCR values

with data of a certain program, interleaved with transport packets with data from other programs. The STC of the program is denoted as STC_P. For the purpose of this example, in tp_2, tp_8 and tp_{12} the PCR field is encoded in the adaptation field of the transport packet header. The encoded values $PCR(tp_2)$, $PCR(tp_8)$ and $PCR(tp_{12})$, indicate the intended arrival time of the final byte of the PCR field. MPEG specifies that the bitrate of a transport stream is constant between two PCR values, so that a transport stream has a piecewise constant bitrate. As a consequence, the intended arrival time of any byte in the transport stream between two consecutive final bytes of PCR fields can be determined from the index of that byte and the encoded PCR values. For each byte in between two subsequent PCRs in a transport stream the transport stream rate R is calculated from the number of bytes in between the previous and the next PCR and the encoded values for the previous and the next PCR, as follows:

$$R = \frac{[index(final\ byte\ next\ PCR) - index(final\ byte\ previous\ PCR)]}{[next\ PCR - previous\ PCR]}$$

Note that in above formula, the rate R is expressed in bytes/s; to find the usual expression in bits/s, multiplication by eight is needed.

For each byte in between the previous PCR and the next PCR the intended arrival time can be determined from its index and R. For example, for byte B in Figure 9.9, the intended arrival time STC[byte B] is calculated as follows:

$$\begin{aligned} STC[byte\,B] &= previous\,PCR + [index(byte\;B) - index(final\,byte\,previous\,PCR)]/R \\ &= PCR(tp8) + [index(byte\,B) - index\{final\,byte\,PCR(tp8)\}]/R \end{aligned}$$

While the above formula applies in all cases, it should be noted that in practice most, if not all, broadcast transport streams have a constant bitrate, which results in the simple linear relationship between the time base of the program and the index of bytes in the transport stream depicted in Figure 9.9.

As each program has its own time base, within a transport stream there may be as many time bases as programs. For decoding a program, the associated time base is applied to control all timing events with mathematical accuracy, including the delivery of the transport stream to the decoder, the input buffering, the decoding and the presentation. For example, in Figure 9.10 a transport stream is depicted carrying transport packets with data of program 1, transport packets with data of program 2 and transport packets with data of other programs. In Figure 9.10, transport packets tp_1, tp_2, . . . , tp_{13} carry data of program 1, while transport packets tp_a, tp_b, . . . , tp_m are carrying data of program 2. When program 1 is decoded, the delivery of a

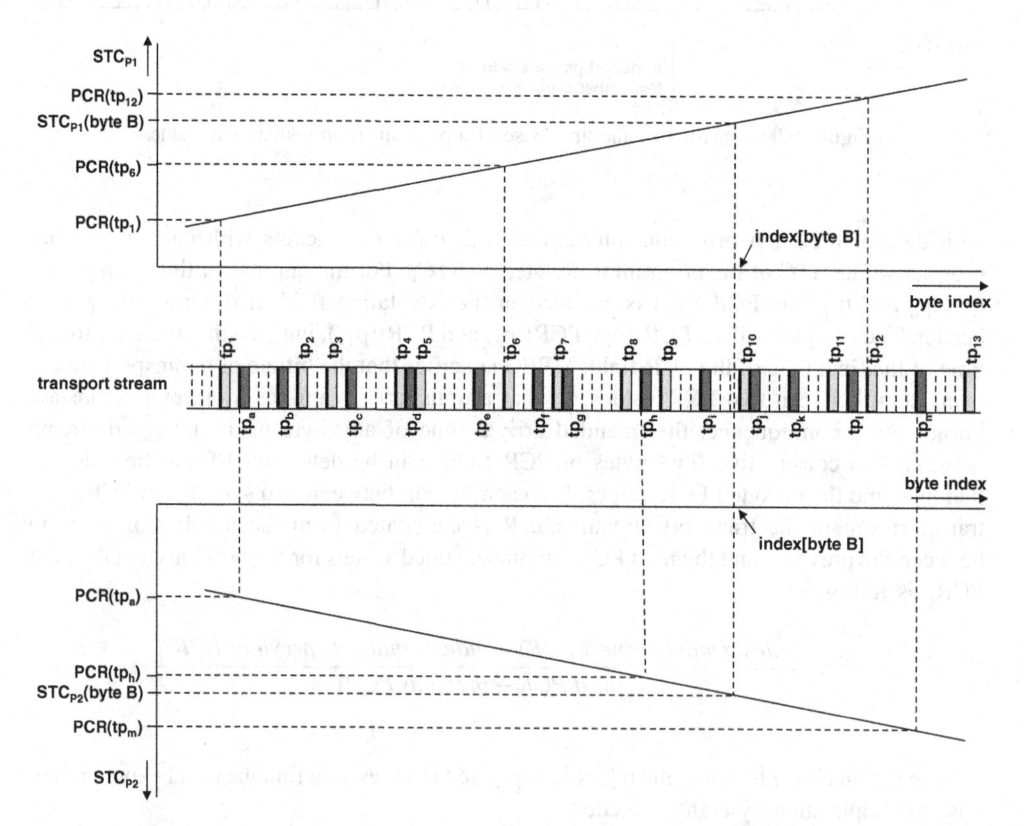

Figure 9.10 Timing of byte delivery dependent on program specific time base

each byte in the transport stream is expressed in units of the time base STC_{P1}, while the delivery of the same bytes is expressed in units of the time base STC_{P2} when program 2 is decoded. See Figure 9.10. At any time, the difference between the time bases of two programs can be any random value within the periodic range of the STC, more than 24 hours. Typically, the program time bases conveyed within a transport streams are mutually unrelated and not locked.

MPEG-2 system streams are fully self-contained; hence all information needed for the decoding of a program within a transport stream must be available in that stream. So as to identify the various programs that may be conveyed in a transport stream, a program-number is assigned to each program. This program-number is used in the so-called Program Specific Information (PSI) to provide a structure for transport stream navigation, a kind of table of content for the transport stream.

The program specific information consists of a set of tables; each table is carried in transport packets identified by a certain PID value. For accessing the programs in a transport stream, two tables are most important: the Program Map Table, PMT, and the Program Association Table, PAT. The Program Map Table provides for each program in a transport stream the contained program elements, such as video or audio streams, as well as for each program element, the stream-type and the elementary-PID value of the transport packets conveying that program element. By providing the PCR-PID, the PMT also specifies which transport packets carry the PCR samples for the program. Furthermore, the PMT conveys (descriptive) information on the program and on each program element. The program map information for a specific program is referred to as the program definition. The Program Map Table is the complete collection of the program definitions of all programs carried in a transport stream.

For each program, identified by its program-number, the program definition is carried in transport packets with a unique PID value selected by the encoder; this program-map-PID value is found in the Program Association Table. The program-number zero cannot be used for program identification, but is reserved to signal the network-PID for carriage of the Network Information Table (NIT); the NIT is discussed further below. The PAT must provide a complete list of all programs contained in a transport stream, with for each program the program-map-PID value. See Figure 9.11. In each transport stream, the PAT must be present and carried in transport packets with a PID value equal to zero, and is therefore easy to find in any transport stream. The PAT also provides the transport-stream-ID, that may be encoded with a value that may serve as a numeric label to identify transport streams within an application domain.

Hence, to start the decoding of a certain program, a receiver first needs to access the PAT to find the program-map-PID of that program. Upon retrieving the program definition from the transport packets with the program-map-PID, the PCR-PID and the PIDs of the transport packets carrying the elementary streams of that program are found, so that the data needed to decode the program can be retrieved from the transport stream. Also descriptive information on the program and on its elementary streams may be found.

Program information and elementary stream information can be included in each program definition by means of descriptors. In MPEG-2 systems, a wide variety of descriptors is defined to extend the definitions of programs and program elements. For instance, a descriptor associated with an audio stream may provide language information. A descriptor may also provide essential information needed by a decoder to determine whether or not the associated stream meets the capabilities of the decoder, without accessing the coded data itself. To ensure a high quality of service, it is desirable to determine upfront whether a stream can be decoded, instead of starting the decoding with a risk of failure after some time, for example because of

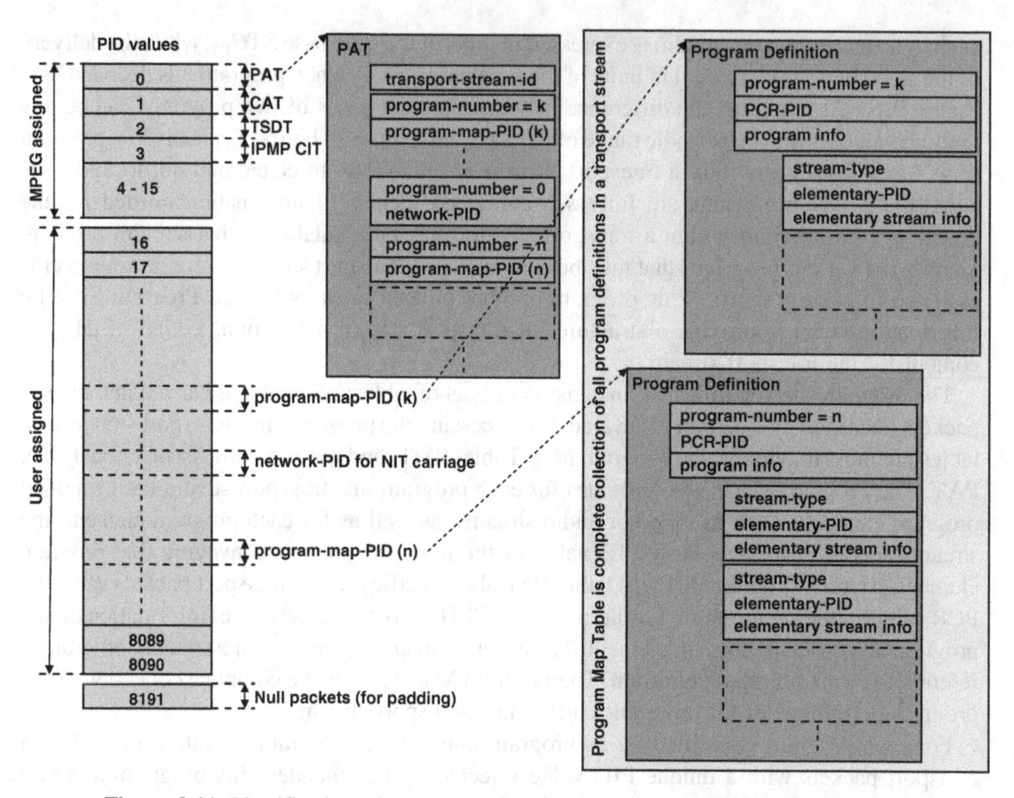

Figure 9.11 Identification of PID values in PAT and in program definitions in PMT

excessive buffer requirements or a too high bitrate. The program definition may also carry private descriptors, for example specified by ARIB, ATSC or DVB.

All descriptors share the same basic format. Each descriptor commences with an 8-bit descriptor tag; hence, in total 256 different tags are available; 64 can be assigned by MPEG-2 systems and the other 192 are left for private usage. The assignment of descriptor tag values is discussed in Section 13.3 and in Chapter 14 for each content format. The descriptor tag is followed by an eight-bit descriptor length field, indicating the number of data bytes in the descriptor remaining after the descriptor length field. As a consequence, a complete descriptor consists of at most 257 bytes.

In the program definition, the program info and the elementary stream info commence with the program-info-length and the elementary-stream-info-length, respectively, so as to specify the total length in bytes of the program info and elementary stream info. If the total length is larger than zero, one or more complete descriptors are carried. Multiple descriptors may be concatenated; the receiver can parse the descriptors by means of the descriptor length. Hence, the encoded value of the info length must be equal to the sum of the encoded values of the descriptor length fields in the included descriptors, plus two bytes for each included descriptor; see Figure 9.12. Further details of the descriptors, including their data formats, are discussed in Section 13.3 and Chapter 14.

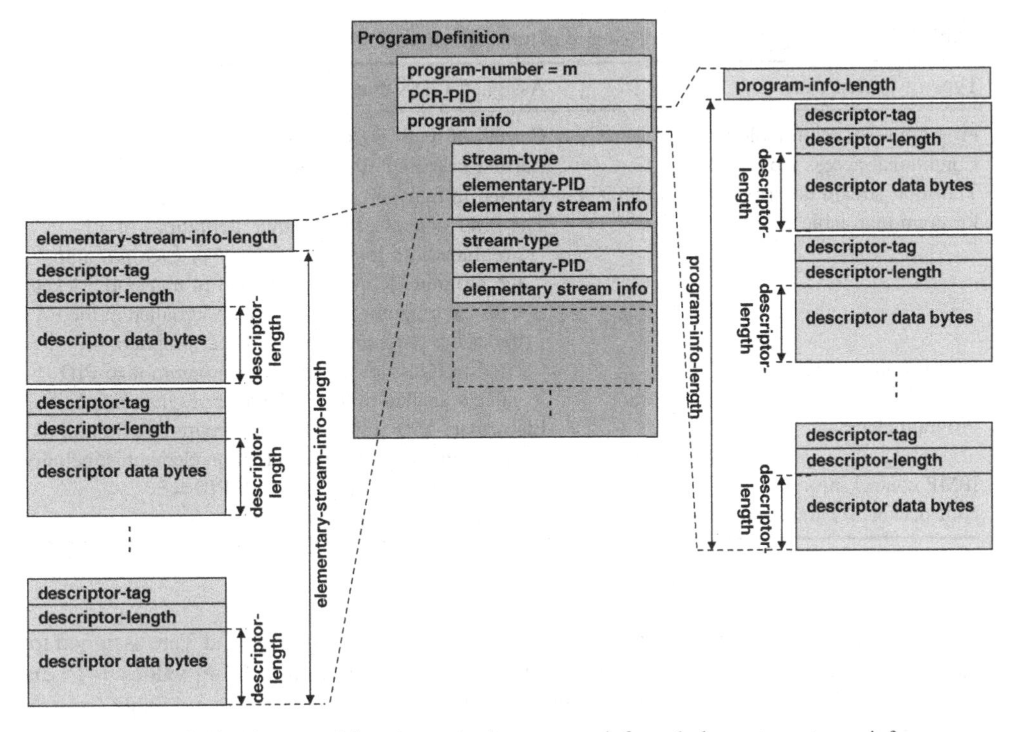

Figure 9.12 The use of descriptors in the program info and elementary stream info

In addition to the PAT and the PMT, the transport stream may carry four more PSI tables, as listed below:

- The Conditional Access Table, CAT, conveyed in transport packets with an encoded PID value equal to one. The CAT is a private table to provide information on the Conditional Access system used for scrambling of elementary streams or other program data. Whenever one or more program elements within a transport stream are scrambled, a complete CAT must be provided. The CAT is discussed in more detail in Section 10.2.
- The Network Information Table, NIT, an optional table that carries private data; the NIT is intended to convey network information, such as the frequencies used to broadcast the transport streams of a service provider, satellite transponder numbers, modulation characteristics and so on. Broadcast standardization bodies, such as DVB, may specify the data format of the NIT.
- The Transport Stream Description Table, TSDT, is an optional table providing descriptive information applicable to the entire transport stream.
- The IPMP Control Information Table (IPMP CIT) is an optional table that provides information when IPMP in MPEG-2 systems is used. The use of IPMP in MPEG-2 systems is specified in ISO/IEC 13818-11 [6].

An overview of the assignment of PID values for carriage of PSI tables and program elements in transport streams is given in Table 9.3. As already indicated in Figure 9.11, the PID

Table 9.3 PID assignments for data carried in transport streams

Type of carried data	Assigned PID value
Program association table (PAT)	Carried in transport packets with PID = 0
Conditional access table (CAT)	Carried in transport packets with PID = 1
Transport stream description table (TSDT)	Carried in transport packets with PID = 2
Program map table (PMT)	The PMT contains the program definitions of all programs in a transport stream. For each program, the program definition is carried in transport packets with the program-map-PID value signalled in the PAT under the program number of that program
Network information table (NIT)	Carried in transport packets with program-map-PID value signalled in the PAT under program number = 0
Program element	Elementary-PID signalled in the program definition of the program in which the program element is included
IPMP control information table	Carried in transport packets with PID = 3
Null packets for transport stream stuffing	Transport packets with PID = 8191

values 0 up to 15 are for assignment by MPEG systems; the values 0, 1, 2 and 3 are assigned for carriage of the PAT, CAT, TSTD and the IPMP CIT, while the remaining values 4–15 are reserved for future use by MPEG systems. The PID values 16–8190 are available for user assignment; for each program element, the assigned elementary-PID is signalled in the PMT. A program element may be used in more than one program; for example, the same video of a sports event may be used in multiple programs, whereby each program provides audio comments in a specific language. However, users may also assign PID values for carriage of data without listing this data in the PMT, but in that case MPEG systems does not specify the type of data carried, nor any relationship between such data and the MPEG specified content carried in transport packets. Consequently, the use of data carried in transport packets with a PID value not listed in any program definition is not addressed in Table 9.3.

The description of the content conveyed in a transport stream provided by a PSI table may change over time. For example, in the PMT, one of the elementary streams may be replaced by another, or the encoding format may change. In other words, the validity of a PSI table is usually limited in time. To accommodate changes in PSI tables, the notion of a table version is introduced when transporting PSI tables.

For the transport of PSI tables, MPEG-2 systems specified the so-called section format as a generic structure to convey tables in transport packets. To conveniently map each PSI table to the section format, a number of important issues were taken into account in the design of the section format:

- The PSI tables are used to retrieve and decode content from a transport stream. For broadcast services, the PSI tables must be repeated regularly to allow random access, but once a receiver has knowledge of a PSI table, it should be conveniently possible for a receiver to determine whether a new table version is available, without processing the entire re-broadcast table data.

- Any change in PSI table data must result in a new table version and the time of transition in to a new table version must be specified accurately, so that receivers can accommodate such transition without noticeable impact.
- The control data carried in PSI tables is often very sensitive to data loss and error; therefore it should be possible to verify from received PSI data whether or not the received data is complete and correct.
- To reduce the sensitivity for data error and loss, it should be possible to constrain the maximum size of some PSI table data to about 1 KB. However, for less sensitive data, a maximum size of conveyed data of about 4 KB is desirable.
- It must be possible to split large tables into multiple fragments and to convey each fragment separately to the receiver. In a receiver it should be possible to conveniently reconstruct the original table from these segments.
- The use of sections should not be restricted to PSI tables, but should also allow transport of other MPEG defined data as well as private data.

The above considerations did lead to the flexible section structure depicted in Figure 9.13. Each section starts with an eight-bit table-ID that identifies the type of data conveyed in the

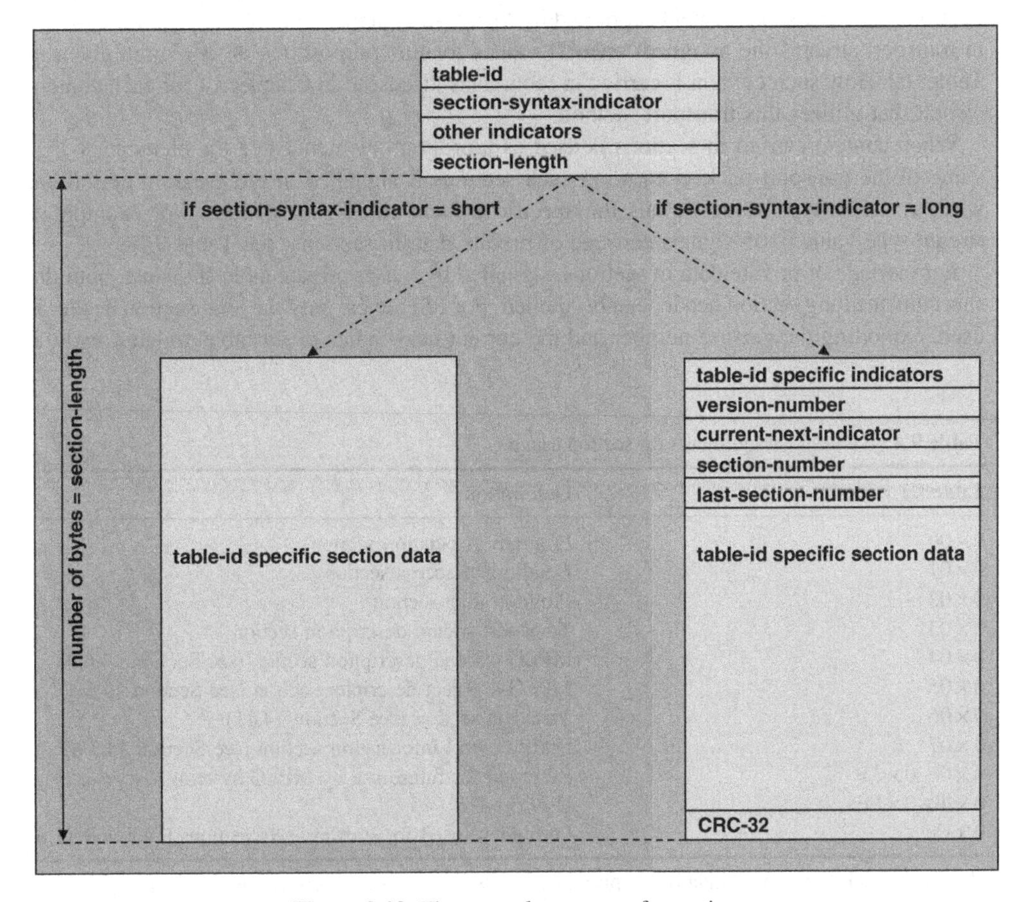

Figure 9.13 The general structure of a section

section, as discussed further below. The table-ID is followed by the section-syntax-indicator by means of which distinction is made between a long and a short section header. Both the short and the long section header have section-length indication, specifying the number of bytes in the section remaining after the section-length field.

In case of the short section header, the section has a very minimal structure: the section length field is followed by section data with a format that depends on the coded value of the table-ID. In case of the long section header, versioning and table fragmentation is supported, while each such section is concluded by a CRC code of 32 bits. By applying a Cyclic Redundancy Check on the entire section (commencing with the table-id field), a decoder can verify the absence of any error in the section, in which case the data can be safely 'trusted'. The CRC decoder is prescribed in Annex A of the MPEG-2 system standard. In case of the long section header, the section-length field is followed by an 18-bit field that is available for table-ID specific indicators. Note that in sections with the short section header, the use of a CRC code is not required by MPEG systems, but instead left to the specification of the section data format.

The table-ID assignments are given in Table 9.4. There are 256 table-ID values; 64 are available for use by MPEG-2 systems and 191 are left for private usage. The use of the value 0xFF is forbidden, so as to allow for stuffing after a section in a transport packet; see Section 9.4.5. For some content supported by MPEG-2 systems, sections are used for carriage in transport streams the assigned table-ID values for this purpose (as of 2013) are given in Table 9.4. How such content is carried in sections is discussed in Chapter 14 for each content format that utilizes this transport method.

When content carried in sections is used as a program element, then the elementary-PID value of the transport packets carrying such sections is signalled in the program definition, while a stream-type value signals the specific content carried in sections. For example, a stream-type value 0x05 signals carriage of private data in sections; see Table 9.2.

For carriage of private data in sections, signalled by a user private table-ID value, both the short and the long section header can be applied. For PSI tables, only the long section header is used, exploiting the version-number and the current-next-indicator for table updates, and the

Table 9.4 Table-ID assignments for section usage

Table-ID value	Description
0×00	Program association section
0×01	Conditional access section
0×02	Program map section[a]
0×03	Transport stream description section
0×04	MPEG-4 scene description section (see Section 14.6.4)
0×05	MPEG-4 object descriptor section (see Section 14.6.4)
0×06	Metadata section (see Section 14.11)
0×07	IPMP control information section (see Section 14.3.6)
0×08–0×3F	Reserved for future use by MPEG systems
0×40–0×FE	User private
0×FF	Forbidden (used for stuffing, see Sections 9.4.5 and 11.6)

[a]In the MPEG-2 system standard the program map section is often referred to as TS program map section or transport stream program map section.

section-number and last-section-number for table fragmentation. Prior to transport, a PSI table can be segmented in at most 256 fragments, whereby each fragment is contained in a single section. The section-number of the first fragment is zero; for each additional fragment, the section-number is incremented by one; the last-section-number specifies the section with the highest section-number and indicates that there are (last-section-number +1) fragments of the same table. By means of this information, receivers can determine that a table with an applicable version-number is completely available, when all sections with data of the table are delivered. Note though that the order in which the sections are delivered is not prescribed by MPEG systems.

The PAT is carried in one or more program association sections, each with a table-ID value of zero. As shown in Table 9.3, these sections are delivered in transport packets with the PID value zero. No other sections are permitted in those transport packets. In program association sections, 16 bits of the table-ID specific indicators depicted in Figure 9.13, are used to signal the transport-stream-ID. After the last section-number, the data bytes follow, providing for each program in the transport stream the program-number followed by the program-map-PID. The program-number signals the program for which the program definition is carried in transport packets with the program-map-PID value, with one exception: for program-number equal to zero, the value of the network-PID is provided. Transport packets with the network-PID carry the NIT. During the continuous existence of a program, the program definition may change regularly, but the program-map-PID must remain the same for each program.

The program definition is delivered by program map sections in transport packets with the program-map-PID specified in the PAT; those transport packets may also convey private sections. This means that along with the program definition, one or more private tables can be transmitted, possibly with a similar structure as the PSI tables defined by MPEG, and that by selecting the long or short section header syntax, these private tables can either exploit or ignore the versioning, section numbering and the data validation by CRC provided by the long or short header syntax.

A program map section must carry a complete program definition. In program map sections, 16 bits of the table-ID specific indicators depicted in Figure 9.13, are used to convey the program-number of the program definition. The same program-map-PID value may be signalled in the PAT for multiple programs; in that case, the transport packets with the program-map-PID carry multiple program definitions, each contained in a separate program map section and identified by its program-number in the header of that program map section. Figure 9.14 shows the format of a program association section and of a program map section, though not in each detail yet.

The NIT is an optional table, intended to convey network information. When present, the NIT is carried in transport packets with the network-PID, as signalled in the PAT under program-number zero. To convey the NIT in transport packets, private sections are used. The format of the NIT may be specified by application standardization bodies, such as DVB, ATSC and ARIB. Application standardization bodies may also define other PSI related tables, (private tables from MPEG perspective) such as Service Information (SI) tables, as deemed necessary.

Figure 9.15 presents an example of segmenting a PAT in two program association sections. In this simple example, the PAT contains 11 pairs of [program-number, PID (program-number)]. For program-numbers 16 up to 25, the program-map-PID is given, while under program-number

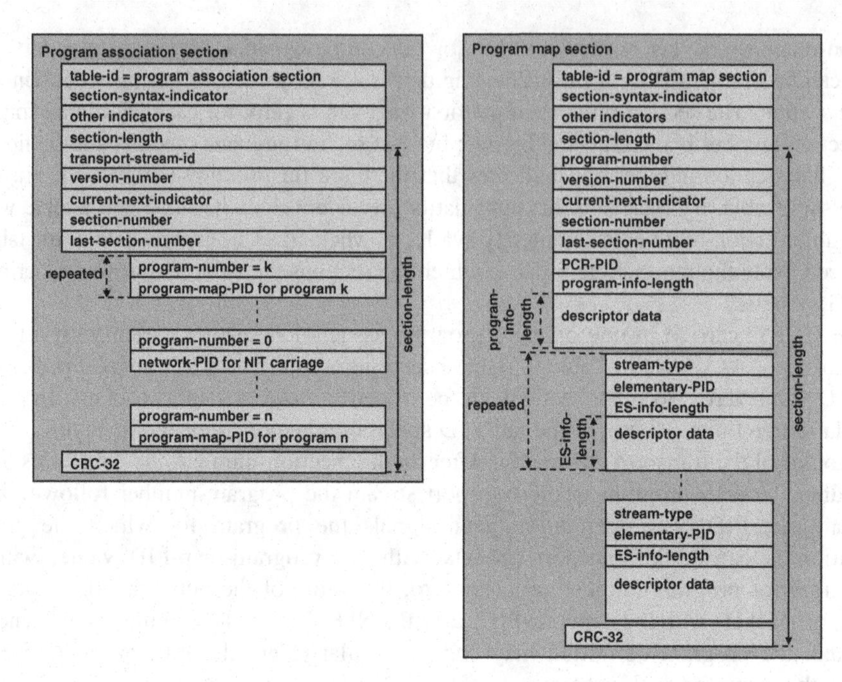

Figure 9.14 Format of a program association section and a program map section

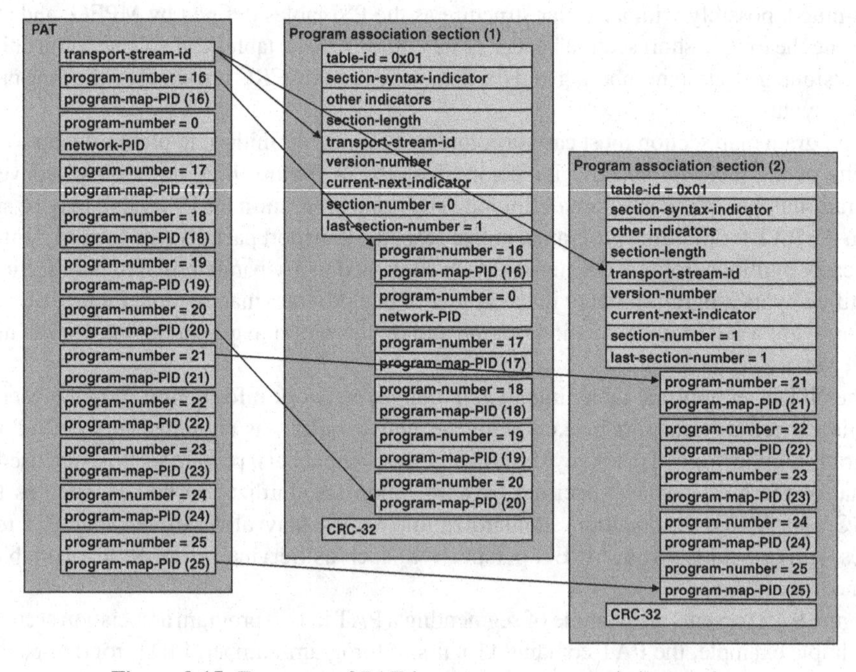

Figure 9.15 Transport of PAT in two program association sections

zero the network-PID is given. Both sections carry the transport-stream-ID of the PAT. Program association section (1) conveys the program-map-PID for program-numbers 16 up to 20 and the network-PID, while program association section (2) provides the program-map-PID for program-numbers 21 up to 25. MPEG systems requires that in the first section with data of the PAT, the section-number is encoded with the value zero, and that the section-number is incremented by one in each additional section. Hence, in program association section (1) the section-number is encoded with the value zero and in program association section (2) with the value one. In both sections, the last-section-number has the value one, indicating that program association section (2) is the last section carrying data of the PAT.

The TSTD table is also optional. When present, the TSTD is carried in transport packets with the PID value two; see also Table 9.3. The TSTD is conveyed in these transport packets by means of transport stream description sections, identified by the table-ID value three; see Table 9.4. The format of a transport stream description section is presented in Figure 9.16. As for all PSI tables, the long section header is used. The table-ID specific indicators depicted in Figure 9.13, following the section-length, are not used for signalling, but instead reserved. After the last-section-number, one or more descriptors follow, that each apply to the entire transport stream.

For table updates, a five-bit version-number is used. For the program association table and the conditional access table, the version applies to the whole table. However, for the program map table, the version applies only to the definition of a single program. The version-number is incremented by one modulo 32, whenever there is a change in the PAT, the program definition, or in the CAT. The transition to the new table data must be accurately specified in time. For that

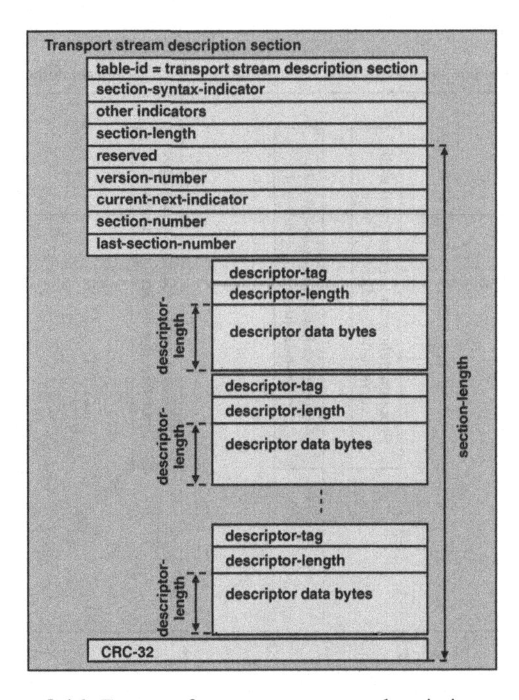

Figure 9.16 Format of a transport stream description section

purpose, the current-next-indicator is used; when this flag is set to 'current', then the PAT in the section is currently applicable. When set to 'next', then the PAT conveyed in the section is not applicable yet, but the next one to become valid. This allows to prepare the transition to the new table data by making this new table data available to the receiver prior to its usage. A receiver can ensure to be ready for the transition by making the new table data fully available in its decoding system, and to wait with making the new table data valid until the current-next-indicator is set to 'current' (see Figure 9.17).

In Figure 9.17, the applicable program definition version (*n*) is delivered at time t_1 and t_3, to allow random access. At t_2, the new version of the program definition is provided with the current-next-indicator set to 'next', to allow receivers to prepare for the transition. At t_4, version $(n+1)$ of the program definition is delivered once again, but now with the current-next-indicator set to 'current', indicating that the transition to version $(n+1)$ is to take place. For random access purposes, the delivery of version $(n+1)$ of the program definition is repeated at t_5. For more details on the time of transition, as specified by MPEG-2 systems, see the discussion on local program insertions in Section 11.3.

The maximum size of a section is equal to 4096 bytes. However, so as to minimize data loss in error conditions, sections carrying data from a PAT, a PMT, a CAT or a TSDT are constrained to have a maximum size of 1024 bytes. For transport in sections, a PAT can be partitioned to occupy multiple sections, but a program definition is required to be conveyed in a single section. As the header of a program map section (up to and including the last-section-number) contains eight bytes, this means that the size of a single program definition can be no more than 1016 bytes. In the exceptional case of a larger program definition, on application level it is in general possible to split the references of the streams, so that they do not all have to be listed together.

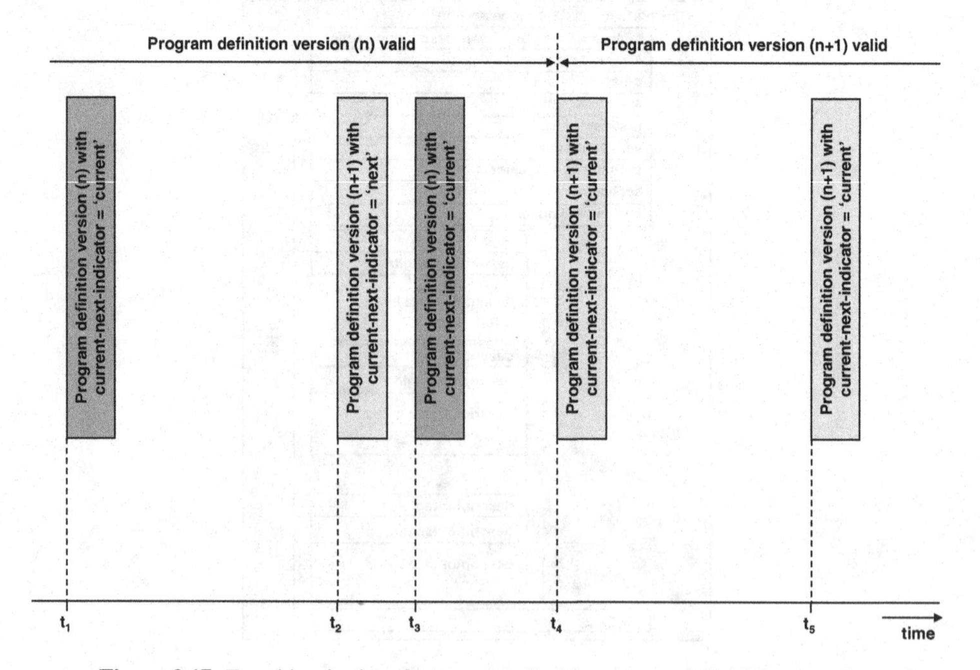

Figure 9.17 Transition in time from one applicable program definition to another

Sections are not only used to convey PAT and PMT information, but also for other PSI tables. The CAT contains private data, as discussed in Chapter 10, and is carried in private sections conveyed in transport packets with the PID value one. The contents of the NIT, if any, is also private, and consequently, the NIT is conveyed in private sections in transport packets with a PID value signalled in the PAT under program-number zero as network-PID. The optional TSDT is carried in transport packets with the PID value three, using TS-description-sections. The IPMP CIT, if any, is conveyed in transport packets with the PID value four, using IPMP control information sections. These section formats are discussed in detail in Section 13.5, except the format of an IPMP control information section, which is defined in ISO/IEC 13818-11.

9.4.5 Conveying Sections in Transport Packets

In the payload of transport packets with a certain PID value, a series of concatenated sections may be transported, without any specific requirement on alignment between the beginning of a section and the first byte of the transport packet payload. There are also no constraints on the header of the transport packets carrying sections; hence the adaptation field may be present as well. Sections may be so small that within the transport packet payload data of multiple sections are present. On the other hand, a section may also span many transport packets; for example, if the used transport packets carry on average a net payload of 182 bytes, then it will take 22 transport packets to convey a section of 4004 bytes.

The general section structure, as depicted in Figure 9.13, is very simple, without any mechanism to find the beginning of a section; however, once the beginning of a section is known, the beginning of the next section can be found from the encoded section-length value. As a consequence of this approach, means external to the section structure needs to ensure that the beginning of a section can be found. For this purpose, MPEG-2 systems requires the following:

- If a section commences in the payload of a transport packet, then an eight-bit pointer field must be present in the first byte of its payload.
- To distinguish transport packets with and without this pointer field, the presence of the pointer field must be signalled in the transport packet header.

An example of carriage of sections in transport packets is depicted in Figure 9.18, where section 1, section 2, section 3 and section 4 are carried in transport packets 1, 2, 3 and 4. Transport packet 1 conveys a trailing fragment of a previous section, after which section 1 commences. The first byte in the payload must therefore be a pointer, indicating the number of bytes prior to the start of section 1, hence the number of bytes of the trailing section fragment. The presence of the pointer field is indicated by one of the fields in the header of transport packet 1, the payload-unit-start-indicator; for further details see Section 13.6. Section 1 is short and therefore also section 2 commences in the payload of transport packet 1; the parser will find the first byte of section 2 by means of section-length (1). Section 2 is relatively large and is conveyed in three fragments in transport packets 1, 2 and 3. Because no section commences in transport packet 2, the pointer field is absent in the payload, as indicated by the payload-unit-start-indicator in its header. In transport packet 3, after the trailing part of section 2, section 3 starts. As in transport

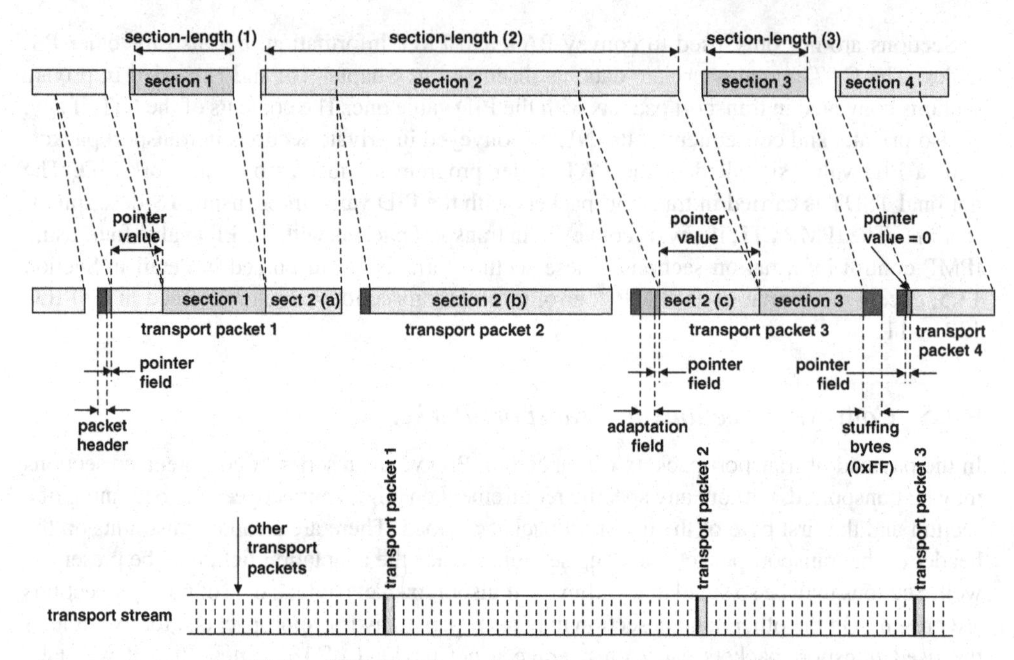

Figure 9.18 Transport of sections in transport packets

packet 1, also in transport packet 3 the pointer field must be present. Section 3 fits entirely in the payload of transport packet 3.

For reasons determined by the encoder, the remaining payload after section 3 is filled by stuffing bytes with a value 0xFF. The receiver will identify those bytes as stuffing bytes, because after the last byte of each section, the parser expects the table-ID of the following section. However, the table-ID value 0xFF is forbidden and if identified, the receiver knows that no new section follows and that stuffing is applied. In case of such stuffing, all remaining bytes in the payload are required to be stuffing bytes with the value 0xFF, as indicated in Figure 9.18. Because of this stuffing, section 4 starts immediately in the payload of the next transport packet with section data. Therefore, also in this packet the pointer field must be present, encoded with the value zero, indicating that the new section commences immediately after the pointer field.

Note: due to the pointer based synchronization mechanism for sections described above, there is no need for restrictions on the occurrence of start codes, sync bytes or other bit patterns in the data contained in sections.

References

1. Bylanski, P. and Ingram, D.G.W. (1980) *Digital Transmission Systems*, IEE Telecommunication Series 4, Revised 2nd edn, Peter Peregrinus Ltd., p. 104.
2. ITU (1996) ITU-T Recommendation H.222.1: Multimedia Multiplex and Synchronization for Audiovisual Communication in ATM Environments; see http://www.itu.int/rec/T-REC-H.222.1-199603-I.
3. Wikipedia (2013) General information on digital modulation techniques can be found at http://en.wikipedia.org/wiki/Modulation.

4. Wikipedia (2013) Background information on QPSK can be found at http://en.wikipedia.org/wiki/Phase-shift_keying.
5. Wikipedia (2013) Background information on QAM can be found at http://en.wikipedia.org/wiki/QAM.
6. ISO/IEC 13818-11 (2004) Information technology – Generic coding of moving pictures and associated audio information – Part 11: IPMP on MPEG-2 systems, see http://www.iso.org/iso/catalogue_detail.htm?csnumber=37680.

10

Conditional Access and Scrambling

Why conditional access is important for many MPEG-2 system applications. Scrambling in transport streams and the use of conditional access sections and descriptors. The role of simulcast in transport stream applications. Scrambling in program streams.

10.1 Support of Conditional Access Systems

For many MPEG-2 system applications, unauthorized access to the content carried in an MPEG-2 system stream is to be prevented. For example, content in a Pay TV service should only be available to subscribers to such a service, while an optical disc may contain free content, as well as content that may be accessed only if an additional payment is made. For this purpose, Conditional Access (CA) systems are used. A conditional access system specifies how to encrypt the content at the encoder side prior to its distribution, so as to make the content unreadable by anyone, unless access to the content is granted explicitly by means of providing the information needed to decrypt the content back into the original format. While the specification of CA systems is beyond the scope of MPEG, it should be possible to apply practical CA systems in MPEG-2 system based applications. Therefore the use of CA systems in a MPEG-2 system stream was an important design consideration.

As an example for the use of a CA system in an MPEG-2 system environment, Figure 10.1 depicts a typical conditional access system for an MPEG-2 transport stream, as often applied by Pay TV applications. In Figure 10.1, at content encoding, transport packets enter the data scrambler, so as to alter the payload of transport packets by a CA system specific scrambling algorithm, based on the use of certain encryption keys. Each key may be valid during a certain period of time only, as frequently changing the key decreases the risk that the CA system can be hacked to achieve non-authorized access to the service. The scrambled payload bytes can obviously no longer be understood by decoders of audio, video and other content. The altering of the payload data, as defined by the scrambling algorithm, is reversible: in the receiver, a descrambler can reconstruct the original payload data when the receiver has access to the required descrambling keys.

To authorize access to the encrypted content at receivers, entitlement messages are conveyed to the subscribers. Two different message types are distinguished: ECMs, Entitlement Control Messages, for controlling access to the scrambled content and EMMs, Entitlement

Fundamentals and Evolution of MPEG-2 Systems: Paving the MPEG Road, First Edition. Jan van der Meer.
© 2014 John Wiley & Sons, Ltd. Published 2014 by John Wiley & Sons, Ltd.

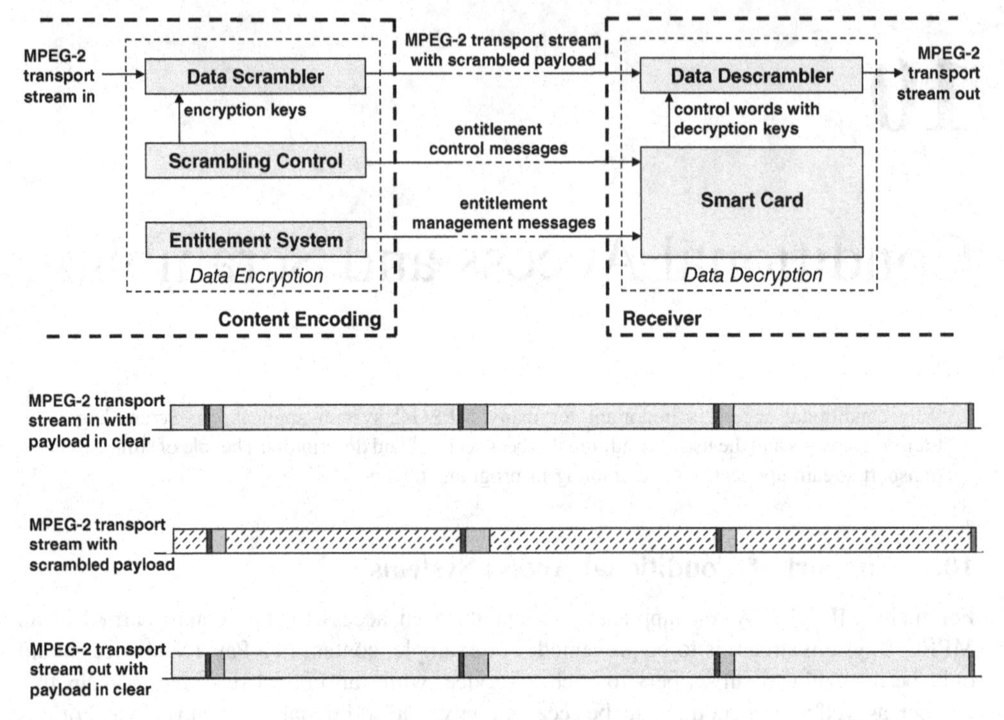

Figure 10.1 Conditional access system for an MPEG-2 transport stream

Management Messages, for managing the authority to access the scrambled content. The scrambling control at content encoding ensures that ECMs are sent to receivers; an ECM carries private CA data, such as encrypted control words with decryption keys and possibly other access control parameters. ECMs may apply to all scrambled program elements of a program, or to a specific program element. Next to the ECMs, also EMMs are conveyed to receivers; an EMM carries the private CA data that specifies the authorization levels of specific receivers to access the content. An EMM may address a single receiver or a group of receivers. Access may be granted for a limited period only; for example, in a Pay-Per-View service, consumers may subscribe to view a film or a sports event transmitted at a certain time.

In the case when a conditional access system is hacked, it is important that the CA system can be updated conveniently. If the security of a CA system cannot be increased easily, for example by more frequently changing keys, a CA system can be updated. Usually, a smart card is used to ensure that, in the case of a serious hack, the CA system can be updated in a cost-effective manner. A smart card can be inserted in the receiver of the subscriber and provides the receiver with a unique identification that, by one way or another, is known to the entitlement system at content encoding. The smart card contains the 'secret' software to grant access to the scrambled content for authorized subscribers. In case the CA system is hacked, it is possible to replace the smart cards of subscribers, without replacing the receivers of subscribers.

The software contained in the smart card compares the entitlements in the smart card with the received EMM(s). If this comparison indicates that access to the service is granted, then the smart card decrypts the control word in the received ECM(s), upon which the decryption key is provided to the descrambler. After the descrambler, the original payload of the transport packets is made available, so that the payload bytes can be processed by the appropriate content decoders.

The delivery of ECMs is often critical in time: for example, when switching from one decryption key to a new one, it must be ensured that the new key is delivered in a timely manner; therefore ECMs are usually provided in-band. The delivery of EMMs is less time-critical: for example, access to the content may be granted in-band at a timing convenient for the service operator, or triggered by a subscriber request across the internet or by telephone. Sometimes also out-of-band delivery of EMMs is possible.

MPEG-2 systems support scrambling of elementary stream data in transport streams and program streams. For example, ECM streams and EMM streams can be conveyed both in program streams and in transport streams. Further details of scrambling in transport streams and program streams are discussed in the following two sections.

The specification of CA systems is private and fully beyond the scope of MPEG-2 systems. Hence, the scrambling algorithm, the format of control words, the format of the ECMs and EMMs as well as any smart card detail is not addressed by MPEG. In practice, applications use many different proprietary CA systems, as considered appropriate by the application.

10.2 Scrambling in Transport Streams

Within a transport stream, the payload of transport packets may be scrambled, but each transport packet header, including the adaptation field, if any, must always be in the clear, irrespective whether the payload is scrambled or not. In a case when the scrambled payload of a transport packet carries a PES packet, the entire PES packet, including its header, is scrambled. The transport packet header is required to be always in the clear, to ensure that scrambling does not impact access to encoded fields in the header of transport packets. As a result, also in cases of scrambling, the stream synchronization discussed in Section 9.4.1, remains possible, as well as transport packet identification. Furthermore, irrespective of any applied scrambling, encoded fields in packet headers can be changed, if deemed necessary during transport.

For scrambling control purposes, the transport packet header contains a two-bit scrambling control field. If coded with the value '00', then the payload of the packet is not scrambled. When coded with another value, the meaning of this field is user-defined; for example, if the CA system dynamically switches from one decryption key to another, this field may be used for that purpose.[1]

To ensure under all circumstances that receivers can properly identify the programs and the associated video, audio and other program elements in transport streams, sections carrying the PAT, the PMT and other PSI data must always be in the clear, that is not scrambled. However, private data bytes carried in sections may be scrambled, also when carried in transport packets in which program map sections are conveyed. However, any field in the private section preceding and following the private data bytes must be in the clear, to prevent impact of the scrambling on the section functionality, such as section length indication in all sections and, in the case of a long section header, versioning, current-next indication and verification by means

[1] Note that in the case of switching keys, accurate synchronization is required between the use of keys for scrambling at content encoding and for descrambling at the receiver. For example, while sending an ECM message with the current key to receivers for descrambling, also an ECM message with the next applicable key may be sent, so that receivers can prepare for its use, as possibly indicated by the scrambling control field.

of the CRC-32 code. Hence, if the private data bytes are scrambled, the Cyclic Redundancy Check, if any, remains performed over the entire section; however, MPEG does not specify whether the CRC is performed over the scrambled or the unscrambled payload.

When scrambling is applied in a transport stream, then for each applied CA system, typically one EMM stream and one or more ECM streams are conveyed in the transport stream. For carriage of each EMM and ECM stream, a specific PID value is assigned. Note that, when simulcrypt is applied, as discussed in Section 10.3, multiple CA systems may be associated to the same scrambled program element. For example, consider a transport stream that carries six programs, each consisting of one audio stream and one video stream; four programmes are in the clear and only programme (1) and programme (2) are scrambled, using the same CA system, without applying simulcrypt. For programme (1) a single ECM stream applies to both scrambled program elements, while for programme (2) a specific ECM stream is conveyed for each scrambled program element. In that case, the following PIDs may be assigned for carriage of the EMM and ECM streams, each with a format specified by the applied CA system:

- Transport packets with PID value (*a*) convey the EMM stream, containing (at some point in time) the authorization of subscribers to access programme (1) and programme (2).
- Transport packets with PID value (*b*) convey the ECM stream, containing the encrypted control words with the keys needed to descramble the audio and video streams of programme (1).
- Transport packets with PID value (*c*) convey the ECM stream, containing the encrypted control words with the keys needed to descramble the audio stream of programme (2).
- Transport packets with PID value (*d*) convey the ECM stream, containing the encrypted control words with the keys needed to descramble the video stream of programme (2).

Hence, to access the scrambled program elements within a transport stream, there is a need to signal which program elements are scrambled, which CA systems are applied, as well as the associated ECM and EMM streams. For this purpose, MPEG-2 systems defined the Conditional Access Table, CAT, and CA descriptors.

For each scrambled program element in a transport stream, a CA descriptor must be included in the PMT of the programme in which such program element is included. This CA descriptor specifies the CA-system-ID of the applied CA system, as well as the PID of transport packets with the associated ECM stream. In addition, the CA descriptor may carry conditional access information, with a format, specified by the applied CA system. In the PMT, the CA descriptor can be included in the program info or in the elementary stream info, thereby signalling whether the CA descriptor applies to an entire programme (i.e. all its program elements) or to the associated program element only.

The CAT provides complete information on the EMMs applicable to all scrambled program elements in the transport stream. Typically, each CA system used in a transport stream, has its own EMM stream; each CA system is identified by means of its CA-system-ID. If an EMM stream is carried in the transport stream for an applied CA system, then the CAT must provide the CA-system-ID and the PID value of transport packets carrying such EMM stream. For this purpose, the CA descriptor is also used in the CAT; hence, the information conveyed in the CA descriptor depends on its context: when carried in the PMT, the CA descriptor signals the PID for ECM carriage and when carried in the CAT, the PID for EMM carriage.

In conclusion, in case a certain program element in a transport stream may be scrambled,[2] a CA descriptor is present in the PMT for each programme containing that program element. The CA descriptor may apply to that program element only or to the entire programme. The CA descriptor signals the CA-system-ID of the applied CA system and the PID value for ECM carriage. If the receiver supports the signalled CA system, then the receiver looks in the CAT for the EMM stream associated to that CA system, if not available already in the receiver. If the EMMs signal that the user is authorized, access to the content is granted by processing the ECMs, followed by descrambling the content. In all other cases, access to the scrambled content is not granted.

As discussed in Section 9.4.4, the CAT is contained in transport packets with a PID value equal to one. A CAT must be transmitted whenever scrambling is applied of one or more program elements within a transport stream. The CAT must be conveyed within transport packets by means of conditional access sections, identified by a table-ID value of one (see Table 9.3). The CAT may be split for carriage in multiple conditional access sections, using the section-number and last-section-number fields discussed in Section 9.4.4.

Within transport packets carrying the CAT, only conditional access sections are permitted. The format of a conditional access section[3] is provided in Figure 10.2. Within the conditional access section, one or more CA descriptors may be carried, while also private descriptors may be contained, possibly specifying proprietary data, such as information on the retrieval of EMMs, associated system-wide or access control management information or related parameters.

Each CA descriptor signals the CA-system-ID of the applied CA system and the CA-PID, identifying the PID value of transport packets with conditional access information for that applied CA system (see Figure 10.3). The type of conditional access information carried in the CA-PID depends on the context of the CA descriptor. When a CA descriptor is carried in a conditional access section, the CA-PID signals an EMM stream, while an ECM stream is signalled when the CA descriptor is contained in a PMT. In the CA descriptor, the CA-system-ID is a 16-bit field, while the CA-PID is encoded with 13 bits; to achieve byte alignment, the CA-PID field is preceded by three reserved bits. The CA-PID field is followed by zero or more private-data-bytes; the number of included private-data-bytes is found from the encoded descriptor-length value. The format of the private-data-bytes is determined by the CA system signalled by the CA-system-ID.

At any time, the CAT must be 'complete': for each scrambled stream, a CA descriptor must be present in the program definition of each program containing that scrambled stream, either in the elementary stream info or in the program info. Any change in scrambling that makes the currently applicable CAT invalid or incomplete,[4] must be described in an updated version of the CAT, using the version-number and the current-next-indicator discussed in Figure 9.17.

Figure 10.4 summarizes how a receiver of a transport stream finds the required data for the decoding of a scrambled program, consisting of a video stream and an audio stream.

[2] Note that the program element is not required to be scrambled; for example, during some periods, the program element may be in the clear.

[3] Note that this format, except the table-ID value, is the same as the format of a transport stream description section.

[4] Note that temporarily switching off the scrambling is possible within the same version of the CAT.

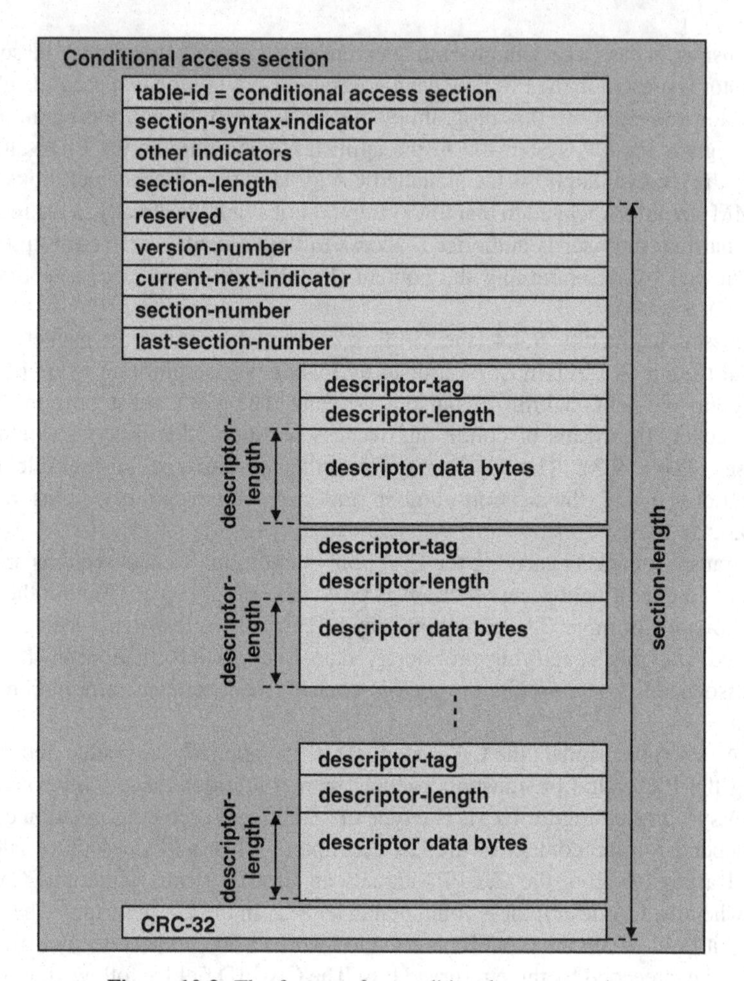

Figure 10.2 The format of a conditional access section

The assigned program-number is k. In case of the example in Figure 10.4, the receiver takes the following steps:

- In the PAT, carried in transport packets with PID = 0, under program-number = k, the PID value is found of transport packets carrying the program definition of program k: program-map-PID (k).
- In the program definition of program k, a CA descriptor is found in the program info, signalling that CA system X is used for the scrambling of the content of program k, and that transport packets with the PID value equal to ECM-X PID carry the ECMs needed for the descrambling of program k. Note that the CA descriptor could also have been included in the elementary stream info; in that case the ECMs carried in ECM-X PID would have applied to that elementary stream only.
- Because a CA descriptor is found in the program definition of program k, signalling the use of CA system X, the receiver looks in the CAT, as carried in transport packets with PID = 1,

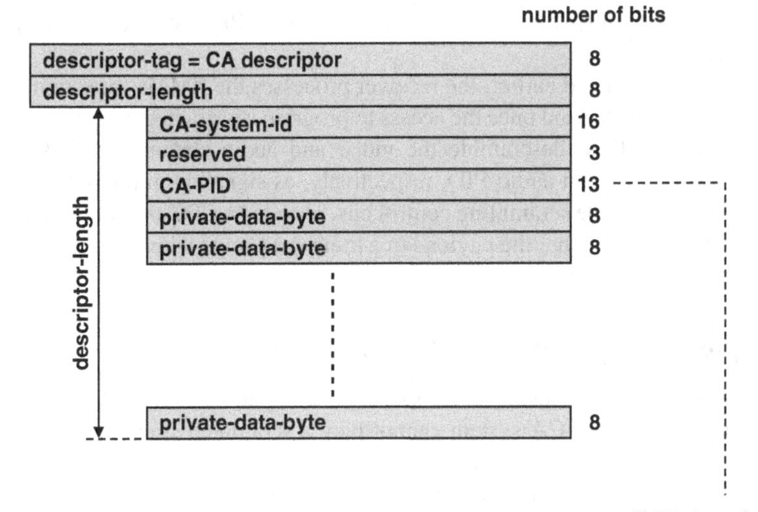

Figure 10.3 The CA descriptor format

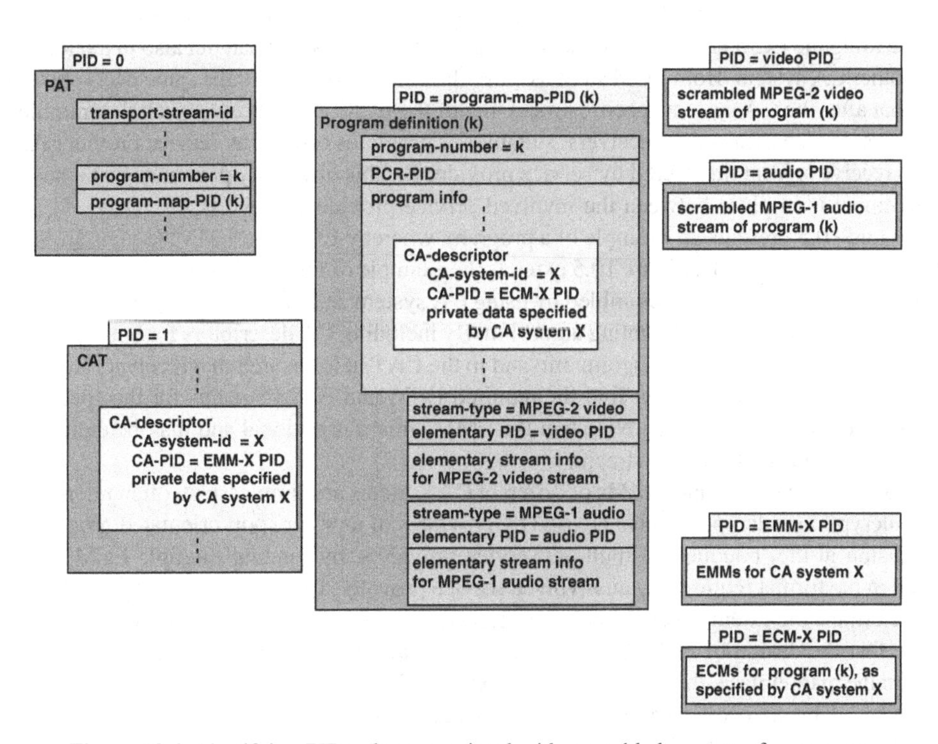

Figure 10.4 Identifying PID values associated with scrambled content of a program

for CA system X and finds that transport packets with PID = EMM-X PID are used to convey the EMMs for CA system X.
- If authorization is not already granted earlier, the receiver processes the EMMs (presuming the receiver supports CA system X), and once the access to program *k* is granted, the receiver uses the ECMs in ECM-X PID to descramble the video and audio streams, carried in transport packets with video PID and audio PID, respectively, as signalled in the program definition of program *k*. Note that the scrambling control bits, encoded as '00' or with a value defined by CA system X, signal whether the payload of a transport packet with video PID or audio PID is actually scrambled or not.

10.3 Improving the Interoperability between CA Systems

The use of different CA systems may result in completely incompatible services, because receivers designed for using a specific CA system cannot play a scrambled stream that uses another CA system. To allow for more compatibility, particularly in transport streams, some application standardization bodies defined a number of CA system elements for (optional) common use, such as a scrambling algorithm and a CA module with a generic interface for usage with smart cards of multiple CA systems.

The use of a common scrambling algorithm, including any use of scrambling-control bits, is one of the essential conditions for achieving compatibility by business agreement. In practice, receivers are designed to support a single CA system, as required by the provider of the service for which the receiver is intended to be used. However, when the receivers designed for two different CA systems apply the same scrambling algorithm, then compatibility can be achieved by sending the ECM and EMM streams not only in the original format, but also in the format of the other CA system. Both ECM streams carry the control words with the same decryption key, so that after the CA system specific processing of the ECM, both receivers can descramble the streams. Which means that receivers, suitable for programs offered by service provider A, can also receive a program offered by service provider B. This obviously presumes that a business agreement is in place between the involved service providers A and B.

Figure 10.5 presents an example of a program, whereby ECM and EMM streams are sent in two CA system formats. Figure 10.5 extends the example of Figure 10.4, in which the program with program-number k is scrambled by using CA system X. In Figure 10.5, the CA systems X and Y apply a common scrambling algorithm. By including CA descriptors for both CA system X and CA system Y in the program info and in the CAT, it is ensured that receivers supporting either CA system X or Y can find the required ECM and EMM streams for the specific CA system the receiver supports. Note that the EMM stream is optional and may be delivered by means beyond the transport stream.

Cases where ECMs and EMMs of different CA systems are simulcast are often referred to as simulcrypt. Simulcrypt can also be used by providers of news or sport orientated programs for inclusion in the 'bouquet' of multiple service providers, by sending multiple ECM streams, each in the format required by an involved service provider. It is beyond the scope of this book to determine the extent at which such arrangements are feasible; however, it is important to note here that the design of MPEG-2 transport streams allows for such arrangements.

On the marketplace, the use of Set Top Boxes (STBs) that are only capable to receive content distributed by a specific service provider is accepted by consumers. However, the use of multiple STBs for various service providers becomes easily cumbersome. For consumers it is

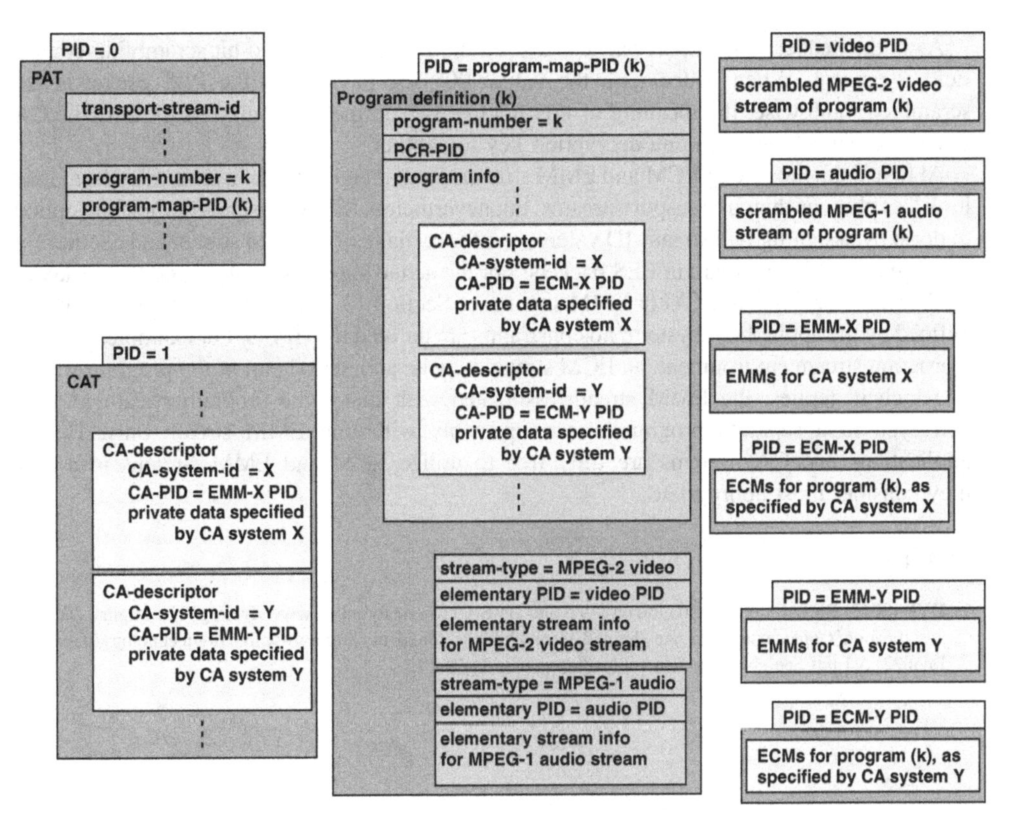

Figure 10.5 Simulcast of ECM and EMM streams for CA systems X and Y

far more convenient to have a single TV set capable of decoding the content from any service provider. An important step to allow integration of STB functionality in TV sets in a manner that is service provider neutral, is defining a (removable) CA module, such as specified by the Common Interface effort in DVB [1]. A CA module allows to separate CA functionality from other STB functionality that can be integrated into a TV set without risking the CA security. The CA module can typically be used with smart cards of multiple service providers. It may be possible to apply more than one CA module; in that case the transport stream with the scrambled data may be passed sequentially through all modules, so as to allow the appropriate CA module to descramble the content.

10.4 Scrambling in Program Streams

Also within a program stream scrambling may be applied. While the payload of a PES packet may be scrambled, the PES header must be in the clear and hence not be scrambled. By requiring that PES headers in program streams are always in the clear, it is ensured that scrambling does not impact the transport mechanism. For example, the stream synchronization discussed for program streams in Sections 9.3 and 7.3, remains possible, as well as PES packet identification. Also, irrespective of any applied scrambling, access to encoded fields in PES packet headers remains possible during transport.

As in the header of transport packets, also in the PES header a two-bit scrambling-control field is present. When coded with the value '00', the payload of the PES packet is not scrambled; otherwise, the meaning of this field is user defined, and may be used by the CA system for switching from one decryption key to another.

At first sight, the role of ECM and EMM streams as an integral part of a program stream may look less obvious than in transport streams, but nevertheless MPEG systems defined the option to do so by assigning two stream-ID values, one for carriage of an ECM stream and another for carriage of an EMM stream in PES packets. For the actual stream-ID values and the syntax of PES packets carrying an ECM or EMM stream see Section 13.2. The format of both streams is defined by the applied CA system; no constraints are defined by MPEG. For example, while in a program stream environment, an ECM stream may be an integral part of the program stream to which it applies, the EMM stream associated with that same program stream may be conveyed in a separate program stream, possibly with one EMM stream only. Hence, applications and CA systems are fully free to utilize ECM and EMM streams in a way they consider most appropriate.

Reference

1. DVB (2013) Background information on DVB and DVB-CI can be found at: www.dvb.org/ The original DVB-CI specification (some extensions are defined meanwhile) is found at: http://www.dvb.org/technology/standards/En50221.V1.pdf See also: http://en.wikipedia.org/wiki/DVB-CI.

11

Other Features of MPEG-2 Systems

Required measures for transport stream usage in error prone environments. Why re-multiplexing of transport streams is needed and what that means for delays and buffering. Commercial breaks in transport stream based broadcast, causing time base discontinuities; associated buffering issues. Statistical multiplexing in broadcast applications. The role of padding and stuffing. The importance of random access and parsing convenience. Options to carry private data. Support of copyright and copy control. Trick mode operation. Issues with single program transport streams and partial transport streams. Carriage of a program stream in a transport stream, the use of PES streams and room created for future extensions.

11.1 Error Resiliency

When transporting an MPEG-2 system stream, errors may occur. Usually, transport systems apply error correction, for example as an integral element of channel coding, as discussed in Figure 9.8, but the error correction capabilities are obviously limited, so that in error prone environments remaining errors in system streams may be expected. Typical errors are packets with a number of uncorrectable erroneous bytes, and lost packets. Video, audio and other elementary stream decoders are sensitive to errors; prior to detecting an error in the received bitstream, these decoders may produce undesirable output signals. To prevent such output, it is desirable to detect data errors at system level and to signal them to the elementary stream decoders, so that the decoder can suitably conceal the error in its output. System data, such as program association tables and program definitions and other PSI data may be even more vulnerable for data errors. Errors in these tables may cause serious malfunctioning of receivers. Hence, both for content decoding and for system processing in receivers it is important to detect data errors and to ensure that errors are signalled to receivers. Note though that the handling of errors is at the discretion of decoder designers and beyond the scope of MPEG.

To enable a reliable detection of data errors, a number of measures have been taken in MPEG-2 system streams. Program streams are intended for use in almost error-free environments, while transport streams must be suitable for use in environments where errors are likely; therefore, the main focus of the measures is on transport streams. The measures are listed below.

Fundamentals and Evolution of MPEG-2 Systems: Paving the MPEG Road, First Edition. Jan van der Meer.
© 2014 John Wiley & Sons, Ltd. Published 2014 by John Wiley & Sons, Ltd.

- Error indication in transport packets. The transport packet header contains a one bit transport-error-indication flag; when set, this flag indicates that this transport packet contains at least one error. This flag, that may be set by a channel (de)coding entity, provides an in-band method for error indication without requiring external means. This straightforward in-band method allows signalling errors through the various transport stages. For example, in a digital TV broadcast application, prior to digital distribution across cable to subscribers, the transport stream may be conveyed to the cable head-end over satellite. If an uncorrectable error occurs in a transport packet during satellite transport, the transport-error-indication flag can be set, and for transport over cable this set flag conveys the presence of the error to receivers. Note though that no indication is provided which data of the packet is in error; payload bytes as well as fields in the header, such as the encoded PID value, may be erroneous.
- Continuity counter in transport packets. To enable detection of packet loss, the transport packet header contains the four-bit continuity-counter field. Each data stream carried in a transport stream has its own continuity counter; the counter is incremented by one with each transport packet carrying data of that stream. More in particular, in each subsequent transport packet with a certain PID, the continuity counter associated to that PID is incremented by one, but only if the transport packet carries one or more payload bytes: if a packet with only the adaptation field and no payload bytes is lost, then control data is lost, but no elementary stream data or other data that is to be decoded. After reaching its maximum value, the continuity counter wraps back to zero. A decoder will detect packet loss, if the comparison of the continuity counter values in two subsequent transport packets with the same PID yields a difference larger than one. Due to the use of four bits for the encoding of the continuity counter, the method will fail, if a multiple of 16 transport packets are lost. For elementary streams, no further measures to cover this unlikely event were deemed necessary, but for PSI data, MPEG systems offers the additional measure on CRC usage discussed further below.

  With respect to the use of the continuity counter for determining packet loss, as described above, there is one exception however. The adaptation field of the transport packet header carries a one-bit flag called the discontinuity-indicator; if this indicator is set to '1' in a transport packet with a PID value other than the PCR-PID of a program, then the continuity counter is allowed to be discontinuous. The same applies for transport packets with the PCR-PID, but only in case that packet carries an encoded PCR. For a discussion on the use of the discontinuity indicator, see Section 11.3.
- Cyclic Redundancy Check, CRC, in sections. Non-identified errors in PSI tables may have serious consequences for receivers, and therefore MPEG systems introduces a cyclic redundancy check in sections carrying PSI tables. Therefore, as already discussed in Figure 9.13, in the case of a long section header, the data of an MPEG-2 section is concluded by a 32-bit CRC code. By means of this code, the decoder can verify whether the received section is free of error. For this purpose, the decoder applies a cyclic redundancy check on the entire section, by using the CRC decoder model specified in Annex A of the MPEG-2 system standard. This CRC model has a 32-bit output. At the encoder, the CRC-32 code is provided so, that when there are no errors, after processing the entire section, each bit of the output of the CRC decoder is zero.
- When the splice-countdown field is used to prepare for splicing (as discussed in Section 11.4 and Figure 11.8) then in subsequent transport packets of the same PID, the encoded value in this field is subtracted by one, excluding duplicate transport packets and packets without

payload data, that is packets with adaptation field only. Hence, when in a sequence of packets with the same PID the splice-countdown is encoded, but one or more values are missing, the loss of packet payload is identified. Note however that this feature is a spin-off from an optional splicing tool and therefore only useful in circumstances that may be rare. Moreover, usage of the splice-countdown is rather redundant with the continuity counter. In other words, usage of the splice-countdown field to identify packet loss is listed mainly for completeness here.

- Optional use of a 16-bit CRC code in PES packets. Also this measure is listed here for completeness only. The PES header contains a PES-CRC-flag, which, if set, signals the presence of a CRC-16 code for the previous PES packet. Hence, if present in a PES header, the CRC-16 code applies to the previous PES packet. The cyclic redundancy check is performed similar as for sections: in this case, at the encoder, the CRC-16 code is provided so, that when there are no errors, after processing all payload bytes of the previous PES packet, each of the 16 bits of the output of the CRC decoder is equal to zero. The calculation is made over the payload bytes only, because during transport the PES packet header may be modified. While the CRC-16 code can be used to verify the absence of errors in the previous PES packet, it is not intended for elementary stream decoding, but for network maintenance in special cases. An example of such network maintenance is isolating the source of intermittent errors.

In addition to measures helpful for error detection, also some measures are taken to assist networks and receivers in the handling of errors. One such measure is transport priority indication in the header of transport packets. The transport-priority is a one-bit flag that, if set, signals that the associated packet is of greater priority than packets with the same PID in which this flag is not set. How to use the transport-priority flag is left at the discretion of transport mechanisms. Depending on the application, the flag may be used regardless of the PID value or within one PID value only. From MPEG systems perspective, transport systems are free to change the encoded value of this flag in packets, but applications and their standardization bodies may impose specific requirements on its usage.

Another relevant flag in this context is the elementary-stream-priority-indicator in the adaptation field of transport packets. If set, within packets with the same PID value, this flag indicates elementary stream data with a higher priority. For MPEG-2 video streams, a set flag means that the payload of the packet contains one or more bytes from a slice with intra coded video data, while for AVC, SVC and MVC video a similar meaning applies. For many other elementary streams, priority is a less obvious notion than for video; therefore the MPEG system experts decided only to specify conditions for the coding of this flag for video and to leave conditions for the use of this flag for other elementary streams to application domains.

Also in the PES header a priority flag is defined; setting this flag indicates a higher priority of the payload in the PES packet. This flag can be used during transport of PES packets; however, transport systems are not allowed to modify this PES-priority flag.

To slightly reduce the probability of losing an important transport packet, a transport packet with payload data may be sent twice; the second one is called a duplicate packet. The bytes in the header and payload of each duplicate packet must be exactly the same as in the previous transport packet with the same PID value, with the exception of the PCR field; if present, the PCR value must be encoded with a valid value. A transport packet may be duplicated only once. A duplicate packet is identified by a continuity counter value that is the same as in the previous packet with the same PID value. However, receivers should take into account that such condition is also met in case a burst of errors or data loss has impact on a multiple of 16 transport packets with the same PID value.

The payload of a duplicate packet is only to be used by the decoder in case the packet it duplicates is not received; note that in such case the duplicate packet is not identified as a duplicate one. However, if both packets are received, the decoder should ignore the payload of one of the packets to avoid incorrect reconstruction of the conveyed data.

11.2 Re-Multiplexing of Transport Streams

An essential requirement in broadcast environments is support for re-multiplexing of transport streams in the network. A wide variety of re-multiplex operations may be applied. A few examples are depicted in Figure 11.1; transport stream 1 carrying n programs is not only broadcast to the receiver of end users, but is also conveyed to three re-multiplexers. In re-multiplexer 1, the programs contained in transport stream 1 are re-multiplexed into two transport streams with a lower bitrate. Re-multiplexer 2 receives next to transport stream 1, three other transport streams. The programs contained in the four transport streams at its input, are re-arranged by re-multiplexer 2 over the five transport streams provided at its output. Re-multiplexer 3 receives transport stream 1 and another transport stream, so as to re-multiplex the contained programs into a single transport stream with a higher bitrate. A re-multiplexer may

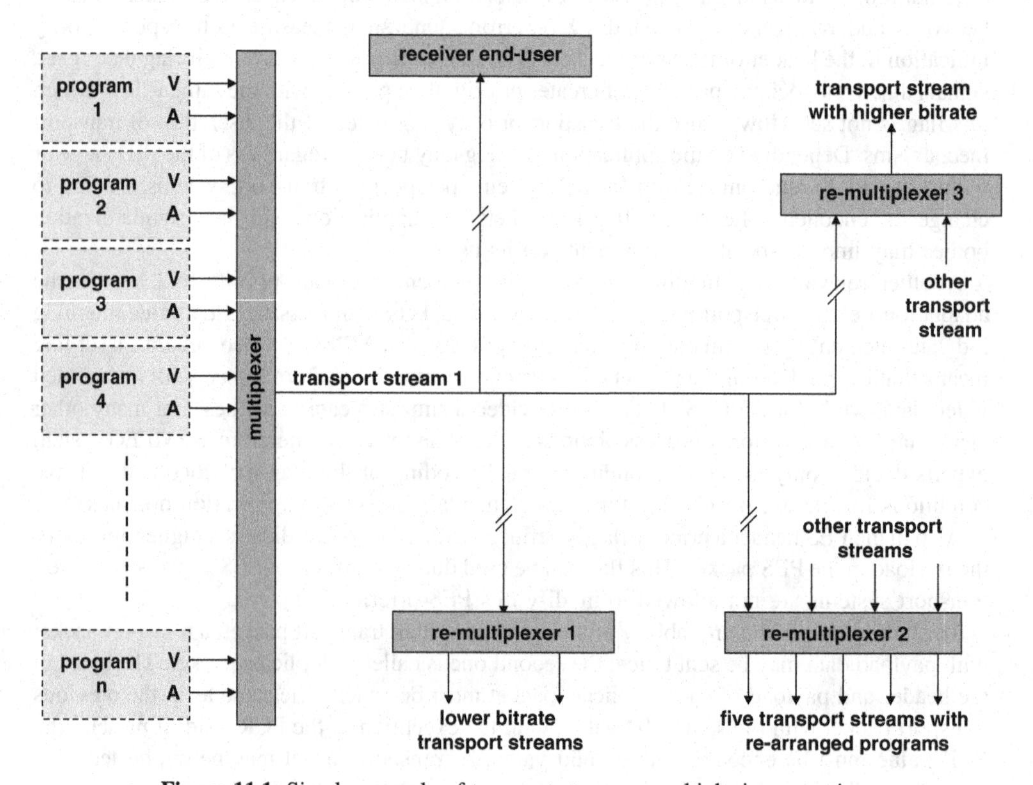

Figure 11.1 Simple example of transport stream re-multiplexing operations

only re-multiplex some of the programs contained in the transport stream(s) at its input, and ignore the other programs. In general, the time bases of the programs involved in the re-multiplex are not locked, and the bitrates of the transport streams at its input and output may not have a known relationship, so that the re-multiplexer may need to compensate for unknown clock behaviours, for example by means of padding.

In any case, the design of practical re-multiplexers, requires knowledge of network characteristics and other issues beyond the scope of MPEG. On the other hand, some requirements are imposed by MPEG to allow for re-multiplexing of transport streams, without violating the compliancy of re-multiplexed transport streams. In this chapter these requirements are discussed by means of the simple, but representative example, depicted in Figure 11.2. This example uses a simple re-multiplexer that selects a few programs conveyed in transport streams 1 and 2 to form a new transport stream 3, intended for delivery across another network. Note that this operation has no impact on transport streams 1 and 2.

In the re-multiplexer of Figure 11.2, the transport packets of the selected programs enter the re-multiplex buffers B_1 and B_2 prior to putting them into transport stream 3. Essential for each re-multiplexer is to maintain the packet order for each elementary stream involved in the re-multiplex, to prevent altering of the elementary streams. With the re-multiplexer in this example, the packets from transport streams 1 and 2 are re-positioned in time when put in

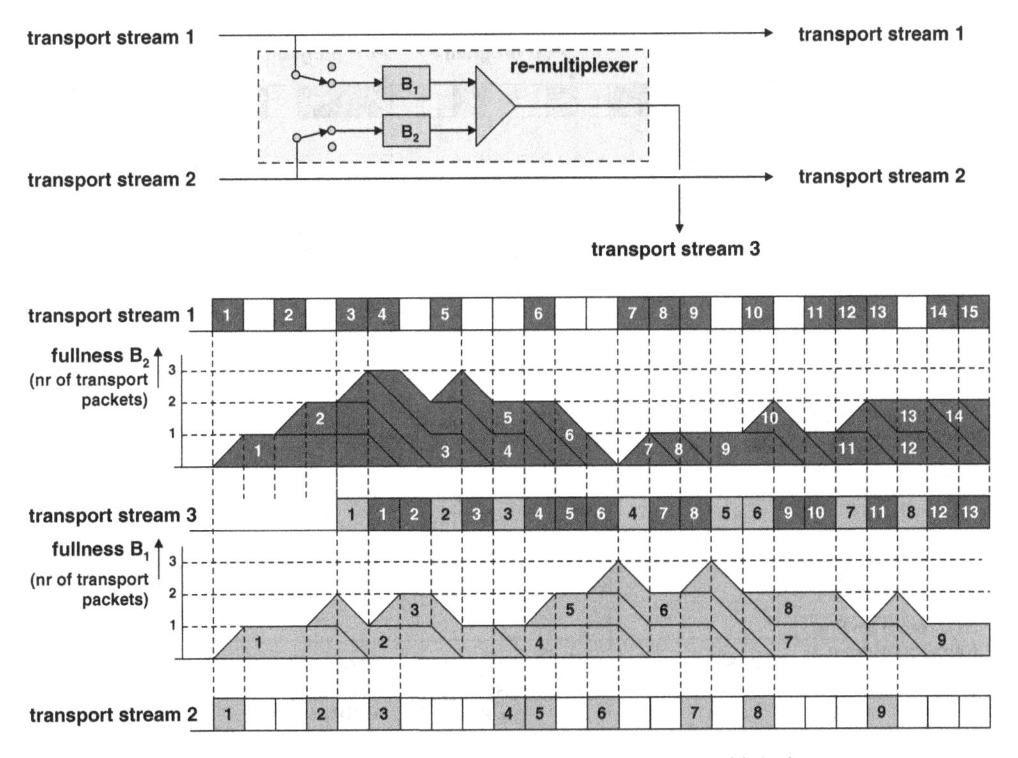

Figure 11.2 Simple example of transport stream re-multiplexing

transport stream 3, while the mutual order of the packets from the same program is maintained. However, in practice it may be required to give priority to one of the elementary streams, for example by inserting a packet with video in the output transport stream prior a packet with audio from the same program, though in the input transport stream the packet with the audio was preceding the packet with the video. As a consequence, a re-multiplexer may change the order in which some of the transport packets with data of a given program arrive at a receiver.

In Figure 11.2, for simplicity reasons, in all three transport streams, the delivery of the transport packets share the same timing, which is only possible if the three transport streams have the same bitrate. The transport packets with data from the selected programs in transport streams 1 and 2 are numbered in the order of arrival. For each selected transport packet, the flow through buffers B_1 and B_2 is depicted. For example, at the start of arrival of packet 3 in transport stream 2, buffer B_1 contains packet 2 only. Upon delivery of the last byte of packet 3, buffer B_1 contains packets 2 and 3, but after putting packet 2 in transport stream 3, only packet 3 is contained. During the time packet 3 is put in transport stream 3, the fullness of buffer B_1 does not decrease, but remains the same, because at the same time packet 4 enters B_1.

As a result of the re-multiplex operation, the packets from transport stream 1 and 2 are put in transport stream 3 with a certain delay that varies per transport packet. Figure 11.3 shows the delay caused by the re-multiplex example in Figure 11.2 for packets 1, 5, 7 and 11 in transport stream 1. For example, the delay D_1 of packet 1, is equal to 5τ, with τ the transport packet duration, while the delay D_7 of packet 7 is equal to τ. The maximum value of this re-multiplex

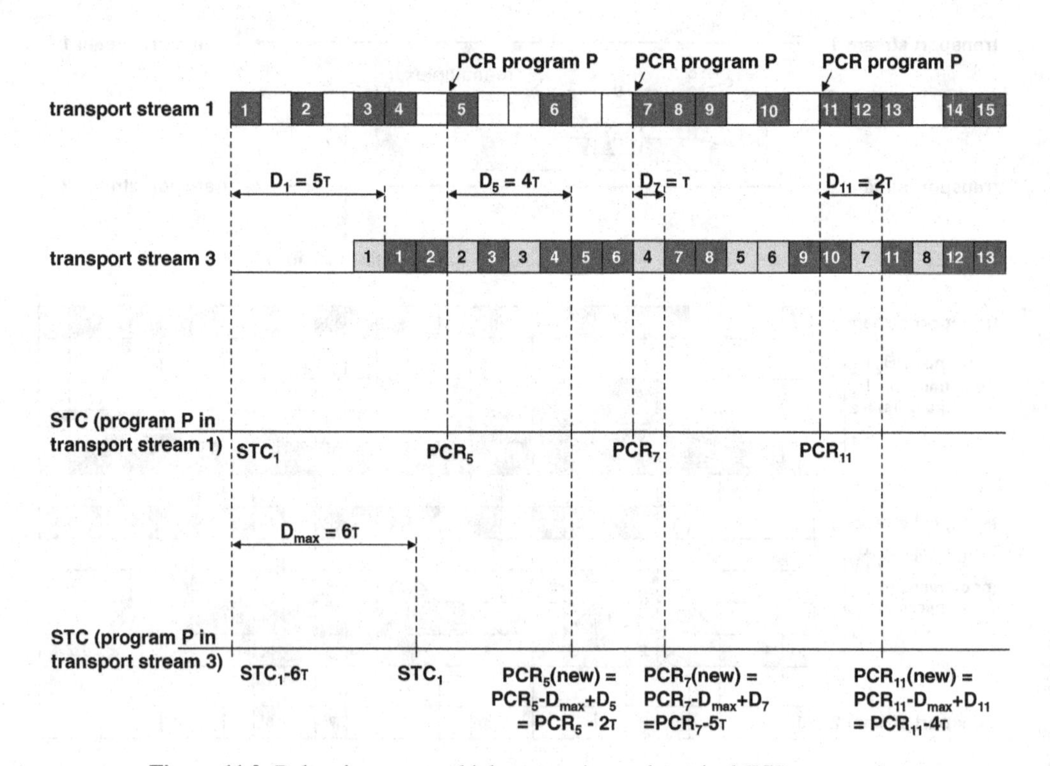

Figure 11.3 Delay due to re-multiplex operation and required PCR re-stamping

delay D_{max} is determined by the re-multiplexer. For the simple re-multiplex example of Figures 11.2 and 11.3, it is assumed that D_{max} is equal to 6τ.

If the data in transport streams 1 and 2 are delivered 'just in time', the data is decoded immediately upon delivery. To prevent underflow when decoding a program from transport stream 3, of each program involved in the re-multiplex, the delay caused by the re-multiplexing must be compensated for in the System Time Clock, STC. For each program in transport stream 3, a constant delay equal to D_{max} must be applied; hence for program P the following must apply: STC(program P in transport stream 3) = STC(program P in transport stream 2) − D_{max} (see Figure 11.3). The time stamps controlling the decoding process remain the same, and therefore the above offset between both STCs ensures that the transport packets from transport streams 1 and 2 can be put in transport stream 3 with any delay between zero and D_{max}.

As a result of the applied offset between the STCs for each program involved in the re-multiplex, additional buffering is needed for the decoding of a program from the transport stream at the output of the re-multiplexer. The required additional amount of buffering for the entire transport stream would be equal to D_{max} times the bitrate of the transport stream. However, for the decoding of a program, there is no need to buffer the entire transport stream; instead it is more efficient to define the additional buffer on a per elementary stream basis, as further discussed in Section 12.3.

Depending on factors, such as the applied networks with their available bandwidth, the number of network switches involved in conveying a transport stream, as well as other constraints and requirements, multiple re-multiplex operations may have been applied in cascade on a transport stream prior to its arrival at the decoder in receivers of users. No constraints are defined for each individual re-multiplexer; instead only a general constraint for all multiplex and re-multiplex operations in an end-to-end delivery system is defined: the maximum amount of delay caused by all multiplex and re-multiplex operations must be less than 4 ms. In most, if not all, practical delivery systems, this constraint leaves more than enough room for re-multiplexing. As an indication, Table 11.1 shows that the 4 ms constraint leaves room for a total delay of about 40 transport packets in case the bitrate of the transport rate is as low as 15 Mb/s; for a bitrate of 45 Mb/s, the delay corresponds to more than 120 transport packets. For the general constraint to be meaningful, the multiplexer that produces the original transport stream should leave room for re-multiplex operations, while the re-multiplexers should use reasonable delays. Which can be ensured in practical end-to-end delivery system specifications, beyond the scope of MPEG systems.

As a consequence of the re-multiplexing delays, the encoded PCR samples of the STC are not valid anymore. For example, in Figure 11.3 it is assumed that packets 5, 7 and 11 in transports stream 1 carry PCR samples of the STC of program P with encoded values PCR_5, PCR_7 and PCR_{11}, respectively. In transport stream 3, these packets are delayed with respectively 4τ, τ and 2τ, while these samples should now refer to the STC of program

Table 11.1 Number of transport packets within 4 ms at two bitrates

Bitrate of transport stream	15 Mb/s	45 Mb/s
Duration of a transport packet ($188 \times 8 = 1504$ bits)	~0.1 ms	~0.03 ms
Number of transport packets within 4 ms	~40	~120

P in transport stream 3, which has an offset of minus D_{max} compared to the STC of program P in transport stream 1. Hence the PCR samples need to be corrected in new PCR values, $PCR_i(new)$, by subtracting D_{max} and adding the actual delay to the original PCR_i value (see also Figure 11.3), as follows:

- $PCR_5(new) = PCR_5 - D_{max} + D_5 = PCR_5 - 2\tau$
- $PCR_7(new) = PCR_7 - D_{max} + D_7 = PCR_5 - 5\tau$
- $PCR_{11}(new) = PCR_{11} - D_{max} + D_{11} = PCR_5 - 4\tau$.

Obviously, the above correction of PCR values is not restricted to delays in terms of τ, but applies in general, as expressed in the following formula:

$$PCR_i(new) = PCR_i - D_{max} + D_{actual},$$

where:

PCR_i is the PCR sample of a program prior to the re-multiplex;
$PCR_i(new)$ is the PCR sample of that program after the re-multiplex;
D_{max} is the applied constant value of the maximum re-multiplex delay;
D_{actual} is the value of the actual re-multiplex delay of the transport packet carrying PCR_i.

Though the PCR re-stamping described above is to be applied on a per program basis, there is no need in the re-multiplexer to reconstruct the STC for each program. The PCR time correction is small, and therefore a free-running 27 MHz clock can be used to measure the D_{actual} value, upon which $PCR_i(new)$ can be calculated and encoded in the PCR field of the transport packet carrying PCR_i. However, the PCR re-stamping may cause a small error in the accuracy of encoded PCR values, in particular when multiple re-multiplexers are cascaded. Therefore the MPEG-2 system specification allows for a tolerance in PCR accuracy of ±500 ns (500×10^{-9} s), corresponding to more than 10 periods of the 27 MHz STC clock. A minor note on this issue: PCR inaccuracies may cause a certain noise on the delivery rate of a transport stream, because this rate is determined from two successive PCRs in the transport stream (see Section 9.4.4).

11.3 Local Program Insertion in Transport Streams

11.3.1 Usage of Local Program Insertions

Local program insertion plays an important role in several digital TV broadcast systems. A typical example is a global news programme that is broadcast countrywide, which is interrupted for some minutes to allow for local insertion of a programme with the local news weather forecast; advertisements, if any, may also be locally orientated. At various areas, a local programme may be inserted, for example each province, state or major city may have its own local programme.

From a quality of service perspective, the local program insertion is required to be as seamless as possible, both when switching to the local program and when switching back to the global program. The latter requires special attention, when from several local program a switch is required to the same global program. In most cases, if not all, a convenient implementation of

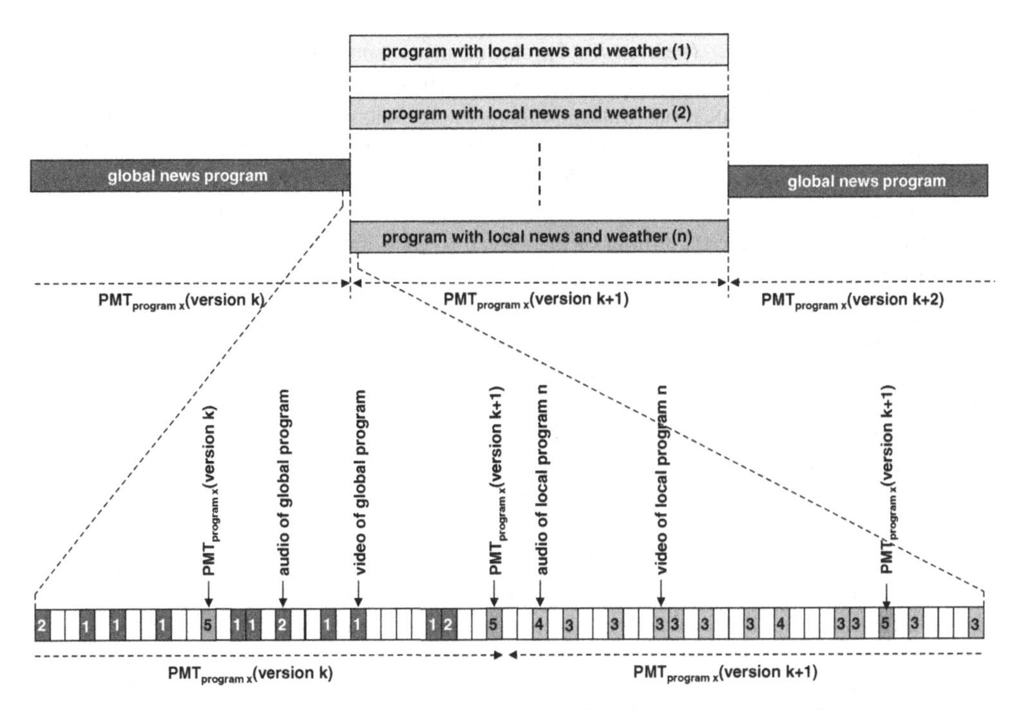

Figure 11.4 Insertion of a local program by changing the program definition

program insertions will require private arrangements in distribution systems. Such arrangements are beyond the scope of MPEG.

11.3.2 Associated PSI Issues

The insertion should be performed so that receivers of the global program continue to receive the local program after the insertion, and return to the global program, once the local program is finished. This can be achieved by ensuring that the insertion does not impact the program-number associated to the program. A straightforward method to implement a local program insertion, is to change the PIDs of the transport packets that carry the elementary streams. This requires switching to another version of the program definition in the PMT.

Figure 11.4 depicts an example for a global program that is interrupted in *n* areas for the local program for that area. In the global program, video and audio are carried in transport packets (1) and (2), respectively. In local program *n*, the video and audio are carried in transport packets (3) and (4). The change is described in a new version of the PMT. The program-map-PID for the local program is the same as for the global program.[1] In Figure 11.4, all versions of $PMT_{program\ x}$ are carried in transport packets (5). For the global program, $PMT_{program\ x}$(version *k*) is valid, and repeated at a certain rate to allow random access to the program. After the last transport packet of the global program, and prior to the first transport packet from local program *n*, the new

[1] Keeping the same program-map-PID value prevents the need for a new version of the PAT as well as the associated timing uncertainty on the validity of the PSI.

PMT version, $PMT_{program\ x}$(version $k+1$), carried in transport packets (5), is inserted in the transport stream, with the current-next-indicator set to '1'. By means of $PMT_{program\ x}$ (version $k+1$) the new locations (and possibly new formats) of the audio and video streams of program x are specified. In case each local program uses the same PID values for carriage of video and audio, the new PMT may be conveyed by the global program with the current-next-indicator set to 'next', and set to 'current' just after the last transport packet with data of the global program, so as to indicate the timing of the switch to each local program.

At the end of the local program a similar switch is performed back to the global program, based on the next PMT version, $PMT_{program\ x}$(version $k+2$). To let this approach work, a private arrangement is needed that the local program will use only a single version of the PMT. Other private arrangements may be needed as well, but they are considered to be beyond the scope of this book.

The approach whereby the program definition is changed by means of a new version of the PMT, has disadvantages; in particular, the time needed to switch from the old to the new program definition will take time. Therefore the option for program insertion without changing the program definition was considered too. This requires the system responsible for the program insertion, to replace the coded video, audio and other elementary streams of the global program by the corresponding streams of the local program and vice versa, and involves time base discontinuities if the local and the global program have different time bases. Note also that a local program insertion without changing the program definition, requires the use of the same stream-type and applicability of the same descriptors in the global and in the local program.

To support local program insertions within the same program definition, MPEG-2 systems specifies some tools for two purposes, as described below.

- The first purpose is signalling to receivers of any discontinuities that may occur, so as to allow proper decoding of transport streams with local program insertions in case such discontinuities occur.
- The second purpose is to assist systems (in the network) to perform program insertions as seamlessly as possible. However, MPEG leaves room for flexibility to implement the program insertions in a manner deemed suitable by the application. This is important because performing 'a perfect edit' may in practice be complicated,[2] and considered too complex to implement, if possible at all.

In any case, to avoid violation of the MPEG-2 system specification, the insertion must be scheduled carefully and in a timely and accurate manner, as further discussed below.

11.3.3 Time Base Discontinuities

Figure 11.5 presents the same program insertion example as Figure 11.4, but without changing the program definition in the PMT. In the example of Figure 11.5, two time base discontinuities are associated to each local program insertion: one at the transition from the global program time base to the local program time base (n), and the reverse one at the transition back to the global program. To signal a time base discontinuity within a program, a one-bit flag, called the discontinuity-indicator, is used in the adaptation field of the transport packets that carry the

[2] A 'perfect edit' would avoid editing problems (as discussed in Figure 6.19 in Section 6.4) and would, for example, require accurate reconstruction of the sampling clocks of audio and video at insertion.

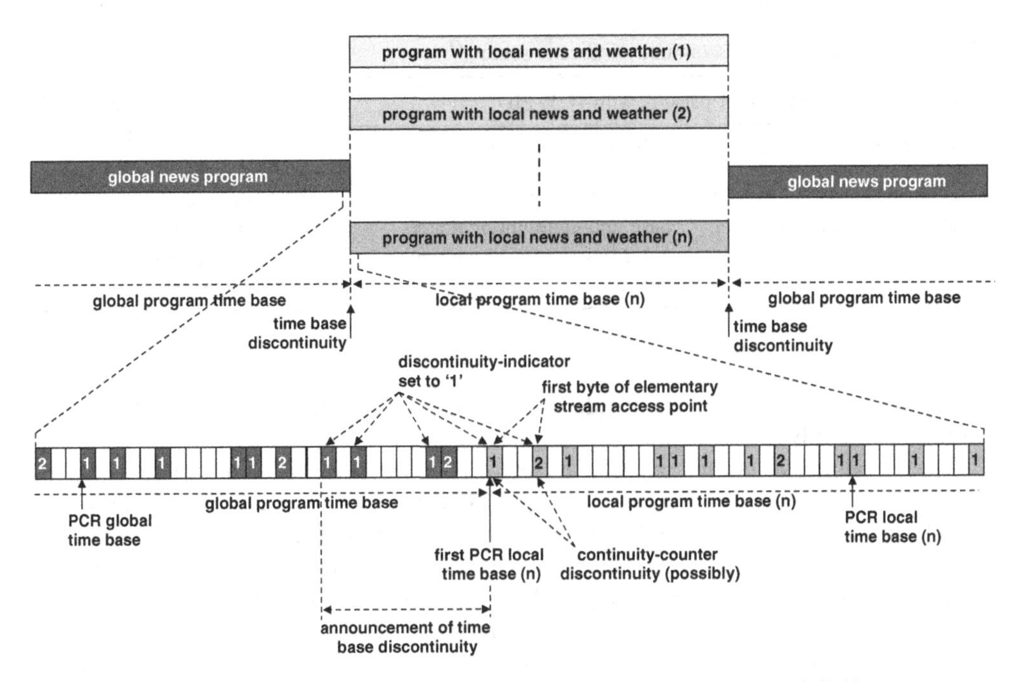

Figure 11.5 Local program insertion without changing the program definition

PCR samples of that program. The discontinuity-indicator must set to '1' in the transport packet that carries the first sample of the new time base. Prior to the arrival of this first sample, the previous time base remains valid, while the new one becomes immediately valid after the arrival of the first PCR of the new time base. A time base discontinuity may be announced by the discontinuity-indicator set to '1' in a series of transport packets with the PCR-PID preceding the packet with the PCR of the new time base. In such packets the PCR field must be absent. However, once set, this indicator must be set in each following packet with the PCR-PID until the PCR of the new time base arrives.

In Figure 11.5, the transport packets (1) and (2) carry the video and the audio of the global and, after the insertion, of the local program, while the same transport stream also carries transport packets of other programs. Transport packets (1) are assigned as the PCR-PID; the first packet (1) in Figure 11.5 carries the PCR of the global time base. Four packets (1) follow; if the adaptation field is present, the discontinuity-indicator is set to '0'. In the next three packets (1), the time base discontinuity is announced by setting the discontinuity-indicator in the adaptation field to '1'. The next transport packet (1) carries the PCR of the new time base; upon its arrival, local time base (*n*) becomes valid.

In the transport packet with the PCR of the new time base, the continuity counter may be discontinuous. At program insertion, MPEG does not require to keep track of the continuity counter values. This allows simple implementations of switching to local programs, and back to the global program. For example, the number of used transport packets for each elementary stream is not necessarily the same for each local program. By allowing continuity counter discontinuities, each local program has freedom to allocate the total amount of available

Table 11.2 Meaning of discontinuity-indicator set to '1'

Transport packets with PCR-PID		Transport packets with a PID other than PCR-PID
No PCR carried in packet	PCR carried in packet	
Announcement of (potential) time base discontinuity	(Potential) discontinuities of time base and continuity-counter No discontinuity of continuity-counter	(Potential) discontinuity of continuity-counter Set to '1' in at most two consecutive packets

transport packets for the carriage of video and audio, while also the local programs do not need to finish at exactly the same time.

The discontinuity-indicator may also be set to '1' in transport packets not designated as the PCR-PID. A discontinuity-indicator set to '1' in such packet, indicates that the continuity counter in that packet may be discontinuous; see the first transport packet (2) following the time base discontinuity in Figure 11.5. Hence, the semantics of the discontinuity-indicator depend on the context, as summarized in Table 11.2. MPEG systems requires that in transport packets not designated as the PCR-PID, the discontinuity-indicator is set to '1' in at most two consecutive packets with the same PID. This prevents compromising the packet loss verification mechanism based on the continuity counter described in Section 11.1.

To ensure a good random access to the inserted program, the first byte of each elementary stream after a time base discontinuity must represent an access point of that elementary stream, carrying sufficient information to start its decoding. Furthermore, a PTS must be present for the video picture or audio frame at this access point. See the first packets (1) and (2) with data associated to the new time base of local program (*n*) in Figure 11.5. In the case of video, the access point data may be preceded by a sequence-end-code, to signal the end of the video sequence associated to the previous time base. Care should be taken regarding the positioning of transport packets; the time base discontinuity serves as a separation point between data associated to the previous and the new time base, as follows.

- Prior to the time base discontinuity, no transport packets should be present with data associated to the new time base. This means that the transport packet with the PCR of the new time base, is the first packet with data associated to the new program. In Figure 11.5, the first packet (2) of local program (*n*) is indeed put in the transport stream after (and not prior to) the first packet (1).
- After the time base discontinuity, no transport packets should be present with data associated to the previous time base. Hence, all required transport packets with data of the previous program should be present in the transport stream prior to the packet with the PCR of the new time base in the transport stream.

However, as discussed in Section 11.2, re-multiplex operations may change the order in which transport packets with data of a given program are received. Hence, when re-multiplexing is applied after the program insertion, the time base discontinuity cannot strictly serve as a separation point, and therefore MPEG systems somewhat relaxed the above requirement towards transport packets that contain a PTS. A transport packet with a PTS referring to the previous time base must be received prior to a time base discontinuity, while a transport packet with a PTS that refers to the new time base must be received after the time base discontinuity.

In practice, after a program insertion, no re-multiplexers may be applied, in which case the time base discontinuity can easily be enforced by applications as a true separation point, if so desired. In case further re-multiplexing may occur, it may prove feasible at program insertion to avoid inserting a transport packet with a PTS within a time window around the time base discontinuity, corresponding to the maximum displacement of transport packets due to re-multiplexing. In a well behaved system, as discussed in Section 11.2, the maximum delay due to re-multiplexing is 4 ms, in which case the transport packets are re-positioned in time no more than 4 ms, which would require a time window of ±4 ms around the time base discontinuity. If the maximum displacement is known to be a smaller value in an application, then the time window can be reduced accordingly. It should be noted however, that moving transport packets away from the time base discontinuity will require displacing some transport packets with data from other programs, which may be modelled as a re-multiplex operation. MPEG-2 systems requires that a PTS is encoded for the first access unit of each elementary stream after a time base discontinuity. Therefore, if a program element is carried in a certain PID, then as long as no packet with a PTS is received after a time base discontinuity, receivers should assume that packets of that PID contain data associated to the previous time base.

When a decoder receives a transport stream with a time base discontinuity, data of two time bases are in the input buffer of the decoder in the period between receiving the time base discontinuity and the instant in time all data associated to the previous time base is decoded. At program insertion it is to be ensured that during this period, no overflow occurs of buffers in the decoder. This issue is discussed in Section 11.4 on splicing.

11.4 Splicing in Transport Streams

Inserting a local program, as discussed in Section 11.3, involves splicing operations, a form of editing of coded audio and video streams within a transport stream, whereby packets with elementary streams of the previous program are replaced by packets carrying the elementary streams associated to the next program. Depending on the requirements of the application, the splicing may or may not be seamless. In case of a seamless splice, discontinuities as discussed in Figure 6.19 are prevented. However, in many cases it is difficult to ensure a seamless splice, if possible at all, while choosing appropriate characteristics of the distributed video and audio can hide undesired effects of non-seamless splices. For example, to hide the transition, the audio may go silent and the video to a dark screen at the end of the previous program. As a consequence, the market relevance of seamless splices may be limited.

Each splice needs to meet a number of requirements, whether the splice is seamless or not: the resulting transport stream must be fully MPEG systems compliant. To ensure this does not only require the appropriate positioning of transport packets with data of the previous and new programs, as discussed in Figure 11.5, but also preventing the overflow of buffers in receivers and complying with timing constraints for elementary stream decoding. MPEG-2 systems specifies these requirements without addressing how to meet them. Though encoders may use tools in transport streams to assist in performing splices, in general, private arrangements may be required to ensure that the splicing can be performed in a manner deemed acceptable by the application.

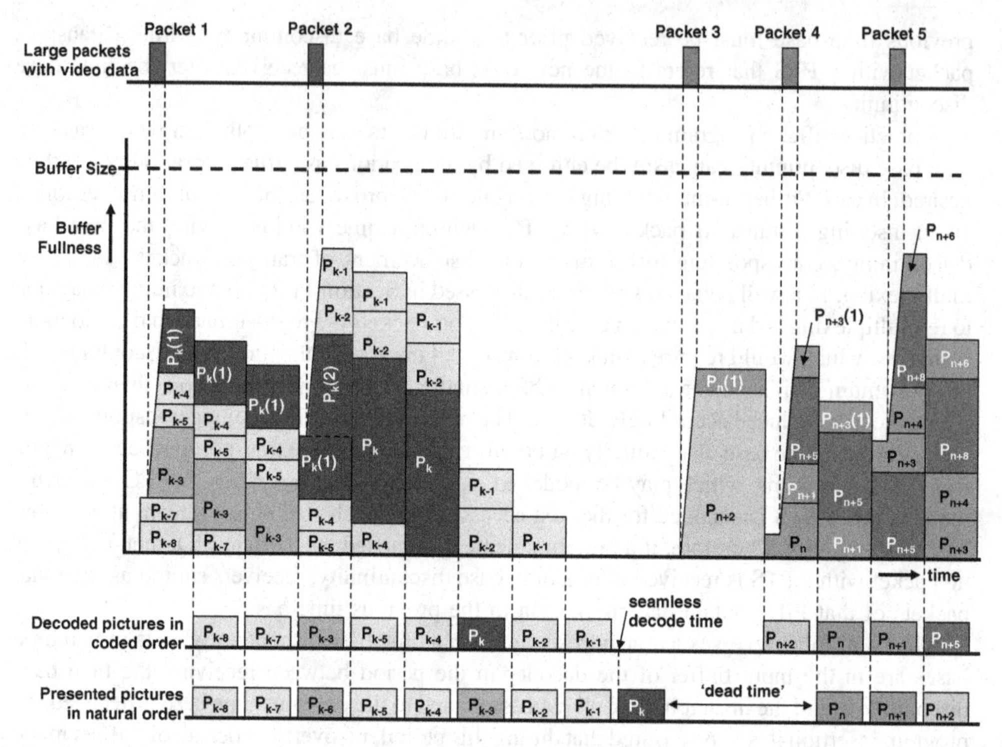

Figure 11.6 Example of a non-seamless video splice

Figures 11.6 and 11.7 discuss some important issues to consider when a video splice is performed. For drawing convenience, both figures presume multiplexing of large packets instead of using 188-byte transport packets. Furthermore, a single input buffer is assumed for storing the video data contained in the large packets prior to the instantaneous decoding of the coded video pictures. By means of this simple decoding model, the issues with buffering are discussed that play a role in the STD model for transport streams, described in Chapter 12. Figures 11.6 and 11.7 show the delivery of the packets with coded picture data, the decoded pictures in coded order and in their natural order for presentation. Also storage of the coded pictures in the input buffer is shown, thereby indicating the fullness of the input buffer over time.

Figure 11.6 shows the performance of a video decoder after a non-seamless video splice. Pictures P_{k-8} up to P_k are from the video prior to the splice, and pictures P_n up to P_{n+8} from the video after the splice. The packets with video data are numbered 1 to 5; when packet 1 arrives the coded data of P_{k-8} and P_{k-7} are present in the buffer. Packet 1 contains the coded data of P_{k-3}, P_{k-5}, P_{k-4} and the first fragment $P_k(1)$ of P_k. Packet 2 contains the remaining video data prior to the splice, and packet 3 the first video data after the splice: P_{n+2} and the fragment $P_n(1)$. To prevent splicing complications, the system taking care of the insertion postpones the delivery of packet 3 until all pictures from the video prior to the splice are decoded. This delay causes a 'dead time' between the presentation of the last picture P_k prior to the splice and the first picture P_n after the splice. The decoder can determine the duration of the dead time after the

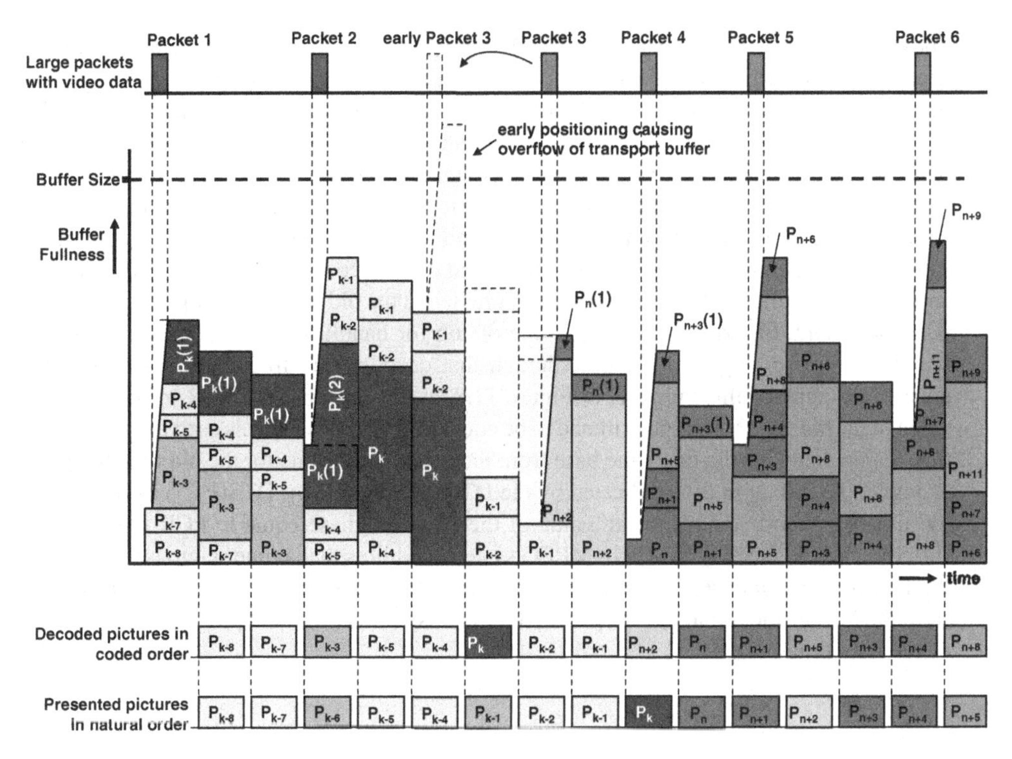

Figure 11.7 Example of a seamless video splice

presentation of P_k from the PTS encoded[3] for P_n. When the splice involves a time base discontinuity, obviously, this PTS relates to the new time base. MPEG does not specify how a decoder needs to perform during such dead time and leaves this to decoder manufacturers or applications; one option is to present P_k during the dead time period as a frozen picture. However, applications may prefer to hide the splice for viewers by requiring that the video just prior to the splice fades out to black, or something similar.

The system responsible for the splicing can perform a seamless splice by ensuring that no decoding discontinuity as discussed in Figure 6.19 occurs. This requires that the first picture after the splice is decoded at the seamless decode time, that is at the same decoding time as the next picture from the video prior to the splice would have had if this video had continued: the seamless decode time as indicated in Figure 11.6. Two conditions are to be taken into account however for any elementary stream involved in the splicing operation:

- The packets with the elementary stream data must be placed in the multiplex so, that at any time, buffer overflow is prevented. For example, if packet 3 in Figure 11.7 is shifted forward to an instance in time, when the coded video data of picture P_k is still in the input buffer, then the fullness of this buffer would become more than the size of this buffer, and hence buffer overflow would occur (see Figure 11.7). As a result, fragments of pictures P_{n+2} and P_n may get lost in receivers.

[3] Encoding of a PTS for the first access unit after this dead time is mandated by MPEG-2 systems; see also the discussion in Section 6.4 and Figure 6.19.

- The decoding time of the first access unit after the splice must not be earlier than the seamless decode time, so as to ensure that the decoding of the stream prior to the splice can be properly finished before the decoding of the stream after the splice commences. This requirement also applies when at the splice a time base discontinuity occurs; in that case the offset between both time bases needs to be taken into account. For example, assume that packet 3 in Figure 11.7 carries the first PCR of the new time base: PCR (packet 3). In that case, the arrival of this PCR signifies a time base discontinuity. After this time base discontinuity, some pictures from the old video stream still need to be decoded and/or presented. The DTS and PTS time stamps of these pictures are expressed in the old time base, though the new one is already valid. Figure 11.8 describes one option for handling the two time bases for the seamless splice of Figure 11.7; note though that decoders are free to apply any strategy deemed suitable for this purpose. In Figure 11.8, at the arrival of the PCR in packet 3, the value of the old time base is determined to be equal to T, and this value is used to calculate the time stamp value for the new time base from the time stamp value for the old time base. For example, the instant in time indicated by the DTS of picture P_{k-1}, DTS(P_{k-1}) in units of the old time base, corresponds to the value of the new time base equal to PCR (packet 3) + [DTS(P_{k-1}) − T] (see Figure 11.8). MPEG systems requires that a time stamp is encoded for the first picture after the time base discontinuity. Hence, the value of DTS(P_{n+2}) is found from the PES header contained in packet 3. Obviously, this time stamp is expressed in units of the new time base.

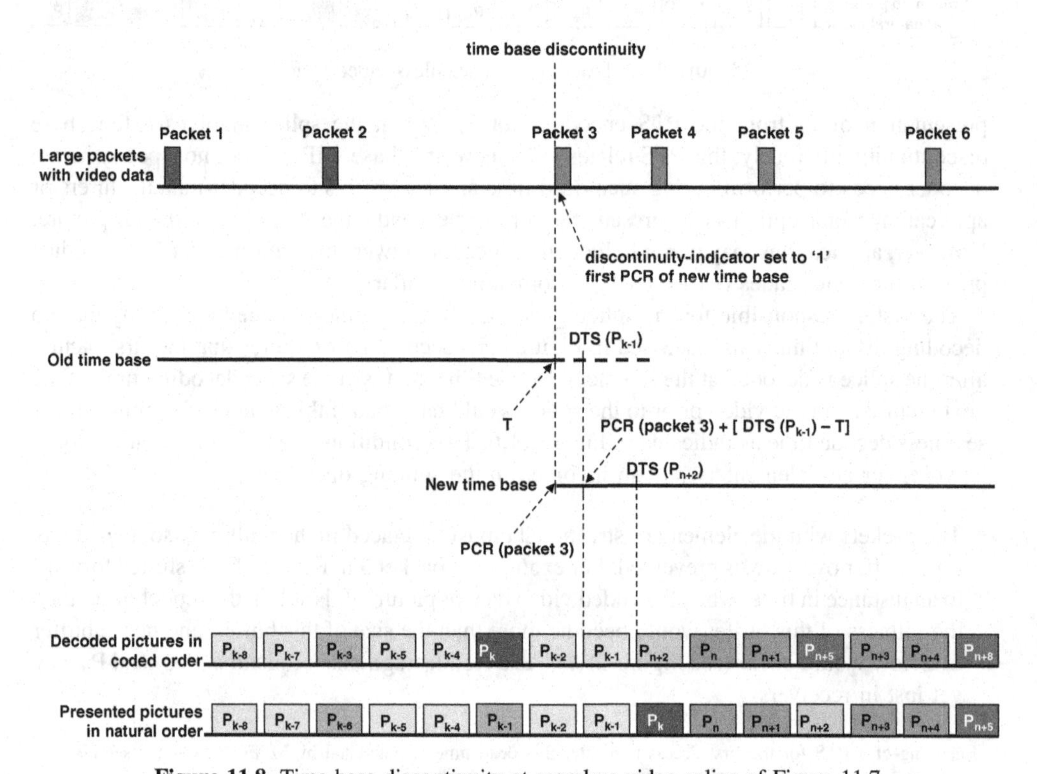

Figure 11.8 Time base discontinuity at seamless video splice of Figure 11.7

To prepare for performing a splicing operations, an eight-bit splice-countdown field may be carried in the adaptation field in transport packets. The presence of this optional field is indicated by the splicing-point-flag. The splice-countdown field is used in packets with the same PID value, and is encoded with a value that may be positive or negative. A positive splice-countdown value indicates the number of transport packets with the same PID until a splicing point is reached, while ignoring duplicate transport packets and packets without payload data (i.e. adaptation field only). The splice point is located immediately after the transport packet with a splice-countdown value of zero; the last byte in the payload of this transport packet must be the last byte of an access unit, such as an audio frame or video picture; to achieve this may require padding in the adaptation field (see Section 11.6).

The first transport packet with the same PID after the splicing point must carry a PES header with a PTS, followed by a PES payload that commences with an elementary stream access point. However, in certain video cases, for example if a different frame rate is applied, the access point may be preceded by an end-of-sequence code. Note that within an MPEG-2 video sequence certain parameters are not allowed to change without starting a new video sequence. In case a splice is performed, the transport packets after the splice point will contain a different elementary stream of the same type. After the splice point, the transport packets with the same PID may carry a splice-countdown field coded with a negative value, specifying the number of transport packets of the same PID after the splice point, excluding duplicate packets and packets without payload (see Figure 11.9).

To perform a seamless splice requires detailed knowledge of the characteristics of the coded elementary streams prior to the splice, such as (decoder input) buffer requirements over time,

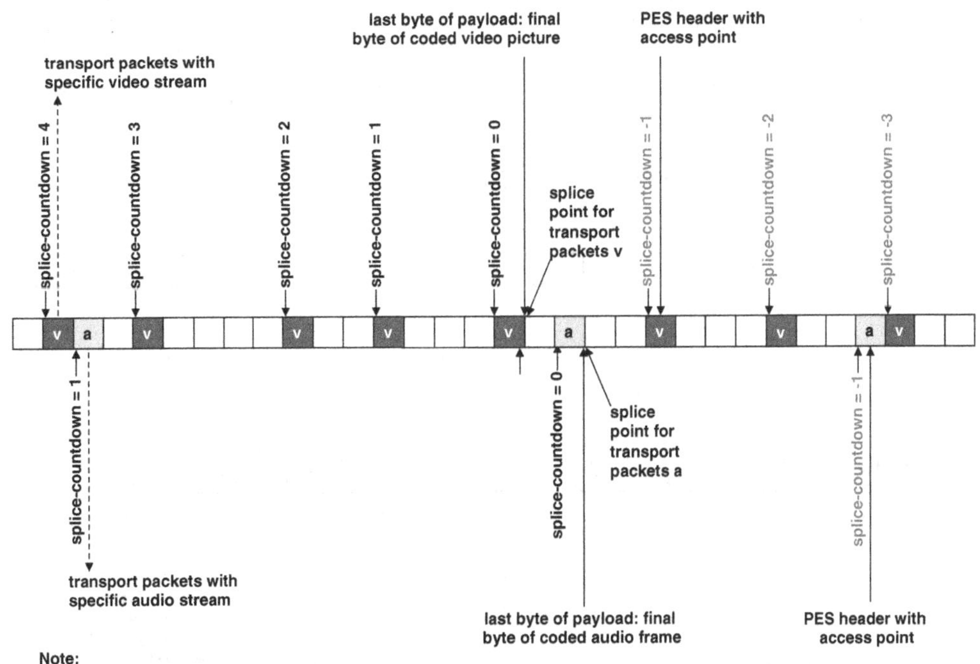

Figure 11.9 The use of splice-countdown

time stamps and video frame rates. This information can be obtained by parsing the elementary streams, but this significantly complicates the system that performs the splicing operation.

Though the market need for seamless splicing is less obvious, several, possibly more research orientated, system experts have put significant effort in designing extensive support for seamless splices in the MPEG-2 system specification. By setting a seamless-splice-flag, the presence of the splice-type and DTS-next-AU fields is indicated. The DTS-next-AU signals the value of the seamless decode time, indicated in Figure 11.6. The use of splice-type is defined for the VBV model of MPEG-2 video only; its objective is to specify constraints on the video stream inserted at the splice point, so as to prevent buffer overflow. For the same purpose, it is possible to specify a so-called legal time window and a piecewise rate for the delivery of transport packets. The use of these fields is complicated and their scope is limited to MPEG-2 video under specific circumstances. In conclusion, these tools may have more scientific than practical value.

The above discussion on splicing, assumes the use of non-low delay video with B-pictures, or the equivalent thereof, whereby at decoding a video delay is caused, as discussed in Section 6.1.2 of this book. The use of non-delay video is common practice in transport stream applications; in broadcast networks to consumers, coding efficiency is commercially far more important than low delay. However, transport stream applications that support low delay video, should be aware that a seamless splice is impossible when switching from low delay to non-low delay video. Figure 11.10 shows the insertion of a low delay video sequence within a non-low delay video stream; two splices are involved, one to low delay video and the other back to non-low delay video. In low delay video, each picture is decoded and presented at the same time: for each picture, the PTS time stamp and the DTS time stamp have the same value. For a seamless first splice, the first picture P_n of the low delay video must be presented immediately after the

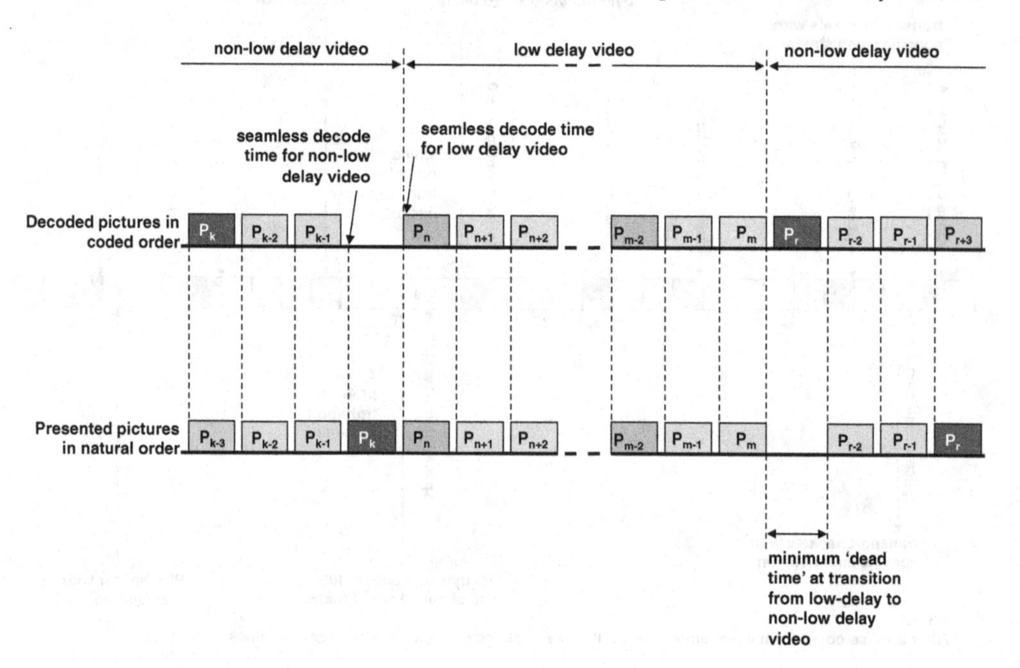

Figure 11.10 Splicing low delay video within a non-low delay video stream

presentation of the last picture of the preceding non-low delay video. This is possible, as shown in Figure 11.10. However, the second splice cannot be seamless, because picture P_{r-2} cannot be presented immediately after the last low delay picture P_m: the I-picture P_r is to be decoded prior to the decoding and presentation of P_{r-2}, and the decoding of P_r can only start after the decoding of P_m, as in the decoding model of MPEG-2 systems the decoding of two pictures at the same time is not permitted. As a consequence there is some minimum 'dead time' of one picture period at this transition (see Figure 11.10). Just for scientific completeness . . .

11.5 Variable Bitrate and Statistical Multiplexing

The coding efficiency of video, audio and other elementary streams depends on the amount of redundancy in the elementary stream that can be exploited by the coding algorithm. The available redundancy within an elementary stream usually varies over time and as a result also the coding efficiency varies over time. To achieve a constant quality for the coded content, it is beneficial to vary the available bitrate for its coding. For example, when coding a movie, to achieve a constant audio and video quality throughout the whole movie, may require for critical scenes a bitrate twice as high as the bitrate for other scenes in the movie. In program streams, the available bitrate for the coding can be adapted conveniently to the compression characteristics of the content by changing the program-mux-rate on a per PES packet basis, as discussed in Section 9.3 and Figure 9.3.

In most, if not all, broadcast applications, transport streams have a constant bitrate, thereby providing 'a pool of bandwidth' for sharing amongst multiple programs. As the contents of the contained programs are usually uncorrelated, there is benefit when a multiplexer can allocate bandwidth to each program depending on the real-time needs of the contained programs. Depending on the complexity of the scenes in the contained programs at a time, the multiplexer aims to assign more bandwidth to complex scenes than to less complex ones. By means of this bandwidth sharing, referred to as 'statistical multiplexing', the best coding quality can be achieved for the contained programs at the available aggregate bandwidth. Figure 11.11 shows a simple example of statistical multiplexing of eight programs contained in a transport stream; the variation of the bandwidth allocation over time is shown. During the time window in Figure 11.11, the bandwidth allocated to program 6 remains constant. At t_1, t_2, . . . , t_{10}, the bandwidth allocation changes; for example, at t_1, more bandwidth is allocated to program 2 and less to program 4, while at t_2, program 5 receives more bandwidth and program 8 less.

Statistical multiplexing is only possible, if the multiplexer can control the amount of bandwidth allocated to each program. However, the content of a program may already be encoded at an earlier stage; for example the compressed data of a program may be provided by a remote encoder or delivered in another transport stream. In such cases, transcoding is needed to benefit from the statistical multiplexing. Note though that transcoding operations contribute significantly to the end-to-end delay.

11.6 Padding and Stuffing

In the design of MPEG-2 systems, byte alignment of all MPEG-2 system defined data structures has been an important requirement. Each transport packet and its header, each PES packet and its header, each section and its header, each descriptor and its header, are all

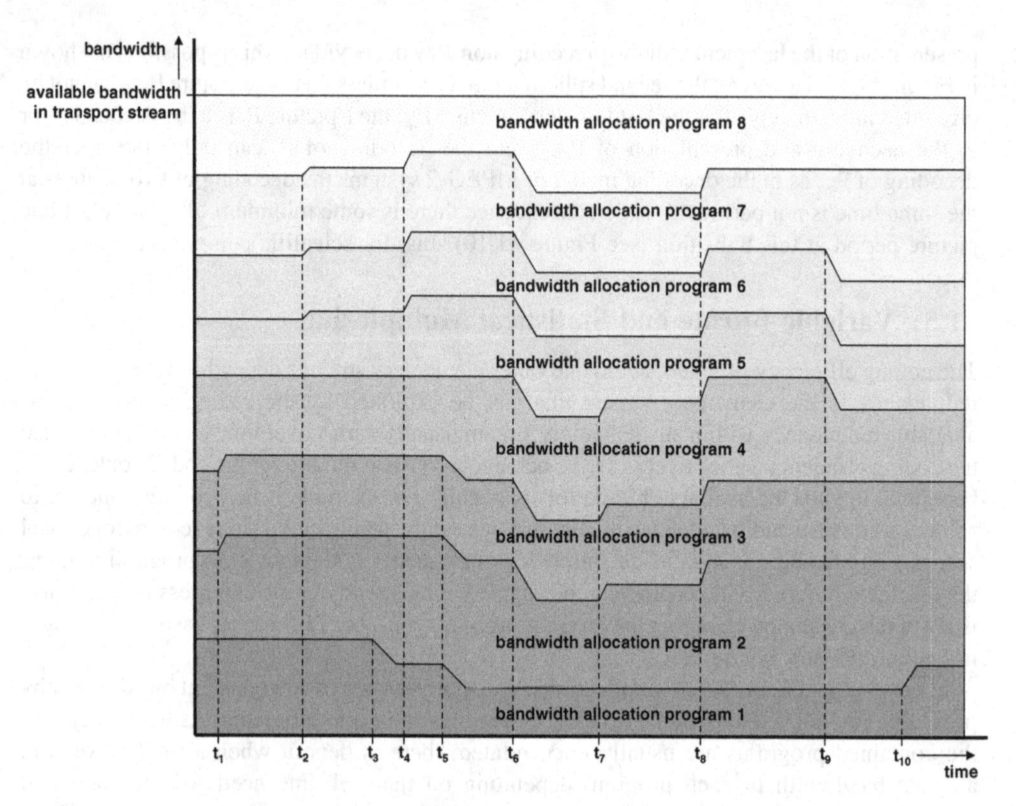

Figure 11.11 Simple example of statistical multiplexing

byte aligned. As a result, also the payload data of transport packets, PES packets, sections and descriptors consist of an integer number of bytes.

Because all data structures defined in MPEG-2 systems are byte aligned, there is no need to support bit stuffing in MPEG-2 systems. However, if access units carried in an MPEG-2 system stream are not byte aligned, then the corresponding elementary stream specification must support bit stuffing, to allow for byte alignment. For example, if an access unit of an elementary stream is required to start in the first byte of the payload of a certain PES packet, than the previous access unit ends in the preceding PES packet. If the last bit of this previous access unit is not byte aligned, then bit stuffing must be applied, for instance at the end of the access unit, to ensure that the PES payload contains an integer number of bytes with valid elementary stream data. For this purpose, each MPEG video and audio specification using non-byte aligned access units, defines how stuffing bits can be added to the coded video picture or audio frame to achieve byte alignment. Note that elementary streams may allow bit and byte stuffing for various reason; for example in audio frames stuffing may also be applied to adjust the average length of an audio frame in time.

In the MPEG-2 system specification, the use of the term padding applies to a mechanism whereby stuffing bytes are added to data structures such as PES packets and transport packets to achieve alignment with other data structures or data rates. MPEG-2 systems defines a variety of options for padding by means of adding stuffing bytes. In each option, the stuffing bytes do not carry useful information and may therefore be discarded by decoders. The use of padding is associated with efficiency loss and its regular usage is therefore preferably prevented. However, under specific circumstances, other considerations may prevail above efficiency, in which the use of padding proves very useful. The available padding options in MPEG-2 systems are discussed below, with some associated use cases. Except in the case of a null transport packet, each stuffing or padding byte has a fixed value equal to '1111 1111', or, in hexadecimal terms, 0xFF.

- Padding PES packets; as in MPEG-1 systems, padding PES packets can be used in MPEG-2 systems; the payload of a padding PES packet contains only padding bytes. A padding PES packet may be useful in program streams. For example, when a program stream is carried in sectors of a fixed size, then a padding PES packet may be needed to conclude the program stream in a sector aligned manner.
- Stuffing bytes in PES header. Padding PES packets cannot be used if the required number of padding bytes is smaller than the minimum size of a padding PES packet. To address the need for adding a small number of stuffing bytes, one or more stuffing bytes can optionally be inserted in the header of PES packets; at most 32 stuffing bytes may be inserted. This padding option is also suitable to provide word (16 bit) or long word (32 bit) alignment in applications where eight-bit alignment is not sufficient.
- Stuffing bytes in pack header; at most seven stuffing bytes can be included in a pack header. In case packs need to be aligned at word or long word level, this option allows alignment without changing PES headers.
- Null transport packets; null packets are intended for padding of transport streams. In cases where the bitrates of some constituent elementary streams are not locked to the bitrate of the transport stream, the bitrate of the transport stream must be chosen high enough to compensate for worst case bitrate inaccuracies. In which case the bitrate of the transport stream must be slightly higher than the total bitrate of the constituent streams; as a compensating measure every now and than a null packet may be inserted. Note however that instead of using null packets, the required stuffing may also be implemented by means of transport packets carrying useful data. For example, a transport packet with a program definition or non-time critical application data can be conveyed slightly earlier than originally scheduled. Null packets may also be inserted or deleted by re-multiplexing processes to compensate for differences in bitrates of the re-multiplexed programs. The data bytes in the payload of null packets may take any value, but their delivery to the decoder cannot be assumed. For this reason, a null packet is not allowed to carry a PCR.
- Stuffing bytes in adaptation field of transport packet header. The adaptation field may contain one or more stuffing bytes. If desired, the entire transport packet can be filled with stuffing bytes, in which case there is no payload in the transport packet. For transport packets carrying PES packets, this is the only available padding option. Each PES packet starts in the payload of a transport packet and therefore the last byte of the payload of the preceding PES packet must be the last byte in the payload of a transport packet. When there are insufficient PES packet data bytes to completely fill the transport packet payload, then stuffing bytes can be inserted in the adaptation field of the involved transport packet, so as to reduce the number of payload bytes as needed to accomplish the required alignment. Note however that PES packets conveyed in transport streams are usually constructed specifically for this purpose; to ensure the most efficient transport, the PES packet dimensions are chosen so that no

stuffing in transport packets is needed. Stuffing bytes in the adaptation field may also be needed in cases where the first byte of an elementary stream in a transport packet needs to be the first byte of an access unit. As a consequence, the last byte in the payload of preceding transport packet must be the last byte of the preceding access unit. However, the remaining last fraction of this preceding access unit is typically smaller that the size of the available payload. For this purpose the adaptation field can be enlarged by including a number of stuffing bytes, so that the payload is reduced with the same number of bytes so as to fit the size of the remaining access unit fraction.

- Stuffing bytes after a section carried in a transport packet; as discussed in Section 9.4, the table-ID value 0xFF (see Table 9.4) is reserved for usage as stuffing byte in the remaining payload of a transport packet after a section. After the last byte of a section, the table-ID value of the next section is found, but when the stuffing byte value 0xFF is identified, then all bytes until the end of the transport packet are stuffing bytes. Hence, once the first stuffing byte is found, there is no need for receivers to continue the parsing of the transport packet. This option can be used in cases where the next section must start in the beginning of the next transport packet with the same PID.

  A special case occurs if in a transport packet without the pointer field, the last byte in the payload is the first byte of a next section; in that case, the pointer field must be included as the first byte in the payload of the transport packet to specify the position in the payload of the first byte of the next section (see Section 9.4.5). However, including the pointer field would move the first byte of the next section to the payload of the next transport packet, in which case the presence of the pointer field in the payload is not allowed.[4] To resolve this conflict, a stuffing byte can be inserted after the last byte of the section in the current transport packet, so that the first byte of the next section is moved to the payload of the next transport packet; in that case no pointer field is required in the current transport packet.

  Note that instead of inserting stuffing bytes after a section, the required stuffing can also be performed using the stuffing bytes in the adaptation field, as discussed above. However, system encoders may prefer to insert stuffing bytes after a section, as this prevents the need to change the header of the transport packet.

11.7 Random Access and Parsing Convenience

It is important to achieve a good random access to an MPEG-2 system stream, for example upon recovery after a transmission error or when zapping from one program to another in the same or in a different transport stream. The time elapsed between starting to read the system stream at the new position and the start of the presentation of the decoded video, audio and other content should be acceptable. In general terms, some random access issues and requirements are already discussed in Section 6.7. Below, issues are discussed in more detail.

A decoder can only play content of a program in an MPEG-2 system stream when it has access to essential system information, such as which streams are part of the program, the time base of the program and time stamps of access units. To retrieve information from a transport stream requires access to the PAT, PMT and other PSI data, which can be found rapidly, when the conveyed PSI data is repeated regularly. MPEG does not specify the PSI repetition rate, as

[4] Presuming that the previous section does not commence in this transport packet.

requirements differ depending on the application. For example, when an application uses rather static or quasi-static program definitions, there may be less need to regularly repeat PSI data. The same may apply in a movie on demand application, where a server transmits the transport stream with the movie to a single user.

To reconstruct the time base and to correctly present decoded content, requires access to the conveyed samples of the time base, the SCR in a program stream and the PCR in a transport stream, as well as the PTS and DTS timestamps. MPEG systems does only specify constraints for the minimum occurrence of SCRs, PCRs and PTSs in program and transport streams. Obviously, applications may enforce to encode these fields more regularly.

To assist random access in transport streams, a payload-unit-start-indicator is provided in the transport packet header; in each transport packet, this indicator must be set to either '1' or '0'. Its meaning depends on the type of data carried in the transport packet, as follows.

- If the transport packet carries PES packet data, then a payload-unit-start-indicator set to '1' indicates that the transport packet payload starts with the first byte of a PES packet header; if set to '0', then no PES packet starts in the transport packet.
- If the transport packet carries section data, then a payload-unit-start-indicator set to '1' indicates that the transport packet payload contains the first byte of a section; in that case the first byte in the transport packet payload is the pointer field, that specifies the position in the payload of the first byte of a section (see Section 9.4.5). If the payload-unit-start-indicator is set to '0', then no section starts in the payload of the transport packet. Hence when transport packets carry sections, the payload-unit-start-indicator operates as a flag signalling the presence of the pointer field.

The payload-unit-start-indicator also provides parsing convenience at random access: when the parser is looking for the start of a PES packet or a section, only the value of this indicator needs to be verified; if set to '0', there is no need to further parse the transport packet header and its payload.

Furthermore, the transport packet header may contain the optional random-access-indicator. This indicator, when set to '1', signals the availability of information in this and possibly subsequent transport packets with the same PID, to further assist with random access. In transport packets with the PCR-PID, the random-access-indicator signals the presence of a PCR sample. In transport packets with a PID other than the PCR-PID, when a video, audio or other elementary stream is carried, the presence is signalled of an elementary stream access point, with an associated time stamp encoded in the PES header.

As discussed in Section 4.1, MPEG-2 systems also supports a framework for carriage of metadata services of various formats. Also for metadata services, random access indication is provided, to indicate an entry point to the metadata service, where decoding is possible without requiring information from previously conveyed metadata. Further details on entry points are left to the specification of the metadata format.

In special playback modes, such as fast forward and fast reverse, it is important for players to rapidly identify elementary stream access points, where enough information is provided to start the decoding of the stream. To prevent the need for extensive parsing of the elementary streams, applications may require to align access units and PES packets so that the first byte in the payload of a PES packet is the first byte of an access unit. For example, in the case of audio, this byte may be a sync word, and in the case of video the first byte of a coded video picture, possibly preceded by a sequence header. The MPEG system specification does not require such alignment, but supports it by providing a data-alignment-indicator in PES headers. If this indicator is set to '1', then the first byte in the PES payload data is a syntax element that depends on the elementary stream type carried in the PES packet. In the program definition, a data stream alignment descriptor may be included for the involved elementary stream, to specify the syntax element in detail; for further details, see Section 13.3.4.2. If this descriptor is absent, a default syntax element applies.

11.8 Carriage of Private Data

Coded video, audio and other coded elementary streams are in many cases capable to convey private data. For example, MPEG-2 video allows carriage of user data within each coded picture; some applications exploit user data to convey closed captioning data. In case the carried data has a direct association with the coded picture, such as a SMPTE time code for the picture, then carriage in the coded picture has some benefits. However, there are also drawbacks; private data included in coded video or audio data may have impact on bitrate and on buffer requirements in decoders, and its inclusion may violate the compliancy of the stream. The captions may be available at the instant in time the video is coded, but may also be produced offline, in which case inclusion as user data in coded picture data becomes more cumbersome.

Carriage of private data by using the system layer provides flexibility without any impact on elementary stream layers. Private elementary streams can be included, using, if required, exactly the same time stamp mechanism as MPEG video and audio streams, but private elementary streams can also be carried without any time stamp. Private subtitles, video and audio can be carried along with MPEG defined video and audio. But also private data not representing an elementary stream, such as application and computer data, can be carried. MPEG-2 systems provides a wide variety of options to carry private data, each with a specific purpose, as discussed below.

- Private data may be carried in the payload of transport packets with a PID value assigned for that purpose. This allows to convey private data in transport streams without using the structures defined by sections and PES packets; instead, private structures may be used.

  Each transport stream must carry the PAT and the PMT. In the program definition a private stream carried in transport packets is signalled by assigning one of the user private stream-type values to the elementary PID of these transport packets. For private usage the stream-type range of 0x80–0xFF is available (see also Table 9.3). A program may contain one or more such private data streams.

  A program may even contain only private streams carried in transport packets, in which case the system time base and the PCR samples thereof may or may not be used. The absence of a system time base must be signalled in the program definition by a PCR-PID value equal to 0x1FFF, corresponding to the PID value of a null packet; note that the encoding of a PCR in a null packet is forbidden, as discussed in Section 11.6. It should be noted though, that in the absence of a system time base, a very minimum usage is made of MPEG-2 systems functionality: the transport stream is used as a transparent data pipe for private streams. The argument for using the transport stream format in this case may be to allow transmission across existing broadcast networks designed for MPEG-2 transport streams.

- A private data stream, such as a video, audio and subtitling stream with a format beyond the MPEG scope, can be transported as a private stream in PES packets. This allows the use of the synchronization mechanism defined in MPEG-2 systems, based on the STC and time stamps. However, it is also possible to use PES packets without any time stamps in the PES header. Therefore, two stream-ID values are assigned for carriage of private data: private-stream-1 and private-stream-2. In case of a private-stream-1 the regular PES header is applied, so that the PTS and DTS time stamps can be used, but in the case of a private-stream-2 a very minimum PES header is applied, only consisting of the packet start code with the stream-ID and the PES-packet-length field. Hence, in the case of a private-stream-2, neither PTS and DTS time stamps can be encoded in the PES header, nor any other field present in the regular PES header. Carriage of private data in PES packets is signalled by a stream-type value 0x06 (see also Table 9.2).

  The stream-IDs for private-stream-1 and private-stream-2 indicate different PES headers for carriage of private streams. For each private stream type only a single stream-ID value is available, and therefore the stream-ID cannot be used to distinguish multiple private streams of the same type in a program stream. Usually, application standardization bodies require a private-stream-1 and a private-stream-2 to include a private header in the payload to uniquely identify each private stream. Note that in the case of a private-stream-2, a private header may also support private synchronization mechanisms.

  However, when new content formats developed in the course of the MPEG-2 systems deployment, the need for additional stream-ID values evolved. At that time only very few stream-ID values were left available for future use, upon which one of these values a 'wild card' was assigned, indicating the presence of an extension of the PES header with a seven-bit stream-ID-extension field, providing 128 more values to uniquely identify elementary streams; for further details see Section 13.2. Of these 128 values, 64 were assigned for private usage, thereby providing another option to distinguish private streams carried in PES packets with a private-stream-1 header. However, as private-stream-2 PES packets do not use the regular header, the stream-ID extension field cannot be used in PES packets with a private-stream-2 stream-ID.
- In a transport stream, private data may also be carried in sections; carriage of private data in sections is signalled by a stream-type value 0x05; see Table 9.2. Each section with a table-ID value in the range 0x40–0xFE is identified as a private section carrying private data. Depending on application requirements, the short or long section header, discussed in Section 9.4.4 and Figure 9.13 can be used. Note that in the case of the long section header, the payload of the section is concluded by a CRC code that may be used for error detection purposes.

  Private sections are a very suitable tool to convey PSI type of data, structured in tables in a way similar to the PAT, PMT and CAT. In this context it should be noted that MPEG-2 systems only defined the basic PSI data important for interoperability, while leaving the definition of other important PSI data to applications and their standardization bodies. Indeed, ATSC, DVB and other broadcast standards did produce extensive PSI related specifications [1,2], addressing network information, service descriptions, time and date information, event information, possibly for multiple transport streams, and other information relevant for the provided services. This PSI related data, often referred to as Service or System Information, SI, is using the same table structure as the MPEG defined PSI.

Private sections may be used in transport packets with the program-map-PID to convey private data, next to program-map sections with PMT data. This provides the option to convey, for example, ATSC, ARIB or DVB defined SI data within a program definition; note that it is also possible to convey private descriptors in the program definition, as discussed in the next bullet.

If elementary stream scrambling is applied within a transport stream, then private sections are conveyed in the CAT. The format of these private sections is usually defined by the applied conditional access system.

Furthermore, it is possible to convey private sections in transport packets with a privately assigned PID. This mechanism is used to convey (P)SI tables defined by broadcast standardization bodies; either a fixed or a user defined PID value is assigned to convey each such (P)SI table in transport packets; sometimes the same PID is assigned to convey multiple tables.

- Private data can also be carried in a descriptor; descriptor-tag values in the range of 64–255 (0 × 40–0 × FF) are reserved for carriage of private data. Descriptors are typically used to convey high level descriptive information on an elementary stream or other program element, on a program, on a transport stream or on a program stream, depending on the context of the descriptor. In a transport stream, descriptors are used in privately or MPEG defined (P)SI tables, and in program streams, descriptors are used in the PSM; for the latter see Section 13.4. The (P)SI tables, as well as the PSM, are rather empty structures in which descriptors form the 'meat' with information associated to the streams or programs. Broadcast standardization bodies defined a large variety of descriptors for usage in (P)SI tables; for example, DVB defined over sixty descriptors. But if deemed necessary, also an application can specify its own private descriptors.

  Next to the use of private descriptors carrying private data only, it is in some descriptors also possible to convey private data in addition to the MPEG defined data.
- Private data in the adaptation field of transport packets. The adaptation field carries the transport-private-data-flag; if set to '1', then the eight-bit transport-private-data-length field is present, specifying the number of private-data-bytes conveyed in the adaptation field. The number of private-data-bytes must fit within the adaptation field. No specific purpose was intended for carriage of private data in the adaptation field; the option may be utilized for any purpose deemed appropriate by applications and their standardization bodies. For example, in DVB, private data in the adaptation field can be used for various purposes, including announcement information, access unit information and Personal Video Recorder (PVR) assistance.
- Private data in PES header, signalled by the PES-private-data-flag; if this flag is set, 16 bytes of private data are present. In this private data, combined with fields preceding and following it, no start-code emulation is permitted. This option to carry private data was agreed as a compromise that was considered weird by some participants. One expert insisted on optional carriage of private data in the PES header, but failed to convince the other experts about its need. After lengthy discussion, a compromise was suggested in the form of a 'minimum parsing' solution of either zero or 16 private bytes, as skipping 16 bytes is easy to implement in a parser. Much to the surprise of the proposer of the compromise, this proposition was accepted. See Note 11.1 for some background on the decision process in MPEG. As a result, there is an option to carry 16 private bytes in the PES header. DVB decided to use it to convey supplementary audio control information, and it can be used for any other appropriate purpose.

Note 11.1 Some Background on the Decision Process in MPEG

MPEG is formally organized in National Bodies (NB); to take a formal decision, such as accepting a draft MPEG standard, requires voting by the NBs. Typically, a small country has a NB with few members, with a simple process to take a position. For example, the NB in the Netherlands is represented by the NNI, the Netherlands Normalization Institute. But at some point in time, there were only two companies from the Netherlands involved in MPEG, KPN and Philips. When Arian Koster, the KPN representative, was asked what NNI means, he responded 'that stands for jaN aNd I'. On the other hand, a NB with many members may need extensive meetings to agree on a NB position.

Despite the NB orientated decision process, many technical decisions in MPEG are taken by consensus, which is very important for building industry acceptance. For many technology discussions often relatively objective criteria can be found to base decisions on. Hence, the usual challenge in MPEG is to convince experts on a certain proposition as the best possible solution to solve a problem, without involving any NB.

However, when an expert is insensitive to all reasonable arguments and maintains an unreasonable position, a NB position can be enforced on the discussed issue. For this purpose the NB of the delinquent expert can call for a 'caucus', a NB meeting to discuss the controversial subject amongst its members. The unreasonable expert is in a minority position, otherwise his position would not be considered unreasonable, and therefore a NB position can be agreed rapidly, upon which the involved NB can inform MPEG about its position, to which the unreasonable expert is formally mandated to comply, due to ISO/IEC rules.

Private data in transport streams and program streams may be specified by any party. It can be decoded correctly only if the receiver has knowledge of its format. Within an application domain, the usage of private data is often specified by the appropriate application standardization body, in which case there is no need to signal the format of the private data, as a receiver has knowledge of the application domain it operates in. However, also private data not specified by the application standardization body may be used within an application domain. For such private data a mechanism is needed to uniquely identify its format within the application domain. For this purpose, the application standardization body may use the private data indicator descriptor or define a private descriptor containing a parameter to identify the type of private data or its specifier. Usually, for assigning values to such parameter, a registration authority is used.[5] Note that a registration authority may also be involved in the registration for the usage of fields left for private usage, such as the transport-stream-ID in the PAT and the CA-system-ID in the conditional access descriptor.

For the use across various application domains of private data (streams), such as non-MPEG defined video and audio streams, a more generic mechanism is desirable. Therefore in MPEG-2 systems, the registration descriptor is defined. This descriptor provides a 32-bit format-identifier to uniquely and unambiguously signal the format of private data. For assigning format-identifier values, SMPTE [3] is appointed by ISO as the registration authority. In the

[5] For example, the registration authority for DVB is found at www.dvbservices.com/.

registration descriptor, the format-identifier field may be followed by additional-identification-info bytes with a format defined by the assignee of the encoded value of the format-identifier.

11.9 Copyright and Copy Control Support

To protect the rights of holders of copyright of content distributed in the market place, systems for copy protection and copy control have been designed. Such systems are beyond the scope of MPEG, but their operation requires the content to carry relevant information, as discussed below. It should be noted that the objective of the provided measures in MPEG-2 systems is not to prevent illegal copying, but instead to provide a suitable framework to undertake legal action when illegal copies are made.

Firstly, it is important to know whether distributed content is protected by copyright. For this purpose, when audiovisual content is carried in an MPEG-2 system stream, one bit in the PES header is assigned to signal whether the carried content is protected by copyright or whether this is unknown, while another bit signals whether the content is an original or a copy. For carriage of further copyright information, the copyright descriptor is defined. In a transport stream, a copyright descriptor may be assigned to an audio or video stream or any other element of an audiovisual program or to an entire program. This is achieved by including the descriptor either in the elementary stream info field or in the program info field in the Program Map Table (see Figures 9.11 and 9.12). In a program stream, the copyright descriptor can be assigned to the content in a similar way (see Section 13.4). Obviously, the copyrights of the holder are to be respected, and therefore a copyright descriptor assigned to content is not allowed to be changed.

In the copyright descriptor, the descriptor data bytes commence with a copyright-identifier field of 32 bits, intended to specify the applied method for labelling the associated content. The copyright-identifier may be followed by additional-copyright-info bytes, if any, with a format that depends on the encoded value of the copyright-identifier. For example, the copyright-identifier may signal a labelling method similar to ISBN for books and the additional-copyright-info may commence with a copyright number, similar to an ISBN number. The additional-copyright-info may also contain information on the copyright holder(s), the actual copyrights and other data deemed necessary.

The original intention was that ISO would appoint a registration authority responsible for assigning copyright-identifier values. However, in the meantime ISAN [4] and V-ISAN [5] were developed for identifying audiovisual content; both which are ISO standards that can be signalled without requiring a registration authority. Furthermore, support was defined for carriage over MPEG-2 system streams of metadata, which may include copyright information or links to such information, and with the option of using ISAN and V-ISAN for content labelling (see also the discussion in Section 4.1). Thereby the copyright descriptor became redundant.

For carriage of metadata, a content labelling descriptor was defined; in this descriptor, the metadata-application-format signals whether ISAN or V-ISAN is used for labelling the content, or another method. One specific value of this parameter, signals the presence of a metadata-application-format-identifier, encoded exactly as the format-identifier field in the registration descriptor for private data (see Section 11.8). Hence, values assigned by SMPTE as the registration authority apply both to the format-identifier and to the metadata-application-format-identifier. However, obviously, only values assigned for usage in the content labelling descriptor should be encoded in the metadata-application-format-identifier field. For further details on the content labelling descriptor see Section 14.11.

To address the need to assign copyright related information of a more dynamic nature to the content, such as the number of (remaining) copies that are allowed to be made, the PES header contains a one bit additional-copy-info-flag, to signal the presence of a byte consisting of a marker bit followed by the additional-copy-info field. From MPEG perspective, this seven bit field contains private data related to copyright information; its usage may be determined by an application standardization body or may depend on encoded parameters in metadata related descriptors.

11.10 Playback Trick Modes

One of the requirements imposed on MPEG-2 systems is to support trick mode playback from Digital Storage Media (DSM). However, trick mode playback comes with several complications, depending on the applied DSM device, its configuration and its location, the data delivery method in trick mode and many other issues. While most of these complications are application specific issues beyond the scope of MPEG-2, it was considered appropriate to include some information in PES headers to signal playback trick modes. Specifically for video decoders this information may be helpful to achieve an appropriate trick mode performance, particularly in the absence of a communication protocol between the player and the DSM. In practice however, often communication is possible between the player and the DSM, in which case there may be no need to include the trick mode information in the PES header. Decoders are therefore not mandated to decode the trick mode operation.

A DSM can include trick mode information in the PES header by setting the DSM-trick-mode-flag to '1'. In that case the PES header carries the three-bit trick-mode-control field, signalling the actual trick mode. This field is followed by five bits of further information, depending on the actual trick mode. The provided information applies to video only and in particular to MPEG-2 video; for example, in the slow motion or slow reverse mode, it indicates the number of times a field or picture is to be presented. For next generation video codecs their semantics are not updated, which may indicate that in practice the insertion of trick mode information in the PES header has a low added value.

Interoperability for Video on Demand (VoD) type of applications does not only require a standardized system stream to deliver the content, but also a standardized interface to control the DSM. In the context of MPEG-2 systems, some initial effort has been performed on a protocol intended to provide the basic DSM control functions in a simple environment with a only a single user and a single DSM. However, this protocol does not support the diverse network environments for VoD applications. It is therefore only included as an informative specification in annex B of the MPEG-2 system standard, while a far more elaborate effort was performed in the context of part 6 of the MPEG-2 specification on extensions for DSM Command and Control (DSM-CC), resulting in ISO/IEC 13 818-6 [6].

11.11 Single Program and Partial Transport Streams

For transport streams, MPEG-2 systems makes distinction between a Multi-Program Transport Stream (MPTS) and a Single Program Transport Stream (SPTS) The use of a SPTS may be practical at production prior to the forming of a MPTS. For example, to exploit available transport stream based infrastructures, for each event the coded video, audio and other data may be stored and conveyed in a single program transport stream prior to multiplexing with other programs into a MPTS. For constructing a SPTS the same constraints apply as for constructing a MPTS. A SPTS may have a variable bitrate; the complications associated to a variable bitrate MPTS do not apply to a SPTS.

The concept of a transport stream with only a single program is very valuable for the recording of one of the programs contained in a transport stream on a Personal Video Recorder (PVR). Significant bandwidth savings can be achieved by recording only that program from the MPTS instead of the entire transport stream. However, to the playback of such recorded program some problems are associated, as discussed in the following paragraphs.

The most straightforward approach to record one of the programs from a MPTS, is to only store transport packets with information relevant to the recorded program, and to delete the other transport packets. In applications, the stored transport packets are usually referred to as a partial transport stream. To playback the recorded program, the partial transport stream is to be delivered to the decoder. One way to do so is to consider the partial transport stream as a single program transport stream. In a SPTS the delivery of the packets is determined by the encoded PCR samples. If STC_p is the system time base for a recorded program p, then upon playback the PVR needs to deliver each PCR sample at the STC_p time encoded in the PCR sample, while the bytes in the single program transport stream between two PCR samples need to be delivered according to the linear interpolation scheme discussed in Section 9.4.4 and Figure 9.10. As a result, for a partial transport stream, the PCR samples are delivered at the same time as in the MPTS, but all bytes in between two PCR samples are delivered at another time than in the MPTS. See Figure 11.12 for an example.

Figure 11.12, depicts transport packets, whereby the packets tp_1, tp_2, . . . , tp_{15} contain information on a program p recorded on a PVR. Packets tp_1, tp_9 and tp_{15} carry the PCR samples PCR(tp_1), PCR(tp_9) and PCR(tp_{15}). At playback from the PVR, these samples are used to deliver the partial transport stream as a SPTS to a decoder. The PCR sample in packet tp_n is delivered at the encoded STC_p time: PCR(tp_n). The transport packet bytes in between two PCR samples are delivered as the consecutive stream of bytes depicted in Figure 11.12, in accordance with the required linear interpolation scheme.

Figure 11.12 shows that the transport packets of the SPTS are delivered at a piecewise constant rate; the rate is constant between two successive PCR samples, but its value may change after each PCR sample, depending on the number of transport packets between two PCR samples and the distance in time between these two samples. More important is that the packets are delivered at another STC_p time than in the original multi-program transport stream. Compared to the original MPTS, the delivery of each transport packet in the SPTS is displaced in time; some packets are delivered earlier and other packets later. In case the transport packets in between two PCR samples in the MPTS are positioned in a rather equidistant manner, such as tp_9 to tp_{14} in Figure 11.12, the displacement is rather small, while the displacement becomes more significant if the packets in the MPTS are delivered in a more bursty manner, such as packets tp_1 to tp_8.

Depending on the value of the caused displacement of the transport packets in the SPTS, the early or late delivery of transport packets may cause buffer under- or overflow in the decoder. For example, the multiplexer forming the MPTS may have postponed the inclusion of tp_2 in the multiplex to prevent buffer overflow. If the buffer in the decoder was indeed almost full at that time, early delivery of this packet (and the subsequent ones) in the SPTS may cause buffer overflow in the decoder, and thereby the delivery of the partial transport stream as a SPTS would be MPEG-2 system non-compliant.

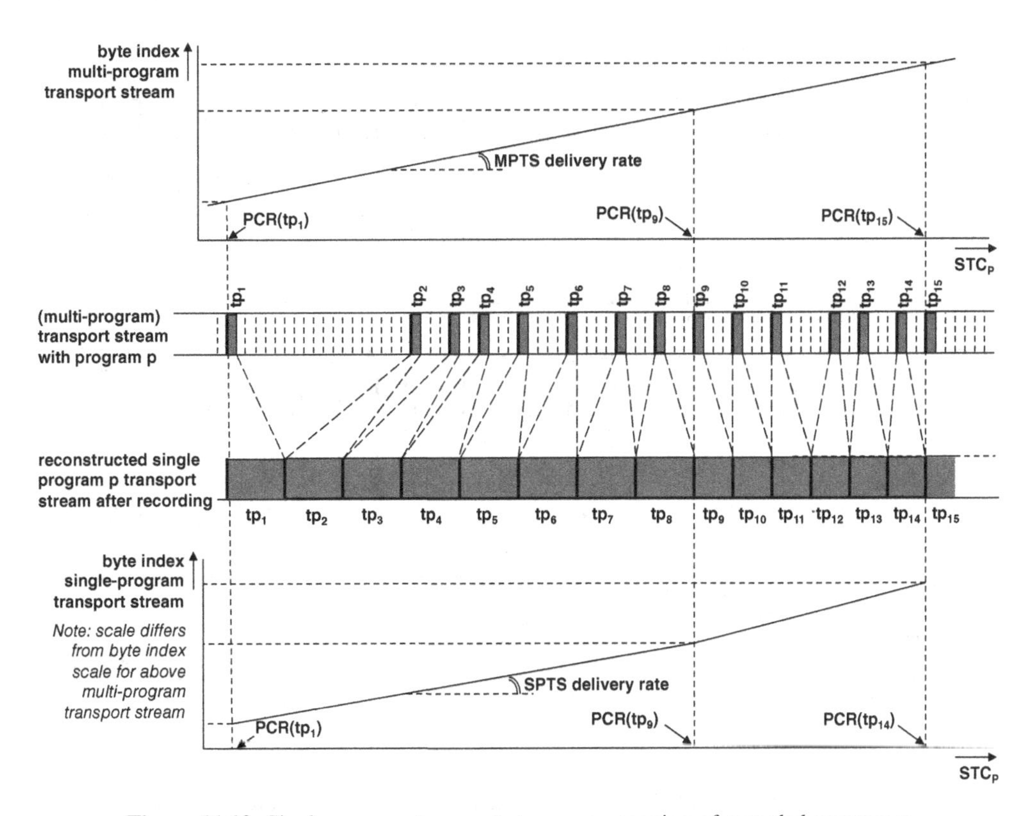

Figure 11.12 Single program transport stream construction of recorded program p

The PVR has no knowledge of buffer status information of a recorded partial transport stream; to ensure that no buffer under- or overflow occurs during the playback of a partial transport stream from a PVR, other playback delivery options are needed. For this purpose, private arrangements can be made, for example by application standardization bodies. In the context of this book, two options need specific attention, as discussed below.

- For each transport packet tp_n of the partial transport stream, the PVR stores the time stamp T (tp_n), specifying the STC_p time of the first byte of the transport packet. The STC_p time may be expressed in units of the 27 MHz STC clock or a derivative thereof, such as 90 kHz. By means of this time stamp, the PVR can deliver each recorded transport packet at the same STC_p time as in the original MPTS, presuming that the PVR knows the rate of the original MPTS. An example is provided in Figure 11.13. With this approach, the PVR provides a bursty transport stream, with gaps at positions where transport packets were deleted from the original MPTS (see Figure 11.13). Formats for such partial transport streams have been defined by IEC for Digital VCRs in IEC 61 883 [7] and for BD-ROM by the Blu-ray Disc Consortium [8].
- When a valid SPTS was used to convey program p prior to multiplexing in a MPTS, then the original value of the PCR sample in the SPTS, the OPCR, may be carried in the MPTS. Each transport packet with a PCR sample may carry the OPCR sample in its adaptation field. If the partial transport stream corresponds exactly to the original SPTS, then the OPCR samples can be used to deliver the partial transport stream during playback (see Figure 11.13).

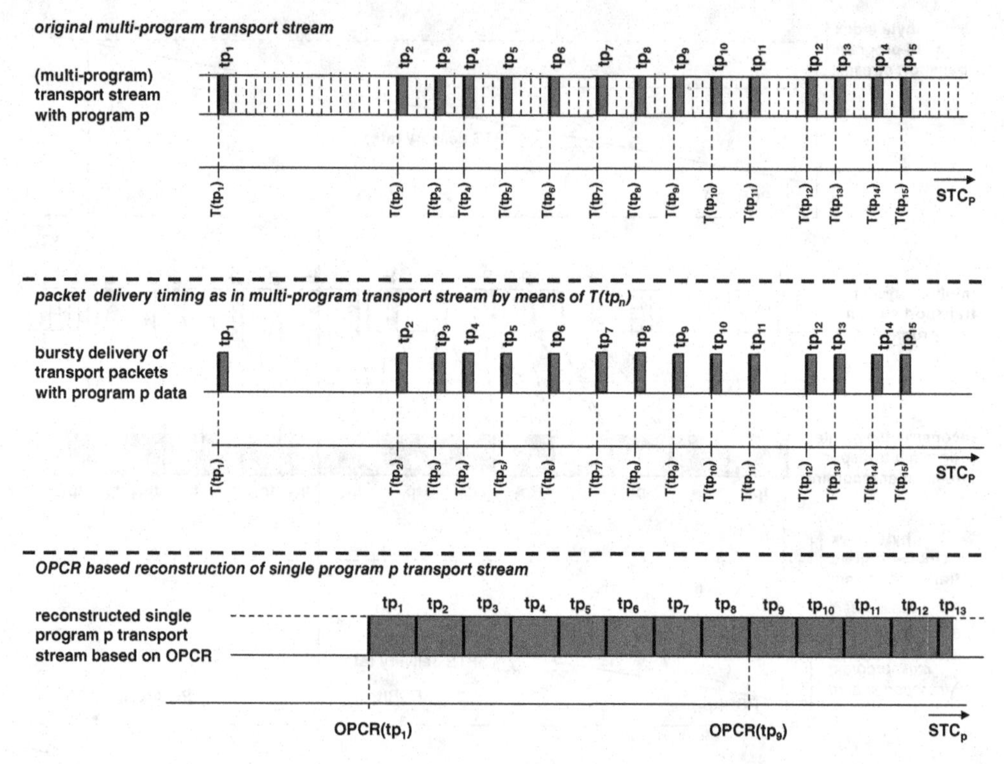

Figure 11.13 Reconstructed single program transport stream

It should be noted that the delivery schedules discussed above are needed in case the PVR must produce a transport stream that can be decoded by any compliant system decoder. In case the PVR is an integral element in a transport stream receiver, then a proprietary hand-shake procedure between the decoder and the PVR may be used for the delivery of the partial transport stream, whereby the timing is determined by the decoder, comparable to the 'pull' delivery method discussed in Section 6.6.2. In that case the decoder is responsible to prevent buffer under- and overflow.

11.12 Program Stream Carriage within a Transport Stream

One of the features provided by MPEG-2 systems is to convey a program stream in a transport stream, in a manner so that the original program stream can be conveniently reconstructed again at the receiver. The same feature is provided for a MPEG-1 system stream. In other words, if so desired, the contents of a DVD or a Video CD can be transparently conveyed over a broadcast channel. This feature is defined in a manner that fits conveniently within the concepts applied in transport streams. For example, carriage of the content of a DVD does not prevent the use of the audio and video from the DVD within a program in a transport stream.

To convey the coded video and audio streams from a program stream or a MPEG-1 system stream, the same packetization is used as in the transport stream, that is in the transport stream the same PES packets are used as in the program stream, though some changes are made to the header of the PES packets. In case of an MPEG-1 system stream, each MPEG-1 packet header is replaced by the equivalent MPEG-2 PES header.

However, not only the coded video, audio and other data streams need to be conveyed, but also the pack header and the system header. Therefore the PES header has been extended with the option to carry a pack and system header, signalled by a pack-header-field flag. The pack header with the system header, if the latter is present, is conveyed in the header of the first PES packet following the pack header in the program stream or MPEG-1 system stream.

For the carriage of a program stream or a MPEG-1 system stream, three more optional fields are defined in the PES header. A program-packet-sequence-counter flag signals the presence of these three fields, described below.

- The program-packet-sequence-counter field; the original program stream or MPEG-1 system stream can only be reconstructed correctly at the receiver, when the original order of the PES packets is known. For this purpose, a seven-bit program-packet-sequence-counter field is inserted in the PES header, encoded with a value incremented by one for each subsequent PES packet in the program stream or MPEG-1 system stream.
- A MPEG1-MPEG2-identifier flag; this one-bit flag signals whether information from a MPEG-2 program stream or from a MPEG-1 system stream is carried.
- The original-stuff-length field; this six-bit field signals the number of stuffing bytes applied in the PES packet header of the original program stream or in the packet header of the original MPEG-1 system stream. Each such stuffing byte is removed, which improves the transport efficiency, but a more important reason is that it allows PES header stuffing in the transport stream fully independent of any PES header stuffing in the program stream.

For transport stream interoperability reasons, the coded audio and video streams, packetized in PES packets, are carried in transport packets in the same way as any coded audio and video. Hence, in the PSI, such audio and video can be referenced in any regular program in a transport stream.

Each PES packet with data from a program stream or MPEG-1 system stream, but with a stream-ID signalling data with little or no relevance to the transport stream, such as the program stream map, the program stream directory, a padding stream, a private stream 2, an ECM, an EMM, or a DSM-CC stream, is carried in the transport stream as an ancillary stream PES packet. In this case, prior to carriage in a transport stream, from each involved original PES packet, the start-code-prefix is removed, upon which all remaining bytes of the original PES packet (i.e. the stream-id field, the remaining header bytes and all payload bytes) are put in the payload of a new PES packet with the stream-ID value of an ancillary stream. Note that this stream-ID value is assigned solely for carriage of data from a program stream or MPEG-1 system stream.

In summary, when the content of a conveyed DVD consists of a video stream, an audio stream, the program stream map and a private stream 2, then in the transport stream, the program definitions may reference four program elements with content from the DVD, one video, one audio and two ancillary streams.

To reconstruct the original program stream or MPEG-1 system stream at the receiver of the transport stream, the following operations are needed. Firstly, from the header of each received PES packet with data from the program stream or MPEG-1 system stream, the data signalled by the pack-header-field flag and the program-packet-sequence-counter flag are removed for use in the reconstruction; both these flags are set to '0'. The pack header and system header, if present, are put in the program stream or MPEG-1 system stream immediately prior to the PES packet in which they are carried. From each ancillary stream PES packet, the header is removed, except the start-code-prefix, which is then concatenated with the bytes in the payload of this PES packet, commencing with the original stream-ID. Furthermore, in the header of each PES packet with data from the program stream or MPEG-1 system stream, stuffing bytes may be inserted or deleted as required by the value of the original-stuff-length field. The encoded values of the program-packet-sequence-counter are used to put the PES packets back in their original order. In case of an MPEG-1 system stream, the header of each PES packet is replaced by the equivalent MPEG-1 packet header. This completes the reconstruction of the original program stream or MPEG-1 system stream.

11.13 PES Streams

MPEG-2 systems also supports stand-alone PES streams, a concept comparable to a program stream with one elementary stream only, but without a pack header, system header and other program stream specific structures. A PES stream is a contiguous stream of PES packets, each containing data of the same elementary stream and with the same stream-ID. PES streams were introduced to enable implementation of separate audio, video and systems encoders without relying on private connections to convey clock references. The delivery timing is defined by two optional fields in the PES header, the ESCR and the ES-rate; the presence of these fields is indicated by the ESCR flag and the ES-rate flag, respectively. The ESCR and ES-rate fields are the equivalent of the SCR and the program-mux-rate. In a PES stream, the contained elementary stream is only multiplexed with the PES header, without involvement of any other elementary stream; therefore the required buffer for the decoding of a PES stream is usually only marginally larger than the input buffer needed for the decoding of the elementary stream itself. For a further discussion see Section 12.2.4.

11.14 Room for Future Extensions

MPEG-2 systems leaves significant room for future extensions by various means, such as the use of reserved bits and bytes in syntactical structures, with a meaning that can be defined by MPEG in future. Existing decoders are required to ignore such bits and bytes, when used in future, to allow forward compatible extensions. Reserved bits and bytes are explicitly defined at selected positions in the syntax. For example, in the adaptation field of transport packets at several positions various reserved bits are available, while the adaptation-field-extension can be concluded with one or more reserved bytes.

Furthermore, when flexibility is required for a parameter, the field for its coding uses more bits than strictly needed, to leave room for extending the semantics in future. For example, for the stream-type parameter, 128 values are available for assignment by MPEG systems, of which about two-thirds are still reserved[6] as of 2013, which leaves significant room for defining new coding formats.

For tags used for syntactical structures such as the stream-ID for PES packets, the table-ID for sections and the descriptor-tag for descriptors a comparable approach is applied as for

[6] For private usage also 128 stream-type values are available.

parameters. However this approach is less essential for tags, because it is possible to define a convenient extension mechanism. For the table-ID, a large number of values is still available for usage in future, but for the stream-ID an extension mechanism is defined already while for the descriptor-tag a extension mechanism may be needed in future. For example one descriptor-tag value can be assigned as extended-descriptor, while the first byte in the payload of the descriptor could be an extended-descriptor-tag byte, thereby creating room for another 256 additional descriptors.

In conclusion, the MPEG-2 system specification is defined so that evolving new content formats can be supported for a long time to come. Most probably, the life time of MPEG-2 systems in the market place will not be restricted by the lack of extension capabilities, but by other relevant factors.

References

1. ETSI (2013) In DVB, Service Information is specified in ETS EN 300 468: 'Digital Video Broadcasting (DVB); Specification for Service Information (SI) in DVB systems', found at: http://www.etsi.org/deliver/etsi_en/300400_300499/300468/01.13.01_60/en_300468v011301p.pdf Implementation guidelines for the use of DVB SI are described in DVB document A005: 'Digital Video Broadcasting (DVB); Guidelines on implementation and usage of Service Information (SI)', found at: http://www.dvb.org/technology/standards/a005_DVB-SI_Imp_Guide.pdf.
2. ATSC (2009) In ATSC, System Information is specified in A/65:2009: 'Program And System Information Protocol For Terrestrial Broadcast And Cable', found at: http://www.atsc.org/cms/standards/a_65-2009.pdf Furthermore, implementation guidelines are described in A/69:2009: 'ATSC Recommended Practice: Program and System Information Protocol Implementation Guidelines for Broadcasters', found at: http://www.atsc.org/cms/standards/a_69-2009.pdf.
3. SMPTE (2013) SMPTE is the Society for Motion Picture en Television Engineers; see: www.smpte.org/ Information on the SMPTE Registration Authority for assigning MPEG-2 system format-identifier values is found at: http://www.smpte-ra.org/mpegreg/mpeg.html.
4. ISO (2002) ISO 15706-1:2002, Information and documentation – International Standard Audiovisual Number (ISAN) – Part 1: Audiovisual work identifier http://www.iso.org/iso/catalogue_detail.htm?csnumber28779.
5. ISO (2007) ISO 15706-2:2007, Information and documentation – International Standard Audiovisual Number (ISAN) – Part 2: Version identifier (V-ISAN) http://www.iso.org/iso/catalogue_detail.htm?csnumber35581.
6. ISO (2000) MPEG-2 Part 6: Extensions for DSM-CC – ISO/IEC 13818-6:2000: http://www.iso.org/iso/catalogue_detail.htm?csnumber25039.
7. ISO (2004) ISO 61883-4:2004-08, second edition – Consumer audio/video equipment – Digital interface – Part 4: MPEG2-TS data transmission. Available at http://webstore.iec.ch/webstore/webstore.nsf/artnum/032987!opendocument.
8. Blu-ray Disc Association (2001) Blu-ray Disc Association– White Paper Blu-ray Disc Read-Only Format – Audio Visual Application Format Specification for BD-ROM Version 2.5 (July 2011). Available at http://blu-raydisc.com/assets/Downloadablefile/BD-ROM-AV-WhitePaper_110712.pdf.

12

The MPEG-2 System Target Decoder Model

The MPEG-2 STD model and differences between the P-STD and the T-STD. Requirements and buffer management for the P-STD; the Constrained System Parameter Program Stream (CSPS). The PES-STD as a special case of the P-STD. Characteristics of the T-STD; the use of transport buffers and other buffers in the T-STD and options for the transfer rates from one buffer to another. General STD constraints and requirements and examples of decoding of audio and video in the STD.

12.1 Introduction to the MPEG-2 STD

The MPEG-2 system target decoder is an essential element of the MPEG-2 system specification, and serves a variety of purposes, as already discussed in Section 6.5. The below bullet points provide a brief summary; for further details see Section 6.5.

- The STD defines the timing of stream delivery and decoding events with mathematical accuracy, based on the instantaneous decoding of access units.
- The STD serves as a mathematical tool to determine compliancy of decoders and streams in an unambiguous manner.
- The STD serves as a hypothetical reference for the design of actual decoders.

In this chapter, the decoding of access units is described in general terms, without addressing for each specific content format all detailed issues on access units, input buffer sizes and the input and output rates of buffers. Therefore in this chapter the decoder of each elementary stream n in the STD is represented by a simple decoder building block D_n, without providing any further information on its structure. In this context it should be noted that the content formats supported in MPEG-2 systems require various elementary stream decoder structures. For several content formats, such as MPEG audio, there is no need to further detail the decoder structure, but for some formats the structure needs extensions to describe impact on MPEG-2 system functionality. For example, MPEG-2 video uses a re-order buffer that impacts system functionality, while AVC uses a decoded picture buffer with more functionality than just

Fundamentals and Evolution of MPEG-2 Systems: Paving the MPEG Road, First Edition. Jan van der Meer.
© 2014 John Wiley & Sons, Ltd. Published 2014 by John Wiley & Sons, Ltd.

re-ordering.[1] For each content format supported in MPEG-2 systems the applicable decoder structure must be provided, as well as the content format specific STD parameters.

In the MPEG-2 STD a STC frequency of 27 MHz is used; the choices for this STC frequency and its tolerance and slew rate are discussed in Section 6.6. The bitrate of each system stream in the STD is measured in terms of the STC; for example, a bitrate of 9 000 000 bits/s indicates that one byte of data is transferred every (27 000 000/9 000 000) × 8 = 24 cycles of the STC. In a transport stream, the STC is specific for each program, though all programs in a transport stream may share the same STC. The frequency of the STC in a transport stream is mandatorily locked to the sampling rate of each program element, such as the audio sampling frequency and the video picture rate. Such locking is common practice in broadcast applications, and simplifies clock regeneration in receivers, as already discussed in Section 6.6. In a program stream, locking of sampling rates to the STC is optional.

For the decoding of MPEG-2 system streams, two different STDs are defined, the P-STD for program streams and the T-STD for transport streams, to appropriately address differences in delivery characteristics of both type of system streams. The P-STD is described in Section 12.2 and the T-STD in Section 12.3. In Section 12.4, general constraints applicable to content decoding in the P-STD and the T-STD are described, while Section 12.5 provides a brief overview of content format specific STD issues.

12.2 The Program Stream STD: P-STD

12.2.1 Description of P-STD

The P-STD, depicted in Figure 12.1, is very similar to the MPEG-1 STD discussed in Section 7.4. Each elementary stream is decoded in a specific branch of the P-STD. However, as discussed in Section 12.1, information on the internal structure of the elementary stream decoders D_n is not given, but instead provided for each content format. As a result, for each elementary stream n the decoding branch in the P-STD depicted in Figure 12.1 consists of an input buffer B_n followed by the decoder D_n. Each elementary stream is decoded in a branch exclusively assigned for that purpose; hence in one branch all PES packets carrying a certain elementary stream are decoded and no other PES packets.

The PES packets that carry a certain elementary stream are identified by a specific stream-ID value. In the program stream map, each stream-ID value used in the program stream is associated with a stream-type value, to signal the elementary stream type carried in PES packets with that stream-ID value. Depending on this information, the system demultiplexer ensures that for each elementary stream, the PES packet data is provided to the input buffer of the appropriate decoding branch. Only the PES packet data enters the input buffer B_n; the PES packet header carrying time stamps and other information is provided to the decoder to control the decoding and presentation of access units in the decoding system.

Not only the headers of PES packets with elementary stream data, but also the pack header, system header and PES packets carrying the program stream map or other system related information are passed as system control data to the decoding system for immediate processing. The STC samples carried in the SCR field in pack headers may be used to synchronize the STC and to determine the timing of the delivery of the program stream. In the P-STD no input buffer

[1] Note though that the control and dimensioning of these buffers is not a systems issue, but instead addressed in the MPEG-2 video and AVC standards.

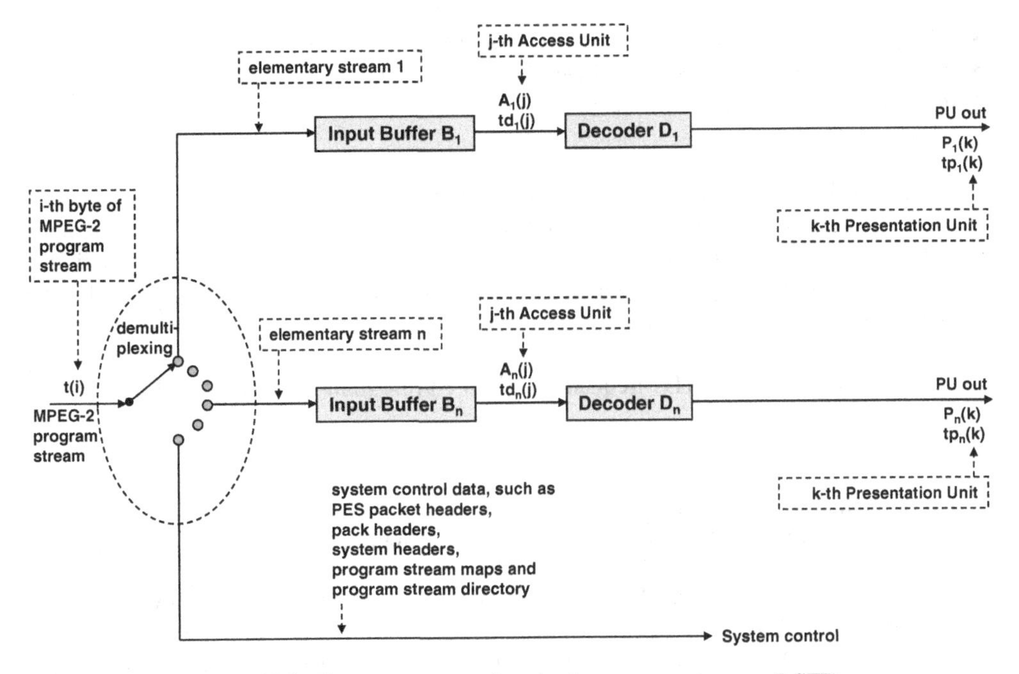

Figure 12.1 The system target decoder for program streams: P-STD

is applied for system control data; hence this data is processed at the rate at which the program stream is delivered to the P-STD. The program stream system header and the program stream map are used to control the demultiplexing and the decoding of the content.

To specify requirements and constraints for the decoding of elementary streams in the P-STD, the P-STD elements and the decoding processes in the P-STD are defined by means of parameters that allow to express the requirements and constraints in generic mathematical formulas. For this purpose the following set of parameters is used:

- n – the index of an elementary stream;
- BS_n – the size of buffer B_n;
- $F_n(t)$ – the instantaneous fullness at time t of the input buffer B_n;
- t – the time measured in seconds;
- i – the index of a byte in the MPEG-2 system stream;
- $t(i)$ – the time of arrival of the ith byte of the MPEG system stream at the input of the STD;
- j – the index in decoding order of an access unit in an elementary stream;
- k – the index in presentation order of a presentation unit;
- SCR(i) – the encoded STC value in the SCR field, where i is the byte index of the final byte[2] of the SCR field that serves as reference for the encoding of the SCR field; in other words, SCR(i) specifies the intended STC delivery time of that byte;
- $A_n(j)$ – the jth access unit in elementary stream n;

[2] Formally, in MPEG-2 systems, this reference byte is the byte containing the final bit of the SCR field (see Section 13.4).

- $Td_n(j)$ – the decoding time in the P-STD of the jth access unit in elementary stream n;
- $P_n(k)$ – the kth presentation unit in elementary stream n;
- $tp_n(k)$ – the presentation time in the P-STD of the kth presentation unit in elementary stream n.

Input to the P-STD is a program stream with a bitrate specified by the program-mux-rate. As discussed in Section 9.3, the delivery time of each byte in the program stream at the input of the P-STD, is determined from the STC sample and the program-mux-rate, both encoded in the pack header. As the program-mux-rate is expressed in units of 50 bytes/s, the following formula applies for the arrival time of each byte contained in a pack of a program stream:

$$t(i) = [\mathrm{SCR}(i')/\mathrm{STC\ frequency}] + [(i - i')/(\text{program-mux-rate} \times 50)],$$

where:

i is the index of any byte in the pack;
i' is the index of the reference byte in the pack for the encoding of the SCR.

Note that the above formula also applies for MPEG-1 systems, discussed in Chapter 7; in Figure 7.5, for simplification, the index of byte i' is set to zero. The program-mux-rate may vary for each pack, but the STC is continuous; time base discontinuities (as discussed for transport streams in Sections 11.3 and 11.4) are not allowed to occur in program streams.

In each elementary stream, access units are concatenated, in the case of video coded pictures and in the case of audio coded audio frames. At its decoding time $td_n(j)$, all data of access unit $A_n(j)$ is removed instantaneously from input buffer B_n and decoded into a presentation unit $P_n(k)$. At its presentation time[3] $tp_n(k)$, $P_n(k)$ is presented at the output of the decoding branch. To avoid ambiguities with respect to the fullness of the input buffer B_n over time, it is to be defined precisely which data belongs to an access unit. Hence, for each elementary stream format supported in MPEG-2 systems an exact definition of access units is required.

The size BS_n of buffer B_n needs to be large enough to accommodate both elementary stream and system level buffering. The fraction of B_n required for elementary stream buffering depends on the size of the access units and may vary over time. The remaining fraction of B_n is available for multiplexing and other system level buffering. As no constraints are defined on the sharing of B_n between these two components, the system encoder needs to ensure that at any time the amount of system level buffering does not exceed the available fraction of B_n left over by elementary stream buffering.

Each program stream contains all timing and other information needed for its delivery to the P-STD and its decoding in the P-STD. This information includes a number of variables selected by the system encoder. The following summarizes the conveyed information and the associated constraints.

- Each pack header carries a coded system clock reference field, SCR(i), and the program-mux-rate. The program-mux-rate value applies to the delivery of the pack and may change at the next pack in the program stream. At least every 0.7 s a SCR(i) value is to be carried in the

[3] Note that the presentation order differs from the decoding order when re-ordering is applied; however, for audio and low delay video, the presentation order is the same as the decoding order.

program stream, and therefore at least every 0.7 s a pack header must occur; the formal requirement is:

$$|t(i') - t(i'')| \leq 0.7\ \text{s},$$

or:

$$|[\text{SCR}(i') - \text{SCR}(i'')]/(\text{STC frequency})| \leq 0.7\ \text{s}$$

for all i' and i'', where i' and i'' are the indexes of the reference byte of two consecutive SCR fields in the program stream.

- The size BS_n of the input buffer B_n applied by the encoder when constructing the program stream is encoded in the P-STD-buffer-size field in the PES header. In the course of a program stream, the value of BS_n may change. The P-STD-buffer-size field is optional, but must be present in the header of the first PES packet of each elementary stream in the program stream; the field must also be present if the value of BS_n changes. In any other PES packet with elementary stream data the P-STD-buffer-size may also be present. A variable BS_n value may be beneficial in some applications; to ensure that a decoder can determine whether its input buffer is large enough to decode the entire elementary stream, the system header contains the largest applied BS_n value for each elementary stream in the program stream.
- The values of $tp_n(k)$ and $td_n(j)$ are conveyed as a PTS or a DTS in thc PES header for some access units; for access units without an encoded PTS or DTS, the decoding and presentation time may be derived from a PTS or DTS conveyed for a previous access unit and from information contained in the elementary stream, such as the frame rate in video or audio. For each access unit the decoding and presentation times must be known; if its value cannot be derived from the PTS or DTS of a previous access unit, then a PTS or DTS must be encoded explicitly for that access unit. This requirement may impact the process of forming PES packets for an elementary stream. See Section 12.4 for constraints on the coding of PTS and DTS fields in the PES header; also content format specific constraints may apply (see Section 12.5 and Chapter 14).

12.2.2 Buffer Management in the P-STD

In the P-STD, buffer overflow is forbidden under all circumstances; buffer underflow is also forbidden, except for low delay video and during trick mode operation. In the case of a low delay video, at its decoding time, the coded data of a picture may not completely be available in the input buffer yet, in which case the encoder skips the next picture. Upon such underflow, the presentation of the picture is postponed until all coded data has arrived in the input buffer. To ensure a unambiguous restart of the decoding process in the P-STD after an underflow, a PTS must be encoded for the coded picture after the underflow.

When the first byte of the program stream enters B_n, the input buffer must be empty; in formal terms:

$$F_n(t) = 0 \text{ instantaneously before } t = t(0).$$

For each elementary stream, the formal requirement for the fullness of the input buffer to prevent overflow is:

$$F_n(t) \leq BS_n \text{ for all t.}$$

For all elementary streams, except low delay video, and during trick mode operation, the formal requirement to prevent underflow is:

$$F_n(t) \geq 0 \text{ for all t.}$$

In other words, for all elementary streams, except low delay video, and during trick mode operation, in the course of a program stream, the fullness of the input buffer may vary between 0 and BS_n:

$$0 \leq F_n(t) \leq BS_n.$$

12.2.3 CSPS: Constrained System Parameter Program Stream

In analogy to MPEG-1 system streams, also for MPEG-2 program streams the concept of a CSPS has been defined. In MPEG-1, the CSPS concept is tightly coupled to a constrained parameter MPEG-1 video stream, as discussed in Section 7.5.2. MPEG-2 video does not use constrained parameter streams, but instead a concept of profiles and levels, thereby reflecting the much wider application scope of MPEG-2. The objective of the CSPS concept in MPEG-2 however is very similar as in MPEG-1, but instead of offering a single 'focus point' to the industry, an MPEG-2 CSPS stream offers multiple 'focus points', so that the concept can be applied for multiple video profiles.

As in MPEG-1, both the PES packet rate and the size of buffer B_n are constrained in a CSPS stream. The CSPS condition on the PES packet rate is depicted in Figure 12.2. The maximum PES packet rate allowed in a CSPS stream, depends on the encoded value of the rate-bound encoded in the system header. The rate-bound is intended to specify the maximum value of the program-mux-rate in the entire program stream; note though that this maximum value of the rate-bound may not be used. For low values of the rate-bound, the maximum PES packet rate is 300 packets/s. For higher rate-bound values, the PES packet rate is bounded by a linear relation to the rate-bound. There are two options, each identified by the packet-rate-restriction-flag in the system header. The maximum PES packet rate of 300 packets/s applies up to 4.5 Mb/s if this flag is set to '1' and up to 2.5 Mb/s if this flag is set to '0'. When the maximum rate of 300 packets/s no longer applies, the linear relations indicated in Figure 12.2 applies. Note that the most restricted option is very similar to the MPEG-1 CSPS constraint, and that the other option allows applications to use CSPS streams with a higher PES packet rate for higher rate-bound values.

The bound on the size of B_n depends on the content format. For audio streams the bound is 4096 byte = 4 KB, except for MPEG-2 AAC audio; to support multi-channel audio, for MPEG-2 AAC with up to eight channels carried in a CSPS, a bound of 8976 bytes[4] has been defined. The available buffer room is shared between elementary stream buffering and multiplex buffering.

[4] For more background on this value see Section 14.3.3.

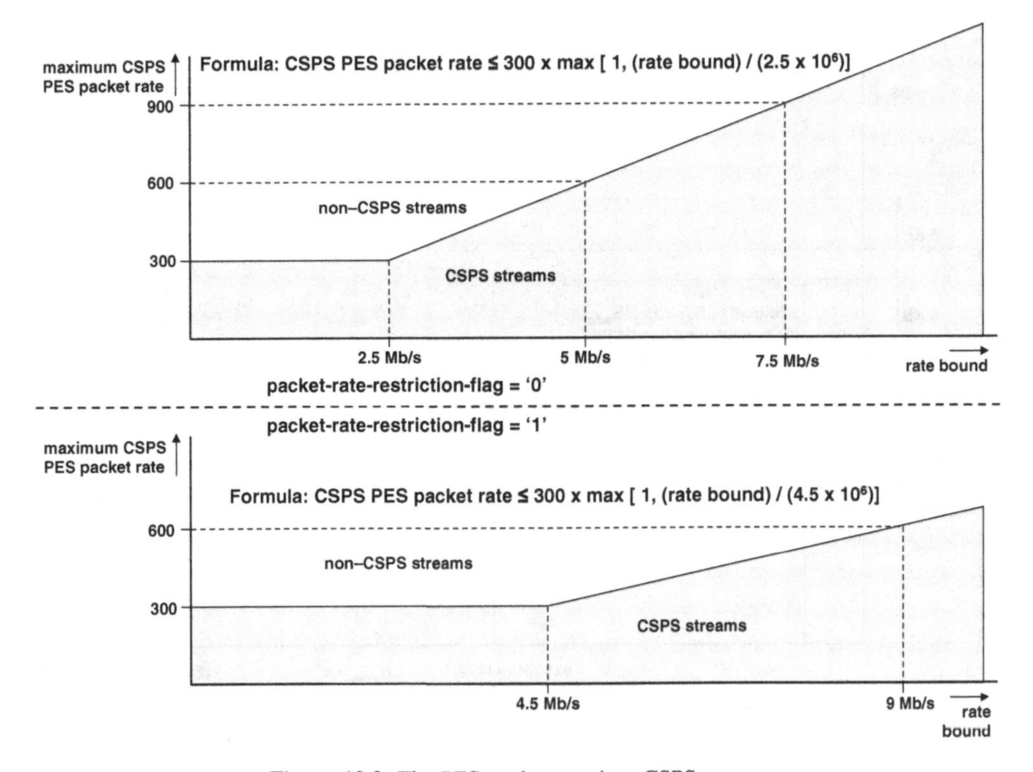

Figure 12.2 The PES packet rate in a CSPS stream

For video streams in a CSPS, bounds have been defined for MPEG-1 video, MPEG-2 video and AVC. The approach is based on the same assumption as for CSPS in MPEG-1: in addition to the input buffer of the elementary video stream, a headroom of 6 KB is sufficient to accommodate for multiplexing. However, this presumes the use of relatively small PES packets; for high video bitrates this means that the PES packet rate may increase beyond the rate bound for CSPS streams. To allow high bitrate video within a MPEG-2 CSPS, the additional buffer room available for multiplexing is defined to increase linearly for higher video bitrates, as follows:

$$BS_{add} = MAX[(6 \times 1024), (R_{vmax} \times 0.001)]\text{bytes},$$

where R_{vmax} is the maximum bitrate of the video stream.

The resulting BS_n constraint for video streams in a CSPS is depicted in Figure 12.3. For R_{vmax} values below $6 \times 1024 \times 8$ kbit/s (about 50 Mb/s), BS_{add} equals 6 KB, while for higher R_{vmax} values BS_{add} increases to 12 KB at R_{vmax} equal to $2 \times 6 \times 1024 \times 8$ kbit/s, about 100 Mb/s. An overview of the resulting maximum BS_n values in a CSPS for audio and video streams is presented in Table 12.1.

Note that the list of content formats in Table 12.1 is not exhaustive. For each content format that is supported by MPEG-2 systems, but for which no maximum BS_n value is specified in Table 12.1, the value of BS_n is unconstrained for carriage in a CSPS.

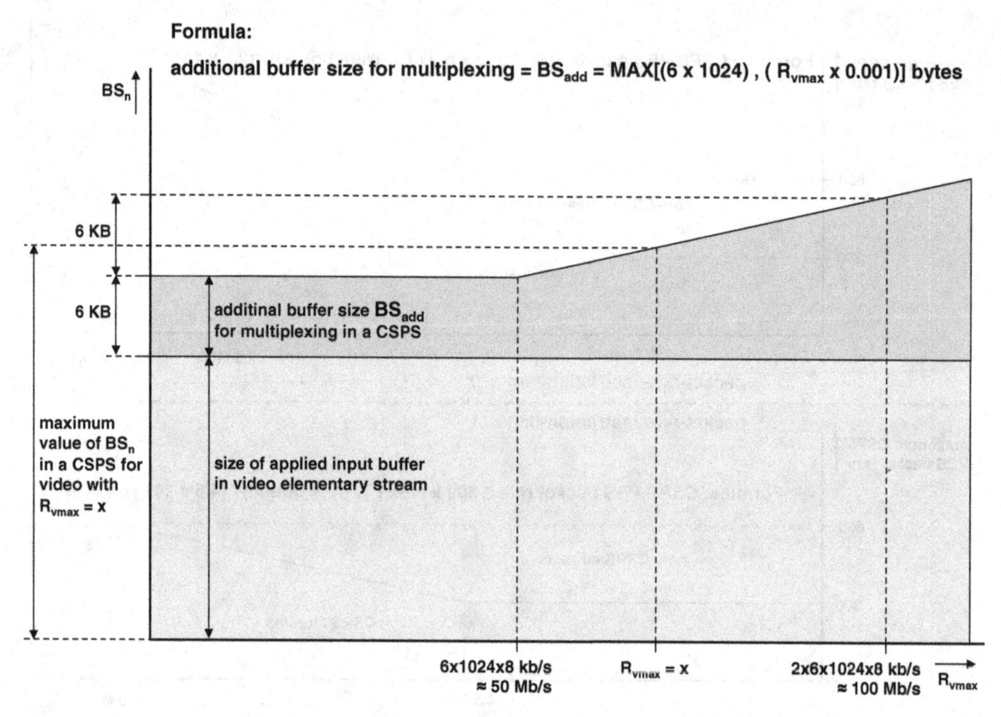

Figure 12.3 The bound on BS_n for video in a CSPS stream

Table 12.1 Constraints on BS_n for audio and video streams is a CSPS

Stream		Maximum value of BS_n in a CSPS
Audio	Except MPEG-2 AAC	4096 bytes
	MPEG-2 AAC (≤ 8 channels)	8976 bytes
Video	MPEG-1	Size of VBV[a] + BS_{add}
	MPEG-2	Size of VBV[b] + BS_{add}
	AVC	CPB-size[c] + BS_{add}

[a]Specified in the sequence header of the MPEG-1 video stream.
[b]Specified in the sequence header of the MPEG-2 video stream.
[c]CPB-size is the CpbSize[cpt-cnt-minus1] size of the CPB signalled in the HRD-parameters() in the AVC video stream; if HRD-parameters() is not present, then CPB-size is (1200 × MAX-CPB) as defined in Annex A of the AVC specification for the applied level.

12.2.4 Usage of P-STD for PES-STD

The model of the P-STD can also be used for PES streams; a PES stream is a contiguous stream of PES packets, each containing data of the same elementary stream and each with the same stream-ID. Some more details on PES streams are found in Section 11.13. While stand-alone PES streams may not be applied widely in practice, the PES-STD provides a convenient tool to discuss the required amount of overhead buffering for PES packetization, which needs to be taken into account when multiplexing elementary streams into transport streams and program streams.

For the decoding of a PES stream, the PES-STD is defined. The PES-STD is described by the P-STD model with only a single decoding branch. At the input of the PES-STD, the demultiplexer ensures that the PES packet payload enters the input buffer B_n of the decoding branch, while providing the PES headers for system control. Buffer B_n needs to accommodate both elementary stream input buffering and overhead buffering due to the insertion of PES headers.

When the encoder inserts a PES packet header within the elementary stream data, the incoming elementary stream data need to be stored in an overhead buffer at the encoder until it can be placed in the PES stream. Thereby the delivery of bytes of the elementary stream data at the decoder is delayed; the delay may vary for each byte. To prevent underflow of buffer B_{dec}, the decoding process is to be postponed by the same amount as the largest delay undergone by any byte in the elementary stream.

Due to postponing the decoding process, more elementary stream data will enter the decoder than anticipated by the decoder buffer and to store the additional data also at the decoder an overhead buffer BS_{oh} is needed. In the PES-STD the size BS_n of the input buffer B_n for elementary stream n is specified to be equal to:

$$BS_n = BS_{oh} + BS_{dec},$$

where:

BS_{oh} is the size of the overhead buffer for PES packetization;
BS_{dec} is the size of the input buffer for elementary stream n.

The delay of the decoding process is demonstrated in Figure 12.4 for an elementary stream with a constant bitrate R_{es} and access units AU_n of variable size. A PES stream is created by inserting a PES header H_n between the last byte of AU_{n-1} and the first byte of AU_n. Hence, each PES packet n consists of header H_n and AU_n. In Figure 12.4, the bitrate of the PES stream is constant during each PES packet and chosen to be exactly equal to the (constant) bitrate R_{es} of the elementary stream plus the additional bitrate due to the inserted PES header bytes of H_n. In other words, the time period consumed by AU_n in the elementary stream is exactly the same as the time period consumed by PES packet n in the PES stream. For each PES packet n both the size of AU_n and the size of H_n may vary, and therefore the bitrate $R_{PES}(n)$ of the PES stream will typically vary for each PES packet n, following a piecewise constant schedule.

In Figure 12.4, the encoder uses a buffer B_{PES} to store the elementary stream to allow the insertion of PES headers. The elementary stream data continuously enters B_{PES} at the rate R_{es}; during the insertion of PES header H_n, no bytes leave B_{PES}, so that its fullness starts to increase at rate R_{es}. If the time consumed by H_n in the PES stream is equal to $t_{PES\text{-}header}(n)$, then the fullness of B_{PES} increases from zero to $t_{PES\text{-}header}(n) \times R_{es}$. In Figure 12.4, for H_5 the fullness increase is maximum. For the duration of PES packet n following the insertion of header H_n, it is assumed that elementary stream bytes are removed from B_{PES} at rate $R_{PES}(n)$ of the PES stream. During this period, elementary stream bytes remain entering B_{PES}, so that the fullness of B_{PES} decreases at a rate equal to $R_{es} - R_{PES}(n)$. At the end of each PES packet n,

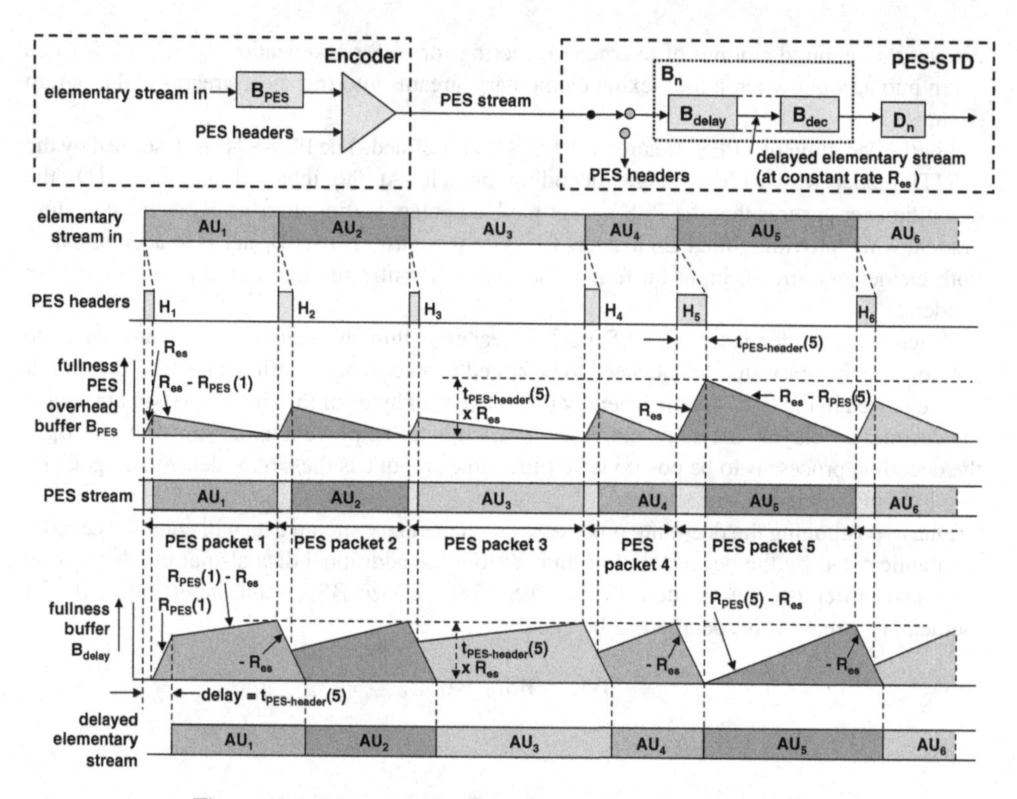

Figure 12.4 Example of PES overhead buffering in the PES-STD

B_{PES} is exactly empty again. Note that the required minimum size of B_{PES} is determined by R_{es} and the maximum delay caused by the PES header insertion: $t_{PES\text{-}header}(5) \times R_{es}$ in Figure 12.4.

Upon the PES stream entering the PES-STD, the PES headers are removed and provided for system control, while the PES payload enters B_n. To compensate for the delayed decoding process, in Figure 12.4 buffer B_{delay} precedes buffer B_{dec}, whereby B_{dec} is the input buffer of the elementary stream decoder. At the output of B_{delay}, it is assumed that the original elementary stream is reproduced with the caused delay of the decoding process. Based on this assumption, Figure 12.4 presents the fullness of B_{delay} and the flow of the access unit data through B_{delay}. Note however that in practice the buffers B_{delay} and B_{dec} are integrated in B_n without any need to reconstruct the original elementary stream; the reconstruction is used here only for explanatory reasons.

When access unit data is present, it leaks out of B_{delay} at the constant rate R_{es}. At the end of each PES packet n, the delivery of AU_n data is completed. Note that with the applied packetization strategy the last byte of each AU_n is delivered at exactly the same time as in the original elementary stream. During the delivery of a PES header no data is entering B_{delay}, so that its fullness decreases at rate R_{es}, but when PES payload bytes enter, its fullness increases by the rate of the PES stream during the PES packet $R_{PES}(n)$ minus R_{es}. Figure 12.4 shows that the fullness of buffer B_{delay} is complimentary to the fullness of buffer B_{PES}. The same applies to the delay through B_{delay} and B_{PES}. The total delay for each byte in the elementary stream is equal to

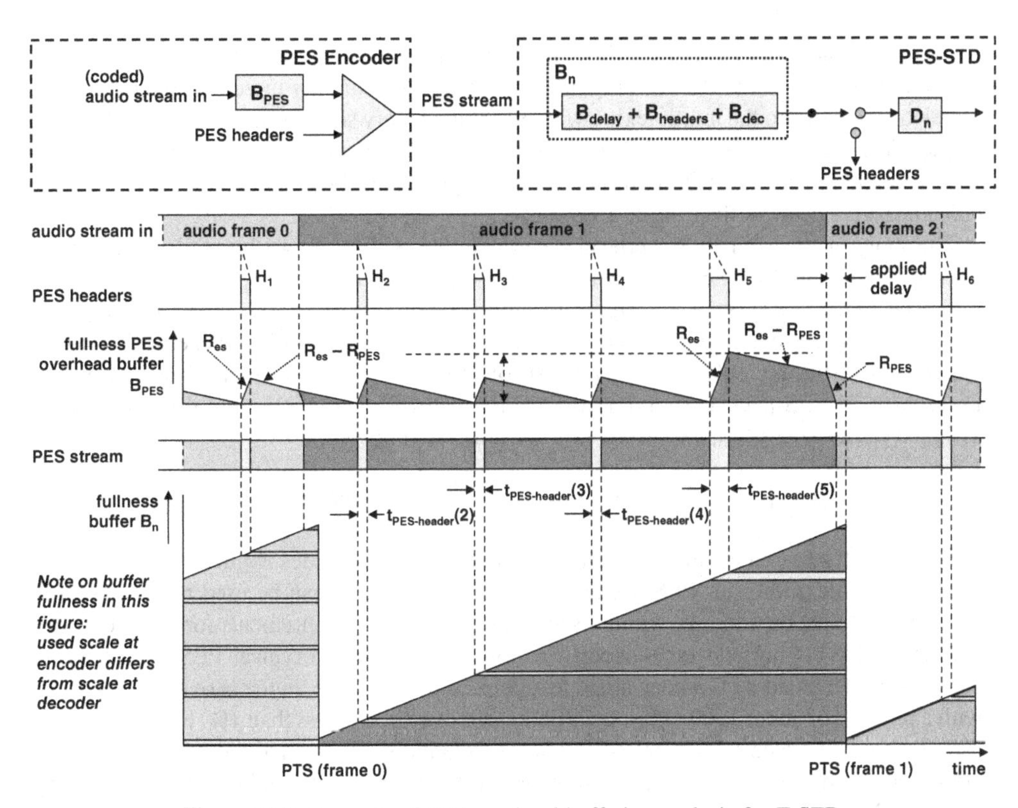

Figure 12.5 Example of PES overhead buffering analysis for T-STD

the longest duration of the PES header $t_{PES\text{-}header}(5)$ and thereby the sum of the fullness of both buffers is equal to $t_{PES\text{-}header}(5) \times R_{es}$ at any time.

In the P-STD, the PES headers do not enter B_n and therefore only the delay caused by the insertion of the PES headers needs to be taken into account. However, as will be discussed in Section 12.3, in the T-STD the PES headers enter the buffers of the decoding branch (e.g. see Figure 12.8). Therefore in the T-STD the buffer overhead due to PES packetization does not only cover B_{delay}, but also a component B_{header} for storage of the inserted PES headers. The buffering for B_{header} can be discussed in the PES-STD by assuming that each PES header enters B_n. Figure 12.5 depicts an example with an audio stream carried in PES packets.

In Figure 12.5, relatively small PES packets are used of variable size that are not aligned with the coded audio frames. As in Figure 12.4, the decoding process is postponed by the maximum duration of the PES header. Figure 12.5 shows the fullness of B_n, but without showing B_{delay} and B_{header} as separate buffers. PES headers enter B_n from the PES stream; when an audio frame is decoded and removed instantaneously from B_n, any included PES headers are removed as well. The room needed for storage of the PES headers depends on the number and size of the PES headers contained in B_n. In Figure 12.5, at the instant in time audio frame 1 is decoded, PES headers H_2, H_3, H_4 and H_5 are contained in B_n. If the maximum number of PES header bytes contained at any time in B_n is equal to sum_{max}(contained PES headers), then in the PES-STD for PES overhead buffering a total buffering room BS_{oh} is needed equal to $t_{PES\text{-}header}(5) \times R_{es} + sum_{max}$(contained PES headers).

One important note: in the PES STD only multiplexing with PES headers takes place and therefore buffer B_n in the PES-STD typically contains only slightly more than a single access unit. However, due to the multiplexing with other elementary streams, in the input buffer in the applicable T-STD branch multiple access units may be stored with any contained PES headers and therefore in the T-STD significantly more room may be required for storage of the contained PES headers than in the PES-STD.

The above analysis for PES overhead is particularly useful to assign the available amount of overhead buffering for PES packetization in transport streams and program streams. Also in this respect MPEG-2 systems is designed to allow for worst case scenarios. The initial approach was, based on the maximum elementary stream bitrate, to allow PES packetization with a maximum overhead of 10% for a delivery rate of 75 PES packets/s, the sector rate for Compact Disc. With 75 PES packets/s, and a 10% overhead for PES packetization, this results in an assigned buffer size BS_{oh}, expressed in a formula, as follows:

$$BS_{oh} = [(1/750) \times (\text{maximum bitrate of carried elementary stream})]$$

For many type of elementary streams the above formula provides applications with such a high level of flexibility for PES packetization that the formula can be used both in program streams and transport streams, without separately addressing requirements for storage of PES headers in T-STD buffers for transport streams. For example, a typical PES packet header containing a PTS and a DTS consists of less than 20 bytes; hence, in the case of a PES packet with a payload of about 2000 bytes, the PES header overhead is less than 1%. Furthermore, the used bitrate of an elementary stream is often substantially lower than the maximum bitrate. However, for some elementary stream formats a different approach is followed, as discussed in Section 12.5 and Chapter 14.

The forming of a PES stream prior to producing a transport stream or program stream is not necessary, but possible. Due to the absence of a pack header, no SCR and program-mux-rate values can be encoded. Instead, the PES packet header carries the ESCR-flag and the ES-rate-flag, to signal the presence of the ES-rate and the ESCR fields in that header. The ESCR carries the STC sample for the PES stream, while the ES-rate signals its delivery rate at the input of the PES-STD. For the encoding of these fields, the same units are used as for the SCR and the program-mux-rate.

The formal requirement for the fullness of the input buffer for elementary stream n in the PES-STD to prevent overflow is:

$$F_n(t) \leq BS_n \text{ for all t.}$$

Except for low delay video and during trick mode operation, the formal requirement to prevent underflow is:

$$F_n(t) \geq 0 \text{ for all t.}$$

Hence, in case underflow is forbidden, in the course of the decoding of a PES stream in the PES-STD, the fullness of the input buffer may vary between 0 and BS_n:

$$0 \leq F_n(t) \leq BS_n.$$

12.3 Transport Stream STD: T-STD

12.3.1 Description of T-STD

A transport stream may carry multiple programs with independent time bases, but in the T-STD only one program is decoded at a time. All timing indications in the T-STD refer to the time base (STC) of that program; this includes not only decoding and presentation times, but also the delivery time of each byte in the transport stream, independent whether the byte is from a transport packet of the decoded program or of any other program. The transport stream is delivered at the input of the T-STD at the piecewise constant bitrate discussed in Section 9.4.4 and Figure 9.10. Note that when another program from the same transport stream is selected, typically another time base applies and other delivery times of the bytes in the transport stream.

At the input of the T-STD, transport packets are provided at the rate of the transport stream. The transport stream demultiplexer ensures that each transport packet carrying system data for the decoded program, such as the PAT, the CAT, the TSDT and the PMT enter the T-STD branch for system decoding. This presumes that the decoding system has knowledge of the contents of the PAT and the PMT of the program to be decoded. From the PMT the PCR-PID is known, so that the PCR with the STC sample of the selected program can be read upon arrival, so as to synchronize the STC and the delivery of the transport stream. The PMT also provides the list of elementary-PIDs for all comprised program elements. The packet with the PCR PID may carry an element of the selected program, but if not, upon reading the encoded PCR value, the involved transport packet is ignored without entering the T-STD.

The demultiplexer ensures that the transport packets for each program element enter the appropriate decoding branch, as in the P-STD. Transport packets carrying privately defined data for the selected program may enter a privately defined branch in the T-STD; for example an application standardization body may specify details of such branch. Transport packets with data irrelevant for the selected program are ignored by the demultiplexer.

The above handling results in a bursty delivery of transport packets to the various branches in the T-STD. The delivery of the bytes is instantaneous: each byte of a transport packet enters the appropriate branch at its delivery time in the transport stream. An example is provided in Figure 12.6, where a transport stream is depicted with transport packets tp_1, tp_2, . . . , tp_{13}, each carrying audio, video or system data of the same program. The demultiplexer selects these transport packets from the stream and ensures that each packet enters the appropriate decoding branch of the T-STD.

The elementary stream specific decoding branches in the T-STD are comparable to those in the P-STD, but for transport streams the use of a single buffer B_n, as used in the P-STD, is insufficiently adequate. A transport stream may have a very high bitrate, in which case the selected transport packets enter the T-STD at a (very) high speed. A demultiplexer must obviously be capable of handling the input rates of the transport stream it is designed for. However, it is undesirable to burden the processing of the conveyed elementary stream data and system data with (very) high transport rates. To achieve this, a transport buffer TB_n of 512 bytes is introduced in each branch of the T-STD. This buffer furthermore ensures that the packets of each program element must be positioned in the transport stream in a reasonably equidistant manner, as discussed in Section 12.3.2 and Figure 12.9.

Input to the transport buffer in the T-STD are transport packets at the delivery rate of the transport stream; under all circumstances a piecewise constant rate applies. MPEG systems requires that for each time base at least two PCRs are encoded, so that for each time base the

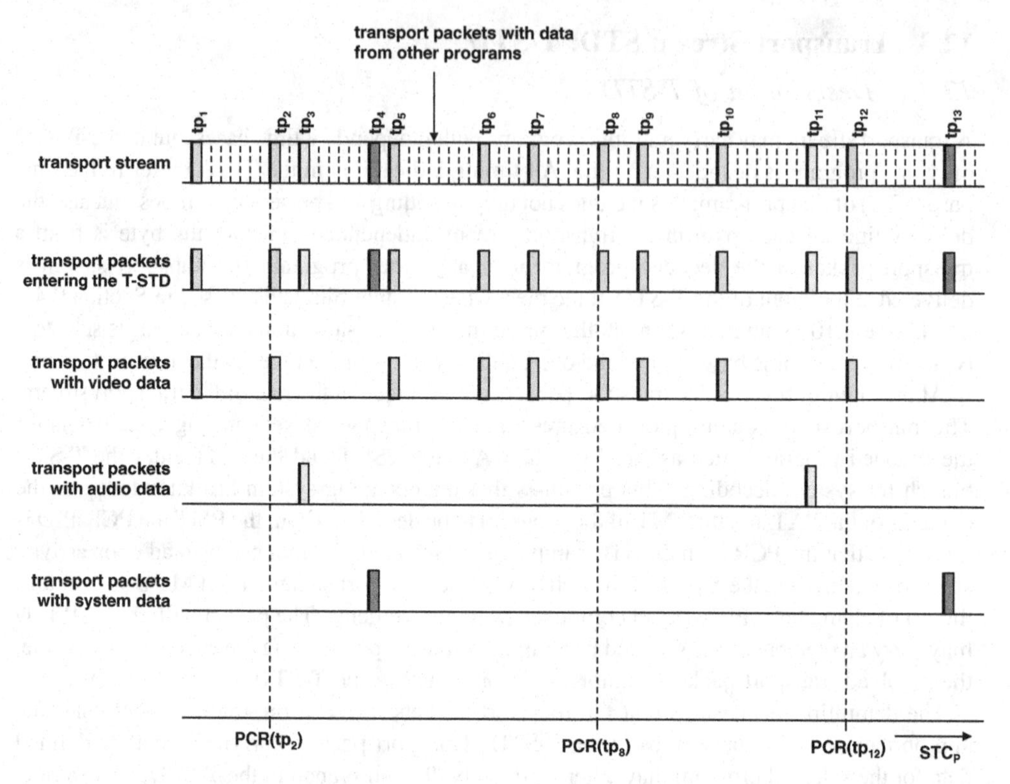

Figure 12.6 Transport packets from a transport stream entering the T-STD

delivery rate can be determined. For bytes in transport packets preceding the transport packet with the first encoded PCR of a time base for a program, the transport rate is extrapolated from the first and the second PCR values.

As an example, in Figure 12.7 a transport stream is depicted in which a new program commences with transport packets tp_1, tp_2, . . . , tp_8. In tp_3 and tp_7 the first and second PCR for this program are encoded. For the delivery of the transport packets tp_1 up to tp_3, the extrapolated transport rate is used based on the rate between PCR(tp_3) and PCR(tp_7). Likewise, for bytes following the last encoded PCR of a time base for a program, the transport rate is determined from the last two PCRs. At the transition from one time base to another, typically a time base discontinuity occurs. For more background on time base discontinuities see Section 11.3 on local program insertion and Section 11.4 on splicing in transport streams. In Figure 12.7, an example is depicted in which a local program, such as the local weather forecast, is inserted in a global program intended for a larger regional area. The time base discontinuity occurs at the first PCR of the local time base, encoded in packet tp_1. In packets tp_a and tp_e the one but last and the last PCR of the global time base are encoded; for the transport packets remaining after the last PCR in tp_e until the time base discontinuity, the transport rate is extrapolated from the calculated rate between PCR(tp_a) and PCR(tp_e).

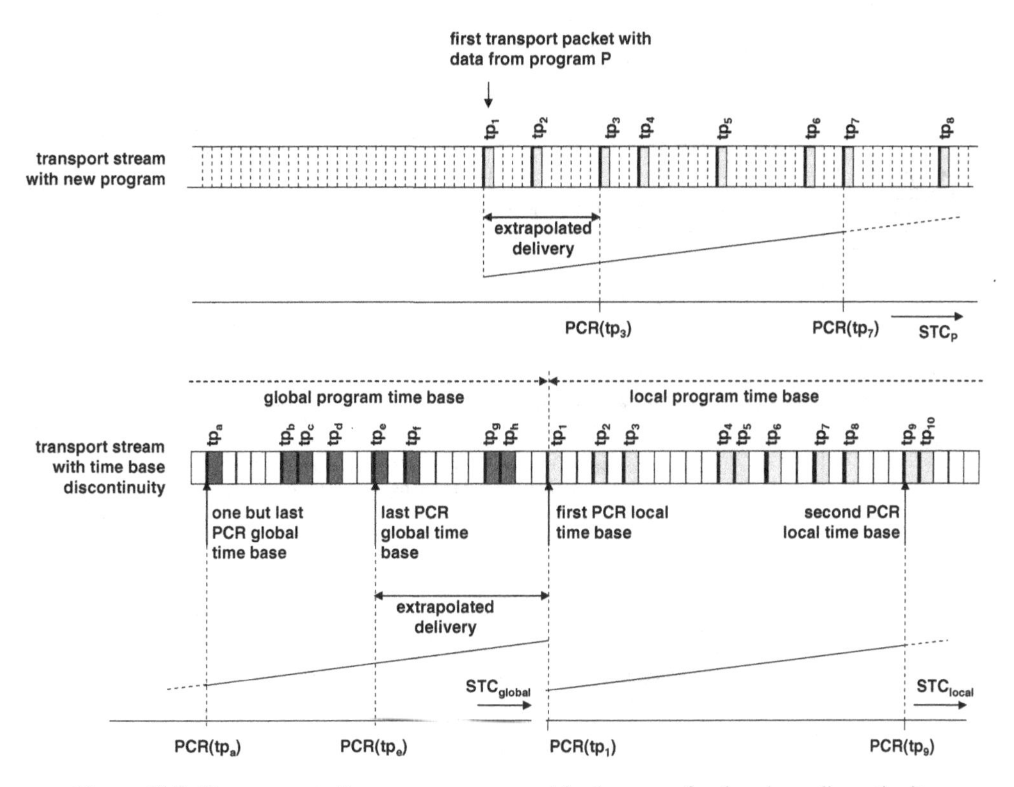

Figure 12.7 Transport rate for a new program and in the case of a time base discontinuity

All bytes that entered the transport buffer TB_n are removed at a rate Rx_n with a fixed value considered 'reasonable' for the type of data that is processed in the branch, so that TB_n empties frequently. For systems data, a single value is chosen, but for all other branches the value of Rx_n depends on the content format. As a rule of thumb, Rx_n is chosen to be 1.2 times the maximum bitrate of the elementary stream. This means that practical decoders are required to process the received elementary stream data at a rate that is only slightly higher than the maximum bitrate of the elementary stream, while sufficient room is left to multiplexers for the positioning in the transport stream of the transport packets carrying the involved elementary stream.

For video streams the value of Rx_n is chosen based on above rule of thumb. For audio streams, usually a single value is selected. The applicable values of Rx_n are discussed in Section 12.5 and in Chapter 14. Overflow of TB_n is forbidden; as each transport packet consists of 188 bytes, fewer than three transport packets can be stored in TB_n at any time. To prevent overflow of TB_n, system encoders must ensure that the transport packets for each branch are positioned in the transport stream in a sufficiently equidistant manner, as discussed in Section 12.3.2 and Figure 12.9. In case the rate of the transport stream is high compared to the value of Rx_n, in the transport stream typically at most two transport packets for the same decoding branch can be conveyed at immediately subsequent positions.

The structure of the T-STD is depicted in Figure 12.8. Compared to the P-STD, not only the transport buffers TB_n are introduced, but there is also another branch type in which the single

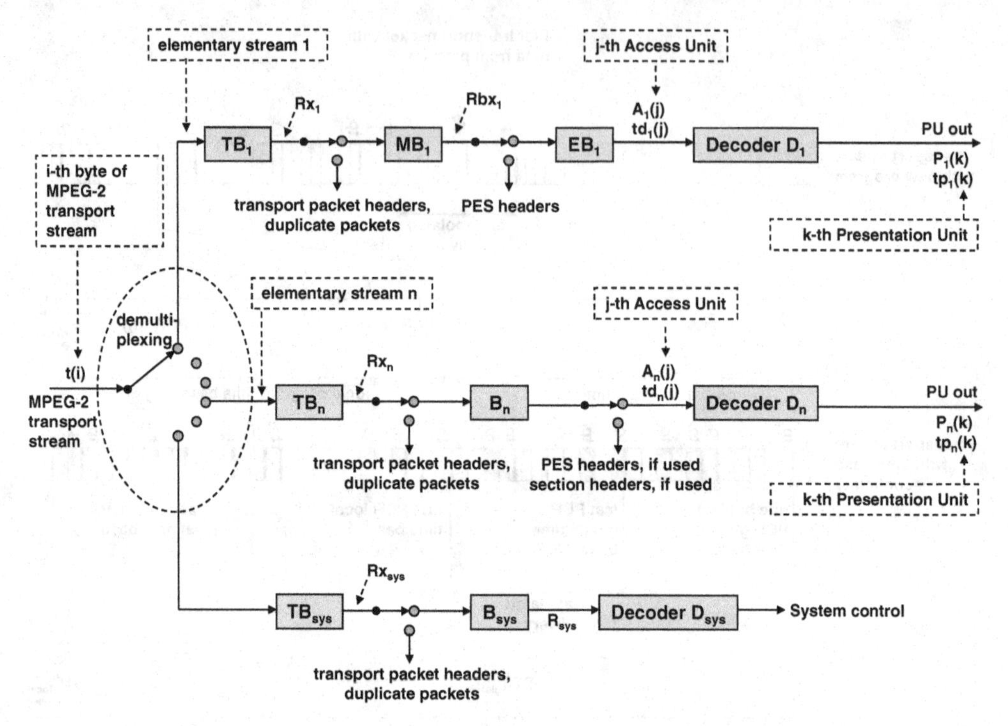

Figure 12.8 The system target decoder for transport streams: T-STD

buffer B_n is replaced by two buffers MB_n and EB_n: a multiplex buffer and a elementary stream buffer. Note the difference in removal of PES headers in the branches. In the branch with MB_n and EB_n, data is transferred from MB_n into EB_n at a rate Rbx_n, whereby two methods may be applied. The use of transport buffers is described in detail in Section 12.3.2, while Section 12.3.3 addresses the processing of the system related data in the branch with B_{sys} and D_{sys}, including constraints for the management of D_{sys}. In Section 12.3.4 the use of both branch types for elementary stream data is discussed, including both transfer methods between MB_n and EB_n. Finally, the constraints for the management of B_n, MB_n and EB_n are discussed in Section 12.3.5.

Due to the introduction of additional buffers and the decoder for system data in the T-STD, some additional parameters are needed, next to those from the P-STD, so as to express the requirements and constraints in generic mathematical formulas. The following parameters are used in the T-STD:

- n – the index of an elementary stream.
- t – the time measured in seconds.
- i – the index of a byte in the MPEG-2 system stream.
- t(i) – the time of arrival of the ith byte of the MPEG system stream at the input of the T-STD.
- j – the index in decoding order of an access unit in an elementary stream.
- k – the index in presentation order of a presentation unit.

- PCR(i) – the encoded STC value in the PCR field, where i is the byte index of the final byte[5] of the PCR field that serves as reference for the encoding of the PCR field; in other words, PCR(i) represents the intended STC delivery time of that byte.
- $A_n(j)$ – the jth access unit in elementary stream n.
- $td_n(j)$ – the decoding time in the T-STD of the jth access unit in elementary stream n.
- $P_n(k)$ – the kth presentation unit in elementary stream n.
- $Tp_n(k)$ – the presentation time in the T-STD of the kth presentation unit in elementary stream n.
- Rx_n – the bitrate of the data leaving TB_n.
- Rbx_n – the bitrate of the data leaving MB_n.
- Rx_{sys} – the bitrate of the data leaving TB_{sys}.
- R_{sys} – the bitrate of the data leaving B_{sys}.
- BS_{sys} – the size of buffer B_{sys} in bytes.
- BS_n – the size of buffer B_n in bytes, specified in Section 12.5 or in Chapter 14 for each content format for which a decoding branch with B_n applies.
- MBS_n – the size of buffer MB_n in bytes, specified in Section 12.5 or in Chapter 14 for each content format for which a decoding branch with MB_n and EB_n applies.
- EBS_n – the size of buffer EB_n in bytes, specified in Section 12.5 or in Chapter 14 for each content format for which a decoding branch with MB_n and EB_n applies.

12.3.2 The Use of Transport Buffers

At the input of the T-STD, a transport stream is delivered at the transport stream rate. Based on the encoded PID values, complete transport packets relevant for a selected program are transferred to the transport buffer TB_n of the appropriate decoding branch or, in the case of system data, to the transport buffer TB_{sys}. The system data transferred to TB_{sys} includes the PAT, CAT and, if present, the TSDT and the IPMP CIT, carried in transport packets with a PID value 0, 1, 2 or 3, respectively, as well as transport packets carrying the PMT of the selected program, as identified in the PAT. The NIT, if present, is not transferred to TB_{sys}. The NIT is a private table; the handling in the T-STD of transport packets carrying the NIT or any other private table is to be specified by private means; for example, an application standardization body may define details of a specific branch for that purpose.

All bytes that enter TB_n or TB_{sys}, leave it at a rate Rx_n or Rx_{sys}, respectively. If no bytes are present in TB_n or TB_{sys}, then Rx_n or Rx_{sys} is equal to zero; otherwise, Rx_n has the value discussed in Section 12.5 or Chapter 14 for each content format, while Rx_{sys} has the fixed value of 1×10^6 bits/s. Transport packet payload bytes leaving TB_n or TB_{sys} are delivered to the subsequent buffer in the branch; this includes the header and payload of PES packets or sections in which the elementary stream or system data is packetized. However, transport packet headers are not delivered to the subsequent buffer, but may be used to control the system. To prevent the decoding of data from duplicate packets, at the output of the transport buffer, it is ensured that also duplicate packets do not enter the subsequent buffer.

[5] Formally, in MPEG-2 systems, this reference byte is the byte containing the final bit of the PCR field (see Section 13.6).

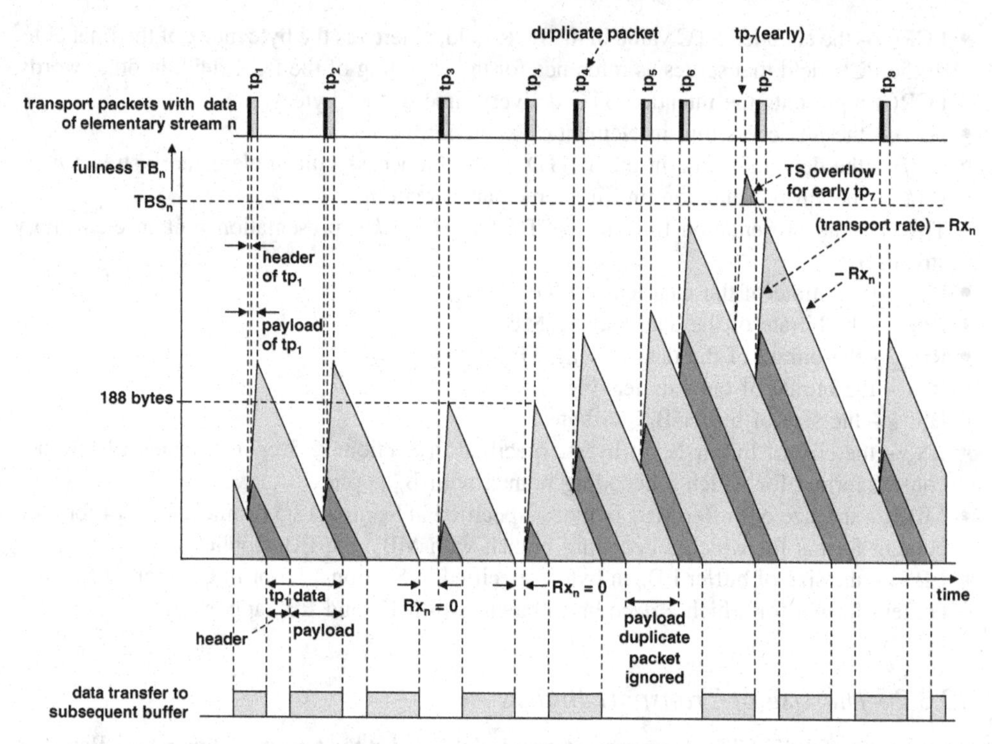

Figure 12.9 Byte delivery into and out of TB_n and the associated fullness of TB_n

MPEG-2 systems defines the following constraints for each transport buffer TB_n and TB_{sys}:

- Buffer overflow is forbidden; at any time TB_n and TB_{sys} contain at most 512 bytes.
- The transport buffer shall be empty at least once every second. Compliancy problems, such as overflow of TB_n, may occur only over (very) long time windows. An empty TB_n provides a unambiguous starting point for testing the compliancy of a carried arbitrary stream. Hence, this requirement facilitates compliancy testing, as testing from its 'start' may be a meaningless concept in the case of continuous broadcasts.

An example of the data delivery into and out of TB_n is provided in Figure 12.9, where transport packets tp_1, tp_2, . . . , tp_8 carry data of elementary stream n. Packet tp_4 is carried twice; the second one is a duplicate packet. During the delivery of a packet with elementary stream data, the fullness of TB_n increases at a rate equal to the transport rate minus Rx_n. When no packet enters, the TB_n fullness decreases at rate Rx_n as long as data is contained in TB_n. Figure 12.9 shows that TB_n is empty just prior to the delivery of tp_3 and tp_4; when TB_n is empty, $Rx_n = 0$. Bytes from the payload of tp_1, tp_2, . . . , tp_8 leaving TB_n are transferred immediately to the subsequent buffer in the decoding branch, except for the second (duplicate) packet tp_4. The transport packet header is not transferred to the subsequent buffer, but used to control the decoding system; for example, based on information from the headers of both tp_4 packets, it is determined that the second one is a duplicate packet.

The requirement that TB_n must not overflow limits how far ahead or behind in time a transport packets can be positioned compared to its 'natural' equidistant position. For example,

tp_6 and tp_7 in Figure 12.9 are delivered so that the TB_n is almost full. If tp_7 is delivered slightly earlier, as indicated by tp_7(early), the fullness of TB_n exceeds its size TBS_n, causing overflow of TB_n.

12.3.3 System Data Processing and Buffer Management

For buffer B_{sys} a size BS_{sys} of 1536 bytes has been selected; while a value in this range is required, this specific value is selected to allow joint implementation of TB_{sys} and B_{sys} within 2048 bytes. The T-STD defines TB_{sys} and B_{sys} as separate buffers with constraints and requirements that must be met by each compliant transport stream. However, designers of practical decoders have freedom to implement these buffers in a way that is deemed most convenient, as long as the practical decoder is capable of decoding the same transport streams as the T-STD.

The rate R_{sys} of system data out of B_{sys} is defined as follows. If B_{sys} is empty, then $R_{sys} = 0$, else:

$$R_{sys} = \text{maximum}[(80\,000\ \text{bit/s}), (\text{transport-rate in bit/s}) \times 0.002]$$

Figure 12.10 depicts the dependency of R_{sys} on the transport-rate. For transport rates up to 40 Mb/s the value of R_{sys} is equal to 80 kbit/s; for higher transport rates the value of R_{sys} increases, so as to allow for higher PSI data rates. At a transport rate of 80 Mb/s, the value of R_{sys} is equal to 160 kbit/s. Note that for transport rates above 500 Mb/s the value of R_{sys} exceeds the fixed value 1×10^6 bits/s of Rx_{sys}, which means that above this transport rate each

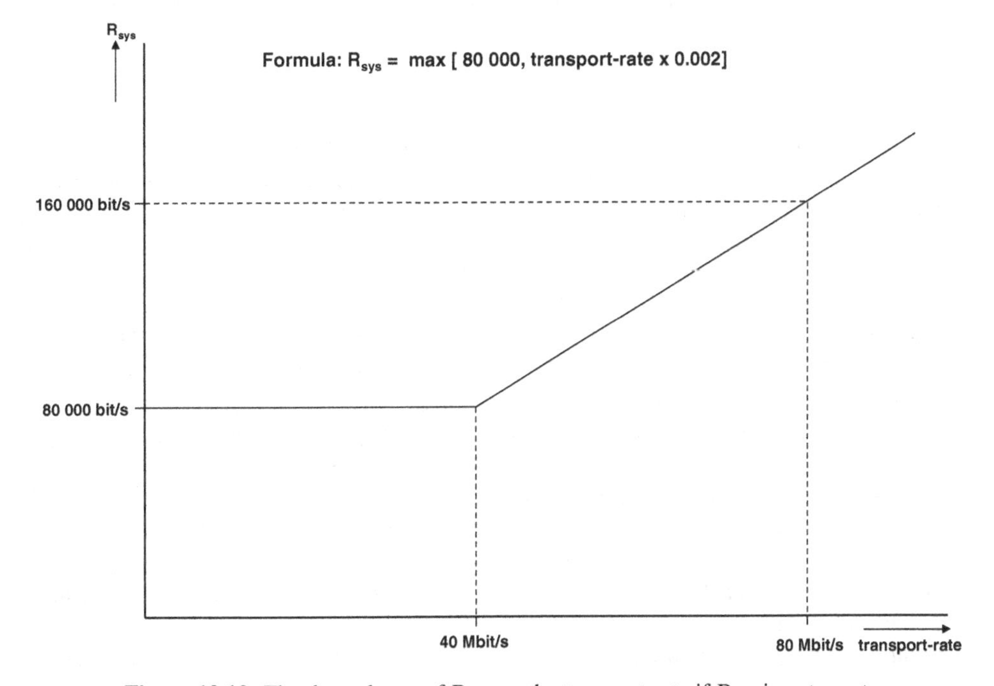

Figure 12.10 The dependency of R_{sys} on the transport rate if B_{sys} is not empty

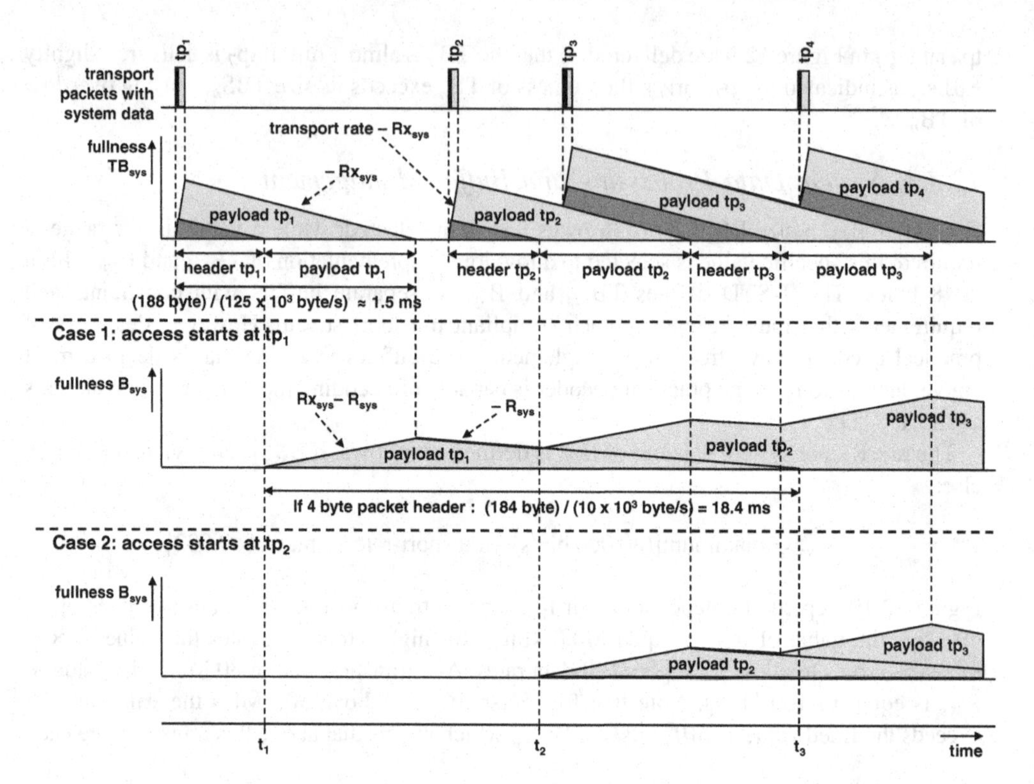

Figure 12.11 Flow of system data through TB_{sys} and B_{sys} in two stream access cases

byte that enters B_{sys} is removed immediately. Hence, the rate R_{sys} has effectively an upper bound equal to 1×10^6 bits/s. In such cases B_{sys} is not useful anymore.

The flow of system data through TB_{sys} and B_{sys} is depicted in Figure 12.11. For the decoded program the transport stream at the input of the T-STD carries the system data in packets tp_1, tp_2, tp_3, and tp_4. At the instant in time packet tp_1 in the transport stream starts entering the empty buffer TB_{sys}, the tp_1 data starts immediately to leak out of TB_{sys} with the specified rate of 1×10^6 b/s, or 125×10^3 bytes/s. In the T-STD it will take $(188/125) \times 10^{-3} \approx 1.5$ ms for the tp_1 data to leave TB_{sys}. In Figure 12.11, TB_{sys} is empty again when the tp_2 data enters TB_{sys}. When about half of the tp_2 data is leaked out of TB_{sys}, packet tp_3 enters and after some more time also tp_4. Figure 12.11 shows the resulting fullness of TB_{sys}. When a transport packet enters TB_{sys}, the fullness increases by a rate equal to the rate of the transport stream minus Rx_{sys}. When no data enters TB_{sys}, its fullness decreases by a rate equal to Rx_{sys}, until TB_{sys} is empty. Note that the leaking of packets tp_1 and tp_2 starts immediately upon their arrival at the input of the T-STD, due to the fact that TB_{sys} is empty when these two packets arrive, but that the leaking of packet tp_3 out of TB_{sys} is postponed until all data of tp_2 is transferred out of TB_{sys}. In other words, the delay of a packet through TB_{sys} depends on the fullness of TB_{sys} at the instant in time the packet arrives.

Figure 12.11 also shows the flow of data through B_{sys} for two different startup cases; in case 1, the decoder starts to access the transport stream at packet tp_1, and in case 2 at tp_2.

At startup, in both cases buffer B_{sys} is empty. Upon leaking out of TB_{sys}, each transport packet payload byte enters B_{sys} instantaneously. The transport packet header bytes that also leak out of TB_{sys} are used to control the system without entering B_{sys}. As soon as system data enters B_{sys}, the data start leaking out of B_{sys}; a transport rate of less than 40 Mb/s is assumed and therefore the leak rate out of B_{sys} is equal to 80 000 bit/s or 10×10^3 bytes/s. Hence if a transport packet header consists of 4 bytes, the 184 payload bytes of a transport packet are transferred out of B_{sys} in 18.4 ms.

In case 1, the payload of tp_1 is transferred out of B_{sys} between t_1 and t_3; the transfer of the tp_2 payload out of B_{sys} has to wait until all payload bytes of tp_1 are transferred. As a result, the transfer of tp_2 payload out of B_{sys} starts at t_3. In case 2, buffer B_{sys} is empty when the tp_2 payload data is transferred from TB_{sys} into B_{sys}, and as a result, the transfer of the payload bytes of tp_2 out of B_{sys} starts already at t_2. In other words, also the delay of packet payload through B_{sys} depends on the fullness of B_{sys} at the instant in time the packet payload arrives. In conclusion, when a transport packet tp_n with system data arrives at the input of the T-STD, the length of the transfer periods of its payload bytes out of TB_{sys} and out of B_{sys} is determined easily from the number of payload bytes and the applicable leak rates. However, the start of the transfer period depends on the number of bytes present in B_{sys} at the time of arrival of the first byte of the payload of packet tp_n.

The timing of transitions of PSI tables is often not known a priori; for example, in the case of local program insertions in a global program, the timing of the transition to the new PMT of the local program is determined by the system in charge of the local insertion. Moreover a time base discontinuity may occur. Therefore time stamps are not used to signal transitions of PSI tables. Instead, MPEG-2 systems defines that a new version of a PSI table becomes valid at the instant in time the last byte of section needed to complete a PSI table leaves B_{sys}. However, this instant in time does not only depend on the delivery time of the transport packet carrying that section, but also on the status of the fullness of the buffers TB_{sys} and B_{sys} as discussed above. In addition, ambiguity may be introduced by re-multiplexing of transport streams. Obviously, system encoders should be aware of the involved complications. For example, if video of a program starts to be carried in another PID, then it must be ensured that all packets with the 'old' PID value are processed and that the first packet with the 'new' PID is not missed. To accommodate the uncertainty in the timing of the PSI change it may be possible to leave a gap in time between the last packet with the 'old' PID and the first packet with the 'new' PID, which can typically be done with some smart buffer control management without any visible impact on the output video. These and similar strategies are beyond the scope of this book; however, it should be noted that for this issue guidelines for the use of PSI were defined by application standardization bodies, such as DVB.

A simple method to prevent complicated tracking of the TB_{sys} and B_{sys} buffer fullness for determining PSI table transitions is to limit the number of packets carrying PSI for the conveyed programs in the transport stream and to position these packets in a rather equidistant manner. Which may be rather straightforward; for example, in a transport stream with eight scrambled programs, for the carriage of the PSI tables the following may often be reasonably assumed:

- the PAT, carried in a single transport packet with PID value 0;
- the CAT, a single transport packet with PID value 1;

- often there are about eight program definitions in the PMT; typically four can be carried in one transport packet, so that two transport packets are required to convey the eight program definitions.

In above case, four transport packets are needed to convey the PSI tables. If for these tables a repetition rate is selected of 10 Hz, to ensure a good random access, then each second 40 transport packets with PSI data enter TB_{sys}; hence on average each 25 ms a transport packet will enter TB_{sys}. In which case both TB_{sys} and B_{sys} will be empty when the first transport packet payload bytes enter it, even if a maximum delivery shift of ±4 ms due to re-multiplexing is taken into account. MPEG-2 systems specifies that overflow of B_{sys} is forbidden; at any time B_{sys} contains at most BS_{sys} bytes.

12.3.4 Processing of Elementary Stream Data

Figures 12.12, 12.13 and 12.14 discuss the decoding branches for elementary streams. In the branches for elementary streams, the decoding times of access units are determined by time stamps, and therefore the timing ambiguity issues discussed above for PSI data do not occur. For explanatory reasons, in the figures the transport packets tp_1, tp_2, tp_3, and tp_4, as well as the leak rate of data out TB_n are copied from Figure 12.11. Note though that the transport packets in

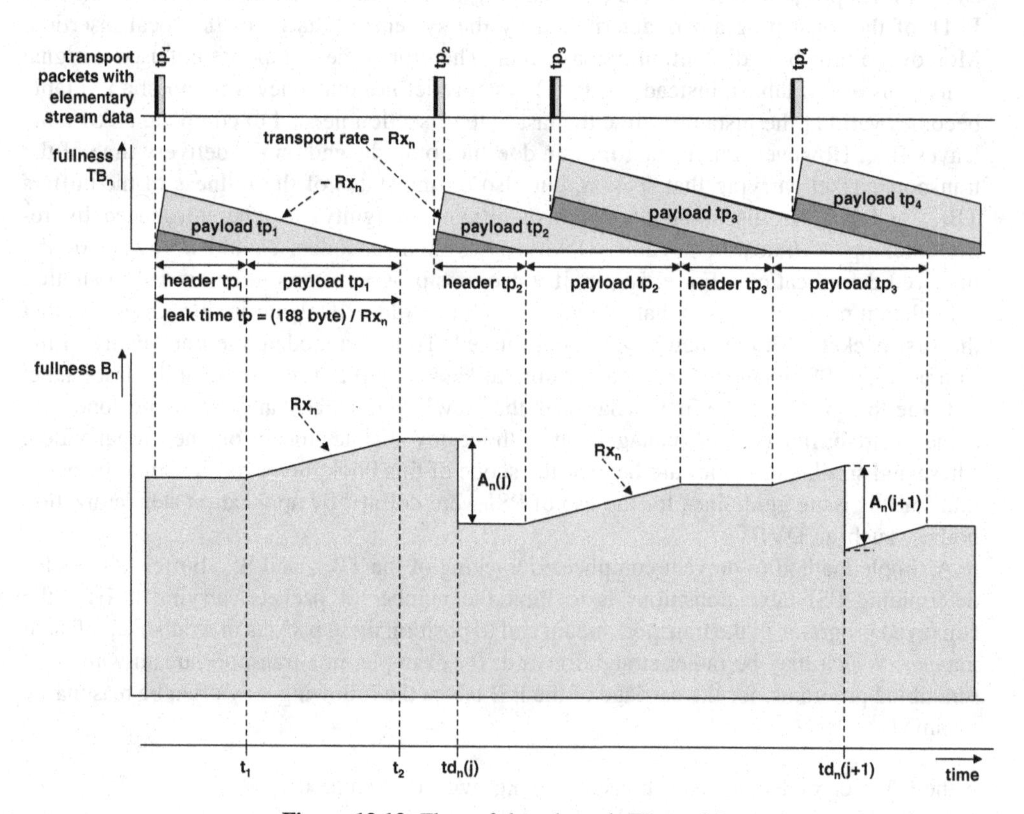

Figure 12.12 Flow of data through TB_n and B_n

each figure carry different elementary streams, and that in practice in each figure the leak rate out of TB_n will differ.

Audio and several other content formats use the decoding branch with the single input buffer B_n. The flow of data through TB_n and B_n is shown in Figure 12.12. When TB_n contains one or more bytes, data leaks out of TB_n at a rate equal to Rx_n. The payload of the transport packets tp_1, tp_2, tp_3, and tp_4 leaks out of TB_n into B_n, and as a result the fullness of TB_n increases at the same rate Rx_n, as long as no data is removed from B_n. The removal of data out of B_n is determined by time stamps: at time $td_n(j)$ and $td_n(j+1)$ access units $A_n(j)$ and $A_n(j+1)$ are decoded, upon which all data of the access unit is removed instantaneously from B_n. All bytes of PES or section headers that are included in the access unit or that immediately precede it, are removed as well and may be used to control the system. If the overhead due to either the PES packet headers or the section headers is ignored, then upon its decoding, the fullness of B_n decreases with the number of bytes contained in the access units $A_n(j)$ and $A_n(j+1)$; see Figure 12.12.

In the branch used for the decoding of MPEG-2 video and AVC two buffers are applied, a Multiplex Buffer MB_n and an Elementary stream Buffer EB_n. Practical system decoder implementations typically apply one buffer that comprises both MB_n and EB_n. Hence, the use of two buffers in this branch is important from MPEG-2 compliancy perspective, but less from implementation perspective.

In the decoding branch with MB_n and EB_n, data leaks out of TB_n into MB_n at the rate Rx_n, while the removal of access units from EB_n is controlled by time stamps, in the same way as the data entry in and removal out of B_n. Each byte that enters MB_n leaves it, but only PES payload bytes are transferred to EB_n. When a PES payload byte is transferred from MB_n to EB_n, all PES packet header bytes in MB_n that immediately precede that payload byte, are removed instantaneously from MB_n and discarded.[6] Hence, prior to transferring a byte from MB_n to EB_n, it is verified whether that byte is part of a PES header, in which case that byte is discarded immediately.

PES payload bytes in MB_n are transferred to EB_n at a rate Rbx_n. Two methods can be used for this transfer: the leak method and the vbv-delay method. With the leak method, as long as MB_n is not empty, PES payload data leaks out of MB_n at rate Rbx_n until buffer EB_n is full. When EB_n is full, no bytes are removed from MB_n. An example of the use of the leak method is depicted in Figure 12.13. With the vbv-delay method, the timing of the original elementary video stream is reconstructed, based on timing information contained in the elementary stream, in the case of MPEG-2 video the coded values of the vbv-delay parameter. An example of the use of the vbv-delay method is depicted in Figure 12.14.

In Figure 12.13, it is shown that transport packet payload bytes leak out of TB_n at rate Rx_n. As long as MB_n contains data, data leaks out of MB_n at rate Rbx_n, unless EB_n is full. The fullness of MB_n increases at a rate equal to $(Rx_n - Rbx_n)$ when data leaks out of TB_n and out of MB_n. If no data enters MB_n, then its fullness decreases at rate Rbx_n, as long as MB_n is not empty and EB_n not full. At times $td_n(j)$ and $td_n(j+1)$, access units $A_n(j)$ and $A_n(j+1)$ are decoded; at these instances in time, these access units are removed from EB_n, so that its fullness decreases instantaneously with the number of bytes contained in these access units. At time t_1

[6] Note though that the information carried in the PES header may be used to control the decoding system.

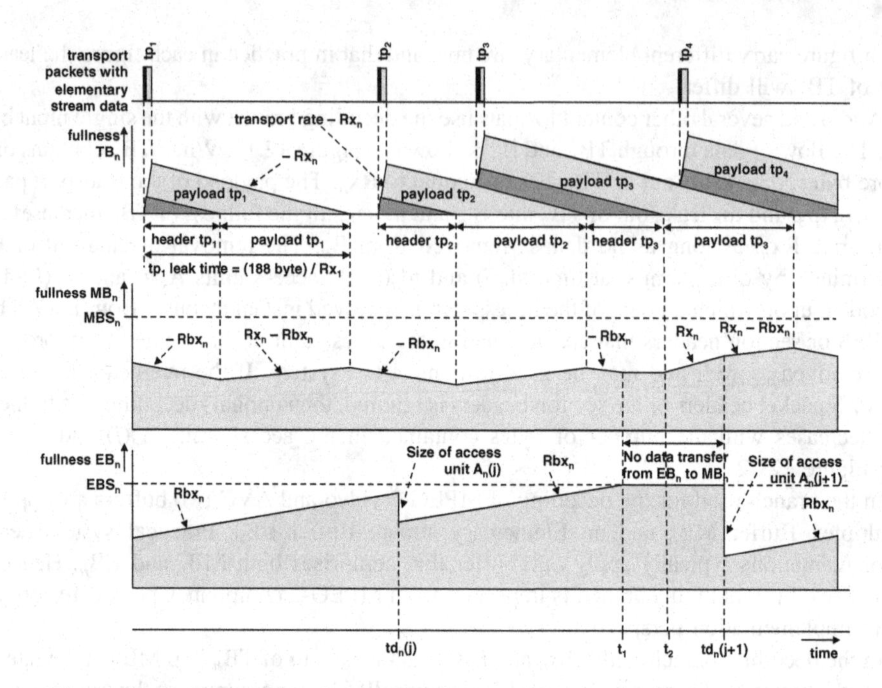

Figure 12.13 Flow of data through TB_n, MB_n and EB_n if the leak method applies

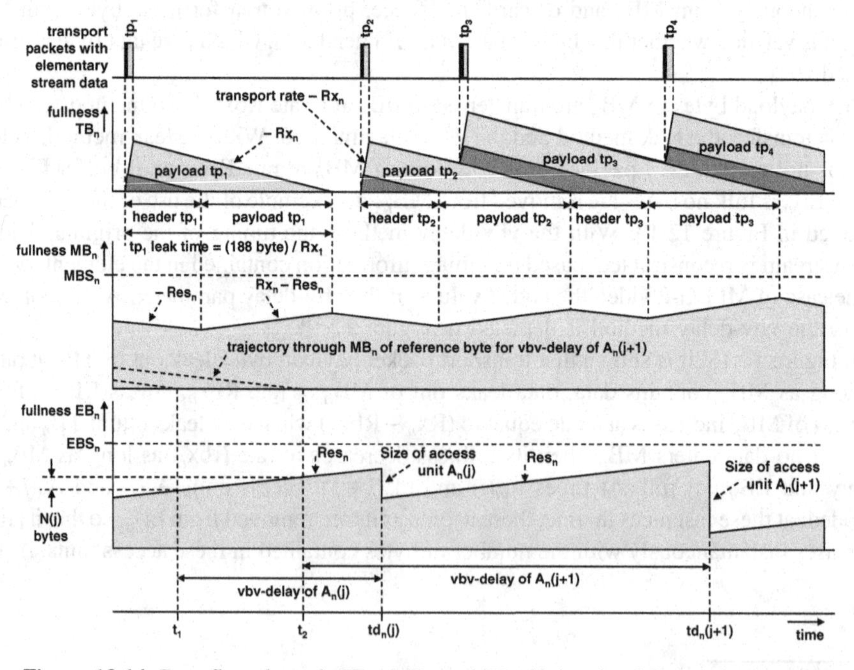

Figure 12.14 Data flow through TB_n, MB_n and EB_n if the vbv-delay method applies

buffer EB_n becomes full, upon which the leaking out of MB_n stops. Buffer EB_n remains full until access unit $A_n(j+1)$ is decoded at time $td_n(j+1)$ and so between t_1 and $td_n(j+1)$ no data is transferred from MB_n to EB_n. At t_1 no data enters MB_n and therefore the fullness of MB_n remains the same until time t_2, when the payload of packet tp_3 starts entering MB_n, upon which the fullness of MB_n starts to increase at the leak rate Rx_n out of TB_n. After the decoding of $A_n(j+1)$, data starts to leak again into MB_n at rate Rbx_n.

When the vbv-delay method is applied for the transfer of data from MB_n into EB_n, then a strict separation is made between the coding of the elementary stream and system coding. With the vbv-delay method, the data transfer between MB_n and EB_n aims to reconstruct the delivery timing of the original video stream. In case of MPEG-1 and MPEG-2 video, the vbv-delay parameter is used for this purpose, and for AVC, an equivalent method is applied. See Section 7.4 and Figure 7.11 for an example of the reconstruction of a video stream from decoding times and vbv-delay values.

Figure 12.14 shows the fullness of MB_n and EB_n in the case of the vbv-delay method; the bitrate of the elementary stream Res_n is assumed to be constant. The vbv-delay indicates the time between the arrival of the reference byte for the vbv-delay for a picture in the coded video stream and the decoding of that picture. Figure 12.14 shows the vbv-delays for pictures $A_n(j)$ and $A_n(j+1)$; at the time indicated by the vbv-delay, the reference bytes enter EB_n at t_1 and t_2 and all bytes in between two reference bytes enter EB_n at the rate Res_n. Hence, elementary stream data continuously leaves MB_n at rate Res_n, while payload bytes of transport packets enter MB_n at rate Rx_n during the periods indicated in Figure 12.14. As a result, the fullness of MB_n either increases at a rate equal to $(Rx_n - Res_n)$ or decreases at a rate Res_n. Note that Figure 12.14 also shows the trajectory through MB_n of the reference bytes for both vbv-delays.

The elementary stream bitrate $Res_n(j)$ between the reference bytes for the vbv-delays of two subsequent coded pictures $A_n(j)$ and $A_n(j+1)$ is found from the number of bytes in the elementary stream between both reference bytes. In Figure 12.14, the number of bytes between the reference bytes of both vbv-delays is indicated as $N(j)$, so that the value of $Res_n(j)$ is equal to $N(j)/(t_2 - t_1)$, and therefore:

$$
\begin{aligned}
Res_n(j) &= N(j)/[td_n(j = 1) - \text{vbv-delay}(j+1) - \{td_n(j) - \text{vbv-delay}(j)\}] \\
&= N(j)/[\text{vbv-delay}(j) + \{td_n(j+1) - td_n(j)\} - \text{vbv-delay}(j+1)]
\end{aligned}
$$

Note that each vbv-delay and td_n value is expressed in seconds.

In Figure 12.14, the value of $Res_n(j)$ is assumed to be constant, but in the general case, rate $Res_n(j)$ is piecewise constant, and may vary for each j. In case the trick mode status is true, the vbv-delay method does not apply; in this status only the leak method is used.

For the transfer of data from MB_n to EB_n, initially only the vbv-delay method was defined by MPEG systems, thereby requiring the encoding of the vbv-delay parameter for each picture. The encoding of vbv-delays allows the reconstruction of the coded video stream with its original timing. Which means that after any multiplex operation and transport, the video stream input to the very first multiplexer can be reconstructed, and thereby all impact of multiplex operations can be undone. Furthermore, in the case of a non-compliant system stream, the encoding of vbv-delays allows to determine whether the video encoder or the system encoder causes the non-compliant behaviour. For example, in the case of buffer overflow in the T-STD when a MPEG-2 video stream is decoded, then the video encoder is to blame if the VBV buffer overflows, else the system encoder causes the non-compliant behaviour.

However, the vbv-delay method limits the available amount of buffering for multiplex operations more than strictly needed. Typically, only a fraction of EB_n is used to store the access units; only in very rare cases EB_n is completely full. The (varying) empty room in buffer EB_n can be used efficiently for multiplexing, and for this purpose the leak method is defined for the transfer of data from MB_n to EB_n. The difference between both methods in available buffer room is demonstrated by the fullness of MB_n in Figures 12.13 and 12.14. With the leak method in Figure 12.13, MB_n is structurally less full, thereby indicating that more buffer room is available for multiplexing.

The leak method does allow the encoding of vbv-delays for MPEG-2 video and the equivalent information in AVC; hence, if this information is present, the delivery timing of the original video stream can also be reconstructed if the leak method is applied. However, in transport stream applications that do not require the coding of valid vbv-delays, the value 0xFFFF may be encoded in the vbv-delay parameter of MPEG-2 video streams, indicating that the vbv-delay for pictures is not specified. Also in AVC, the coding of vbv-delay equivalent information is optional.

The conditions for the use of the vbv-delay method and the leak method depend on the content format, as discussed in Chapter 14. For each content format applying the branch with MB_n and EB_n, the used method for the transfer between MB_n and EB_n for the decoding in the T-STD is specified unambiguously. Which is very important for determining whether the produced transport stream is compliant. However, practical decoder implementations do not need to know which method is used, as long as the implementation took into account that any compliant transport stream can be decoded, for example by using additional memory to compensate for differences between the specific implementation and the hypothetical STD model. The choice between both methods may depend on the encoding architecture and on the need for the capability to undo all multiplex operations, as discussed above.

12.3.5 T-STD Buffers for Elementary Stream Decoding

12.3.5.1 Buffering Components and Buffer Sharing

For the decoding of each elementary stream in the T-STD, MPEG-2 systems defines the size of each input buffer in the applicable decoding branch. The size depends on the content format of the elementary stream. It needs to be ensured that for each elementary stream the input buffers can accommodate the three buffer components listed below.

- The input buffer required for the decoder of elementary stream *n*, without any multiplexing, assuming the instantaneous decoding of access units; this buffer component is referred to as B_{dec} with size BS_{dec}. For MPEG-2 video this is the VBV buffer. The buffer room reserved for BS_{dec} depends on the content format of elementary stream n and on format specific issues, such as the level and profile of video and the number of conveyed audio channels.

- The overhead buffering needed for PES packetization; this buffer component is referred to as B_{oh} with size BS_{oh}. In Section 12.2.4 the buffer room reserved for BS_{oh} is discussed in general terms; for most, but not all elementary stream types the following applies:

$$BS_{oh} = [(1/750) \times (\text{maximum bitrate of carried elementary stream})]$$

Which BS_{oh} value applies depends on the content format; and this is discussed in more detail in Section 12.5 and in Chapter 14.
- The buffering required by the multiplexing to form the original transport stream and for any subsequent re-multiplexing of the transport stream; this buffer component is referred to as B_{mux} with size BS_{mux}. For all multiplex and re-multiplex operations a maximum delay is assumed of 4 ms (see Section 11.2). The assigned room for this purpose in buffer B_{mux} is determined by the following rule of thumb: $BS_{mux} = 4$ ms times the maximum bitrate of the elementary stream.

For each of the above components a certain buffer room is reserved, in a manner so that cost-effective decoders can be built, while at the same time applications sufficient flexibility is offered to form and convey transport streams. To achieve the required flexibility in (re-) multiplex operations for the dimensioning of the buffer components a rather worst case approach is followed. However, in practice often the reserved buffer room for a component is not used completely, in which case more room can be made available for other components. Therefore in the T-STD, the total amount of buffering can be shared in a flexible manner: when one component uses less room than reserved for it, the left-over room is available to the other components. For example, if a small B_{dec} is used, then B_{mux} can be made equivalently larger to create even more flexibility for multiplexing.

12.3.5.2 Size BS_n of Buffer B_n

The decoding branch in the T-STD with input buffer B_n is used for audio streams and for various other content formats. Buffer B_n needs to accommodate all three buffer components discussed in Section 12.3.5.1. Hence, in a formula, the following applies:

$$BS_n = BS_{dec} + BS_{oh} + BS_{mux}$$

Detailed information on the applied value(s) of BS_n for the content formats supported by MPEG-2 systems is provided in Section 12.5 and in Chapter 14.

12.3.5.3 Sizes MBS_n and EBS_n of Buffers MB_n and EB_n

The decoding branch with buffers MB_n and EB_n are used for MPEG-2 video and for AVC. Buffer EB_n accommodates the elementary stream buffer, while MB_n needs to accommodate PES overhead and multiplex buffering. Hence, the following formulas apply:

$$EBS_n = BS_{dec}$$
$$MBS_n = BS_{oh} + BS_{mux}$$

For each content format supported by MPEG-2 systems using MB_n and EB_n, either Section 12.5 or Chapter 14 provides the applicable values of MBS_n and its components.

12.3.6 Buffer Management for Elementary Stream Data

Below the constraints for overflow and underflow of the buffers B_n, MB_n and EB_n are discussed. Overflow of a buffer occurs when more data enters the buffer than can be stored in the buffer; for example, at any time B_n must contain at most BS_n bytes. Underflow occurs when removal of data from the buffer is required that has not arrived yet; for example, to prevent underflow of B_n, at the decoding time $td_n(j)$ of each access unit $A_n(j)$, all data belonging to $A_n(j)$ must be present in B_n; however note that exceptions apply for low delay video or in trick mode operation. The following buffer constraints apply:

- Overflow of B_n is forbidden.
- Underflow of B_n is forbidden, except during trick mode operation and for low delay video.
- Overflow of MB_n is forbidden.
- If the leak method is in effect, MB_n must be empty at least once every second.
- If the vbv-delay method is in effect, underflow of MB_n is forbidden.
- Overflow of EB_n is forbidden.
- Underflow of EB_n is forbidden, except during trick mode operation and for low delay video.

12.4 General STD Constraints and Requirements

For the decoding operation in the P-STD and T-STD, the availability of STC samples and time stamps is crucial. Therefore minimum requirements have been defined for their presence in program streams and transport streams. The following requirements apply:

- The PTS and DTS are optional fields in the PES packet header. If a PTS is encoded, a DTS is only permitted to be present if its value differs from the PTS value. Due to the instantaneous decoding in the STD, for many access units the decoding time is the same as the presentation time. If a PTS is absent, also a DTS is absent.
- For each elementary stream in a program stream or transport stream, a PTS must be present for the first access unit.
- When in a program stream or a transport stream a decoding discontinuity exists for an elementary stream, as discussed in Sections 6.4, 11.3 and 11.4, then a PTS must be encoded for the first access in that elementary stream after the decoding discontinuity.
- When a time base discontinuity occurs in a transport stream, as discussed in Sections 11.3 and 11.4, then after the time base discontinuity a PTS must be encoded for the first access unit in each elementary stream of the program.
- In an elementary video or audio stream carried in a program stream or in a transport stream, a PTS must be encoded at least every 0.7 s. The following formal requirement applies:

$$|tp_n(k) - tp_n(k')| \leq 0.7\ \text{s},$$

for all n, k and k', where k and k' are the indexes of two consecutive[7] presentation units with an encoded PTS.

The above constraint does not apply for still pictures, nor for an AVC video stream with a very low frame rate, in which the presentation times of subsequent pictures differ by more than 0.7 s.

- In each program stream, a SCR must be encoded at least every 0.7 s. The following formal requirement applies:

$$|t(i') - t(i'')| \leq 0.7 \text{ s},$$

for all i' and i'', where i' and i'' are the indexes of the reference bytes for two consecutive SCRs.

- In each transport stream, a PCR must be encoded at least every 0.1 s. The following formal requirement applies:

$$|t(i') - t(i'')| \leq 0.1 \text{ s},$$

for all i' and i'', where i' and i'' are the indexes of the reference bytes for two consecutive PCRs.

- Between two time base discontinuities in a transport stream, at least two PCRs must be encoded to allow determining the delivery timing of bytes in the transport stream between both time base discontinuities.
- If a program stream or transport stream contains PES packets of an elementary stream with a header in which the ESCR is encoded, then an ESCR must be encoded in the PES headers of this elementary stream at least every 0.7 s.

As discussed in Section 6.7, for a high QoS, MPEG-2 systems defines a maximum delay of coded data through the STD input buffers. The typical value of this delay is 1 s, which means that each access unit must be decoded within 1 s after it starts entering the input buffer. However, there are some exceptions, as discussed below.

- For still picture video data (signalled in a descriptor associated to the involved video stream, see the discussion on video formats in Chapter 14) the maximum delay is increased to 60 s. This allows delivery of coded still picture data at a very low bitrate. For example, if a coded still picture contains 25 000 bytes, and if that still picture is to be presented 50 s after the previous one in a kind of slide show, then that still picture data can be delivered at an average rate of 200 000/50 = 4000 bit/s. If the 1 s constraint also applies to still picture video, then all coded still picture data is to be delivered within the 1 s just prior to its presentation at an average rate of 200 000 bits/s.
- To allow efficient coding under all circumstances, the maximum delay value for MPEG-4 video, AVC, SVC and MVC is increased to 10 s.

The general requirement for the maximum delay not to exceed N seconds is formally expressed as:

$$[td_n(j) - t(i)] \leq N \text{ (seconds)}$$

for all j, and each byte i in each access unit $A_n(j)$ in each elementary stream n.

[7] This condition applies to the presentation order, which may differ from the coding order.

Hence, in summary the following applies:

In the case of still picture video data:	$N = 60$ s; else
In the case of MPEG-4 video, AVC, SVC and MVC:	$N = 10$ s; else
	$N = 1$ s.

Note: a maximum delay of 10 s may also be applied when in future support in MPEG-2 systems is defined for carriage of next generation video coding formats, such as HEVC, that also exploit temporal redundancy to improve the compression efficiency.

12.5 Content Format Specific STD Issues

To fully define the decoding of an elementary stream in the STD, all STD details need to be specified, such as the relevant building blocks of elementary stream decoder D_n in the STD, the structure of the decoding branch and the values of buffer sizes and transfer rates applicable for the elementary stream. Furthermore, in addition to the general STD requirements and constraints discussed in Section 12.4, some content format specific requirements may apply. As an introduction, below some general considerations are discussed for MPEG audio and video elementary streams. Chapter 14 discusses support of content formats in MPEG-2 systems, including details on STD parameters and conditions as well as the data structure of the associated descriptors.

12.5.1 Decoding of MPEG Audio Streams in STD Model

12.5.1.1 Structure of STD Decoding Branches

Figure 12.15 shows that the decoding branch for an MPEG audio stream in the T-STD and in the P-STD consists of a single buffer B_n, preceded by transport buffer TB_n in the T-STD, and followed by decoder D_n. The decoder for an MPEG audio stream does not contain any further buffers or other building blocks next to the audio decoder. Figure 12.15 shows also the table with applicable STD parameters for several MPEG-1, MPEG-2 and MPEG-4 audio formats up to eight channels. For MPEG audio formats not listed in Figure 12.15, the applicable STD parameters are discussed in Chapter 14.

Typically, the decoding of content in the STD takes place within a single decoding branch. However, in some cases, such as scalable video, more than a single decoding branch may be involved. This may also apply to MPEG-2 audio: to decode MPEG-2 audio in the STD, two parallel decoding branches may be needed. As discussed in Section 3.3.2, in the case of MPEG-2 audio coding, the multichannel information may be carried in two elementary streams: the base stream with a format that is backward compatible with the MPEG-1 audio format, and an optional extension stream that is used if the multichannel bitrate exceeds the maximum bitrate of MPEG-1 audio. Therefore, if an extension stream is present, for the decoding of MPEG-2 audio in the STD two decoding branches are needed, one for the base stream and the other for the extension stream. In such case there is one branch with buffer B_n and decoder D_n for the audio frames of the base stream and another branch with buffer $B_{n'}$ and decoder $D_{n'}$ for the extension frames of the extension stream. Obviously, in practice, both decoders D_n and $D_{n'}$ will be tightly integrated to produce a single multichannel output signal. The delivery of the base stream and the extension stream must be so that the STD requirements

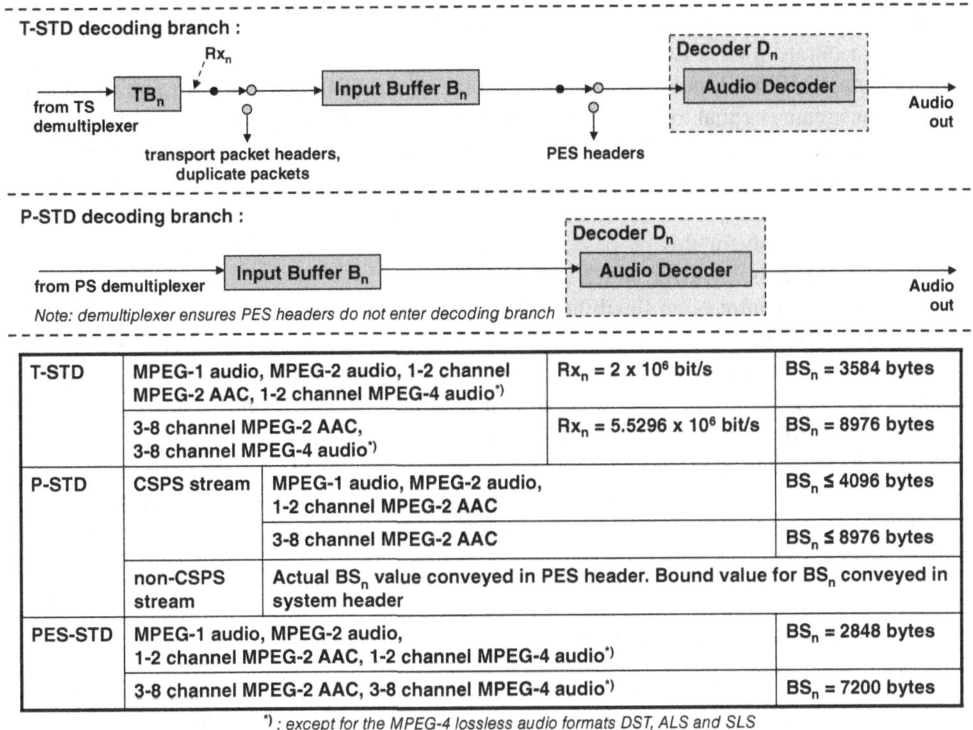

T-STD	MPEG-1 audio, MPEG-2 audio, 1-2 channel MPEG-2 AAC, 1-2 channel MPEG-4 audio*)	$Rx_n = 2 \times 10^6$ bit/s	$BS_n = 3584$ bytes
	3-8 channel MPEG-2 AAC, 3-8 channel MPEG-4 audio*)	$Rx_n = 5.5296 \times 10^6$ bit/s	$BS_n = 8976$ bytes
P-STD	CSPS stream: MPEG-1 audio, MPEG-2 audio, 1-2 channel MPEG-2 AAC		$BS_n \leq 4096$ bytes
	CSPS stream: 3-8 channel MPEG-2 AAC		$BS_n \leq 8976$ bytes
	non-CSPS stream	Actual BS_n value conveyed in PES header. Bound value for BS_n conveyed in system header	
PES-STD	MPEG-1 audio, MPEG-2 audio, 1-2 channel MPEG-2 AAC, 1-2 channel MPEG-4 audio*)		$BS_n = 2848$ bytes
	3-8 channel MPEG-2 AAC, 3-8 channel MPEG-4 audio*)		$BS_n = 7200$ bytes

*) : except for the MPEG-4 lossless audio formats DST, ALS and SLS

Figure 12.15 Decoding in STD model of MPEG audio formats up to eight channels

are met for each decoding branch. Moreover, corresponding frames in the base stream and the extension stream must have identical time stamps. Further details are discussed in Chapter 14.

For decoding MPEG audio formats in the P-STD, the size of B_n is specified in PES headers, while the delivery rate of the data entering B_n is determined by the rate of the program stream; the maximum BS_n values for an MPEG audio stream carried in a CSPS stream are already provided in Table 12.1, but for completeness these values are listed also in Figure 12.15, as the chosen values for a CSPS stream have a close relationship to the BS_n values specified for the T-STD.

12.5.1.2 Leak Rate from TB_n into B_n in the T-STD

As discussed in Section 12.3.1, the value of Rx_n in the T-STD for an elementary stream type is chosen to be reasonably higher than the maximum bitrate of the elementary stream, with a general rule of thumb that Rx_n is equal to $1.2 \times$ this maximum bitrate. However, for most MPEG audio formats this general rule of thumb is not followed, but instead Rx_n is chosen to be equal to a fixed 'easy' value: $Rx_n = 2 \times 10^6$ bits/s. This value is sufficiently low to enable cost effective implementation in software at the time MPEG-2 systems was developed and sufficiently high to ensure proper T-STD operation for an MPEG audio stream as long as its maximum bitrate is reasonably below 2×10^6 bits/s.

The value $Rx_n = 2 \times 10^6$ bits/s is very suitable for MPEG-1 and MPEG-2 audio: the maximum bitrate for MPEG-1 audio is equal to 448 kbit/s, while for backward compatible multichannel MPEG-2 audio the maximum total bitrate of the base stream and the optional extension stream is equal to 1130 kbit/s; hence, for the extension stream, the maximum bitrate is 682 kbit/s. However, for MPEG-2 AAC, the maximum bitrate R_{max} is 576 kbit/s per channel, and hence with three or more channels the bitrate may exceed or get close to 2×10^6 bit/s. To allow decoding in the T-STD of MPEG-2 AAC with three or more channels, a larger Rx_n value is to be determined; for this purpose, the general rule of thumb for Rx_n is used: if N is the number of channels, then $Rx_n = 1.2 \times R_{max} \times N$.

Applications require often flexibility with respect to the number of channels; multichannel applications for consumers typically use a certain number of channels between three and eight. To keep usage of various channel configurations simple from T-STD perspective, a single Rx_n value is defined for any MPEG-2 AAC stream with three to eight channels, based on $N_{max} = 8$: $Rx_n = 1.2 \times 576\,000 \times 8$ bits/s $= 5.5296 \times 10^6$ bits/s.

The Rx_n approach for MPEG-2 AAC is also applied for MPEG-4 audio formats, except for the lossless DST, ALS and SLS formats. For example, in the case of MPEG-4 audio, except DST, ALS and SLS with one or two channels, $Rx_n = 2 \times 10^6$ bits/s, and in the case of three to eight channels, $Rx_n = 5.5296 \times 10^6$ bits/s (see Figure 12.5).

12.5.1.3 The size BS_n of Buffer B_n in the T-STD

For the decoding of a MPEG audio stream in the T-STD, not only for Rx_n, but also for BS_n, fixed values are defined. The room in buffer B_n is to be shared between access unit storage, PES overhead and multiplexing. For the latter two, the rules discussed in Section 12.3.5.1 for B_{oh} and B_{mux} would result in a buffer room $BS_{oh} + BS_{mux} = 455$ bytes: the maximum audio bitrate of 682 kbit/s multiplied by 1/750 yields 114 bytes, while 682 kbit/s multiplied by 0.004 yields 341 bytes. The size of B_{dec} must be chosen so that the largest audio frame can be stored; for a MPEG-2 extension stream, the maximum size of an frame is equal to 2048 bytes. Hence, based on the above, a total buffer size BS_n would be needed of $2048 + 455 = 2503$ bytes.

However, compared to video, this BS_n value provides relatively little opportunity for buffer sharing, and thereby for flexibility in performing multiplex operations. The multiplex flexibility can for example be expressed in the number of transport packets that can be placed 'early' in the transport stream. If the audio frames are known to be smaller than 2048 bytes, the remaining room in B_{dec} can be used for early placement of transport packets, but when (in the case of an audio stream with extension frames of the maximum size) the entire 4 ms room is needed for multiplexing, then the only option for early positioning is to use PES packets in a manner so that the overhead for PES packetization is minor. In that case, the remaining B_{oh} room can be used for early placement, but this room is obviously less than the size of the payload of one transport packet. Hence, if such audio is carried in transport packets, there is only very little opportunity for early placement.

The MPEG systems group decided that it is desirable to create some more flexibility for multiplex operations by defining the size of B_n larger than discussed above. For the decoding of MPEG-1 and MPEG-2 audio streams in the T-STD, the size BS_n was chosen so that the total buffering in the branch can be implemented conveniently by means of a buffer with the size of 4096 bytes. Note that in the P-STD the available size BS_n for a CSPS stream is also equal to 4096 bytes. In the T-STD, the total buffering equals $TBS_n + BS_n$ and therefore BS_n was defined to be equal to 4096 – 512 = 3584 bytes in the T-STD. Which means that even in the worst case situation that all 2503 bytes calculated above are required for B_{dec}, B_{oh} and B_{mux}, there are still 1081 bytes available for early placement of transport packets: with this value the multiplexer may place at least five transport packets early in the transport stream. Obviously, if so desired, the additional buffer room can also be exploited by the system encoder to implement (a part) of the delay for audio as needed to compensate for a different video delay; see Section 6.1 for a discussion on this subject.

Applications are free to exploit the additional room available in B_n in the T-STD within certain constraints. A portion of 736 bytes from B_n is allocated to multiplexing; the remaining 3584 – 736 = 2848 bytes are available for sharing between B_{dec}, B_{oh} and additional multiplexing. If the latter is not used, then 2848 – BS_{dec} bytes are available for PES overhead. In case of the largest audio (extension) frame with a size of 2048 bytes, at most 800 bytes can be spend for PES overhead, while with smaller audio frames even more bytes become available. The requirement $BS_{oh} \leq (2848 - BS_{dec})$ prevents outrageous PES packetization strategies, for example using very short PES packets with very large headers. However, this BS_{oh} constraint still provides a huge flexibility for PES packetization.

For MPEG-2 AAC with three or more channels a size of 3584 bytes for BS_n is no longer suitable, due to the maximum bitrate per channel of 576 kbit/s and the input buffer of 6144 bit (768 bytes) per channel, as defined in the MPEG-2 AAC specification. To allow flexibility in the channel configuration, for any MPEG-2 AAC stream with three to eight channels, a single BS_n value is defined, based on $N_{max} = 8$, as for Rx_n. To specify the size BS_n in the T-STD for MPEG-2 AAC with three to eight channels the following approach is followed.

- $BS_{dec} = 768 \times N_{max} = 768 \times 8 = 6144$ bytes;
- $BS_{oh} = 528$ bytes; this value allows for storage of two PES headers of the maximum size of 264 bytes in B_n; hence even with this very large size, two PES headers can enter B_n, so that the requirement to insert a PTS every 0.7 s can be met, given that at most audio data of one second can be in B_n. Furthermore, the BS_{oh} value of 528 bytes allows for storage of 37 PES headers with a size of 14 bytes, which is the size of a typical PES header for MPEG audio with a PTS and no other optional fields, thereby allowing an accurate coding of PTS time stamps.
- $BS_{mux} = 0.004 \times R_{max} \times N_{max}$, with R_{max} the maximum MPEG-2 AAC bitrate per channel; note that this formula is according the rule of thumb discussed in Section 12.3.5.1 for BS_{mux}. Hence $BS_{mux} = 0.004 \times 576\,000/8 \times 8 = 2304$ bytes.

In conclusion, for any MPEG-2 AAC stream with three to eight channels, the value of BS_n is equal to 6144 + 528 + 2304 = 8976 bytes. The same value is applied for any MPEG-4 audio stream with three to eight channels. For an overview see Figure 12.15.

12.5.2 Decoding of MPEG Video Streams in STD Model

In Figure 12.16 the decoding branches in the T-STD and in the P-STD are depicted for MPEG-2 video. In the P-STD a single buffer B_n is used, followed by a decoder D_n, while in the T-STD the transport buffer TB_n is followed by multiplex buffer MB_n and elementary stream buffer EB_n. Decoder D_n consists of a video decoder and a picture re-ordering buffer, as specified in the MPEG-2 video standard.

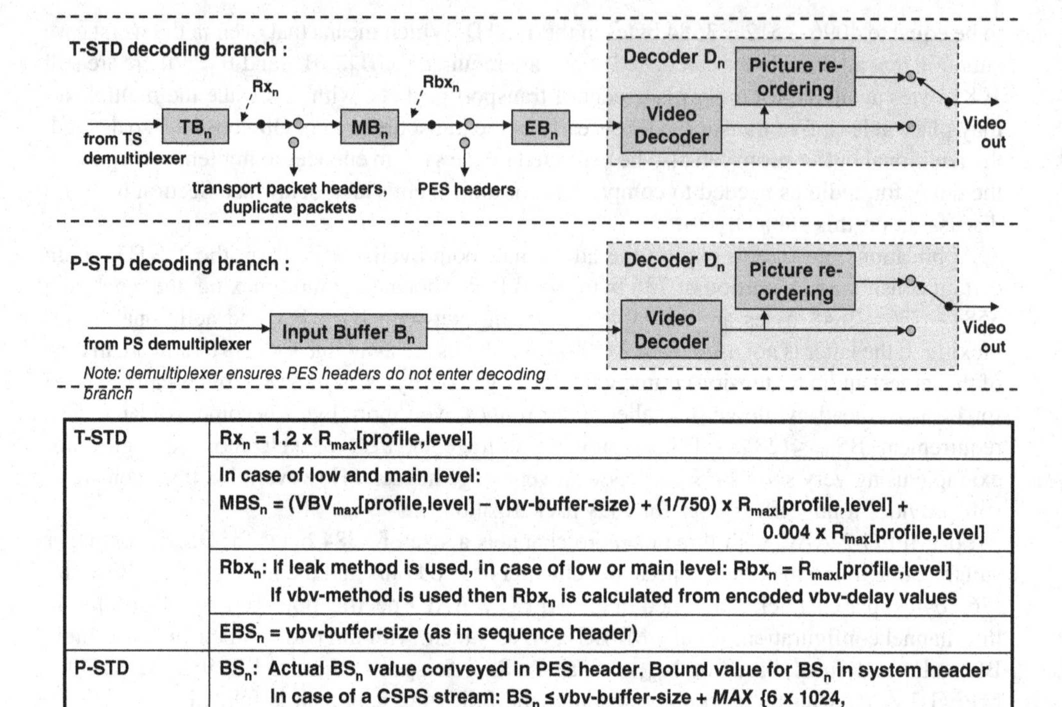

T-STD	$Rx_n = 1.2 \times R_{max}[profile,level]$
	In case of low and main level: $MBS_n = (VBV_{max}[profile,level] - vbv\text{-}buffer\text{-}size) + (1/750) \times R_{max}[profile,level] + 0.004 \times R_{max}[profile,level]$
	Rbx_n: If leak method is used, in case of low or main level: $Rbx_n = R_{max}[profile,level]$; If vbv-method is used then Rbx_n is calculated from encoded vbv-delay values
	EBS_n = vbv-buffer-size (as in sequence header)
P-STD	BS_n: Actual BS_n value conveyed in PES header. Bound value for BS_n in system header; In case of a CSPS stream: $BS_n \leq vbv\text{-}buffer\text{-}size + MAX\{6 \times 1024, R_{max}[profile,level] \times 0.001\}$
PES-STD	$BS_n = VBV_{max}[profile,level] + (1/750) \times R_{max}[profile,level]$

Figure 12.16 STD model and parameters for MPEG-2 video

Figure 12.16 also shows applicable STD parameters for MPEG-2 video. The values of the STD parameters depend on the applied MPEG-2 video profile and level. For Rx_n the general rule of thumb is used: Rx_n is equal to $1.2 \times$ the maximum bitrate for the MPEG-2 video profile and level. In a formula: $Rx_n = 1.2 \times R_{max}[\text{profile, level}]$.

The elementary stream buffer size EBSn is equal to the value of the vbv-buffer-size parameter carried in the sequence header of the MPEG-2 video stream. The used vbv-buffer-size may be smaller than the maximum size $VBV_{max}[\text{profile, level}]$ of the VBV buffer allowed for the applied level and profile. In that case the left-over room is made available as additional buffer room for multiplexing in MBS_n. Note that decoder implementations designed to decode a certain profile and level need to be able to accommodate the maximum VBV buffer size anyway, so the left-over buffer room can also be made available in a straightforward manner in practical decoders. Hence, the total available room in MB_n is equal to:

$$MBS_n = (VBV_{max}[\text{profile, level}] - \text{vbv-buffer-size}) + BS_{oh} + BS_{mux}.$$

The general rules of thumb are followed for BS_{mux} and BS_{oh} and therefore:

$$MBS_n = (VBV_{max}[\text{profile, level}] - \text{vbv-buffer-size}) + (1/750) \times R_{max}[\text{profile, level}] + 0.004 \times R_{max}[\text{profile, level}].$$

Usage of the left-over room for additional multiplexing is only applied when the low or main level are used in the decoded MPEG-2 video stream. For the high-1440 and high levels, the maximum bitrates are significantly higher than for the low and main level, so that the rules of thumb for BS_{mux} and BS_{oh} provide enough flexibility for multiplex operations. Hence, for the high-1440 and high levels:

$$MBS_n = (1/750) \times R_{max}[\text{profile, level}] + 0.004 \times R_{max}[\text{profile, level}]$$

For Rbx_n two methods are defined: the leak method and the vbv-delay method, as discussed in Section 12.3.4. The method selected by the encoder is signalled in the transport stream; for a further discussion on this subject see Chapter 14. If the leak method is used, then for the low and main level $Rbx_n = R_{max}$[profile, level]. However, due to the (very) high maximum video bitrates for the high 1440 and high levels, for these levels the value of Rbx_n is constrained to at most 5% above the highest video bitrate R_{es} applied in the video stream, which value is encoded in the sequence header: $Rbx_n = \text{minimum}\{1.05 \times R_{es}, R_{max}[\text{profile, level}]\}$.

In comparison, Figures 12.17 shows decoding branches for AVC in the T-STD and the P-STD. The structure of both decoding branches is the same as for MPEG-2 video, but decoder D_n is build up differently; it consists of a video decoder and the Decoded Picture Buffer, DPB, as defined by the AVC specification. The equivalent of the VBV input buffer in MPEG-2 video

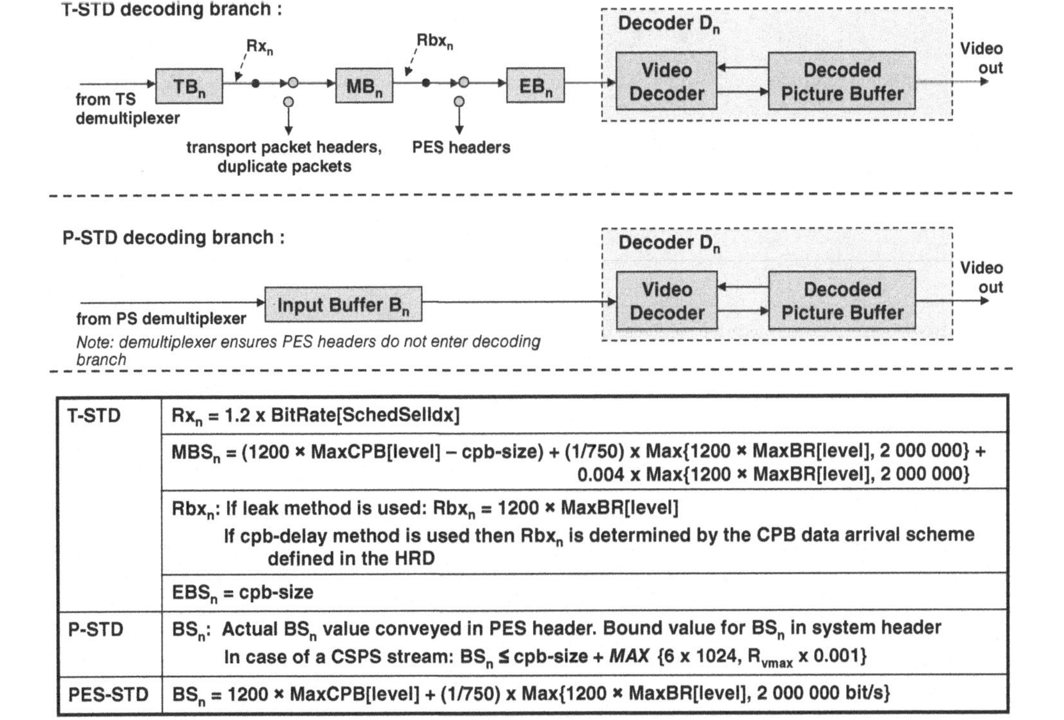

T-STD	Rx_n = 1.2 x BitRate[SchedSelIdx]
	MBS_n = (1200 × MaxCPB[level] – cpb-size) + (1/750) x Max{1200 × MaxBR[level], 2 000 000} + 0.004 x Max{1200 × MaxBR[level], 2 000 000}
	Rbx_n: If leak method is used: Rbx_n = 1200 × MaxBR[level] If cpb-delay method is used then Rbx_n is determined by the CPB data arrival scheme defined in the HRD
	EBS_n = cpb-size
P-STD	BS_n: Actual BS_n value conveyed in PES header. Bound value for BS_n in system header In case of a CSPS stream: $BS_n \leq$ cpb-size + *MAX* {6 x 1024, R_{vmax} x 0.001}
PES-STD	BS_n = 1200 × MaxCPB[level] + (1/750) x Max{1200 × MaxBR[level], 2 000 000 bit/s}

Figure 12.17 STD model and parameters for AVC

for AVC is the CPB, the Coded Picture Buffer; the size EBS_n is equal to the cpb-size specified in the AVC stream for the byte stream format of AVC; this format is applied when AVC is carried over MPEG-2 systems; for further details see Section 14.7.

In the T-STD, the cpb-size may be smaller than the maximum CPB size allowed for the applied AVC level. The left-over room may be used for multiplexing. For BS_{mux} and BS_{oh} the general rules of thumb are followed; however, to ensure sufficient flexibility for multiplex operations in the case of a low maximum bitrate for a level, the used value for the maximum bitrate in the formula is constrained to be at least 2×10^6 bits/s. For decoding of an AVC stream in the T-STD two methods are defined for the transfer of data from MBn into Bn: the leak method and the cpb-delay method, similar as for MPEG-2 video. For further details on the parameters and formulas related to the support of AVC in MPEG-2 systems see the discussion in Section 14.7.

13

Data Structure and Design Considerations

How STC samples and PTS and DTS time stamps are coded. What the data structure of a PES header is. What descriptors are, how their data is structured and which types of descriptors are defined. The purpose and specific data structure of the various system descriptors and general content descriptors. The data structure of a program stream; the pack header with optional system header; the program stream map. Data structure of sections in general and of program association, program map and conditional access sections in particular. The data structure of transport packets with the adaptation field.

13.1 System Time Clock Samples and Time Stamps

Compatibility with MPEG-1 systems was one of the requirements when MPEG-2 systems was designed. For the use of PTS and DTS time stamps such compatibility was easy to achieve, as both 90 kHz resolution and the use of 33 bits for time stamp coding were considered suitable for MPEG-2 systems. However, for the STC (the System Time Clock) instead of 90 kHz, a resolution of 27 MHz was decided. For more background on this issue see Section 6.6.

On the other hand, there are also applications for which a 90 kHz STC resolution would have been sufficient. To allow applications to conveniently apply 90 kHz as the time base in their decoder implementations, as well as for compatibility with MPEG-1, it was decided to split the coding of STC samples in the PCR, OPCR, SCR, and ESCR fields in two parts. One part provides the STC base value in units of 90 kHz (to be precise 27 MHz/300), and the other part provides the STC extension value in units of 27 MHz. Note that all the frequencies above represent nominal values.

The STC base value is coded with 33 bits, equivalent to the time stamp coding. The STC extension part is coded with nine bits and counts during each cycle of the 90 kHz base component, from zero up to 299 (see Figure 13.1).

Fundamentals and Evolution of MPEG-2 Systems: Paving the MPEG Road, First Edition. Jan van der Meer.
© 2014 John Wiley & Sons, Ltd. Published 2014 by John Wiley & Sons, Ltd.

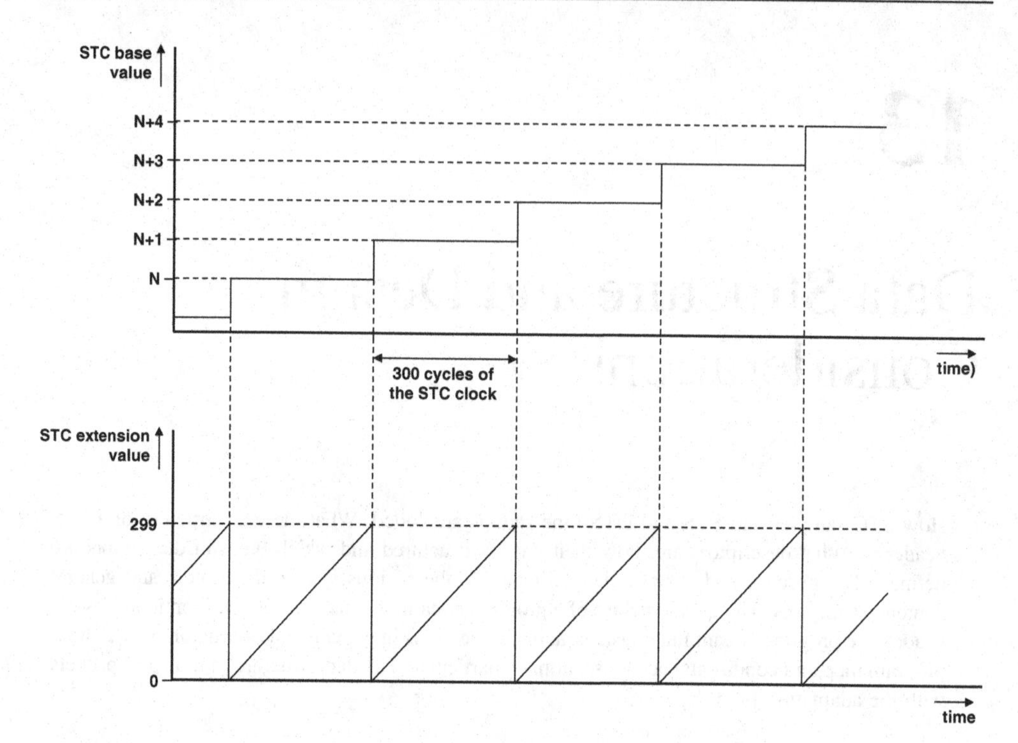

Figure 13.1 Splitting the STC counter into an STC base value and an STC extension value

As a result of the above approach, the value of the conveyed STC sample is found as follows:

$$\text{STC-sample-value} = [\text{STC-base-value}] \times 300 + \text{STC-extension-value}$$

Due to the use of a 90 kHz base value, it can be verified whether the STC reaches the value of a time stamp by simply comparing STC-base-value with the time stamp value, thereby avoiding the need to divide values of a 27 MHz STC counter by 300. The conveyed time in a STC sample time represents the STC time at which the byte containing the last bit of the STC-base-value is intended to arrive at the input of the STD.

Fields of 48 bits are used to convey a STC sample in the system stream, but by means of two slightly different data structures. In the fields for the PCR and the OPCR, the 33 bits of the STC-base-value are followed by six reserved bits, followed by the nine bits of the STC-extension-value. The PCR and OPCR are used in transport streams only, but the SCR and the ESCR are also used in program streams, in which avoiding the emulation of start-codes (23 times '0' followed by a '1') is important. Therefore in the fields to convey the SCR or ESCR in the system stream, the 33 bits of the STC-base-value are interleaved with marker-bits with the value '1'; furthermore, to prevent start-code emulation with surrounding fields, some more marker-bits are put at the positions depicted in Figure 13.2. Both the STC-base-value and the STC-extension-value are unsigned integers with the most significant bit occurring first in the system bitstream.

Figure 13.2 depicts also the data structure of the fields that carry either the PTS only or the PTS and the DTS. As in the SCR and ESCR fields, marker-bits are used at appropriate positions

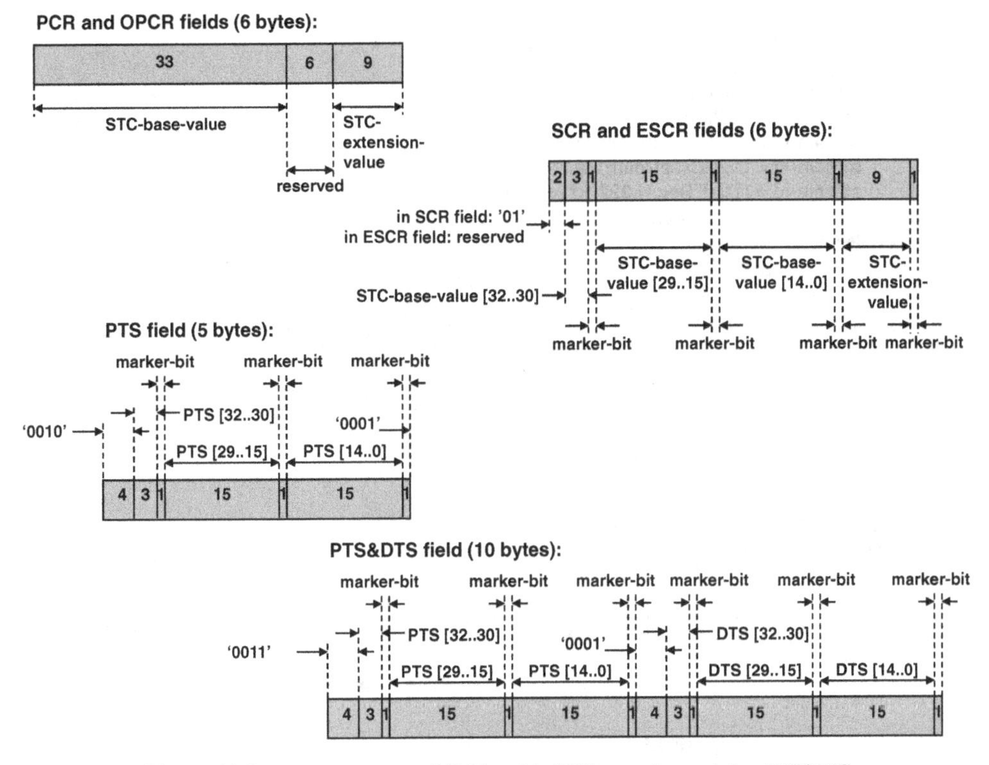

Figure 13.2 Data structures of fields with STC samples and the PTS/DTS

to prevent start-code emulation. The data structure of the time stamp fields is copied from MPEG-1 systems (see Figure 7.14); however, the semantics are slightly different. In MPEG-1 the first four bits of the field following the STD-buffer-size field indicate whether a PTS is encoded, or a PTS and a DTS, or no time stamp at all. With the more complex PES header in MPEG-2 this construction was less suitable, and therefore a two-bit PTS-DTS flag was explicitly included in the PES header to signal the presence of a PTS, or a PTS and a DTS or the absence of both.

13.2 PES Packets

Each PES packet commences with a start-code-prefix followed by an eight-bit stream-ID value and a 16-bit packet-length field. The stream-ID identifies the PES packet, while the packet-length specifies the number of bytes in the PES packet immediately following the packet-length field, very similar to a MPEG-1 packet. The data structure of a PES packet after the packet-length field depends on the encoded value of the stream-ID. Two data structures are used; for most stream-ID values the packet-length is followed by a field with the remaining-PES-header, after which PES-packet-data-bytes conclude the PES packet. However, for some stream-ID values the packet-length is followed immediately by the PES-packet-data-bytes, in which case the header does not contain any further fields, so that for instance no time stamps can be encoded. Both data structures are depicted in Figure 13.3.

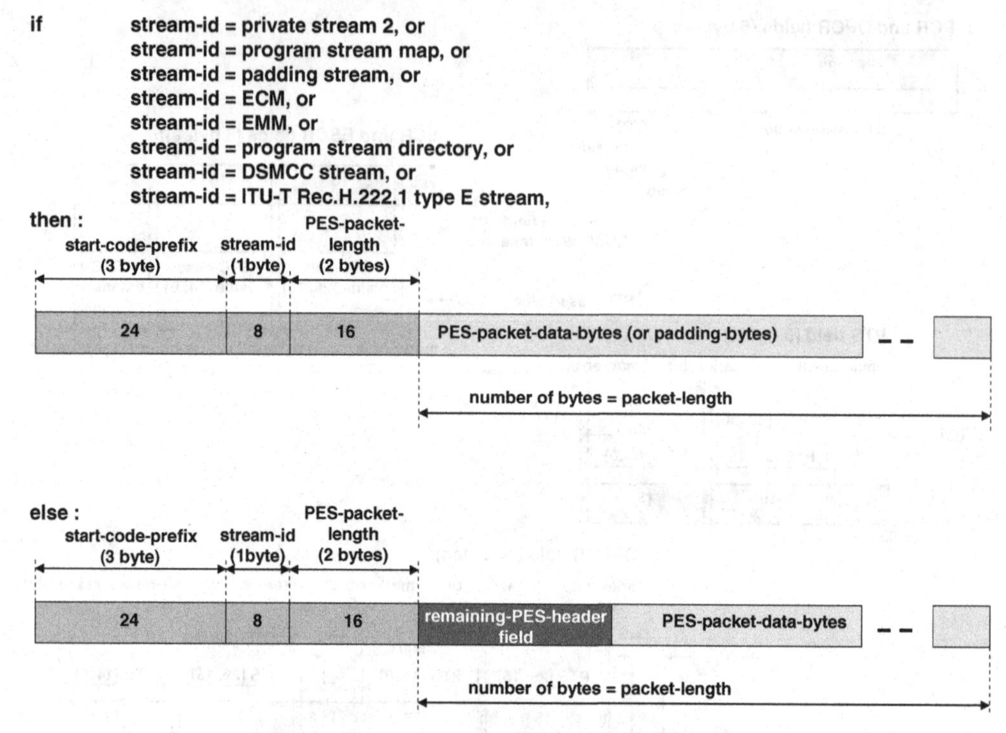

Figure 13.3 The two data structures of a PES packet

The usage of stream-ID values in MPEG-2 systems, as listed in Table 13.1, is very similar to the usage in MPEG-1 systems (see Tables 7.2 and 7.3). The differences are as follows:

- When more video and audio formats evolved in MPEG, MPEG-2 systems generalized the stream-ID values assigned to MPEG-1 audio and video into 32 MPEG-defined audio streams and 16 MPEG-defined video streams. The actual audio or video format is specified in the PMT of a transport stream or in the program stream map by means of the stream-type, while additional information may be provided in audio and video descriptors. In Chapter 14 the stream-type values and descriptors for each content format are discussed.
- In MPEG-2 systems, new type of streams were assigned to values that were reserved for future use in MPEG-1 systems; as of 2013, in MPEG-2 systems only the value 254 (0xFE) was left reserved for future use.
- The stream-ID value 253 (0xFD) is defined in MPEG-2 systems as an escape, that is the PES packet is not identified by this stream-ID value, but by the stream-ID-extension value encoded in an extension of the PES header and signalled by the stream-ID-extension-flag (see Figure 13.7).

Further details on the use of stream-ID value(s) for content formats are discussed in Chapter 14.

There are major differences between the header of a MPEG-1 system packet and the header of a PES packet; to address the MPEG-2 systems requirements appropriately, the packet header

Table 13.1 Usage of stream ID values for the type of streams in MPEG-2 systems

Type of stream	Stream ID (decimal)	Stream ID (hexadecimal)
Program stream map	188	0xBC
Private stream 1	189	0xBD
Padding stream	190	0xBE
Private stream 2	191	0xBF
MPEG defined audio stream 0 through 31	192 through 223	0xC0 through 0xDF
MPEG defined video stream 0 through 15	224 through 239	0xE0 through 0xEF
ECM stream	240	0xF0
EMM stream	241	0xF1
DSMCC stream	242	0xF2
MHEG stream	243	0xF3
H.222.1 type A stream (ITU defined video)	244	0xF4
H.222.1 type B stream (ITU defined audio)	245	0xF5
H.222.1 type C stream (ITU defined data)	246	0xF6
H.222.1 type D stream (ITU defined format)	247	0xF7
H.222.1 type E stream (ITU reserved)	248	0xF8
Ancillary stream	249	0xF9
MPEG-4 SL packetized stream	250	0xFA
MPEG-4 FlexMux stream	251	0xFB
Metadata stream	252	0xFC
Extended stream-ID	253	0xFD
Reserved data stream	254	0xFE
Program stream directory	255	0xFF

is extended significantly. The remaining-PES-header data field, depicted in Figure 13.4, commences with PES-header-flags taking two bytes; some flags provide information on the data carried in the PES packet, while some other flags signal the presence or absence of a number of optional fields. The PES-header-flags are followed by a one-byte PES-header-data-length field that specifies the immediately following number of remaining bytes in the PES header. If this number is N larger than the number of bytes included by the optional fields, then the PES header is concluded by N stuffing-byte fields. At most 32 stuffing bytes are permitted.

The PES-header-flags start with two bits with the fixed value[1] '10', followed by two two-bit flags and ten one-bit flags. In the following paragraphs for each flag the semantics are discussed briefly, including any associated optional fields.

The two-bit PES-scrambling-control field provides information on scrambling of the payload of the PES packet; the value '00' indicates that no scrambling is applied. Any other value indicates a scrambling mode defined by private means, for example by the used CA system (see also Section 10.4). The PES-priority bit signals the priority of the payload of the PES packet. A PES-priority set to '1' signals a higher payload priority than this bit set to '0'. A multiplexer may use this bit to prioritize its data within an elementary stream, but transport mechanisms are not allowed to change the encoded value of this bit. The meaning of the

[1] This distinguishes the syntax of a PES packet after the packet-length field from the syntax of a MPEG-1 system packet after the packet-length field; see the remaining-packet-header depicted in Figure 7.14 for a MPEG-1 packet.

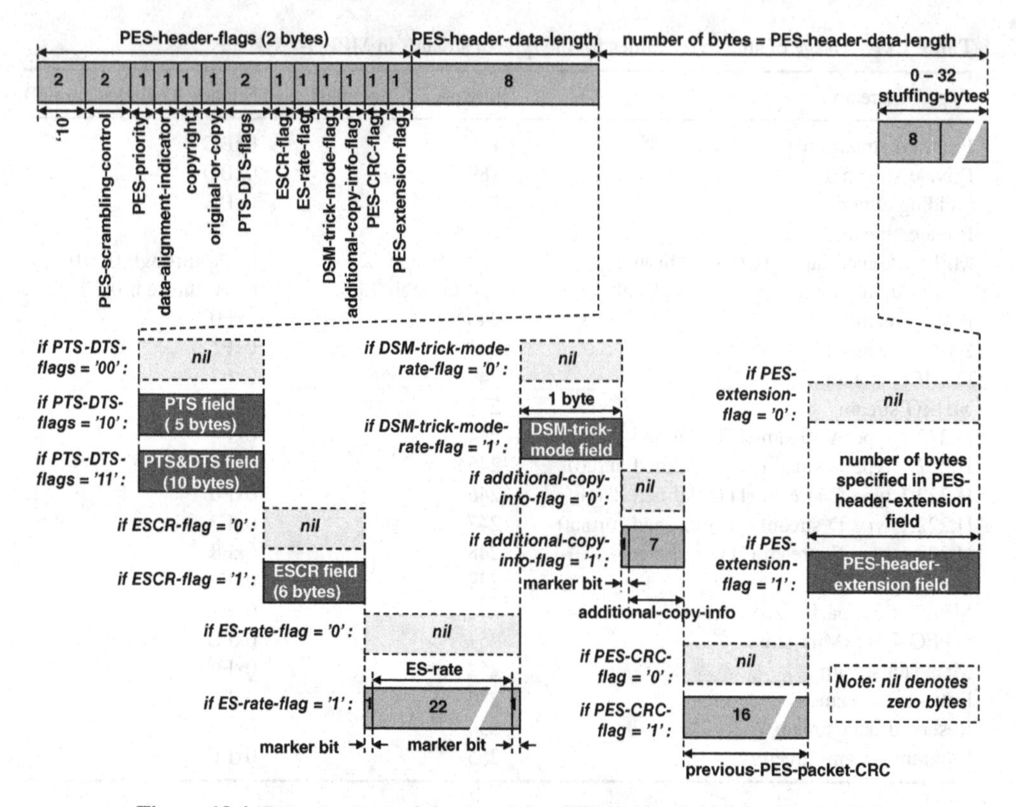

Figure 13.4 Data structure of the remaining-PES-header field from Figure 13.3

signalled priority depends on the type of elementary stream carried in the PES packet and is not defined by MPEG-2 systems, but is left to applications and their standardization bodies.

The data-alignment-indicator bit signals for audio and video elementary streams whether alignment is applied to convey elementary stream data in the payload. If this bit is set to '1', then the first byte in the payload is a certain syntax element, such as an audio sync word or a picture start code. Which alignment is applied is either specified by the data-stream-alignment-descriptor, if present, or by default (see Section 13.3.4.2). If the data-alignment-indicator is set to '0', it is not specified whether any alignment is applied.

The copyright bit indicates whether copyright applies to the content carried in the PES packet, while the original-or-copy bit can be set to '1' or '0' to signal that the content is an original or a copy, respectively. In addition, in the seven-bit additional-copy-info field private copyright information can be provided; the presence of this optional field is signalled by the additional-copy-info-flag. Note though that also a copyright descriptor is defined to carry copyright related information (see Section 13.3.3.8).

The presence or absence of the time stamps is indicated by the two-bit PTS-DTS-flags. The value '00' indicates that the PTS and DTS time stamps are both absent. The value '10' signals that only the PTS is present and the value '11' that both the PTS and the DTS are present. The value '01' is forbidden. If a PTS is encoded in the header of a PES packet, then the payload of

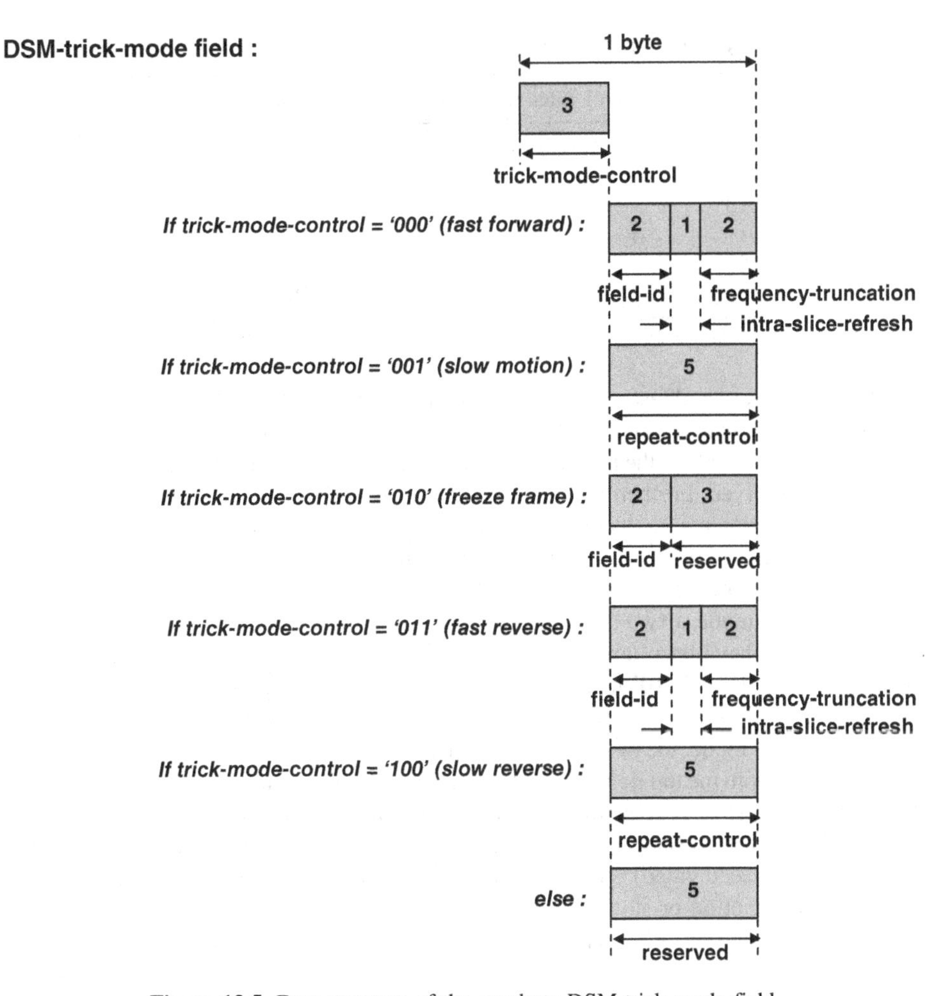

Figure 13.5 Data structure of the one-byte DSM-trick-mode field

that PES packet must contain the first byte of an access unit. A DTS is encoded only, if for the associated access unit the value of the DTS differs from the PTS value.

By means of the ESCR-flag and the ES-rate-flag, the presence of the optional ESCR and the ES-rate fields is indicated. For the use of these fields in PES streams see Sections 11.13 and 12.2.4. If the PES-CRC-flag is set to '1', the presence is signalled of a CRC-16 code calculated over the previous PES packet (for more information see Section 11.1).

The DSM-trick-mode-flag signals the presence of a one-byte DSM-trick-mode field, and the PES-extension-flag the presence of the PES-extension-field. These fields are depicted in Figures 13.5 and 13.6, respectively. The server responsible for the playback operation can use the first three bits of the DSM-trick-mode field to signal the current trick mode; see Table 13.2. The meaning of the remaining five bits depends on the trick mode signalled by the trick-mode-control, as discussed in the next paragraph.

Table 13.2 Trick-mode-control values

Trick-mode-control value (binary)	Current trick mode
'000'	Fast forward
'001'	Slow motion
'010'	Freeze frame
'011'	Fast reverse
'100'	Slow reverse
'101' – '111'	Reserved

In the fast forward and fast reverse mode, the two-bit field-ID signals from which field video is to be displayed, which can be useful in the case of interlaced video. A value '00' signals the top field only, a value '01' the bottom field only and a value '10' the complete video frame; the value '11' is reserved. Furthermore the 1-bit intra-slice-refresh, when set to '1', indicates that there may be missing macroblocks between coded slices in the payload of the PES packet. How to conceal such errors is at the discretion of the video decoder; for example co-sited macroblocks may be used from previously decoded pictures. Finally, in the fast forward and fast reverse mode a two-bit frequency-truncation field indicates that a restricted set of coefficients may have been used. The value '00' signals that only DC coefficients are coded and the values '01' and '10' that only the first three or six coefficients are used for coding, respectively, while the value '11' signals that all coefficients may be non-zero.

In freeze frame mode, the two-bit field-ID is used as well, to signal whether the still picture is to be displayed from the top field, the bottom field or from the complete video frame. Finally, in slow motion and slow reverse mode, a five-bit repeat-control field provides an integer that specifies the number of times each field (in the case of interlaced video) or each frame (in the case of non-interlaced video) is to be displayed. The value zero is forbidden. By changing this value, the slow motion or slow reverse play-back speed is changed.

Figure 13.6 shows the PES-header-extension field, starting with a number of one-bit flags. The first flag is the PES-private-data-flag; if set to '1', it signals the presence of the 128-bit PES-private-data field (see also Section 11.8). The second flag is the pack-header-field-flag; if set to '1', then a pack header of a MPEG-1 system stream or a MPEG-2 program stream is carried in the pack-header field. This is only permitted in a transport stream, so as to allow the reconstruction of the original MPEG-1 system stream or the MPEG-2 program stream (for more details see Section 11.12). Note that the pack-header field is preceded by a one-byte pack-field-length to signal the number of bytes carried in the pack-header field.

If the program-packet-sequence-counter-flag is set to '1', a two-byte field is present to further support the reconstruction of an original MPEG-1 system stream or MPEG-2 program stream conveyed over a transport stream. The first bit of this two-byte field is a marker bit, followed by the seven-bit program-packet-sequence-counter, yet another marker bit, the one-bit MPEG1-MPEG2-identifier and the six-bit original-stuff-length. A discussion on these fields and their usage is found in Section 11.12.

The P-STD-buffer-flag, if set to '1', signals the presence of a two-byte field to specify the size of the P-STD buffer for the elementary stream carried in the payload of the PES packet, in case the PES packet is contained in a program stream. The size BS_n of input buffer B_n in the P-STD is specified by the 13-bit P-STD-buffer-size field and is either expressed in units of 128

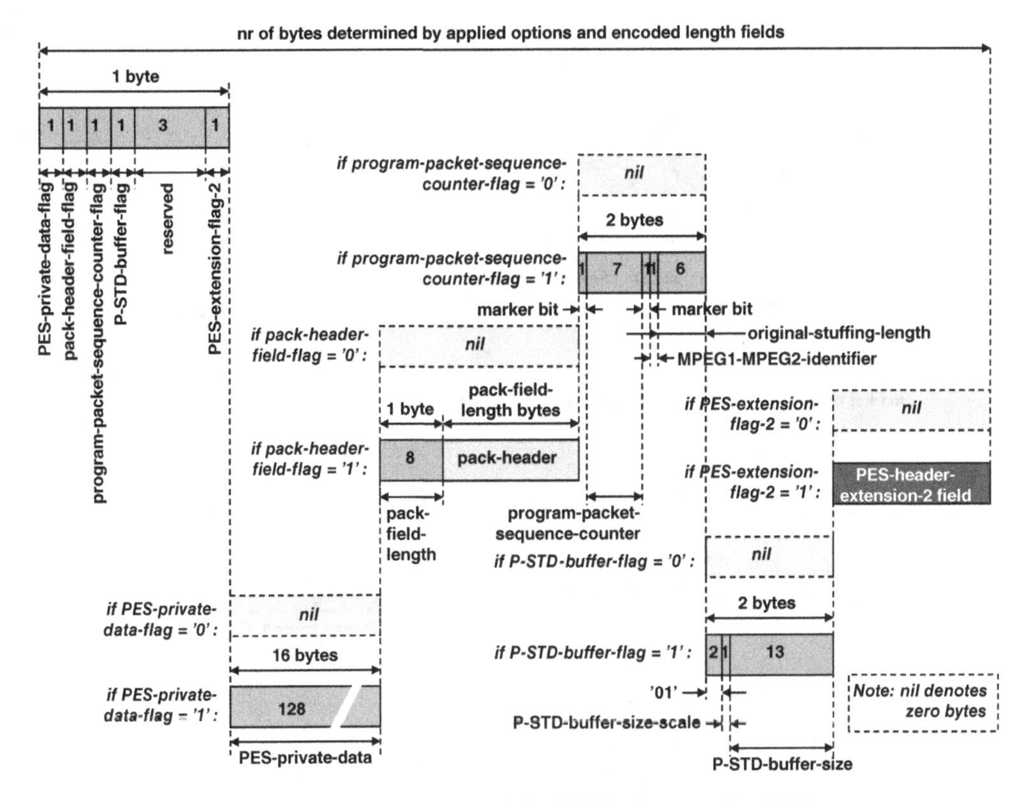

Figure 13.6 Data structure of the PES-header-extension field

bytes or in units of 1024 bytes. The used resolution is signalled by the one-bit P-STD-buffer-size-scale field that precedes the P-STD-buffer-size field. For audio streams, BS_n must be expressed in units of 128 bytes and for video streams in units of 1024 bytes; for all other streams either 128 or 1024 may be used. The value of BS_n may change in the course of a program stream; the encoded P-STD-buffer-size value takes effect immediately upon arrival in the P-STD.

In the PES-header-extension field, the P-STD-buffer-flag is followed by three reserved bits; these three bits cannot be used to define in future further optional fields in the PES extension field, as decoders without knowledge of the new optional fields would not know how to skip these fields. To specifically allow for future extensions of the PES header, the PES-header-extension-2 field is defined; the number of bytes in this field is specified by the PES-extension-field-length that allows a parser in a decoders to skip over unknown fields. The presence of the PES-header-extension-2 field is signalled by the PES-extension-flag-2, the last bit of the PES-extension-flags field.

In the first edition of MPEG-2 systems, the PES-header-extension-2 field did exist, but contained only reserved bytes. Meanwhile this field has been used for stream-ID-extension and for adding a time stamp for scalable video as defined in SVC. In future, more extensions may be defined. MPEG-2 systems requires that each bit that is reserved (for future use) has the value '1'; this applies also to the bits in a reserved byte. Bitstream parsers need to

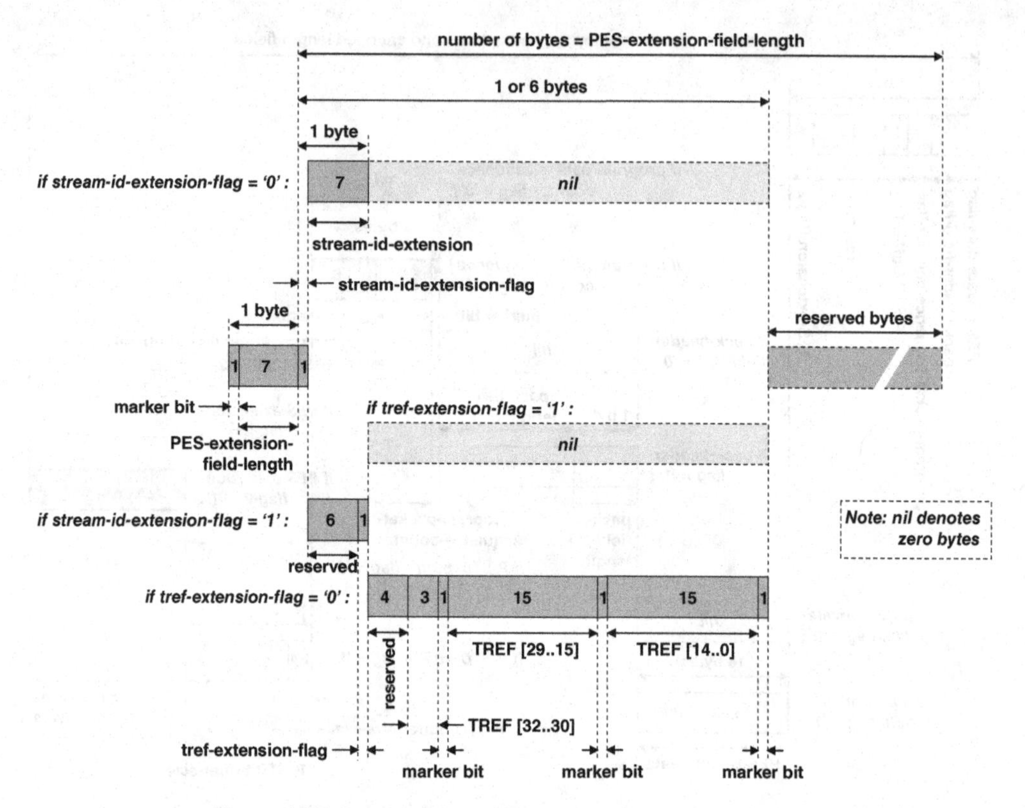

Figure 13.7 Data structure of the PES-header-extension-2 field

distinguish a reserved bit from a flag signalling for example the presence of a certain field. Which means that the usual method with the value '1' signalling the presence of a field cannot be used, and that the flag value '0' is to be used for this purpose. Hence in the PES header extension-2 field, flags use the value '0' to signal the presence of the associated field.

The data structure of the PES-header-extension-2 field is presented in Figure 13.7. The first byte of the PES-header-extension-2 field contains a marker-bit, followed by the seven-bit PES-extension-field-length discussed above. The first bit of the second byte is the stream-ID-extension-flag; if set to the value '0', then the seven-bit stream-ID-extension follows; if set to '1', then a field with six reserved bits and the one-bit tref-extension-flag follows. If the tref-extension-flag is set to '0', then a reference time stamp TREF for SVC usage is encoded. The format of the TREF is very similar to the PTS field; the only difference is that the first four bits of the five-byte field are reserved. If the encoded value of the PES-extension-field-length is larger than the number of bytes of the included optional fields, then the PES-header-extension-2 field is concluded by one or more reserved bytes.

In Table 13.3 the use of stream-ID-extension values is given. Two values are assigned for usage by IPMP, specified in part 11 of MPEG-2 systems; one value signals an IPMP control information stream and another value an IPMP stream (see Section 14.3.6 for further information on MPEG-2 IPMP). Furthermore 14 stream-ID-extension values are assigned to signal carriage of a MPEG-4 text stream and 16 values for a video depth or parallax stream as

Table 13.3 Assigned stream-ID-extension values

Type of stream	Stream-ID-extension (decimal)	Stream-ID-extension (binary)
IPMP control information stream	0	'000 0000'
IPMP stream	1	'000 0001'
MPEG-4 text stream	2–15	'000 0010' – '000 1111'
ISO/IEC 23002-3 auxiliary video stream	16–31	'001 0000' – '001 1111'
Reserved data stream	32–63	'010 0000' – '011 1111'
Private streams	64–127	'100 0000' – '111 1111'

specified in ISO/IEC 23002-3. Note that 64 stream-ID-extension values are made available for private usage.

13.3 Descriptors of Programs and Program Elements

13.3.1 General Format of Descriptors

The general data structure of descriptors is very simple. Each descriptor starts with a eight-bit descriptor-tag, followed by a eight-bit descriptor-length field and the number of descriptor-data-bytes indicated by the descriptor-length; see Figure 13.8. This data structure allows to concatenate multiple descriptor in the descriptor loop in Figure 9.12. In general, it allows to add more fields in future, while maintaining compatibility, and for this purpose, decoders should take into account that the descriptor-length may indicate more descriptor-data-bytes than specified in the version of the MPEG-2 system standard the implementation is based upon, and

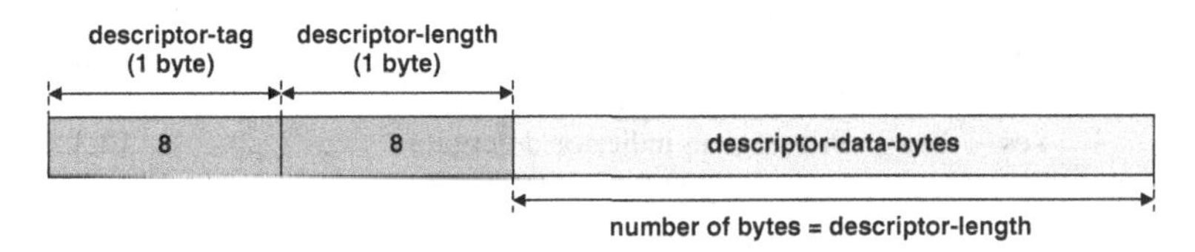

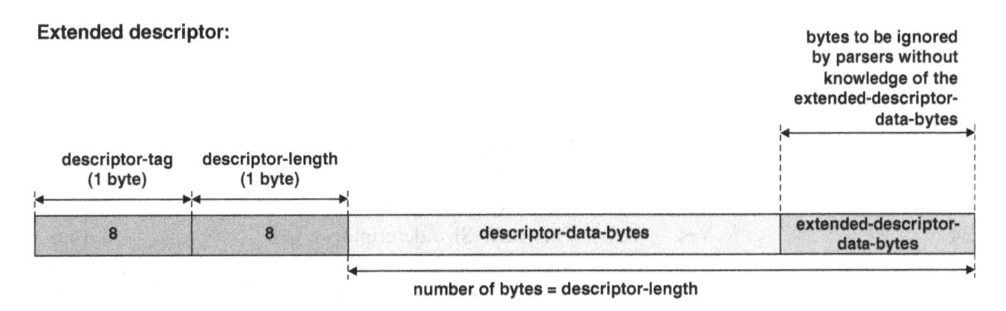

Figure 13.8 General data structure of a descriptor and compatible extensions

that the parser should skip these additional bytes. However, for some descriptors, the syntax does not allow to add more fields.

In total 256 descriptor tags can be assigned; however, only 64 values are available to MPEG; the other 192 are for private usage. As of 2013, MPEG systems assigned over 50 descriptors; see Table 13.4. Which means that the number of remaining descriptor-tags for MPEG system usage is rather limited; as a consequence, most likely in future an extension mechanism is needed, for example one descriptor tag value that signals the presence of an eight-bit extended-descriptor-tag as the first byte in the descriptor-data-bytes field.

Table 13.4 also indicates for each descriptor whether it can be used in a transport stream and in a program stream. Furthermore for each descriptor the chapter is given in which the descriptor is discussed in this book; however, when the content of a descriptor is largely beyond the scope of this book, then such descriptor may be addressed only marginally.

Table 13.4 Usage of descriptor tag values

Descriptor-tag (decimal)	TS	PS	Type of descriptor	Section
0	—	—	Reserved	
1	—	—	Forbidden	
2	Yes	Yes	(MPEG-1/2) Video stream descriptor	14.3.1
3	Yes	Yes	(MPEG-1/2) Audio stream descriptor	14.3.2
4	Yes	Yes	Hierarchy descriptor	13.3.4.1
5	Yes	Yes	Registration descriptor	13.3.3.7
6	Yes	Yes	Data stream alignment descriptor	13.3.4.2
7	Yes	Yes	Target background grid descriptor	13.3.4.4
8	Yes	Yes	Video window descriptor	13.3.4.5
9	Yes	Yes	Conditional access descriptor	13.3.3.6
10	Yes	Yes	ISO 639 language descriptor	13.3.4.3
11	Yes	Yes	System clock descriptor	13.3.3.1
12	Yes	Yes	Multiplex buffer utilization descriptor	13.3.3.2
13	Yes	Yes	Copyright descriptor	13.3.3.8
14	Yes	No	Maximum bitrate descriptor	13.3.3.3
15	Yes	Yes	Private data indicator descriptor	13.3.3.9
16	Yes	Yes	Smoothing buffer descriptor	13.3.3.4
17	Yes	No	STD descriptor	13.3.3.5
18	Yes	Yes	(MPEG-1/2/4) IBP descriptor	14.3.1
19–26	Yes	Yes	Defined in ISO/IEC 13818-6 (DSM-CC)	
27	Yes	Yes	MPEG-4 video descriptor	14.6.1
28	Yes	Yes	MPEG-4 audio descriptor	14.6.2
29	Yes	Yes	IOD descriptor	14.6.3
30	Yes	No	SL descriptor	14.6.3
31	Yes	Yes	FMC descriptor	14.6.3
32	Yes	Yes	External ES ID descriptor	14.6.3
33	Yes	Yes	MuxCode descriptor	14.6.3
34	Yes	Yes	FmxBufferSize descriptor	14.6.3
35	Yes	No	Multiplex buffer descriptor	14.6.3
36	Yes	Yes	Content labelling descriptor	14.11
37	Yes	Yes	Metadata pointer descriptor	14.11

38	Yes	Yes	Metadata descriptor	14.11
39	Yes	Yes	Metadata STD descriptor	14.11
40	Yes	Yes	AVC video descriptor	14.7
41	Yes	Yes	Defined in ISO/IEC 13818-11 (MPEG-2 IPMP)	
42	Yes	Yes	AVC timing and HRD descriptor	14.7
43	Yes	Yes	MPEG-2 AAC descriptor	14.3.3
44	Yes	Yes	FlexMuxTiming descriptor	14.6.3
45	Yes	Yes	MPEG-4 text descriptor	14.6.4
46	Yes	Yes	MPEG-4 audio extension descriptor	14.6.2
47	Yes	Yes	Auxiliary video descriptor	14.9.2
48	Yes	Yes	SVC extension descriptor	14.8
49	Yes	Yes	MVC extension descriptor	14.9.3
50	Yes	—	J2K video descriptor	14.10
51	Yes	Yes	MVC operation point descriptor	14.9.3
52	Yes	Yes	MPEG-2 stereoscopic video format descr.	14.9.1
53	Yes	Yes	Stereoscopic program info descriptor	14.9.1
54	Yes	Yes	Stereoscopic video info descriptor	14.9.3
55–63	—	—	Reserved for future use by MPEG systems	
64–255	—	—	User private	

13.3.2 Types of Descriptors

There are various types of descriptors; some are related to system performance, while others provide information either on one specific content format, or on contents decoding more in general, without being specific to a certain content format. In Section 13.3.3 the descriptors related to system performance are discussed, while the descriptors on general content decoding are addressed in Section 12.3.4. The descriptors that are specific for a certain content format are discussed in Chapter 14.

13.3.3 System Orientated Descriptors

13.3.3.1 System Clock Descriptor

By means of the system clock descriptor the accuracy can be signalled of the applied system clock (STC) that was used to generate the PTS and DTS time stamps. For more background on this issue, see Section 6.6.3. The applied STC accuracy is conveyed by means of the clock-accuracy-integer and the clock-accuracy-exponent parameters, as follows:

$$\text{STC accuracy} = \text{clock-accuracy-integer} \times 10^{-(\text{clock-accuracy-exponent})}\text{ppm}$$

The clock-accuracy-integer and the clock-accuracy-component are integers coded with six and three bits, respectively, see Figure 13.9. If the clock-accuracy-integer has the value zero, then an STC accuracy of 30 ppm is signalled. Note that the value of the clock-accuracy-integer is in the range of 1–63 and the value of the clock-accuracy-component in the range of 0–7. As a consequence, a STC accuracy value can be signalled in the range of 1×10^{-7} to 63 ppm.

System clock descriptor

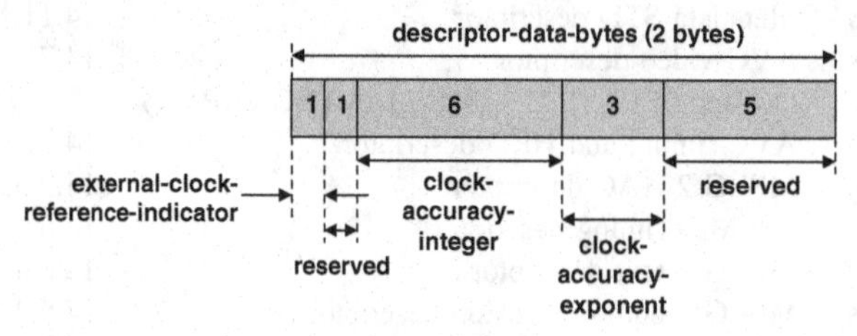

Multiplex buffer utilization descriptor

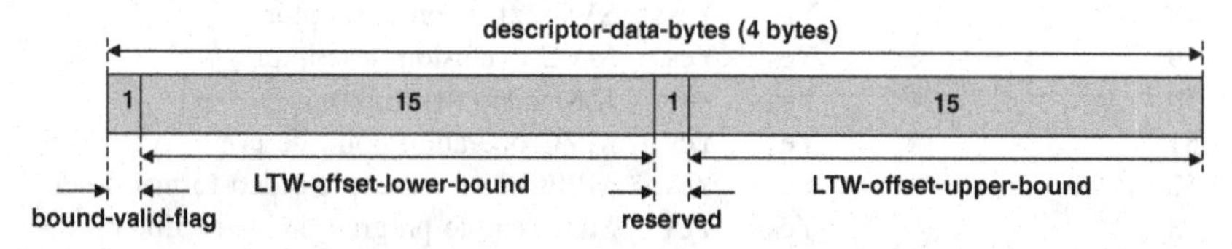

Maximum bitrate descriptor

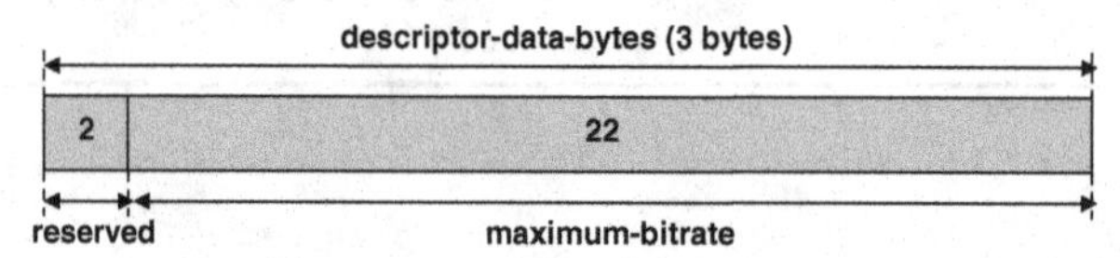

Figure 13.9 Structure of the descriptor-data-bytes of the system clock descriptor, the multiplex buffer utilization descriptor and the maximum bitrate descriptor

By means of setting the external-clock-reference-indicator to '1', the use of an external reference clock can be signalled; in this case, the signalled STC accuracy signals the accuracy of such external clock; the decoder may have access to the same reference clock.

13.3.3.2 Multiplex Buffer Utilization Descriptor

The multiplex buffer utilization descriptor provides information for transport streams on the fullness of the buffer MB_n and is closely related to the legal time window (ltw) concept that was designed to assist devices in the network to implement seamless splices; see also Section 11.4. The fields in this descriptor are depicted in Figure 13.9. The first bit is the bound-valid-flag that, if set to '1', signals that the bound values provided in this descriptor are valid. After a re-multiplex operation this flag may be set to '0'. This flag is followed by two bound fields of 15 bits, separated by a reserved bit. These ltw-offset-lower-bound and the ltw-offset-upper-bound fields specify upper and lower bounds to the encoded value of the ltw-offset field in the adaptation-field of transport packets, independent whether the ltw-offset field is present or absent.

13.3.3.3 Maximum Bitrate Descriptor

By means of the maximum bitrate descriptor, the maximum bitrate applied for the associated program element or entire program in a transport stream can be signalled. This information may be useful for decoders and re-multiplexers. Its data structure is depicted in Figure 13.9. The maximum-bitrate field is a 22-bit integer that specifies an upper bound for the value of the

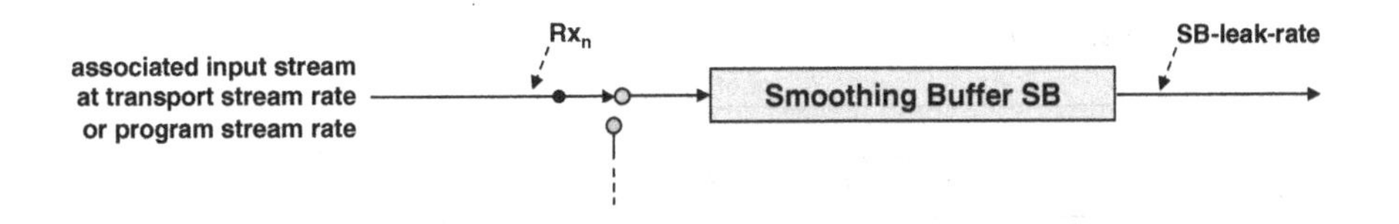

Smoothing buffer descriptor

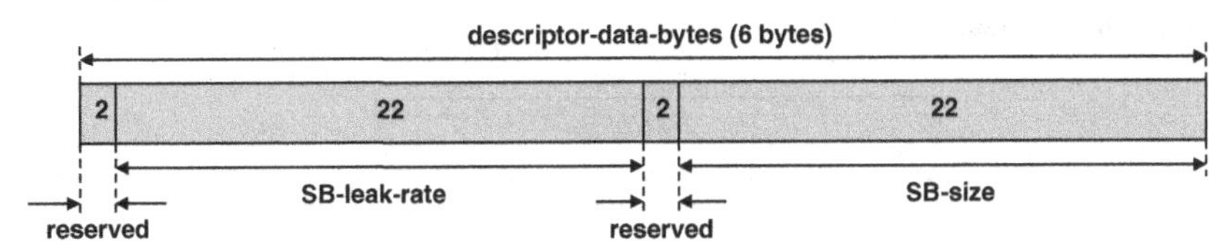

STD descriptor

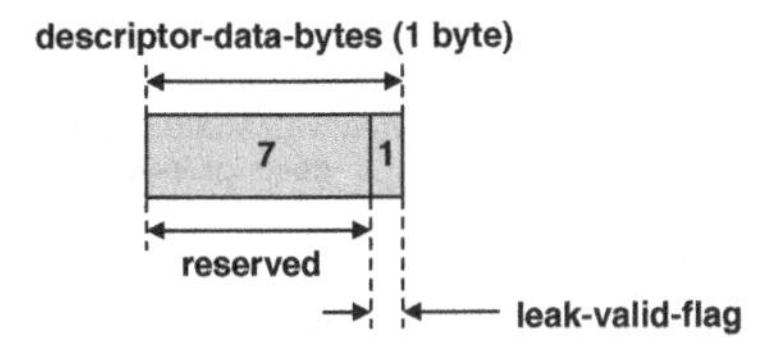

Figure 13.10 The model of the smoothing buffer SB and the structure of descriptor-data-bytes of the smoothing buffer descriptor and the STD descriptor

bitrate of the associated stream, including any transport overhead, in units of 50 bytes/s. Note though that it is not specified how this bitrate is measured.

13.3.3.4 Smoothing Buffer Descriptor

The smoothing buffer descriptor can be used to introduce an additional constraint on the bitrate of a stream, next to the constraints defined by the STD; the descriptor can be associated to an elementary stream, to a program stream and to a program within a transport stream. For the purpose of the constraint, it is presumed that the associated stream is input to a Smoothing Buffer SB. At the input of SB a complete transport stream or program stream is available. However, only bytes of the associated stream enter SB, as follows; see also Figure 13.10.

- In case of a program stream, if the descriptor is associated to the program stream, then the entire program stream enters SB; otherwise only the PES packets of the elementary stream to which the descriptor is associated enter SB.
- In case of a transport stream, if the descriptor is associated to one program element then all transport packets that carry that program element enter SB, while all other transport packets are ignored.
- In case of a transport stream, if the descriptor is associated to a program, then the following transport packets enter SB: each transport packet that carries data of an element of the program, each transport packet with the PCR-PID, and each transport packet with the PMT-PID.

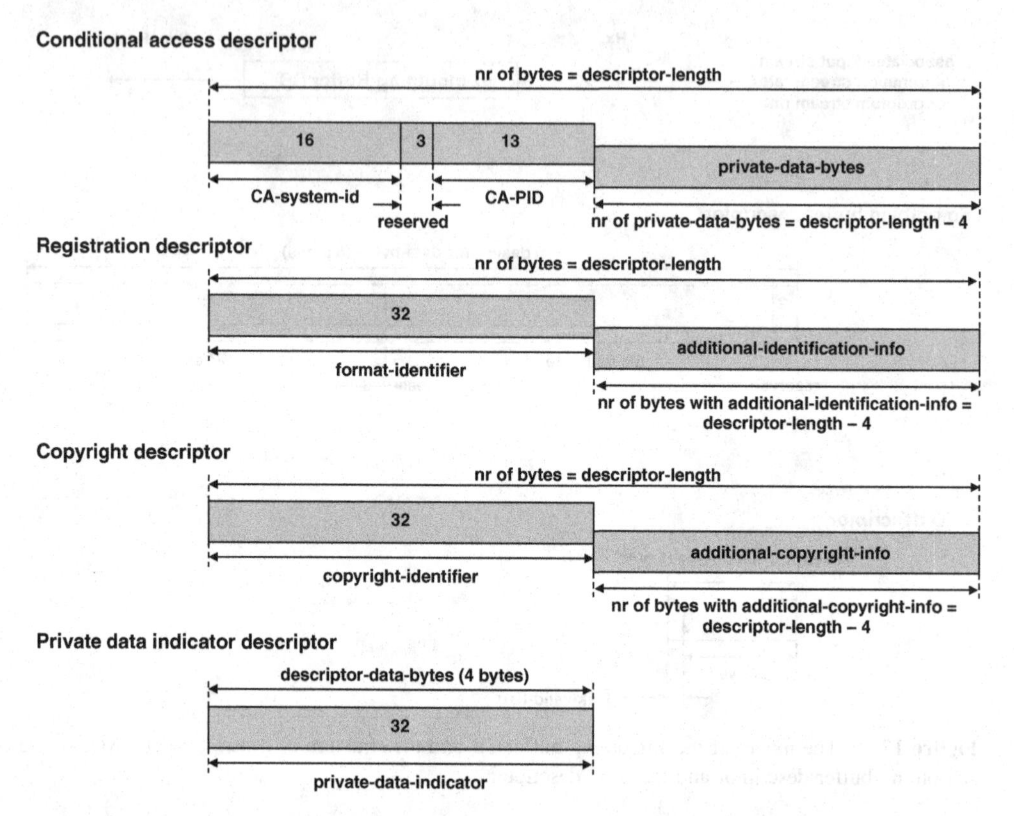

Figure 13.11 Structure of descriptor-data-bytes of the conditional access descriptor, the registration descriptor, the copyright descriptor and the private data indicator descriptor

Each byte that enters SB, leaves it; when there is data in SB, then bytes are removed at the SB-leak-rate, specified in the descriptor in units of 400 bits/s. The descriptor also specifies the size of SB, SB-size in units of one byte. For each stream to which the smoothing buffer descriptor is associated, the transport stream or program stream must be constructed so, that SB never overflows.

13.3.3.5 STD Descriptor

The STD descriptor must be used when the leak method is applied in the T-STD for a MPEG-2 video stream. For more background on the leak method see Section 12.3.4. The STD descriptor carries next to a seven-bit reserved field only the one-bit leak-valid-flag; see Figure 13.10. When set to the value '1', the leak method is applied for the transfer of data from MB_n to EB_n. If the flag has the value '0', and the vbv-delay field is encoded with a valid value (that is unequal to 0xFFFF), the transfer of data from MB_n to EB_n uses the vbv-delay method, as discussed in Section 12.3.4.

13.3.3.6 Conditional Access Descriptor

The format and usage of a conditional access (CA) descriptor is discussed in Section 10.2, but for completeness included here as well; see Figure 13.11. The conditional access descriptor carries the CA-system-ID and the CA-PID, as well as private-data-bytes with a format that depends on the encoded value of the CA-system-ID. If the descriptor is found in the CAT, then the CA-PID signals the PID in which the EMM stream is conveyed; if the descriptor is found in the PMT, then the CA-PID signals the ECM stream. For further details see Section 10.2.

13.3.3.7 Registration Descriptor

By means of the registration descriptor, the format of private data can be identified in a unique and unambiguous manner. This is particularly beneficial for private data used across multiple application domains; for a further discussion on this subject see Section 11.8. The data structure of the registration descriptor is depicted in Figure 13.11. The descriptor-data-bytes start with the 32-bit format-identifier field. Entities using private data can request SMPTE, the nominated registration authority, for one or more values of the format-identifier to signal specific private data. The assigned format-identifier value is then 'owned' by the requesting entity. This entity may decide that the format-identifier field is followed by zero or more private-data-bytes; the format of these bytes, if any, is determined by the requesting entity.

13.3.3.8 Copyright Descriptor

The data structure of the copyright descriptor is very similar to that of the registration descriptor: the format-identifier field is replaced by the copyright-identifier field of the same length (see Figure 13.11). The copyright descriptor and its usage, as well as alternative descriptors for the same purpose are discussed in Section 11.9.

13.3.3.9 Private Data Indicator Descriptor

The private data indicator descriptor provides a method to distinguish private data formats uniquely within one application domain. The entity responsible for standardization within that domain may assign values to the 32-bit private-data-indicator field in this descriptor to signal a specific private data format; see Figure 13.11. Note however that a standardization entities operating in another application domain may assign the same value to another private data format. For an elaborate discussion on the carriage of private data see Section 11.8.

13.3.4 General Content Descriptors

13.3.4.1 Hierarchy Descriptor

By means of the hierarchy descriptor, elementary streams can be identified that are a component of hierarchically coded video, audio and private streams. Hierarchical coding is characterized by multiple layers, whereby the higher layer(s) depend(s) on the lower layer(s); for example a higher layer may be predicted from a lower layer. Below, three examples are given of hierarchical coding.

- A MPEG-2 audio encoder may produce two elementary streams: a base stream and an extension stream; see Section 3.3.2. Decoding of an extension stream requires that also the base stream can be accessed.
- In very error-prone environments, the coding of high resolution video may be split in two layers, a base layer coded at standard resolution and a higher layer coded at high resolution,

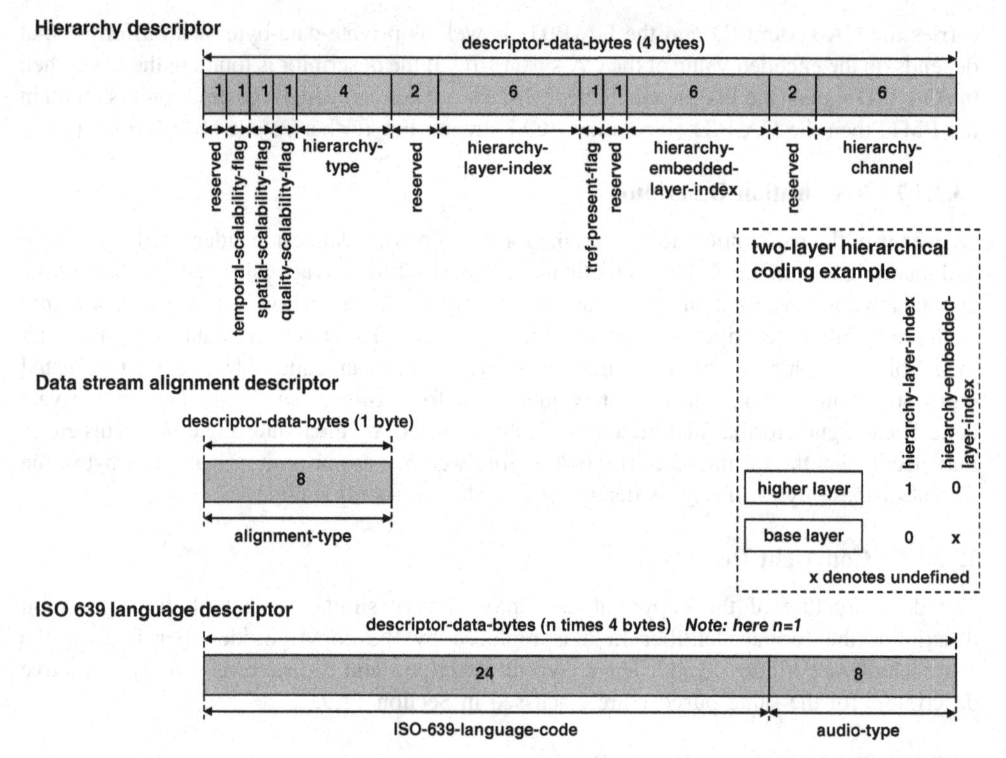

Figure 13.12 Structure of descriptor-data-bytes of the hierarchy descriptor, the data stream alignment descriptor and ISO 639 language descriptor

using prediction from the base layer. For transport of the base layer a more robust channel may be used, so as to improve the probability that the base layer remains available when the higher layer gets lost under severe error conditions. Note though that the business case for such scalable coding may be hard to justify.

- When using the MVC standard for the coding of a right view and a left view, one of these views is the base layer, from which the other view is predicted.

It is possible to use more than two layers and therefore the hierarchy descriptor provides a mechanism to specify the dependencies of the various components of the hierarchically coded content. For this purpose the hierarchy descriptor assigns a unique index to each component; a higher layer stream that is predicted from a lower layer one is referencing the index of this lower layer stream, so as to indicate that this lower layer stream needs to be accessed prior to the decoding of the higher layer stream. Furthermore the hierarchy descriptor assigns an intended channel number to an elementary stream in an ordered set of transmission channels, whereby the lowest number with respect to the transmission hierarchy indicates the most robust transmission channel. Note that the same channel number may be assigned to more than one elementary stream.

The data structure of the hierarchy descriptor is given in Figure 13.12. One reserved bit is followed by three flags, the temporal-scalability-flag, the spatial-scalability-flag and the quality-scalability-flag. When set to '0', each of these flags signals a type of applied scalability.

Table 13.5 Hierarchy-type values for hierarchically coded content

Hierarchy-type value (decimal)	Hierarchy type
0	Reserved
1	Spatial scalability
2	SNR scalability
3	Temporal scalability
4	Data partitioning
5	Extension bitstream
6	Private stream
7	Multi-view profile
8	Combined scalability
9	MVC video sub-bitstream
10–14	Reserved
15	Base layer or base view used for MVC or AVC sub-bitstream of MVC

In the first editions of the MPEG-2 system standard, these bits were reserved, and hence coded as '1'; consequently, the value '1' of these flags cannot be used for signalling a specific type of scalability. More than one type of scalability may be applied at the same time, and therefore all three flags may be set to '0' in the hierarchy descriptor. The three scalability flags are followed by the four-bit hierarchy-type field, coded as indicated in Table 13.5. If multiple scalability types are applied, then the hierarchy-type signals combined scalability, while the scalability flags signal which types are actually used. If the hierarchy-type is encoded with the value 15, then the base layer is indicated or the base view used for MVC or an AVC sub-bitstream of MVC (for MVC see also Sections 4.5.3 and 14.9.3).

The hierarchy-type field is followed by two reserved bits and the six-bit hierarchy-layer-index field encoded with a value that assigns a unique index to the elementary stream to which the descriptor is associated. Next is a one-bit tref-present-flag that, if set to '0', indicates that a TREF field may be present in the PES headers that carry the associated elementary stream. After one reserved bit, the six-bit hierarchy-embedded-layer-index follows, so as to provide the index of the elementary stream that must be decoded prior to the decoding of the elementary stream to which the descriptor is associated. Figure 13.12 provides also an example of a two-layer hierarchical coding system. The embedded-hierarchy-layer-index value of the higher layer references the hierarchical-layer-index value 0 of the base layer.

The descriptor-data-bytes are concluded by a two-bit reserved field and the six-bit hierarchy-channel field that indicates the intended robustness of the transmission channel for the elementary stream to which the descriptor is associated.

13.3.4.2 Data Stream Alignment Descriptor

The transport of access units in PES packets may be fully asynchronous, but also some alignment may be applied between access units and PES packets. By setting the data-alignment-indicator bit in the PES header to '1', it is signalled that some form of alignment is applied (for background on this bit see Section 13.2 and Figure 13.4). For example, each PES packet payload may commence with the first byte of an access unit.

Table 13.6 Alignment-type values and applied alignment

Alignment-type value (decimal)	MPEG-1, or MPEG-2, or MPEG-4 part 2 video	AVC video (includes SVC and MVC)	Audio
00	Reserved	Reserved	Reserved
01 (default)	Slice or video access unit	AVC slice or access unit	Sync word
02	Video access unit	AVC access unit	Reserved
03	Group of pictures (GOP) or sequence (SEQ) header	SVC slice or dependency representation	Reserved
04	Sequence (SEQ) header	SVC dependency representation	Reserved
05	Reserved	MVC slice or view-component subset	Reserved
06	Reserved	MVC view-component subset	Reserved
07–255	Reserved	Reserved	Reserved

By means of the data stream alignment descriptor the type of applied alignment can be specified. The descriptor-data-bytes consist of a single byte only: the eight-bit alignment-type field; see Figure 13.12. The alignment-type value signals the applied alignment; depending on the content format, one or more types of alignment are possible. The descriptor does not need to be present; if the data-alignment-indicator bit is set to '1', and the data stream alignment descriptor is absent, then the default type of alignment corresponding to the alignment-type value 01 is applied. Table 13.6 shows the relationship between the alignment-type value and the applied alignment for various content formats.

Alignment implies that in a PES packet with the data-alignment-indicator bit in its header set to '1', the payload commences with the first byte of the data structure indicated in Table 13.6. In case of MPEG-1, MPEG-2 or MPEG-4 video, this means that the payload starts with the first byte of a start code prefix; see also Figure 3.15 and Table 3.2 in Section 3.2.2. Applications and application standardization bodies may require that the data-alignment-indicator bit is set to '1' in each PES header, thereby requiring alignment in each PES packet. For more details on the referenced data structures for AVC, SVC and MVC see Sections 14.7, 14.8 and 14.9.3, respectively.

13.3.4.3 ISO 639 Language Descriptor

By means of the ISO 639 language descriptor, one or more languages are associated to an audio, subtitling or other type of elementary stream. The ISO 639 language descriptor carries one or more occurrences of a three-byte ISO-639-language-code field containing the three-character code as specified in ISO 639, part 2 [1], followed by the eight-bit audio-type field that provides further specifics on the stream to which the descriptor is associated (see Figure 13.12). In Table 13.7 the coding of the audio-type is given. In case of a multi-lingual audio stream the sequence of language code and audio type fields reflects the contents of the audio stream.

Note: for the ISO 639 language descriptor the extension mechanism discussed in Figure 13.8 does not apply. The loop of n times a three-byte field followed by a one-byte field is only constrained in length by the descriptor-length field. Hence, the number of bytes is not fixed and consequently, the ISO 639 descriptor cannot be extended with new fields.

Table 13.7 Audio-type values

Audio-type value (decimal)	Audio type
00	Undefined
01	Clean effect, that is language not applicable, for example music only
02	Audio for the hearing impaired
03	Commentary for visually impaired viewer
04–127	User private
128–255	Reserved

13.3.4.4 Target Background Grid Descriptor

To allow simple positioning of multiple decoded videos in different video windows on specific positions at the same screen, the target background grid and the video window descriptors have been defined. An example is a picture-in-picture video to be displayed in a small window on top of full screen video. Note though that there are many methods to control the position of a video window. The target background grid descriptor defines a (background) grid of pixels projected on to the display area, while the video defines the position of the video window for the associated video stream. Multiple video windows may be displayed at the same screen.

The descriptor-data-bytes of the target background grid descriptor contains two 14-bit fields; the first specifies the value of the horizontal size and the second the value of the vertical size, both in pixels, of the background grid (see Figure 13.13). The two 14-bit fields are followed by a four-bit aspect-ratio-information field. This field is encoded in the same manner as the aspect-ratio-information field in the sequence header of MPEG-2 video. The aspect-ratio-information specifies either that square pixels are used or provides the display aspect ratio; for further information see ISO/IEC 13818-2.

13.3.4.5 Video Window Descriptor

The video window descriptor is used in conjunction with the target background grid descriptor, and defines the position of the video window for the associated video stream on the target background grid. The horizontal-offset and the vertical-offset are provided in two 14-bit fields; see Figure 13.13. Both parameters provide the distance in pixels between the top-left pixel of the background grid and the top-left pixel of the video window, as indicated in Figure 13.13. The two 14-bit fields in the descriptor-data-bytes are followed by a four-bit field that carries the value of the window-priority that specifies the priority of the window in the case of overlap with one or more other windows. The window-priority value of '0' has the lowest priority and the value '15' the highest priority, which means that a video window with priority 15 is always displayed.

13.4 Program Streams

The general structure of program streams is build on MPEG-1 systems, using packs and PES packets. Figure 13.14 depicts an example similar to the example of the MPEG-1 system stream structure given in Figure 7.2. Like a MPEG-1 system stream, a program stream consists of a concatenation of one or more packs, concluded by a MPEG program end code (see also Section 9.3 and Table 9.1). A program stream system header is used as well, but as an optional

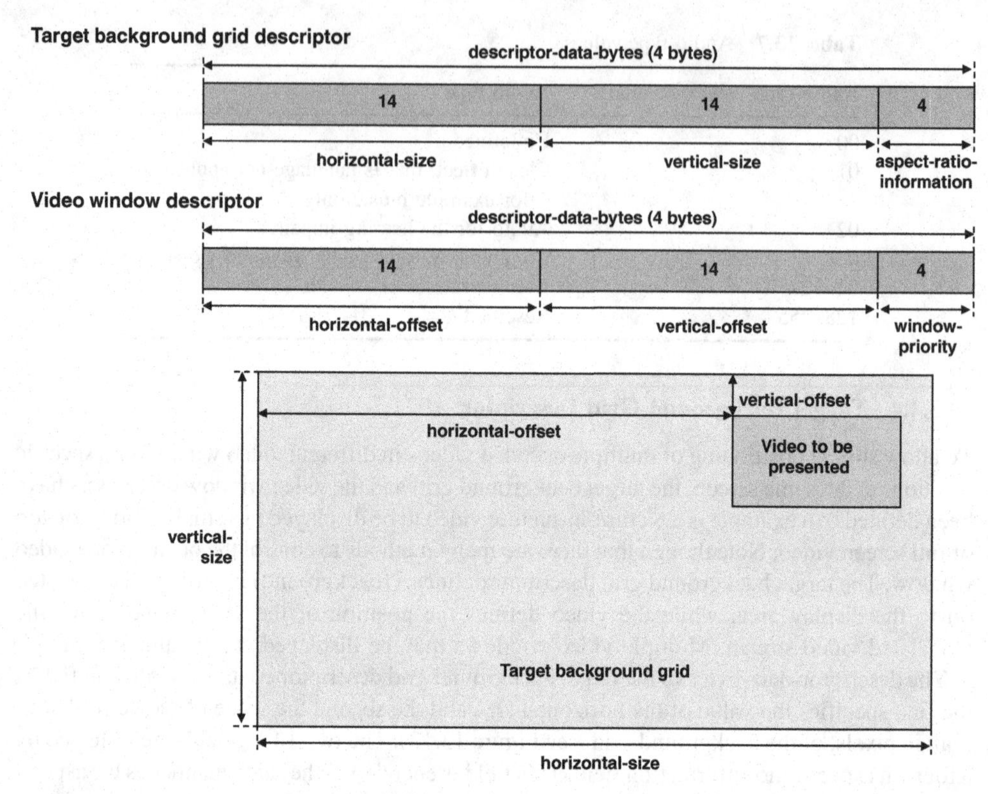

Figure 13.13 Structure of the descriptor-data-bytes of the target background grid descriptor and video window descriptor

element within the pack header; see Figure 13.14. Furthermore a program stream map is defined, the program stream equivalent of a program definition in the PMT of a transport stream. The program stream map lists all stream-ID values currently used in the program stream and specifies for each stream-ID value the content format carried in the PES packets with that stream-ID value. Within a program stream multiple versions of the program stream map may occur. For this purpose, the program stream map uses the same versioning mechanism as the PMT in transport streams with the version number and the current-next-indicator (see Section 9.4.4 and Figures 9.13 and 9.17). This versioning mechanism allows to produce a continuous program stream in which the used stream-ID values and the content formats change over time. However, the program stream system header must remain the same within a program stream, which imposes a constraint on the maximum size BS_n of the input buffer B_n in the P-STD for each stream-ID value.

The structure of a program stream pack header with the optional program stream system header is depicted in Figure 13.15. After the pack-header-start-code the six-byte SCR field, discussed in Figure 13.2, follows; for the pack-header-start-code see Table 9.1. Next is the 22-bit program-mux-rate field and two marker bits; the program-mux-rate value signals the bitrate of the program stream in units of 50 bytes/s. For a discussion on the bitrate of program streams see Section 9.3 and Figure 9.3. The remaining pack header after the two marker bits consists of a five-bit reserved field, followed by the three-bit pack-stuffing-length field that

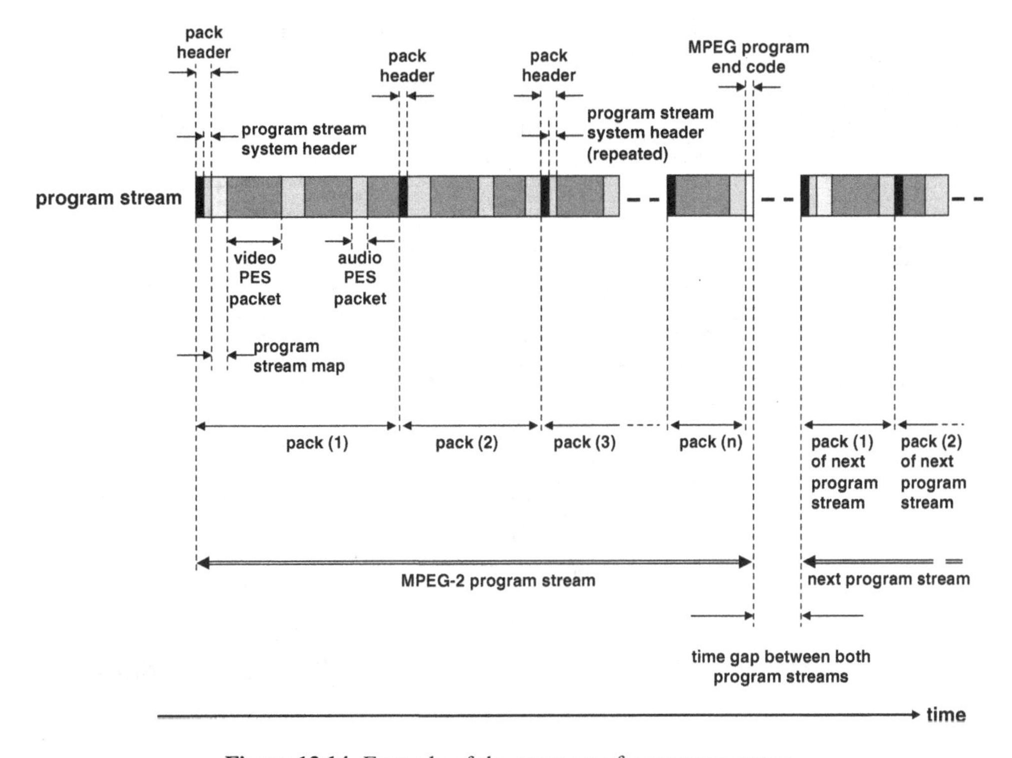

Figure 13.14 Example of the structure of a program stream

specifies the number of immediately following stuffing-bytes; at most seven stuffing-bytes can be applied. The pack header may be concluded by the (optional) system header; see Figure 13.15.

An included system header in the pack header is identified by a system-header-start-code; for the value assigned to a system-header-start-code the see Table 9.1. Note that, if no system header is included, the pack header is followed by either a PES packet header, or another pack header, or a MPEG program end code. The system-header-start-code is followed by the 16-bit system-header-length that specifies the number of immediately following bytes in the system header.

As in MPEG-1, the program stream system header carries important characteristics of the elementary streams carried in the program stream. In particular, it summarizes bounds of system parameters defined in the program stream, so that a system decoder can verify whether its capabilities are suitable for decoding the entire program stream. The program stream system header provides the following information. For details of its data structure see Figure 13.15.

- The value of the rate-bound specifies the maximum value of the bitrate of the program stream. The encoded values in the program-mux-rate field in the pack headers will not exceed the rate-bound value.
- Audio-bound: the maximum number of audio streams that are to be decoded simultaneously. Note though that a program stream may for example contain one video stream with multiple audio streams in different languages to chose from. These audio streams are not intended to be decoded at the same time.

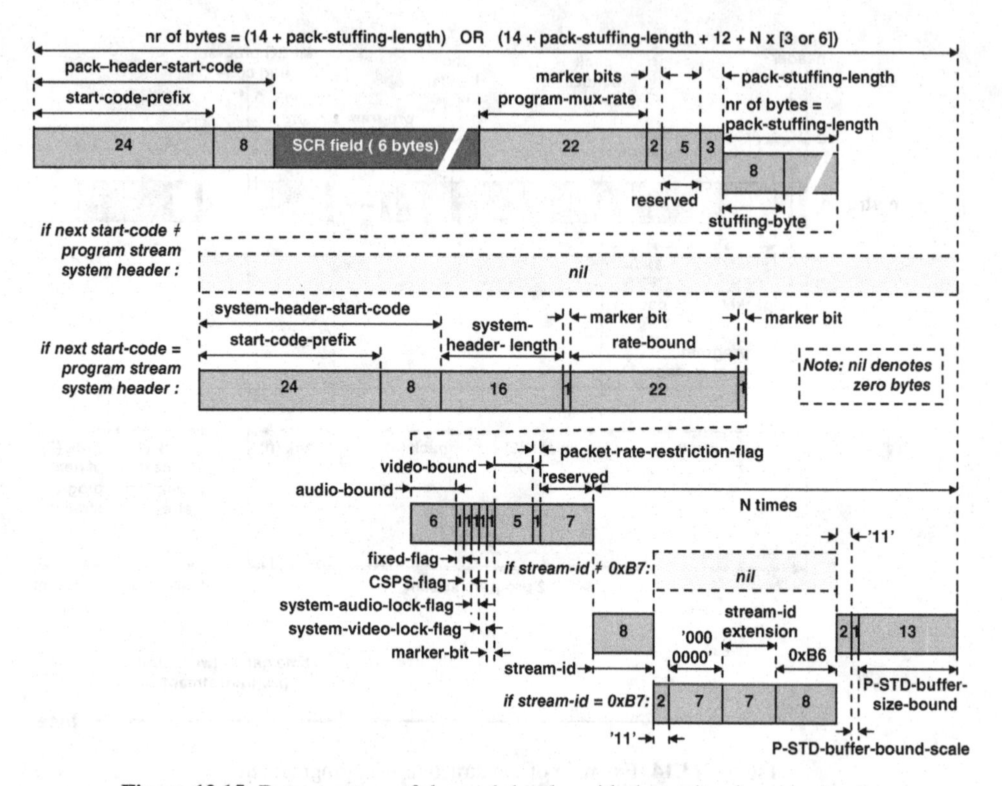

Figure 13.15 Data structure of the pack header with the optional system header

- Video-bound: the maximum number of video streams that are to be decoded simultaneously.
- Fixed-flag: if set to '1', this indicates constant bitrate operation; in other words, in each pack header the same value of the program-mux-rate is encoded.
- CSPS flag: if set to '1', this indicates that the program stream meets the constraints for a constrained system parameter program stream, as discussed in Section 12.2.3.
- System-audio-lock-flag: if set to '1', this indicates that for each audio frame in each carried audio stream, the audio sampling rate in the P-STD is locked to the STC frequency, that is that the division (STC frequency)/(audio sampling rate in the P-STD) yields the same result as this division for the nominal audio sampling frequency.
- System-video-lock-flag: if set to '1', this indicates that for each video picture in each carried video stream, the picture rate in the P-STD is locked to the STC frequency, that is that the division (STC frequency)/(picture rate in the P-STD) yields the same result as this division for the nominal video picture rate.
- Packet-rate-restriction-flag: if set to '1', this indicates that the packet rate is restricted for a CSPS program stream (see Section 12.2.3 and Figure 12.2).
- By means of the stream-ID, STD-buffer-bound-scale and STD-buffer-size-bound, an upper bound is specified for the input buffer B_n for each elementary stream carried in the program stream, as follows.
  - o The stream-ID field indicates to which elementary stream(s) the following STD-buffer-size-scale and STD-buffer-size-bound refer, either by providing the stream-ID value that

is assigned to an elementary stream, or by providing the wildcard values 0xB8 and 0xB9, signalling all audio streams and all video streams carried in the program stream, respectively, with a stream_ID value in the inclusive range between 192 and 239. Note that the values for these wild cards are not available for stream-ID usage in the PES header, see Table 13.1. A stream-ID value 0xFD signals that the provided upper bound applies to all streams in the program stream identified by a stream-ID-extension value. For a stream with an extended stream-ID, an individual bound can be specified by using the stream-ID value 0xB7, which is then followed by a compatible data structure with the applicable stream-ID-extension value.

- The STD-buffer-bound-scale indicates whether values of the following STD-buffer-size-bound are encoded in units of 128 or 1024 bytes.
- The STD-buffer-size-bound specifies a value greater than or equal to the maximum size of BS_n encoded in any PES packet carrying elementary stream n.

The three bytes with the stream-ID, the STD-buffer-bound-scale and the STD-buffer-size-bound are required to be present for each elementary stream present in the program stream. For each elementary stream the STD-buffer-size-bound must be specified exactly once by this mechanism. The first (most significant) bit of the stream-ID field is a '1' for each stream-ID value, which is used in the system header syntax as a flag signalling the presence of the stream-ID, the STD-buffer-bound-scale and the STD-buffer-size-bound fields.

MPEG left room in the system header for extension in future by allowing trailing bytes to follow the final three byte combination of stream-ID and STD buffer bound fields, but the first of these following bytes is required to have the first (msb) bit set to the value '0', so as to indicate that this byte is not a stream-ID field. The number of trailing extension bytes is equal to the remaining number of bytes from the number indicated by the system header length. In this context it may be interesting to note that the system header is followed by either a PES packet, a pack header or a MPEG program end code, all with a first byte that has its first (most significant) bit set to '0'. Note that the option for adding trailing extension bytes is not indicated in Figure 13.15.

The program stream map specifies the formats of the elementary streams carried in the program stream and provides information on these elementary streams, as well as on the program stream itself. Multiple versions of a program stream map may exist in the same program stream; each program stream map describes all elementary streams that occur in the program stream during the time the program stream map is valid. Hence, when a new elementary stream not yet described in the program stream map is included in the program stream, then prior to such inclusion a new version of the program stream map is to be inserted.

In Figure 13.16 the data structure of the program stream map is depicted. The program stream map uses the most simple PES header with the program-stream-map-start-code and the program-stream-map-length, after which the PES-packet-data-bytes follow, representing the program stream map. The current-next-indicator is followed by the single-extension-stream-flag, one reserved bit and the five-bit program-stream-map-version field; the value of the program-stream-map-version is incremented by one, modulo 32, with each new version of the program stream map. The transition procedure from one version to another is the same as for the program definition in a transport stream. See Section 9.4.4 and Figure 9.17. Next, a seven-bit reserved field, a marker bit and the 16-bit program-stream-info-length follows; the latter specifies the number of bytes contained by the immediately following program descriptors, if any.

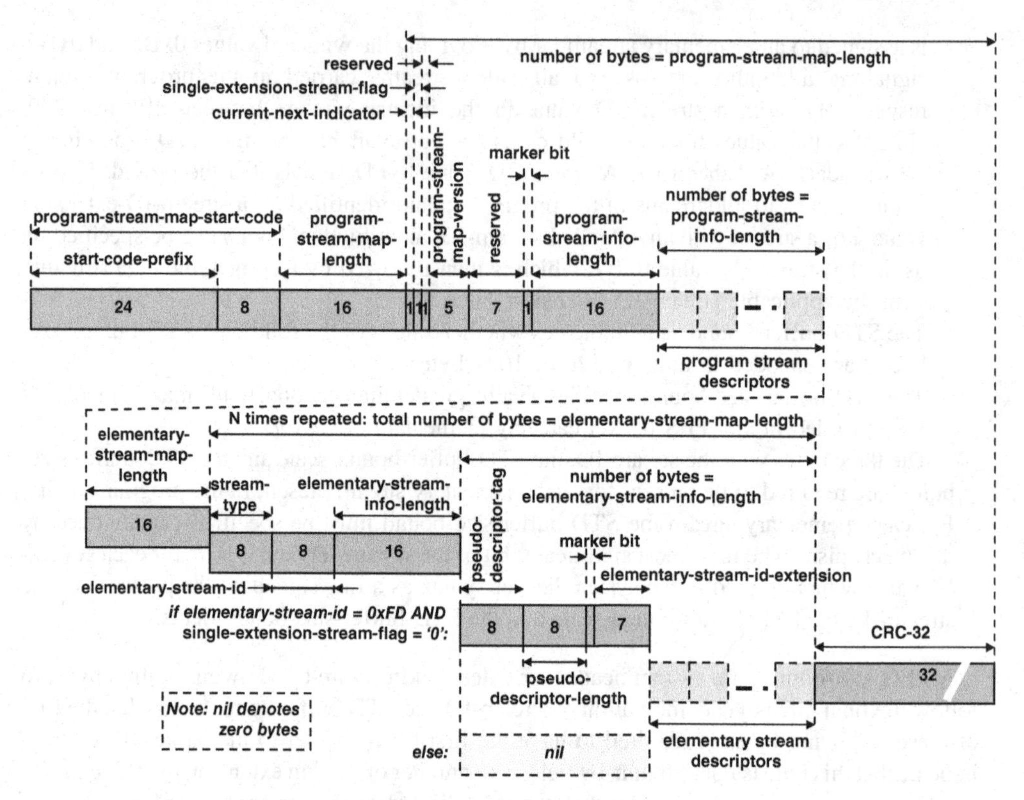

Figure 13.16 Data structure of the program stream map

The program descriptors are followed by the 16-bit elementary-stream-map length that specifies the number of following bytes with elementary stream data. Next, for each elementary streams described in the program stream map elementary stream data follows. The first fields are the stream-type and the elementary-stream-ID, specifying the format of the elementary stream carried in the PES packets with the stream-ID equal to the elementary-stream-ID value. Next, the 16-bit elementary-stream-info-length specifies the remaining number of bytes for this stream, including zero or more elementary stream descriptors. In case there are multiple streams with an extended stream-ID, then to identify each such stream, prior to any descriptors, (for compatibility reasons) an eight-bit pseudo-descriptor-tag is present, coded with the 'forbidden' value 1 (see Table 13.4), followed by the pseudo-descriptor-length with the value 1, a marker-bit and finally the seven-bit elementary-stream-ID-extension field coded with the applied stream-ID-extension for the stream. After describing the elementary streams, the program stream map is concluded by a 32-bit CRC-32 field, containing the CRC value that gives a zero output of the registers in the CRC decoder defined in Annex A of the MPEG-2 system standard, so that the integrity of the received program stream map can be verified.

The program stream directory is intended to provide references to elementary stream access points at regular distances, for example every 0.5 s. The directory for an entire program stream is made up of all directory data carried by program stream directory packets. The directory may be very useful in, for example, fast forward and fast reverse playback modes. However, the directory defined by MPEG systems represents an initial effort that was later considered

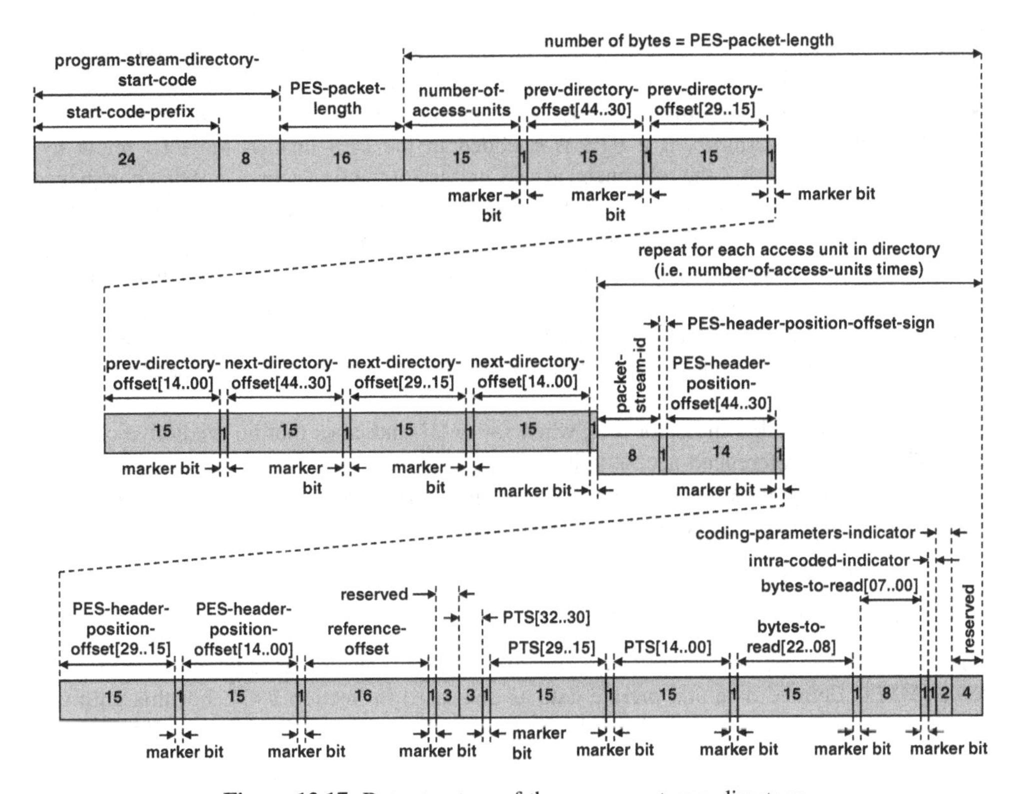

Figure 13.17 Data structure of the program stream directory

insufficient by the DVD Forum: instead, they defined their own (MPEG private) directory. As a consequence, the use of the MPEG program stream directory is not mandated by MPEG.

For completeness the data structure of the program stream directory is briefly described below; see also Figure 13.17. After the program-stream-directory-start-code and the PES-packet-length field, the number-of-access-units specifies the number of access units referenced in this packet. The 45-bit fields prev-directory-offset and next-directory-offset specify the positions of the previous and next directory packets in the program stream; in particular, the byte address offset is specified between the first byte[2] of the current directory packet and the first bytes of the previous and the next directory packet. Note that both offset fields are split in multiple fragments, interleaved with marker bits to avoid start-code emulation. For each referenced access the following information is provided:

- An eight-bit packet-stream-ID, specifying the stream-ID value of the PES packet that carries the referenced access unit.
- A one-bit PES-header-position-offset-sign and a 45-bit PES-header-position-offset, defining the byte offset address between the first byte of the PES packet that carries the referenced access unit and the first byte of this directory packet; if the PES-header-position-offset-sign is set to '0', the offset is positive, if set to '1', the offset is negative.

[2] The first byte of a packet is the first byte of the start-code of the packet.

- A 16-bit reference-offset, indicating the position of the first byte of the referenced access unit with respect to the first byte of the PES packet that carries this access unit.
- A 31-bit PTS, specifying the PTS of the referenced access unit; a PTS may be absent in the PES header; furthermore, if a PTS is encoded in the PES header, the PTS refers to a preceding access unit if the referenced access unit is not the first access unit that commences in the PES packet; encoding the PTS value in the directory packet prevents potentially complicated processing of elementary stream and systems data.
- A 23-bit bytes-to-read, specifying the number of bytes in the program stream that are needed to completely decode the referenced access unit; this includes all bytes in the program stream between the first and the last byte of the referenced access unit and may therefore include bytes from other PES packets containing information from different streams that are to be ignored.
- A one-bit intra-coded-indicator that, when set to '1', indicates that no predictive coding is applied for the referenced access unit.
- A two-bit coding-parameters-indicator, providing information on the availability of coding parameters needed to decode the referenced access unit.

13.5 Sections

MPEG-2 sections are used in transport streams only. Sections are used to convey PSI tables, other MPEG defined data and private data as discussed in Section 9.4.4. For this purpose a generic section data structure is defined with short and long section headers; see Figure 13.18.

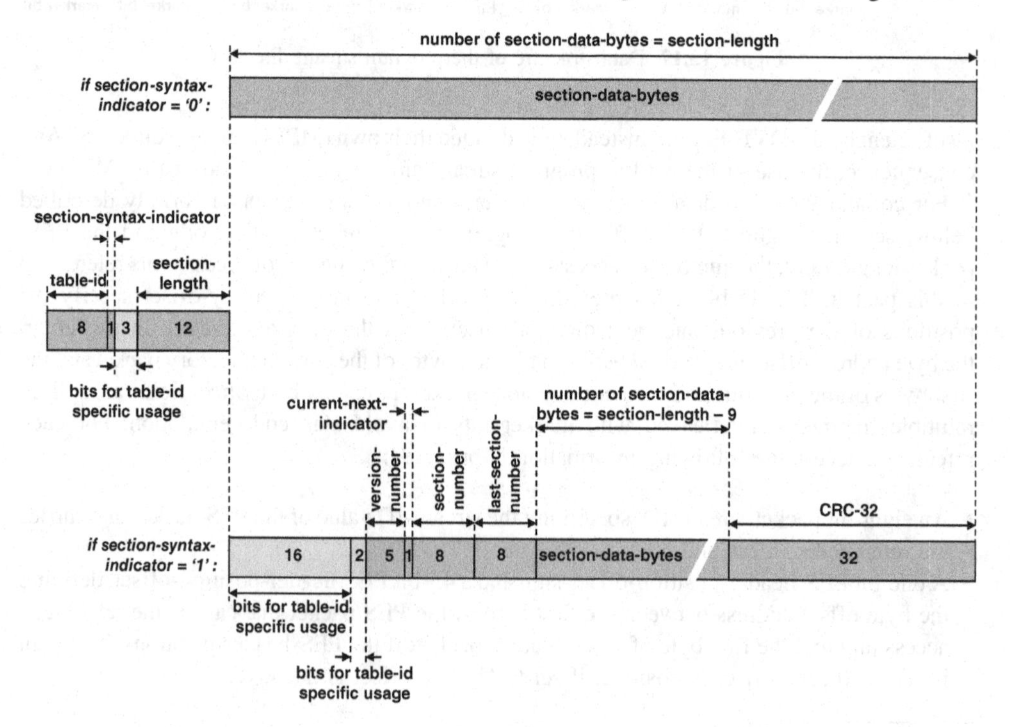

Figure 13.18 Generic section data structure with short and long headers

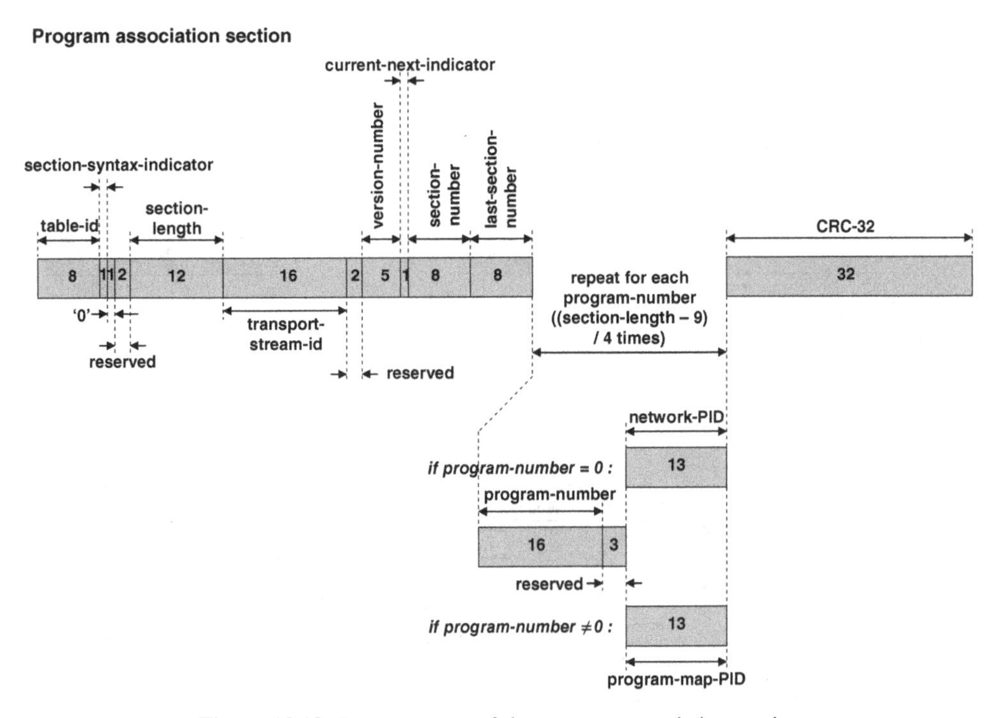

Figure 13.19 Data structure of the program association section

The eight-bit table-ID, specifying the type of data carried in the section (for assignments see Table 9.3), is followed by the one-bit section-syntax-indicator, three bits that are used in a way depending on the encoded table-ID value and the 12-bit section-length field that specifies the remaining number of bytes in the section. If the section-syntax-indicator is set to '0', the short section header is used, in which case the section-length is followed immediately by section-data-bytes until the end of the section. If the section-syntax-indicator is set to '1', then the long section header applies, in which case the section-length is followed by a 16-bit and a two-bit reserved field, the five-bit version-number and the one-bit current-next-indicator. By means of both latter fields a PSI or other table can be renewed. After the current-next-indicator, the eight-bit section-number and last-section-number fields conclude the long section header. These two fields are used for table fragmentation. The last-section-number field is followed by zero or more section-data-bytes, while in the case of the long section header the section is concluded by a CRC-32 code. The use of the fields in the long section header and the CRC-32 code is discussed in Section 9.4.4.

Without further discussion, in Figures 13.19 and 13.20 the detailed data structure of the program association section and of the program map section is presented (for further background see Section 9.4.4 and Figure 9.14), while Figure 13.21 depicts the data structure of three sections: the conditional access section, the transport stream description section and the private section. For the latter the short header can be used, but also the long header is available, in which case a 16-bit table-ID-extension can be privately used to distinguish different private sections.

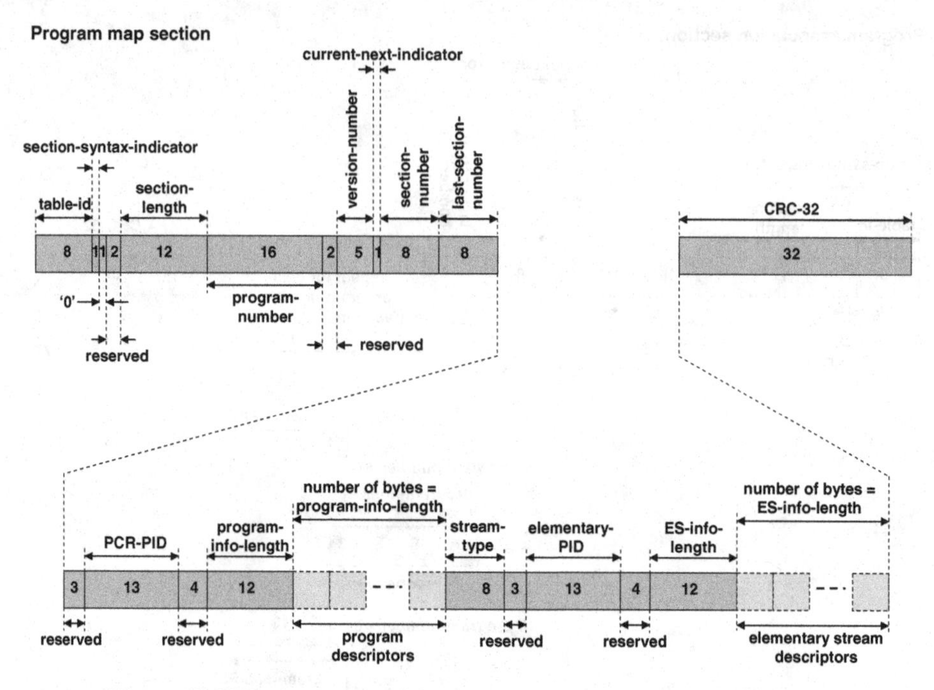

Figure 13.20 Data structure of the (transport stream) program map section

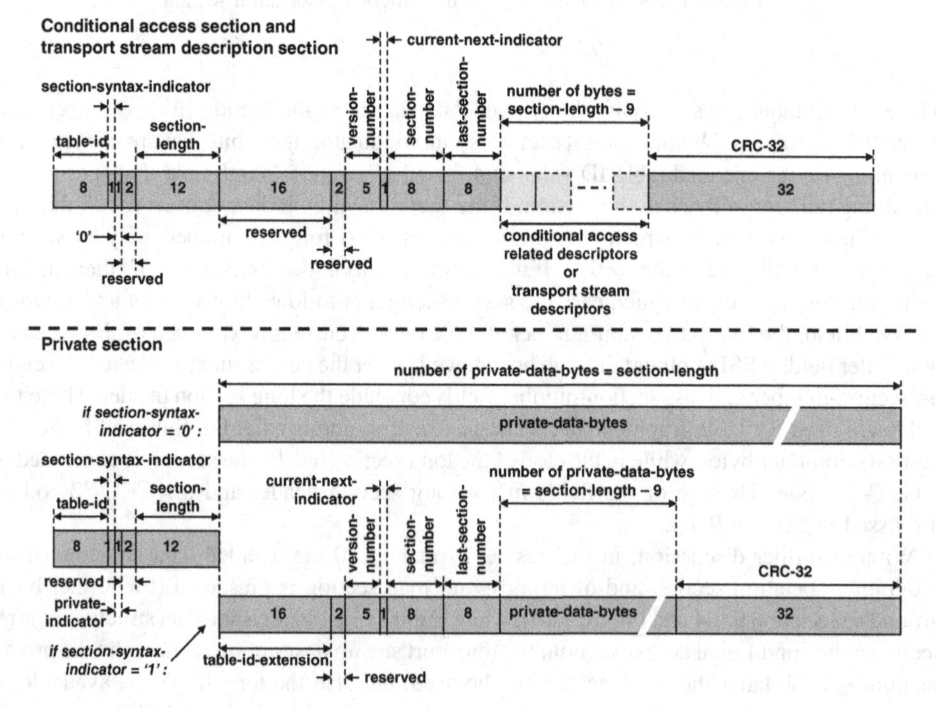

Figure 13.21 Data structure of the conditional access section, the transport stream description section and the private section

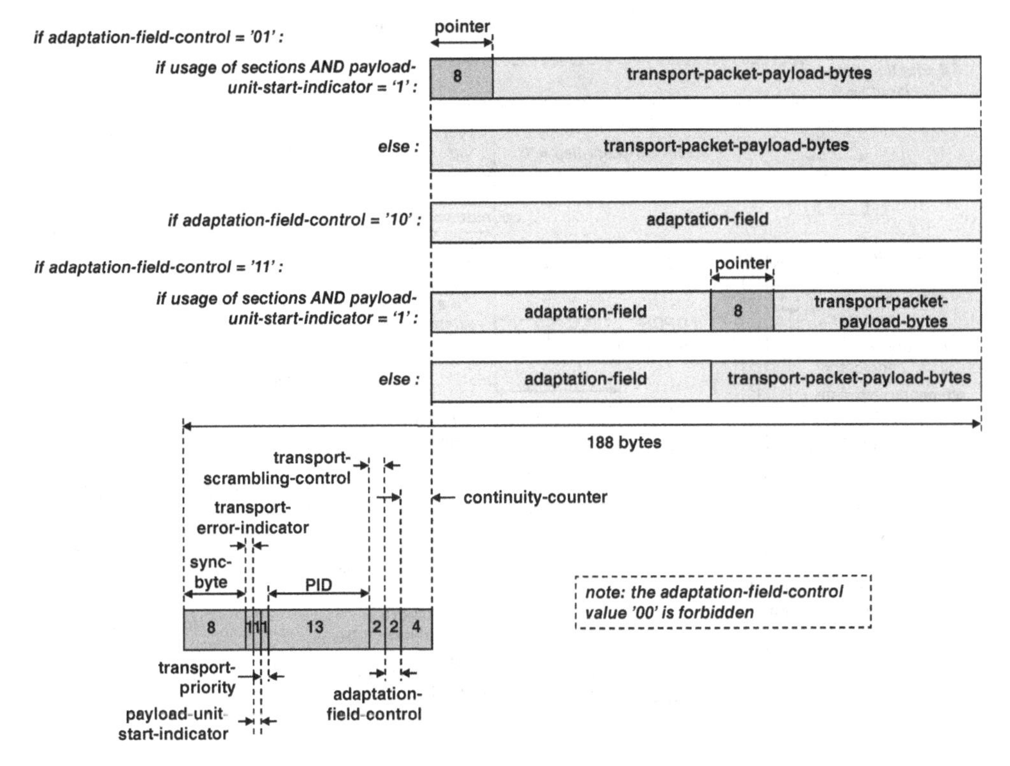

Figure 13.22 Data structure of a transport packet

13.6 Transport Streams and Transport Packets

A transport stream is a concatenation of transport packets. The data structure of a transport packet is depicted in Figure 13.22. Each transport packet commences with the eight-bit sync-byte field; each sync-byte field is encoded with the value '0100 0111' (0x47); the choice of this value is discussed in Section 9.2.1. The sync byte is followed by three one-bit fields: the transport-error-indicator discussed in Section 11.1, the payload-unit-start-indicator discussed in Section 11.7 and the transport-priority discussed in Section 11.1. These three bits are followed by the 13-bit PID field that indicates the type of data carried in the transport packet (for assignment of PID values see Section 9.4.4 and Table 9.2).

The PID field is followed by the two-bit transport-scrambling-control field (for the use of this field see Section 10.2), the two-bit adaptation-field-control field and the four-bit continuity-counter field discussed in Section 11.1. The adaptation-field-control field specifies whether the continuity-counter field is followed by an adaptation-field only, or an adaptation-field followed by transport-packet-payload-bytes or by transport-packet-payload-bytes only (see Figure 13.22). When set to '1', the first bit of the adaptation-field-control specifies the presence of the adaptation-field and the second bit the presence of transport-packet-payload-bytes; as a consequence, the value '00' of the adaptation-field-control is forbidden.

The data structure of the adaptation-field, shown in Figure 13.23, commences with the eight-bit adaptation-field-length, specifying the number of bytes in the adaptation-field following the

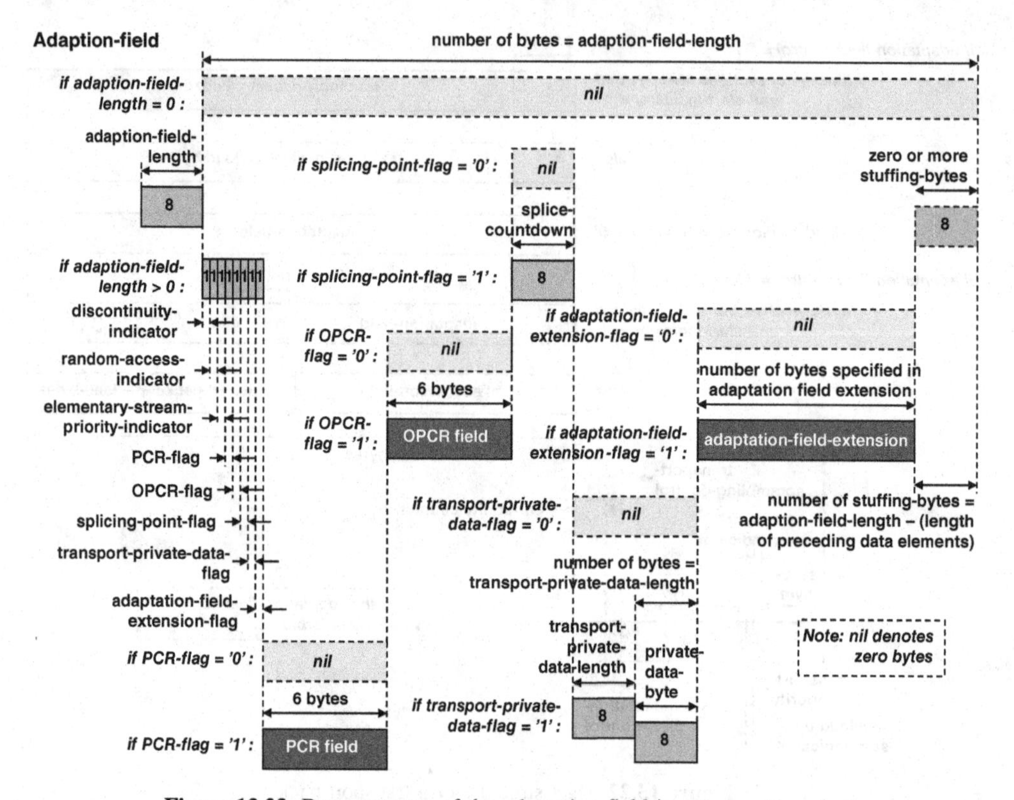

Figure 13.23 Data structure of the adaptation-field in a transport packet

adaptation-field-length. The value of the adaptation-field-length may be zero; in that case the adaptation-field contains only a single byte: the adaptation-field-length. This is useful when stuffing with a single byte is necessary. If the adaptation-field contains more than one byte, then the adaptation-field-length is followed by three one-bit indicators and five one-bit flags. The first bit is the discontinuity-indicator discussed in Section 11.3 and Table 11.2. The second bit is the random-access-indicator; for the use of this bit see Section 11.7. The third bit is the elementary-stream-priority-indicator, the use of which is discussed in Section 11.1. The three indicators are followed by the PCR-flag and the OPCR flag, indicating the presence of the PCR and the OPCR fields in the adaptation-field, respectively. The data structure of these two fields is discussed in Section 13.1 and Figure 13.2. Usage of the PCR is addressed in Section 9.4.4 and Figures 9.9 and 9.10, and of the OPCR in Section 11.11. The coding of an OPCR is only permitted in transport packets in which a PCR field is present; hence, if the OPCR flag is set to '1', then also the PCR-flag must be set to '1'.

In the adaptation-field the PCR and OPCR flags are followed by the splicing-point-flag, the transport-private-data-flag and the adaptation-field-extension-flag. These flags indicate the presence of the eight-bit splice-countdown field (discussed in Section 11.4 and Figure 11.9), the presence of the eight-bit transport-private-data-length that specifies the number of immediately following private-data-bytes, and the presence of the adaptation-field-extension, respectively (see Figure 13.23).

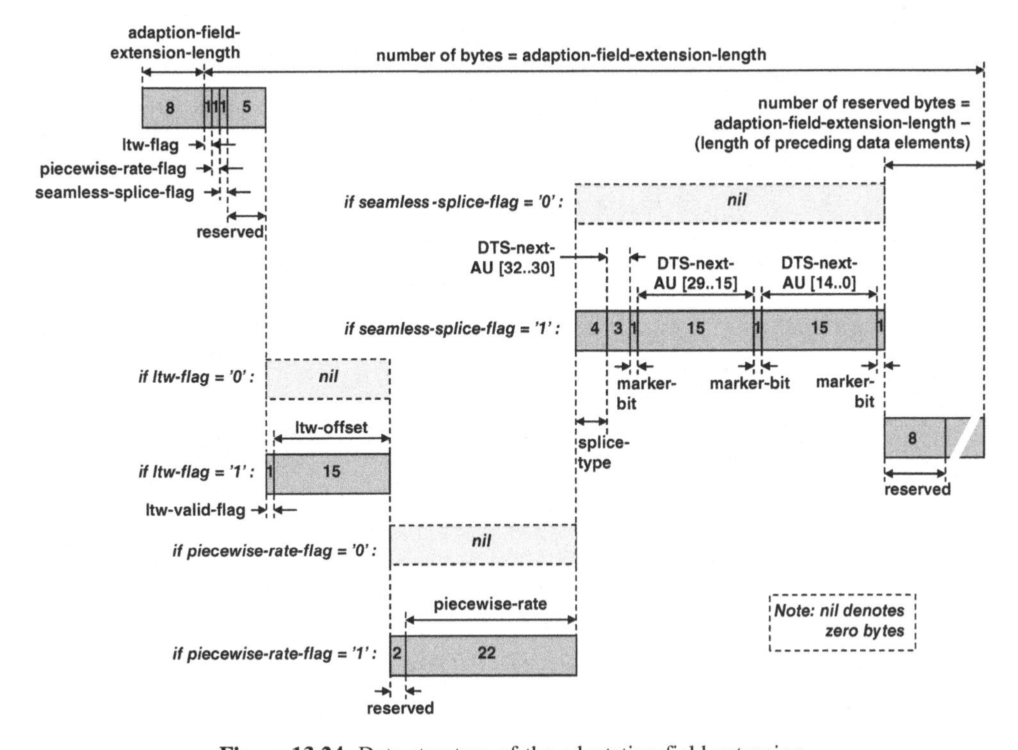

Figure 13.24 Data structure of the adaptation-field extension

The data structure of the adaptation-field-extension, depicted in Figure 13.24, starts with the eight-bit adaptation-field-extension-length, specifying the number of bytes in the adaptation-field-extension. This length field is followed by three flags, and five reserved bits. All three flags are signalling optional fields intended to support splicing. The ltw-flag signals the presence of the one-bit ltw-valid-flag combined with the 15-bit ltw-offset field, while the piecewise-rate-flag signals the presence of two reserved bits followed by the 22-bit piecewise-rate field. If the seamless-splice-flag is set to '1', the four-bit splice-type is present, followed by the 33 bits of the DTS-next-AU parameter, separated by some marker-bits, similar to the PTS and the PTS and DTS field data structures presented in Figure 13.2. The usage of these fields is described briefly in Section 11.4. The adaptation-field-extension may be concluded by one or more reserved bytes; this is the case if the encoded value of the adaptation-field-extension-length exceeds the length of the various fields in the adaptation-field-extension. Note that the reserved fields may be used by MPEG to extend the functionality of MPEG-2 systems in future.

Reference

1. ISO (1998) 639-2. Codes for the representations of names of languages – Part 2: Alpha-3 code, found at: http://www.iso.org/iso/catalogue_detail?csnumber4767 See also: http://en.wikipedia.org/wiki/ISO_639.

14

Content Support in MPEG-2 Systems

How the various content formats are supported in MPEG-2 systems. Packetization of the content in PES packets and sections, the associated signaling parameters and required descriptors. Decoding of each format in the STD model and the structure of the involved decoding branch in the STD.

14.1 Introduction

For a wide variety of content formats, carriage over MPEG-2 system streams has been specified in the MPEG-2 system standard and in amendments thereto, resulting in the evolution of transport capabilities discussed in Section 8.3 and Figures 8.1 and 8.2. It should be noted however that other standardization bodies, such as DVB, ATSC and SMPTE, also specified carriage of content over a MPEG-2 system stream, mostly formats beyond the scope of MPEG. This book discusses only those content formats for which carriage is specified in the MPEG-2 system standard. For each such format, transport over a MPEG-2 system stream needs to be specified in detail; this includes the following:

- The available transport methods in PES packets and sections, with assigned stream-ID, stream-type and table-ID value(s).
- Unambiguous access unit definition, if not provided in the specification of the content format.
- Constraints to be applied of the elementary stream data prior to packetization in PES packets or sections.
- Rules for packetization in PES packets or sections.
- The use of fields in the header of a transport packet, a PES packet or a section and the use of a descriptor, for example, the use of the elementary-stream-priority-indicator and the random-access-indicator in the adaptation field of a transport packet, the meaning of the data-alignment indicator in the PES header and the use of the alignment-type table in the data stream alignment descriptor.
- Any additional data structures that may be needed prior to packetization in a PES packet or a section.

Fundamentals and Evolution of MPEG-2 Systems: Paving the MPEG Road, First Edition. Jan van der Meer.
© 2014 John Wiley & Sons, Ltd. Published 2014 by John Wiley & Sons, Ltd.

- The decoding in the STD; the structure of the decoding branches in the P-STD and the T-STD, the size of each buffer and the value of each transfer rate.
- Any content format specific descriptor(s) that may be needed, as well as any other relevant system issues for a content format.

This chapter provides an overview of the system issues related to carriage of content in MPEG-2 system streams without providing each detail. For the latter the MPEG-2 system specification should be checked. For some content formats supported by MPEG-2 systems, but considered insufficiently relevant for usage by MPEG-2 system applications, only a brief description is provided. On the other hand, a more extensive description of a content format in this chapter does not necessarily mean that its market relevance is proven.

14.2 MPEG-1

14.2.1 MPEG-1 Video

The support of MPEG-1 video [1] in MPEG-2 systems is very similar to the support of MPEG-2 video, discussed in Sections 12.5.2 and 14.3.1. For transport of MPEG-1 video only PES packets are used; Table 14.1 presents the stream-ID and stream-type values used for MPEG-1 video. For constrained parameter video the same stream-type value is used as for MPEG-2 video. This reflects the normative requirement that each MPEG-2 video decoder must be capable of decoding a constrained parameter MPEG-1 video stream. Further details on a carried MPEG-1 video stream may be provided in a video stream and a IBP descriptor with formats discussed in Section 14.3.1. Figure 14.1 provides the STD decoding branches for MPEG-1 video and the values of the STD parameters. See Figure 12.16 for similarities with MPEG-2 video.

14.2.2 MPEG-1 Audio

The decoding branches and STD parameters used for MPEG-1 audio [2] are discussed in Section 12.5.1. Only PES packets are used for carriage of a MPEG-1 audio stream; Table 14.2 presents the used stream-ID and stream-type values. Further details on a carried MPEG-1 audio stream may be provided in an audio stream descriptor with a format discussed in Section 14.3.2.

14.2.3 MPEG-1 System Stream

The format of a MPEG-2 program stream is intended to be forward compatible with the MPEG-1 system stream format [3]. However, there are some minor differences between both formats.

Table 14.1 Signalling parameters for carriage of MPEG-1 video

Stream-ID	User-selectable from the range 0xE0 to 0xEF
Stream-type	0x01: MPEG-1 video in general
	0x02: constrained parameter MPEG-1 video (same stream-id value as for MPEG-2 video)
Descriptor-tag	0x02 (02): video stream descriptor
	0x12 (18): IBP descriptor

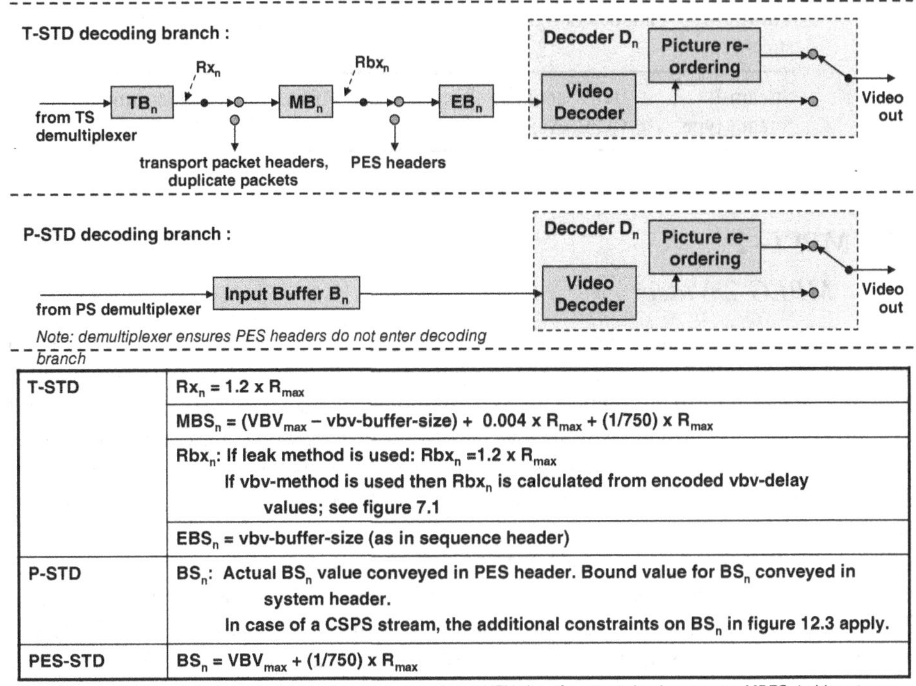

T-STD	Rx_n = 1.2 x R_{max}
	MBS_n = (VBV_{max} – vbv-buffer-size) + 0.004 x R_{max} + (1/750) x R_{max}
	Rbx_n: If leak method is used: Rbx_n =1.2 x R_{max} If vbv-method is used then Rbx_n is calculated from encoded vbv-delay values; see figure 7.1
	EBS_n = vbv-buffer-size (as in sequence header)
P-STD	BS_n: Actual BS_n value conveyed in PES header. Bound value for BS_n conveyed in system header. In case of a CSPS stream, the additional constraints on BS_n in figure 12.3 apply.
PES-STD	BS_n = VBV_{max} + (1/750) x R_{max}

Note: R_{max} and VBV_{max} refer to the maximum bitrate and maximum VBV size of a constrained parameter MPEG-1 video stream

Figure 14.1 STD decoding branches and STD parameters for MPEG-1 video

Therefore, to achieve the required forward compatibility, a program stream decoder is mandated to be capable of decoding MPEG-1 system streams as well.

Carriage of a MPEG-1 system stream across a MPEG-2 transport stream is discussed in Section 11.12. By replacing the header of each MPEG-1 packet by a PES packet header, without changing the coded audio, coded video data or private data carried in the PES payload, and some other measures discussed in Section 11.12, it is ensured that the contained content can be decoded in a transport stream decoder. Pack and system headers are carried in PES packet headers (see Section 11.12). Each MPEG-1 system packet with padding data is carried in a PES packet with the stream-ID indicating ancillary data, according to the procedure discussed in Section 11.12. See Table 14.3 for the assigned stream-ID and stream-type values. The assigned stream-ID value is exclusively for carriage in a transport stream of data from a MPEG-1 system stream or from a program stream. Hence, this stream-ID value may also signal a program stream directory or a program stream map from a MPEG-2 program stream (see Section 14.3.5).

Table 14.2 Signalling parameters for carriage of MPEG-1 audio

Stream-ID	User-selectable from the range 0xC0 to 0xDF
Stream-type	0x03: MPEG-1 audio
Descriptor-tag	0x03 (03): audio stream descriptor

Table 14.3 Parameters signalling padding data from a MPEG-1 system stream

Stream-ID	0xF9: ancillary data, such as a MPEG-1 padding stream
Stream-type	0x0E: Auxiliary data

14.3 MPEG-2

14.3.1 MPEG-2 Video

MPEG-2 video [4] is transported across MPEG-2 systems in PES packets; the signalling parameters are provided in Table 14.4. The applied decoding branch and the associated STD parameters are discussed in Section 12.5.2. Important characteristics of a carried MPEG-2 video stream may be provided by means of several descriptors, for example some general content descriptors discussed in Section 13.3.4, such as the data stream alignment descriptor, the target background descriptor, the video window descriptor and the hierarchy descriptor. Also descriptors specific for MPEG-2 video may be used, in particular the video stream descriptor, the IBP descriptor and the STD descriptor, as discussed in the following paragraphs.

Figure 14.2 depicts the data structure of the video stream descriptor, the IBP descriptor and the STD descriptor. The name video descriptor reflects the situation at the time this descriptor was defined: it applies to MPEG-1 video and MPEG-2 video. The same applies to the STD descriptor, while the IBP descriptor can also be used for MPEG-4 video. These descriptors are not used for video streams coded according to the next generation video codecs.

The video descriptor and the IBP descriptor can be used to inform receivers on certain characteristics of the coded video stream, based on which a receiver may determine whether the associated video stream can be decoded. For example, a receiver may be designed to only support either 50 or 60 Hz based video of a certain profile and level. Without accessing the video stream itself, the receiver may conclude from the information provided in the descriptor that the associated video is beyond the decoding capabilities of the receiver, upon which the receiver can, if so desired, inform the user accordingly. This prevents undesired complications; for example in case a received video stream has a much higher bitrate than can be processed by the decoder, parsing the video stream for finding information on the profile and level may fail.

The first bit encountered after the descriptor-length is the multiple-frame-rate-flag that signals whether a single or more than a single frame rate is applied in the video stream. This flag

Table 14.4 Parameters signalling carriage of MPEG-2 video

Stream-ID	User-selectable from the range 0xE0 to 0xEF
Stream-type	0x02: MPEG-2 video (same value as for constrained parameter MPEG-1 video)
Descriptor-tag	0x02 (02): video stream descriptor
	0x12 (17): STD descriptor
	0x11 (18): IBP descriptor

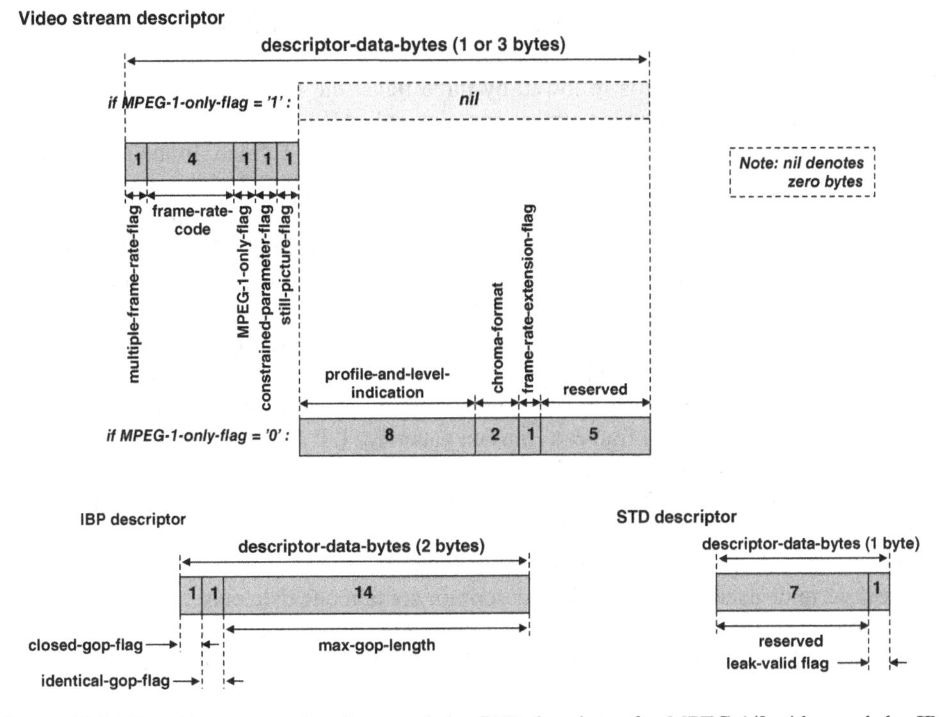

Figure 14.2 The video stream descriptor and the STD descriptor for MPEG-1/2 video and the IBP descriptor for MPEG-1/2/4 video

is followed by the frame-rate-code that specifies the applied frame rate in the case of a single frame rate, or a series of allowed multiple frame rates (see Table 14.5). For example, video source material may be partly shot with a video camera and partly from a movie; in such case the video stream may be a concatenation of a series of video sequences, each with the frame rate

Table 14.5 Frame-rate-code values

Encoded frame-rate-code value (binary)	Indicated frame rate	Other permitted frame rate(s) if multiple-frame-rate-flag '1'
0000	Forbidden	
0001	23.976	—
0010	24.0	23.976
0011	25.0	—
0100	29.97	23.976
0101	30.0	23.976, 24.0, 29.97
0110	50.0	25.0
0111	59.94	23.976, 29.97
1000	60.0	23.976, 24.0, 29.97, 30.0, 59.94
1001–1111	Reserved	

of either the camera or the movie. The frame-rate-code is encoded in the same way as the same field in the MPEG-2 video sequence header.

The frame-rate-code field is followed by three flags; the first flag is the MPEG-1-only-flag, signalling whether the video stream contains only MPEG-1 coded video or possibly a mixture of MPEG-2 video and constrained parameter MPEG-1 video. In the latter case, the three flags are followed by a byte specifying the applied MPEG-2 video profile and level, chroma format and usage of the frame-rate-extension fields. The second flag signals the absence or potential presence of unconstrained MPEG-1 video, while the still-picture-flag signals whether the video stream contains still picture data only or possibly a mix of moving video and still picture data.

The data structure of the descriptor-data-bytes of the IBP descriptor is also presented in Figure 14.2. This descriptor provides information on the usage of some parameters for the coding of the video coding relevant for random access, such as the presence of a group of picture headers prior to each I frame, a constant pattern of I, P and B frames and the maximum number of B and P frames between any two consecutive I frames. The usage of I, P and B frames in MPEG-1, MPEG-2 and MPEG-4 video is very similar; therefore the IPB descriptor can also be used for MPEG-4 video. In some cases a constant IBP pattern may be beneficial to implement the fast forward play back mode.

The descriptor-data-bytes of the STD descriptor contain one byte only, with seven reserved bits followed by the one-bit leak-valid flag. The STD descriptor is used for signalling the transfer mode between the buffers MB_n and EB_n in the T-STD. If the STD descriptor is present with the leak-valid flag set to '0', and if the vbv-delay fields in the associated video stream do not have the value 0xFFFF,[1] then the vbv-delay method is used; otherwise the leak method is applied.

14.3.2 MPEG-2 (BC) Audio

An MPEG-2 audio [5] encoder produces a base stream and an optional extension stream, as discussed in Section 3.3.2. In the STD, the base stream and the extension stream are each processed in a specific decoding branch of the STD.

Depending on application requirements, the base stream may be signalled by a stream-type value as a MPEG-1 or a MPEG-2 audio stream. The extension stream is signalled as a MPEG-2 audio stream (see Table 14.6). The hierarchical relationship between both streams can be expressed by means of hierarchy descriptors, as discussed in Section 13.3.4.1. The hierarchy

Table 14.6 Parameters signalling a MPEG-2 audio base or extension stream

Stream-ID	User-selectable from the range 0xC0 to 0xDF
Stream-type	0x03: MPEG-1 audio (compatible)
	0x04: MPEG-2 audio
Descriptor-tag	0x03 (03): audio stream descriptor

[1] The vbv-delay value 0xFFFF signals that the vbv-delay is undefined. The MPEG-2 video standard requires the vbv-delay for each picture throughout an entire video sequence to be either valid or undefined.

Audio stream descriptor

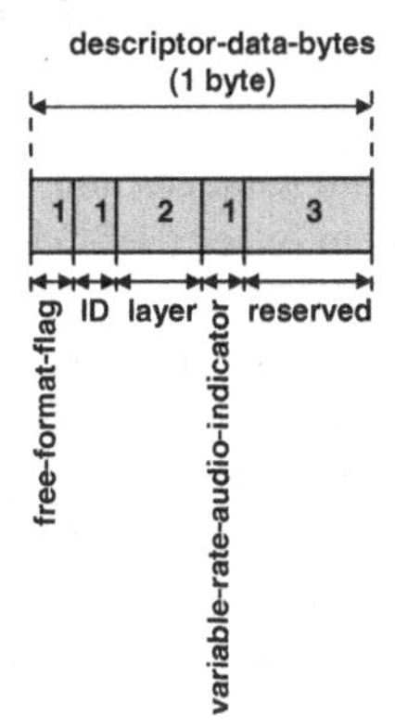

MPEG-2 AAC audio descriptor

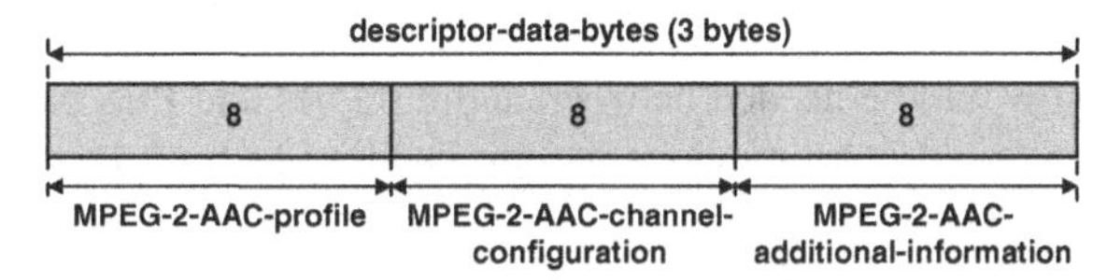

Figure 14.3 Structure of the descriptor-data-bytes of the MPEG-1/2 audio stream descriptor and MPEG-2 AAC audio descriptor

descriptor associated to the base stream signals the hierarchy-type 15 (base layer) and assigns a unique hierarchy-layer-index number to the base stream. The hierarchy descriptor associated to the extension stream signals the hierarchy-type 5 (extension bitstream), while encoding the hierarchy-layer-index number of the base stream as the hierarchy-embedded-layer-index value, thereby signalling that the base stream must be accessed prior to decoding the extension stream. See also Table 13.5 and the description of these index fields in Section 13.3.4.1.

Further information on the MPEG-1 or MPEG-2 audio stream may be provided by the audio stream descriptor. Despite its generic name, the audio descriptor does not apply to any audio format other than MPEG-1 and MPEG-2 audio. Figure 14.3 depicts the data structure of the audio stream descriptor. Three flags indicate: (1) whether the free-format is applied, (2) the ID of the coding algorithm (a '1' signals MPEG-1 audio compatible frames, a '0' signals use of the lower sampling frequencies) and (3) whether the bitrate may change between consecutive audio frames. Furthermore, the two-bit layer specifies the highest layer applied in the audio stream.

When audio is coded using a base stream and an extension stream, the corresponding audio frames in both layers are required to have identical PTS time stamps. Each such time stamp may be an explicitly coded PTS time stamp for an audio frame or a time stamp value that is implicitly derived from the coded PTS value for a previous audio frame, the audio frame duration and the number of audio frames since the last coded PTS value.

Note 14.1 The Evolving Need for an Additional Transport Syntax for MPEG Audio

When the initial work on the MPEG-2 system standard and the associated compliancy document was finished, most MPEG-2 system experts left MPEG for other challenges, such as designing MPEG-2 system based products. At the same time, the MPEG systems group with its new participants focussed on their next objective, the design of a system suitable for the object orientated approach within the MPEG-4 work item. Within that context the MPEG systems group discussed the interface between the elementary stream layer and the systems layer and concluded that framing structures for access units, such as the sync word of an audio frame and the start codes in video, if needed at all, should be defined in the systems layer instead of in the elementary stream layer.

Upon discussing this proposal, the video group disagreed: start codes in video are not only used at access unit level, but also within access units, for example at slice level. However, the audio group agreed to refrain from defining sync words for audio frames in MPEG-2 AAC and in MPEG-4 audio, the standards the MPEG audio group was working on at that time. This decision of the audio group may have been good from a MPEG-4 system perspective, but not for usage in MPEG-2 systems, as without a framing structure, alignment would be needed between audio frames and PES packets. To achieve such an alignment would in many cases require padding and results thereby in an unacceptable loss of efficiency.

To resolve this problem, the audio group decided to define a framing structure for MPEG-2 AAC and for MPEG-4 audio. For use in MPEG-2 systems and other system environments, two optional transport formats were defined:

- The ADTS transport syntax for carriage of MPEG-2 AAC audio; ADTS is specified in the MPEG-2 AAC specification.
- The LATM transport syntax for carriage of MPEG-4 audio; LATM is specified in the MPEG-4 audio standard. Note however that LATM is not always used; for some MPEG-4 audio formats, such as the lossless formats DST, ALS and SLS, instead of using LATM, alignment can be applied between audio frames and PES packets. For very large audio frames such alignment causes only a marginal efficiency loss.

14.3.3 MPEG-2 AAC

A MPEG-2 AAC [6] audio stream is transported across MPEG-2 systems in PES packets. Prior to PES packetization, the ADTS transport syntax specified in the MPEG-2 AAC standard is applied to provide a framing structure for the MPEG-2 AAC audio frames; see Note 14.1 for background on the use of ADTS and other transport syntaxes for MPEG audio. A MPEG-2 AAC audio stream with the ADTS transport syntax is signalled by the parameters provided in Table 14.7. MPEG-2 systems does not support carriage of MPEG-2 AAC streams without the ADTS transport syntax.

For MPEG-2 AAC, the applied decoding branch is discussed in Section 12.5.1, as well as, for up to eight channels, the associated STD parameters. For MPEG-2 AAC with more than

Table 14.7 Parameters signalling a MPEG-2 AAC audio stream

Stream-ID	User-selectable from the range 0xC0 to 0xDF
Stream-type	0x0F: MPEG-2 AAC audio with the ADTS transport syntax
Descriptor-tag	0x2B (43): MPEG-2 AAC audio descriptor

Table 14.8 STD parameters for MPEG-2 AAC with nine or more channels

	T-STD		PES-STD
	Rx_n (bit/s)	BS_n (bytes)	BS_n (bytes)
MPEG-2 AAC with 9–12 channels	8.2944×10^6	12 804	10 800
MPEG-2 AAC with 13–48 channels	33.1776×10^6	51 216	43 200

eight channels, distinction is made between two configurations, one with 9–12 channels and another with 13–48 channels. For both channel configurations the assigned values of Rx_n and BS_n are provided in Table 14.8 without further discussion.

Important information on the coding parameters in a carried MPEG-2 AAC stream can be provided in the MPEG-2 AAC audio descriptor, depicted in Figure 14.3. The descriptor-data-bytes field consists of three bytes, specifying the applied MPEG-2 AAC profile, the applied channel configuration and additional information. The first two are coded as specified in ISO/IEC 13818-7. An encoded value 0x01 of the MPEG-2-AAC-additional-information signals that bandwidth extension data, as specified in the 2006 release of the MPEG-2 AAC specification, is embedded in the stream. An encoded value 0x00 of the MPEG-2-AAC-additional-information signals absence of such data, while all other values are reserved.

14.3.4 MPEG-2 DSM-CC

In MPEG-2 systems some support is defined for playback trick mode operation (see Section 11.10). However, there are many types of digital storage media, many potential locations for such media and many possible networks to connect to those media. To improve the interoperability between decoders and storage media, a standardized interface for communication between decoders and storage media is desirable. An initial step towards defining such interface was undertaken in (the informative) Annex B of the MPEG-2 system specification, called 'Digital Storage Media Command and Control', DSM-CC [7]. However, this annex addresses only some basic DSM control functions for very simple environments, without addressing the various configurations and diverse network environments used in practice.

MPEG clearly recognized the importance of a DSM-CC specification. Annex B was only included as an informative part of the MPEG-2 systems specification, while a much more elaborative effort was started that resulted in MPEG-2 part 6, called 'Extensions for Digital Storage Media Command and Control', also referred to as DSM-CC [8]. In ISO/IEC 13838-6 a large variety of protocols are defined in a manner that abstracts from the underlying network, thereby providing a solution suitable for many network technologies and configurations. ISO/IEC 13838-6 can be applied to browse, select, download and control content, as well as to manage network and application resources. For example, when a MPEG-2 system stream is played from a server in a Video on Demand application, ISO/IEC 13838-6 can be used to

Table 14.9 Parameters values signalling DSM-CC data in MPEG-2 system streams

Stream-type	0x08: DSM-CC according to Annex B of MPEG-2 systems
	0x0A: DSM-CC multi-protocol encapsulation
	0x0B: DSM-CC U-N messages
	0x0C: DSM-CC stream descriptors
	0x0D: DSM-CC sections, any type, including private data
	0x14: DSM-CC synchronized download protocol
Table-ID	0x38: ISO 13818-6 reserved
	0x39: ISO 13818-6 reserved
	0x3A: DSM-CC sections containing multi-protocol encapsulated data
	0x3B: DSM-CC sections containing U-N messages, except download data messages
	0x3C: DSM-CC sections containing download data messages
	0x3D: DSM-CC sections containing stream descriptors
	0x3E: DSM-CC sections containing private data
	0x3F: ISO 13818-6 reserved
Descriptor-tag	0x13 (19): DSM-CC carousel identifier descriptor
	0x14 (20): DSM-CC association tag descriptor
	0x15 (21): DSM-CC deferred association tag descriptor
	0x16 (22): ISO 13818-6 reserved
	0x17 (23): NPT reference descriptor
	0x18 (24): NPT endpoint descriptor
	0x19 (25): Stream mode descriptor
	0x1A (26): Stream event descriptor
Stream-ID	0xF2: DSM-CC data according to annex B of the MPEG-2 system standard or according to ISO 13818-6

control the playback mode, with the use of NPT, the normal play time concept, whereby the time progresses in a way synchronized with the speed of the playback. Furthermore, in a MPEG-2 transport stream that is broadcast to a large population of receivers, a DSM-CC object carousel can be implemented to download data and content of interactive applications to the receivers. Typically, the data is repeated regularly in the carousel, so as to ensure that receivers can access the objects conveniently upon random access. DSM-CC also provides a consistent set of protocols for communication between the receiver and a broadcast server across a return channel, if so desired.

ISO/IEC 13818-6 is deployed widely on the market place, but its specifics are beyond the scope of this book. The MPEG-2 system specification allows to carry DSM-CC defined data in a MPEG-2 system stream in sections, in descriptors and in PES packets. In the MPEG-2 system standard various table-ID, stream-type, descriptor-tag and stream-ID values are reserved for usage by DSM-CC (see Table 14.9). For further information see ISO/IEC 13818-6.

14.3.5 MPEG-2 System Stream

As discussed in Section 11.12, it is possible to carry the content of a program stream across a transport stream. While the PES packets with audio and video are carried in a transport stream

Table 14.10 Parameters signalling carriage in a transport stream of a stream with data from a program stream that cannot serve as a meaningful program element within a transport stream

Stream-ID	0xF9: ancillary data, such as the following MPEG-2 streams that may occur in a program stream: program stream map, padding stream, ECM stream, EMM stream and program stream directory
Stream-type	0x0E: auxiliary data

Table 14.11 Parameters values signalling IPMP data in MPEG-2 system streams

Stream-type	0x1A: IPMP stream, as defined in ISO/IEC 13818-11
	0x7F: (MPEG-4) IPMP stream
Table-ID	0x07: IPMP control information section defined in ISO 13818-11
Descriptor-tag	0x29 (41): IPMP descriptor, as defined in ISO/IEC 13818-11
Stream-ID	0xFD: extended-stream-ID
Stream-ID-extension	'000 0000': IPMP control information stream defined in ISO 13818-11
	000 0001': IPMP stream defined in ISO 13818-11

as regular program elements, PES packets in the program stream that cannot be conveyed in a meaningful way as a program element, can be carried in a transport stream as an ancillary stream, as discussed in Section 11.12. Such streams are signalled, as indicated in Table 14.10. Note that the same mechanism can be used to convey a padding stream from a MPEG-1 system stream (see Section 14.2.3).

14.3.6 MPEG-2 IPMP

Within the MPEG-4 and MPEG-21 work items, MPEG has been working on an architecture for Digital Right Management, called IPMP (Intellectual Property Management and Protection). To offer an alternative for the proprietary conditional access systems discussed in Section 10, the architecture and concept of the MPEG-4 IPMP Extension (MPEG-4 IPMP-X) was mapped to MPEG-2 system. The result is specified in MPEG-2 part 11, ISO/IEC 13818-11 [9]. The use of MPEG-2 IPMP requires the transport of MPEG-2 IPMP data and IPMP control information data within MPEG-2 system streams, either in PES packets or in sections; further information may be provided in an IPMP descriptor. ISO/IEC 13818-11 specifies the format of the IPMP descriptor, the syntax and semantics of the IPMP and IPMP control information data, and requirements for the processing of the IPMP data. Carriage of IPMP data is identified by the parameter values listed in Table 14.11. When carried in PES packet, the stream-ID-extension field is used, as indicated by a coded stream-ID value of 0xFD.

14.4 (ITU-T Rec.) H.222.1

At the time of the joint development by ISO/IEC and the ITU-T of the MPEG-2 system standard, the intention was to apply MPEG-2 system streams in ATM networks. As discussed in Section 9.4.3, the transport packet size was chosen so that each transport packet can be conveyed conveniently and efficiently in ATM cells. Within the ITU, the MPEG-2 system

Table 14.12 Parameters values signalling a ITU-T H.222.1 defined stream

Stream-type	0x09: ITU-T H.222.1 defined
Stream-ID	0xF4: Rec. ITU-T H.222.1 type A
	0xF5: Rec. ITU-T H.222.1 type B
	0xF6: Rec. ITU-T H.222.1 type C
	0xF7: Rec. ITU-T H.222.1 type D
	0xF8: Rec. ITU-T H.222.1 type E

specification is referenced as ITU-T Recommendation H.222.0. The use of MPEG-2 systems in ATM environments is specified in ITU-T Recommendation H.222.1 [10].

In ATM environments it should not only be possible to carry MPEG defined content formats, but also ITU defined content. To allow such carriage, in the MPEG-2 system standard room is created to convey five types of so-called Rec. ITU-T.222.1 streams; the format of each type is defined in H.222.1. For example, type A indicates a stream of class video with a format, such as H.261 or H.263, defined by a ITU descriptor specified in H.222.1. Similarly, a G.711, G.222, G.723 or G.728 audio stream can be conveyed as a Rec. ITU-T.222.1 type B stream of class audio. Type C indicates a stream of class data, such as a H.245 or a T.120 sub-channel. Finally, type D indicates a stream of any class, while the use of type E is reserved for future use. Carriage of such streams in a MPEG-2 system stream is signalled by the parameters provided in Table 14.12. Further information on the carriage of ITU-T H.222.1 defined streams is provided in H.222.1.

14.5 MHEG

During the development of the MPEG-1 and MPEG-2 specifications, the MHEG specification was developed by the Multimedia and Hypermedia Experts Group working under the same ISO/IEC umbrella as MPEG. The resulting MHEG standards address the presentation of multimedia information, and are published as various parts of ISO/IEC 13522 [11]. For example, MHEG-5, ISO/IEC 13522-5, specifies a declarative programming language which can be used in applications to describe a presentation of text, images, audio and video. Within MPEG-2 systems the option has been created to convey MHEG streams in PES packets. For the signalling of a MHEG stream see Table 14.13.

Requirements and constraints for the decoding of a carried MHEG stream are not specified by MPEG-2 systems and may be an issue for applications or application standardization bodies. In this context it should be noted that also other protocols, such as DSM-CC may be involved in the implementation of MHEG applications.

Table 14.13 Parameters values signaling a MHEG stream

Stream-type	0x07: ISO/IEC 13522 (MHEG) defined
Stream-ID	0xF3: ISO/IEC 13522 (MHEG) stream

14.6 MPEG-4

As shown in Section 8.3, the MPEG-4 standard consists of many parts, whereby each part may be a suite of standards in itself. For example, part 3 of MPEG-4 specifies a series of audio compression standards, as discussed in Section 3.3.2. As several MPEG-4 standards are irrelevant for usage in MPEG-2 systems, this chapter discusses only the MPEG-4 standards for which support is defined in the MPEG-2 system standard. Sections 14.6.1–14.6.3 discuss the support of MPEG-4 visual, audio and timed text; and Section 14.6.4 discusses the support of MPEG-4 systems. It should be noted that support for MPEG-4 systems also includes support of elementary streams, but carried by the MPEG-4 system layer, as discussed briefly in Section 14.6.4.

14.6.1 MPEG-4 Visual

Due to the object orientated approach followed in MPEG-4, there is a slight difference in terminology: access units may be referenced as objects. For example, instead of a video access unit, the term visual object is used. For carriage over MPEG-2 systems, each MPEG-4 visual [12] stream is required to be self-contained, that is, it contains all information required for its decoding. Consequently, the stream must contain (so-called) Visual Object Sequence Headers, Visual Object Headers and Visual Object Layer Headers.

Carriage of a MPEG-4 visual stream over MPEG-2 systems is very similar to carriage of a MPEG-2 video stream, using PES packets only. However, in the T-STD model there is an important difference. Instead of the input buffers MB_n and EB_n, only a single buffer B_n is used (see Figure 14.4). As a consequence, when a visual object is instantaneously removed from B_n

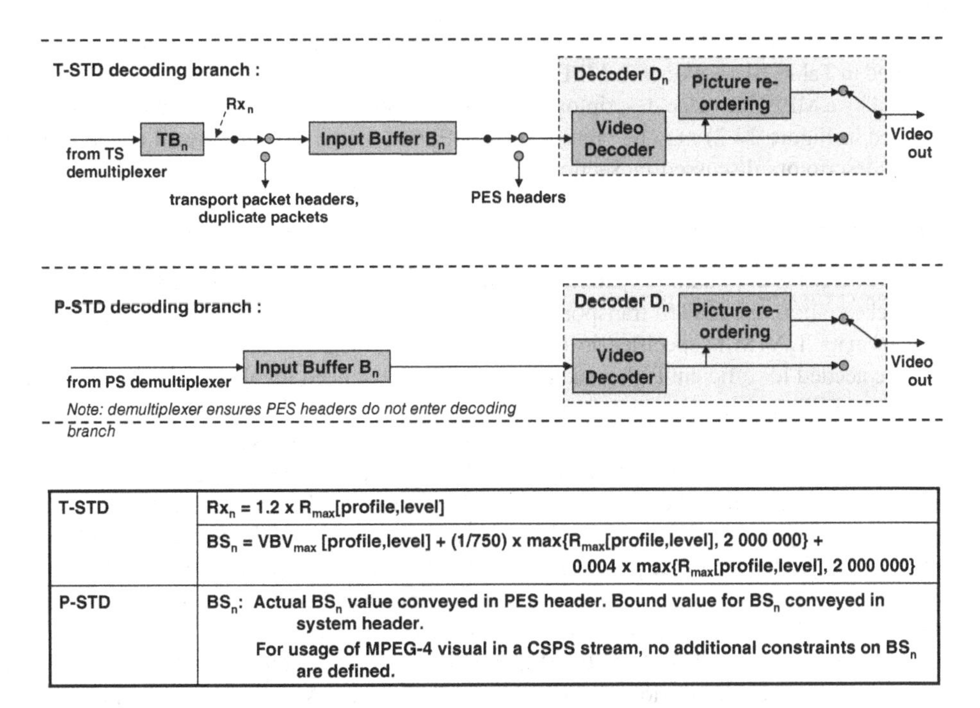

T-STD	Rx_n = 1.2 x R_{max}[profile,level]
	BS_n = VBV_{max} [profile,level] + (1/750) x max{R_{max}[profile,level], 2 000 000} + 0.004 x max{R_{max}[profile,level], 2 000 000}
P-STD	BS_n: Actual BS_n value conveyed in PES header. Bound value for BS_n conveyed in system header. For usage of MPEG-4 visual in a CSPS stream, no additional constraints on BS_n are defined.

Figure 14.4 STD decoding branches and STD parameters for MPEG-4 visual

Table 14.14 Parameters signalling carriage of MPEG-4 visual streams

Stream-ID	User-selectable from the range 0xE0 – 0xEF
Stream-type	0x10: MPEG-4 visual
Descriptor-tag	0x1B (27): MPEG-4 visual descriptor

and decoded, all PES header bytes included in the visual object or immediately preceding it are removed from B_n at the same time.

The STD parameters are determined in a way very similar to those for MPEG-2 video and AVC (see Section 12.5.2). BS_{dec} is defined to be equal to the VBV_{max} value for the used profile and level. For MPEG-4 visual profiles and levels for which no VBV_{max} value is specified, the value of BS_n is not defined by MPEG and is user defined instead. For BS_{oh} and BS_{mux} the general rules of thumb are followed, but to ensure sufficient buffer room in the case of low bitrates, the maximum bitrate is constrained to be at least 2×10^6 bits/s. In formulas:

$$\begin{aligned} BS_n &= B_{dec} + B_{oh} + B_{mux} \\ &= VBV_{max}[\text{profile, level}] + (1/750) \times \max\{R_{max}[\text{profile, level}], 2 \times 10^6\} \\ &\quad + 0.004 \times \max\{R_{max}[\text{profile, level}], 2 \times 10^6\} \end{aligned}$$

Note that due to the use of a single buffer B_n, the left-over room VBV_{max}[profile,level] minus the size of the actually applied VBV buffer is fully available for multiplexing. For carriage in a PES stream of a MPEG-4 visual stream no explicit BS_n value is defined and the PES-STD is therefore not included in Figure 14.4.

The available stream-ID and stream-type values for signalling a MPEG-4 visual stream are presented in Table 14.14. For each MPEG-4 visual stream the applied profile and level must be provided by a MPEG-4 video descriptor (depicted in Figure 14.5). In addition, a IBP descriptor (depicted in Figure 14.2) may be present for a MPEG-4 visual stream, as well as the general content descriptors discussed in Section 13.3.4.

14.6.2 MPEG-4 Audio

To convey a MPEG-4 audio [13] stream in MPEG-2 systems, PES packets are used. Prior to PES packetization, the LATM transport syntax must be applied for most, but not all MPEG-4 audio formats. LATM is specified in the MPEG-4 audio standard and provides the framing structure needed for efficient PES packetization without the need for alignment between audio frames and PES packets. For some more background on the need for an additional transport syntax see Section 14.3.3 and Note 14.1. For transporting the lossless MPEG-4 audio formats DST, ALS and SLS in PES packets, the use of LATM is not mandated. If LATM is not applied, then the audio frames must be aligned with the PES packets; in that case, fragmenting an audio frame over multiple PES packets is allowed, but the first byte of an audio frame must be the first byte of a PES payload, and the presence or absence of such first byte must be signalled by the data-alignment-indicator in the PES header. The stream-ID and stream-type values assigned to usage of MPEG-4 audio with and without LATM are presented in Table 14.15.

The decoding in the STD of MPEG-4 audio streams up to eight channels is discussed in Section 12.5.1, except for the lossless MPEG-4 audio formats. All MPEG-4 audio formats use the type of decoding branch with input buffer B_n depicted in Figure 12.15, but the parameter values for decoding in the STD obviously differ.

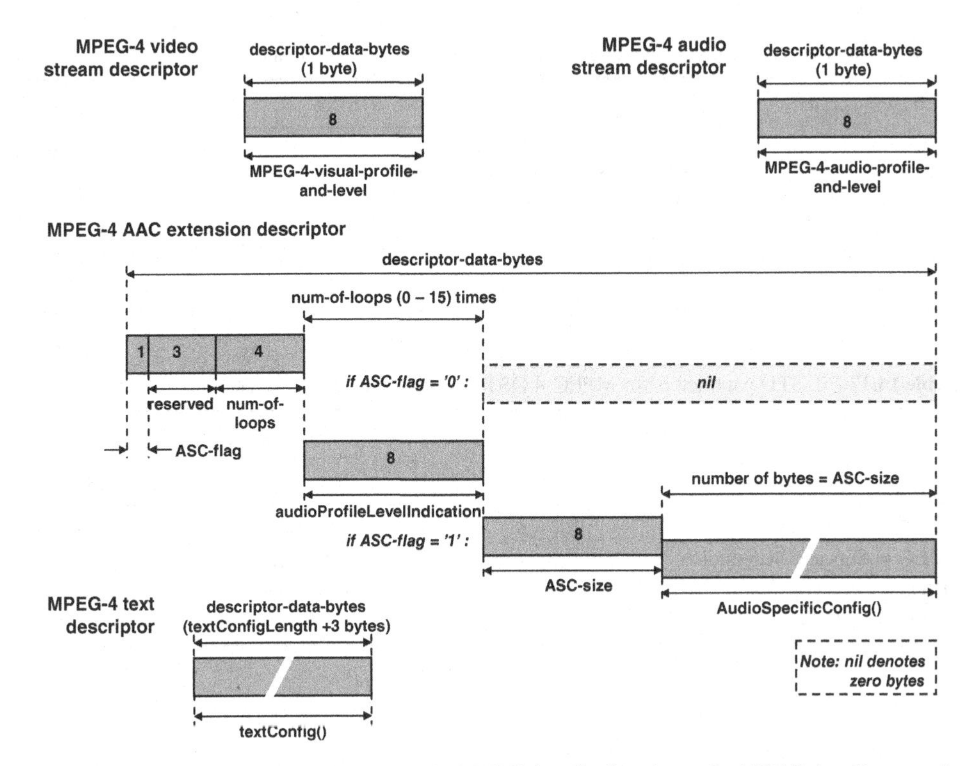

Figure 14.5 The MPEG-4 video descriptor, the MPEG-4 audio descriptor, the MPEG-4 audio extension descriptor and the MPEG-4 text descriptor

The STD parameters for MPEG-4 audio formats with nine or more channels are given in Table 14.16; the values are the same as for MPEG-2 AAC with nine or more channels. Without further comments the values specified in the MPEG-2 system standard for the lossless MPEG-4 DST, ALS and SLS audio formats with N channels are presented in Tables 14.17 and 14.18 for the T-STD and the PES-STD, respectively.

For each MPEG-4 audio stream the MPEG-4 audio descriptor must be present, but also the MPEG-4 audio extension descriptor may be present. Figure 14.5 presents the structure of the descriptor-data-bytes of both descriptors. The MPEG-4 audio descriptor contains the eight-bit MPEG-4-audio-profile-and-level field of which the coding is specified in the MPEG-2 system specification. The following MPEG-4 audio profiles can be signalled by this field:

- Main profile at four levels,
- Scalable profile at four levels,

Table 14.15 Parameters signalling a MPEG-2 audio base or extension stream

Stream-ID	User-selectable from the range 0xC0 to 0xDF
Stream-type	0x11: MPEG-4 audio with the LATM transport syntax defined in ISO/IEC 14496-3
	0x1C: MPEG-4 audio without any additional transport syntax
Descriptor-tag	0x1C (28): MPEG-4 audio descriptor

Table 14.16 STD parameters for MPEG-4 audio with nine or more channels

	T-STD		PES-STD
	Rx_n (bits/s)	BS_n (bytes)	BS_n (bytes)
MPEG-4 audio with 9–12 channels, except DST, ALS and SLS	8.2944×10^6	12 804	10 800
MPEG-4 audio with 13–48 channels, except DST, ALS and SLS	33.1776×10^6	51 216	43 200

Table 14.17 T-STD parameters for MPEG-4 DST, ALS and SLS with *N* channels

		T-STD	
		Rx_n (bits/s)	BS_n (bytes)
MPEG-4 DST-64 audio and	$N \leq 8$	30×10^6	1 600 000
MPEG-4 ALS and SLS audio	$N > 8$	120×10^6	$200\,000 \times N$
MPEG-4 DST-128 audio	120×10^6	$400\,000 \times N$	
MPEG-4 DST-256 audio	120×10^6	$800\,000 \times N$	

Table 14.18 Remaining STD parameters for MPEG-4 DST, ALS and SLS with *N* channels

	PES-STD
	BS_n (bytes)
MPEG-4 DST-64 audio	$5\,000 \times N$
MPEG-4 DST-128 audio	$10\,000 \times N$
MPEG-4 DST-256 audio	$20\,000 \times N$
MPEG-4 ALS and SLS audio	$33\,000 \times N$

- Speech profile at two levels,
- Synthesis profile at three levels,
- High quality audio profile at eight levels,
- Low delay audio profile at eight levels,
- Natural audio profile at four levels,
- Mobile audio internetworking profile at six profiles,
- AAC profile at four levels,
- High efficiency AAC profile at four levels,
- High efficiency AAC v2 profile at four levels.

Each profile and level combination is signalled by a specific value of the MPEG-4-audio-profile-and-level field. The value 0xFF signals that the audio profile and level is not specified in this MPEG-4 audio descriptor, and in that case the MPEG-4 audio extension descriptor is required to be present as well. See Note 14.2 for some background on this approach.

Note 14.2 Use of the MPEG-4 Audio Extension Descriptor

The coding of the MPEG-4-audio-profile-and-level field is specified in the MPEG-2 system standard. As a consequence, for each new MPEG-4 audio profile and level combination evolving over time a new MPEG-4-audio-profile-and-level value was to be defined in an amendment to MPEG-2 systems. This was rather impractical and therefore the MPEG-4 audio extension descriptor was specified to carry the audioProfileLevelIndication and the AudioSpecificConfiguration(), the coding of which is specified in the MPEG-4 audio standard for each new audio profile or level, without requiring a MPEG-2 systems update.

The MPEG-4 audio extension descriptor contains the audioProfileLevelIndication and optionally the AudioSpecificConfiguration(), two data structures used in MPEG-4 audio for signalling the audio profile and level, and for providing additional basic information, such as the use of embedded MPEG Surround data. The presence of the AudioSpecificConfiguration() field is indicated by the ASC-flag, and its length in bytes by the value encoded in the ASC-size field (see Figure 14.5). A single MPEG-4 audio stream may comply to more than one audio profile and level; therefore the descriptor is designed to carry at most 15 audioProfileLevelIndication fields, if so indicated by the ASC-flag, followed by the AudioSpecificConfiguration().

14.6.3 MPEG-4 Timed Text

MPEG-4 Timed Text, specified in part 17 of the MPEG-4 standard [14] is carried over MPEG-2 systems in PES packets. As discussed in Section 4.2, a MPEG-4 timed text stream consists of a concatenation of TTUs. Each MPEG-4 timed text stream is carried in PES packets, whereby the TTUs are aligned with the PES payload, that is the first byte of the PES payload must be the first byte of a TTU header. No alignment with text samples is required: the first TTU in the PES payload may contain a non-first fragment of a text sample.

Table 14.19 presents the assigned stream-type value and the 14 stream-ID-extension values that can be used to convey MPEG-4 timed text streams, The use of stream-ID-extension is indicated by an encoded stream-ID value 0xFD (see also Section 13.2).

The STD decoding branches for MPEG-4 timed text in the T-STD and P-STD are depicted in Figure 14.6. The Hypothetical Text Decoder HTD defined in MPEG-4 part 17 and discussed in Section 4.2 and Figure 4.4 operates as decoder D_n. Both branches use a single buffer B_n; the data removal out of B_n follows the data delivery schedule to the HTD: the data enters the HTD at a rate of R bits/s. If the text sample buffer, B_{ts} in Figure 14.6, is full, then $R=0$, else

Table 14.19 Parameters values signalling a MPEG-4 timed text stream

Stream-type	0x1D: MPEG-4 text	
Stream-ID	0xFD: extended-stream-ID	
	Stream-ID-extension	User-selectable from the range '000 0010' – '000 1111'
Descriptor-tag	0x2D (45): MPEG-4 text descriptor	

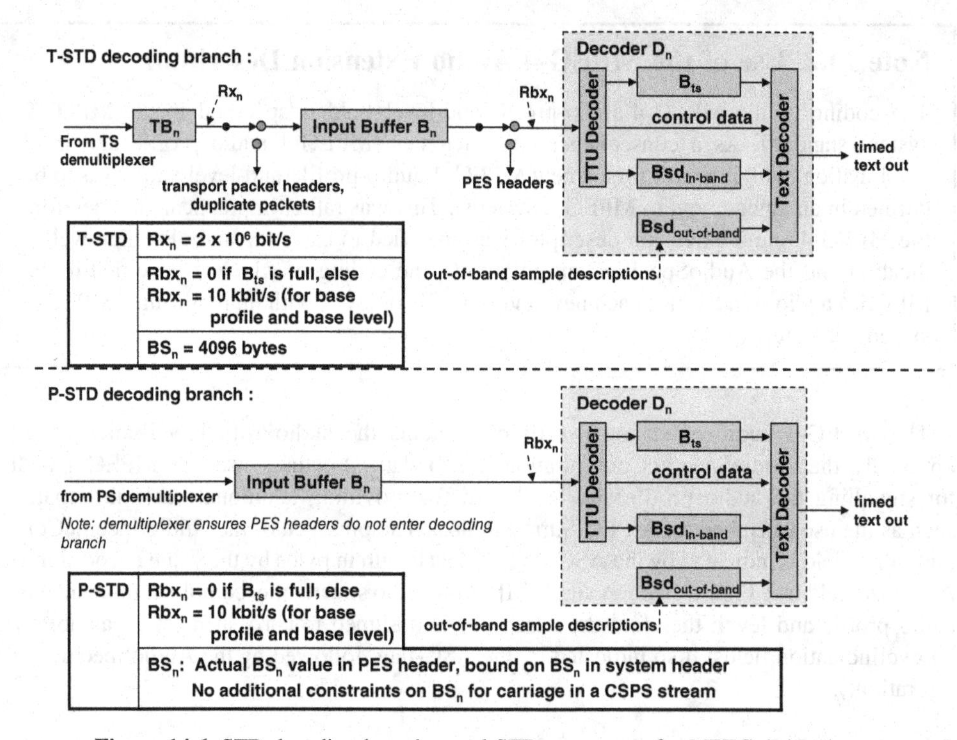

Figure 14.6 STD decoding branches and STD parameters for MPEG-4 timed text

$R = 10$ kbits/s for the base profile and base level. Hence, the instantaneous decoding is not performed at the output of D_n, but instead by the text decoder within the HTD.

Information on the MPEG-4 timed text stream must be provided in the MPEG-4 text descriptor. Figure 14.5 shows that the MPEG-4 text descriptor carries the textconfig() structure defined in MPEG-4 part 17, consisting of an eight-bit textFormat field, a 16-bit textConfig-Length, followed by the formatSpecificTextConfig() structure also defined in part 17 of MPEG-4. Hence, the number of bytes of textConfig() is equal to the coded textConfigLength value + 3. Obviously, as long as (in future) no additional fields are defined for use in MPEG-4 text descriptors, then the same number of bytes is found from the descriptor-length field.

14.6.4 MPEG-4 Systems

The MPEG-4 work item addresses coding of audio-visual objects. An object may contain for example video, still images, audio, speech, text, or synthetic content produced by a computer, and may need streaming data, conveyed in one or more elementary streams. Within this work item the MPEG systems group developed a framework for creating, transporting, decoding and presenting interactive object orientated applications, in which the objects are composed in audiovisual scenes. The user may interact with the scene by positioning and controlling the media objects, depending on the flexibility provided by the application. Important objective is to ensure that the applications are fully independent of the underlying transport media.

For the above purpose, MPEG-4 systems specified in part 1 of the MPEG-4 standard [15] a variety of tools, such as a method to describe the scenes and the associated objects, a language for composing objects within a scene and a method to transport and synchronize objects. Initially also the MPEG-4 file format was specified in part 1, but this specification is moved later to other parts of MPEG-4 (see Section 16.3).

To transport and synchronize objects, MPEG-4 system defines the Synchronization Layer, using SL packets to convey data in SL packetized streams; SL packets can be concatenated into a FlexMux stream. Also a MPEG-4 object time base is defined. Carriage of a scene and its objects does not only require to convey the objects, but also a number of MPEG-4 defined structures, such as an initial object descriptor, a variable number of object descriptor streams and scene description streams.

To fully achieve transport layer independency, all MPEG-4 scene information and all content used in a MPEG-4 scene must use the MPEG-4 timing and synchronization methods and must be contained in SL packets. Hence, when an elementary stream, carried over MPEG-2 systems, is used within a MPEG-4 scene, then that elementary stream must be carried in SL packets. SL packets are similar to PES packets but differ in important aspects, so that a SL packet cannot be defined as a compatible extension of a PES packet. Hence, this requirement means that two methods will be needed to convey an elementary stream, one in the payload of PES packets, and another in SL packets as an intermediate layer, using signalling methods from MPEG-4 systems.

The requirement for the use of MPEG-4 SL packets prevents a backward compatible introduction of MPEG-4 scenes with its associated content in MPEG-2 systems. Any stream carried in PES packets within a program in a transport stream or within a program stream cannot be used in a MPEG-4 scene, while an elementary stream used in a MPEG-4 scene cannot be decoded in a legacy MPEG-2 system decoder lacking knowledge of the SL layer. Which looks like a rather doubtful strategy for market introduction of MPEG-4 system technology in MPEG-2 system based applications.

Nevertheless, it was decided to support carriage of 'full fledge' MPEG-4 systems over MPEG-2 system streams, but carriage of individual MPEG-4 elementary streams in the payload of PES packets was strictly separated from MPEG-4 system support. Two methods were defined for carriage of SL and FlexMux packets: (1) in the payload of a PES packet and (2) in the payload of a section. The used stream-ID, table-ID, stream-type and descriptor-tag values are listed in Table 14.20.

The integration of MPEG-4 system technology in MPEG-2 systems was not straightforward. After lengthy and challenging discussions, often between pragmatic and more idealistic system experts, the data structure of PES packets and sections carrying SL and FlexMux packets was defined and associated descriptors were specified. In the payload of a PES packet or a section either a single SL packet is carried, or an integer number of FlexMux packets, indicated by the presence of an associated SL descriptor or FMC descriptor.

Furthermore the MPEG-2 STD models and the System Decoding Model specified in MPEG-4 systems were integrated to specify the decoding of SL packetized streams and FlexMux streams conveyed in a MPEG-2 system stream, taking into account the use of MPEG-4 object time bases, which are required to be locked to the MPEG-2 STC. Under certain conditions, the relationship between the STC time and the time of the object time base can be conveyed by means of a coded PTS in the PES packet carrying a SL packet or FlexMux packets. Several complications are addressed; for example, due to the use of single stream-ID

Table 14.20 Parameter values signalling carriage of MPEG-4 system data

Stream-type	0x12: MPEG-4 SL packetized stream or FlexMux stream carried in PES packets
	0x13: MPEG-4 SL packetized stream or FlexMux stream carried in MPEG-4 sections
Table-ID	0x04: MPEG-4 section with scene description data contained in SL or FlexMux packets
	0x05: MPEG-4 section with object descriptor data contained in SL or FlexMux packets
Descriptor-tag	0x1D (29): IOD descriptor
	0x1E (30): SL descriptor
	0x1F (31): FMC descriptor
	0x20 (32): External-ES-ID descriptor
	0x21 (33): MuxCode descriptor
	0x22 (34): FmxBufferSize descriptor
	0x23 (35): MultiplexBuffer descriptor
	0x2C (44): FlexMux-timing descriptor
Stream-ID	0xFA: MPEG-4 SL packetized stream
	0xFB: MPEG-4 FlexMux stream

values for signalling a SL packetized or a FlexMux stream, in a program stream only one SL packetized stream and one FlexMux stream can be carried.

Describing the support of MPEG-4 systems in detail, requires extensive introduction of MPEG-4 system technology. However, at the market place the MPEG-4 system technology did not became the success the MPEG-4 system experts were hoping for. Business cases for interactive applications in MPEG-2 system environments evolved only slowly, while alternative internet technology with similar features as MPEG-4 systems became available at the same time, and with less concerns on the conditions under which the technology would be licensed. Also the initial perception of MPEG-4 technology did not help (see Note 14.3). In other words, the incompatibility of the MPEG-4 system approach discussed above was not the only obstacle for MPEG-4 system usage. Therefore in this book the data structures of the MPEG-2 descriptors defined for MPEG-4 systems and listed in Table 14.20 are discussed only briefly, while further details, including the integrated models for the decoding of SL and FlexMux streams, are considered out of scope. Figure 14.7 presents the MPEG-4 descriptors.

The IOD descriptor carries the InitialObjectDescriptor(), the structure that allows access to the scene and object description streams, which provide further information about the scene and its content. Also an IOD label is provided (multiple IOD descriptors may be associated to a program) as well as an indication whether the label is unique within a program stream, or within a program in a transport stream, or within a transport stream.

The SL and FMC descriptors assign a ES-ID value to a SL packetized stream and to each channel in a FlexMux stream, respectively. The assigned ES-ID value identifies a stream within a scene; a stream external to the scene can be referenced by means of the external-ES-ID descriptor, for example an elementary stream can be associated with an IPMP stream. The Muxcode, FmxBufferSize and FlexMuxTiming descriptors convey for a FlexMux stream: (1) the MuxCodeTableEntry() that carries information on the configuration of the FlexMux stream, (2) the FlexMux buffer sizes carried in DefaultFlexMuxBufferDescriptor() and by zero or more FlexMuxBufferDescriptor() structures and (3) information on the timing parameters and the size of timing parameter fields in a FlexMux stream. Finally, the

Note 14.3 The Initial Up-Hill Battle on MPEG-4 Technology

After finishing the MPEG-2 system standard and the associated compliancy effort, most MPEG-2 system experts left MPEG to assist in implementing MPEG-2 system based products. At the same time other experts, mainly from research entities, started to participate in MPEG systems, eagerly looking for the next challenge that should make equally much sense to the market place as MPEG-2 systems. But what to do next after the tremendous success of MPEG-2?

The MPEG-4 discussions initially took place as an activity separate from MPEG-2; several potential work items were reviewed, many of which were significantly more research than market orientated, thereby causing MPEG-2 experts a 'virtual headache'; some of them characterized MPEG-4 as a 'dream driven' scientific exercise with much intellectual and little market value. Several jokes were made.

Chad Fogg, email 30 November 1994 on creating critical MPEG-2 video bitstreams for testing video decoders:

"Warning: Delinquent members who fail to timely submit their bitstreams will be exposed to public spanking by Mr Convenor before the next plenary and be forced to attend MPEG-4 all week."

Chadd Fogg, email message 19 August 1994 with Top 10 alternative names for MPEG-4:

Nr 4: "Generic Coding of nearly-still pictures and associated noises for job security at up to about five more years."

Chad Fogg, email message 24 August 199 with Top 10 ways MPEG-4 will improve compression another 8:1 (over MPEG-2):

Nr 10: "With the help of the two M's of Compression: Marketing and Mass-hypnosis."

Chad Fogg, email message 20 August 1994 with Top 10 requirements for MPEG-4:

Nr 3: "Non-causal in every sense: both linear filter theory and market demand!"

When the work on MPEG-2 neared completion, the MPEG-4 effort was turned into actual task forces on coding of audio-visual objects. The MPEG video group started to focus within MPEG-4 on improving the efficiency for lower bitrates and resolutions, as discussed in Section 3.2.3. But the market was busy to exploit MPEG-2 video and there was the risk that a next generation codec, defined a few years after MPEG-2 video, would be perceived as a successor, which would undermine the potential of MPEG-2 video. The activity was therefore not really welcomed by many MPEG-2 video experts. Amongst them was one of the most influential video experts who stated in the press: '*Most of the people I know believe MPEG-4 is a solution in search of a problem. In fact it isn't much of a solution at all.*'

Source of text from emails: Chad Fogg. Reproduced with permission from Chad Fogg.

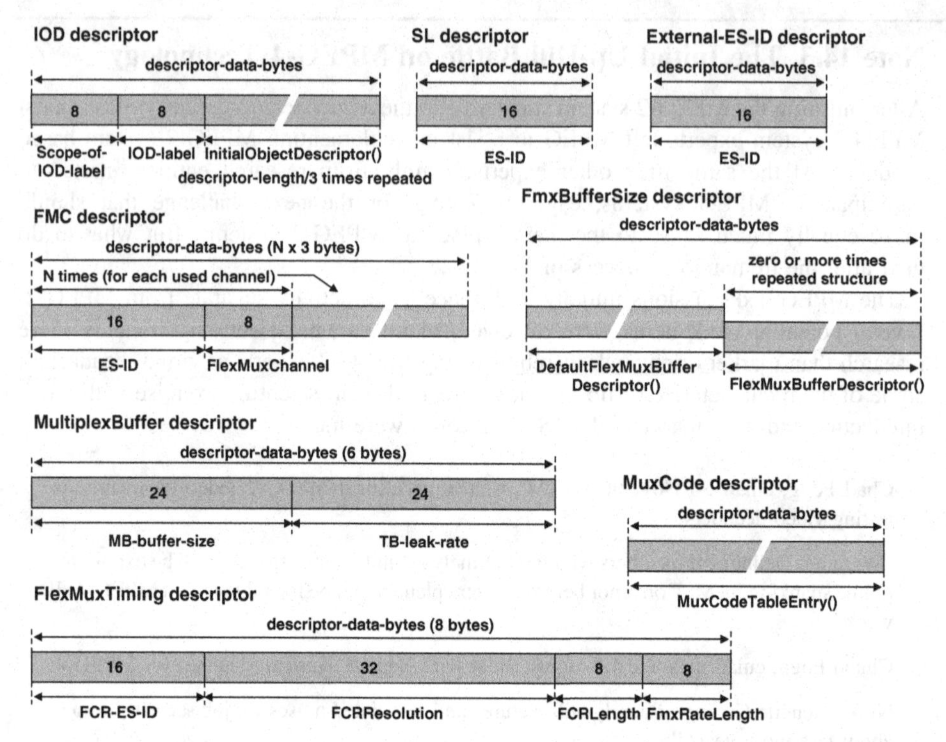

Figure 14.7 The MPEG-2 system descriptors related to MPEG-4 system support

MultiplexBuffer descriptor provides the size of the multiplex buffer MB_n and the leak rate Rx_n of the data out of TB_n into MB_n, as defined in the integrated models for the decoding of SL and FlexMux streams carried in a MPEG-2 system stream. The MultiplexBuffer descriptor must be included for each SL and FlexMux stream, but may be used for other purposes as well.

14.7 AVC

About 10 years after MPEG-2 video, the AVC standard [16], also known as H.264, became available as its successor, as discussed in Section 3.2.3. Due to a typical gain in coding efficiency of 50%, new services started to apply AVC, while existing services continued using MPEG-2 video to serve the installed base of receivers. To upgrade or to replace existing receivers for AVC support is often neither practical nor cost-efficient.

Compared to MPEG-2 video, the AVC standard offers a huge amount of flexibility, not only with respect to the tools to compress the video signal, but also regarding applicable constraints and used parameters. For instance, a wide range of clock frequencies may be used to express AVC internal timing. While most MPEG-2 system applications may not need the provided flexibility, various measures are needed to handle it. For example, ambiguity in the coding of DTS and PTS time stamps is to be prevented for pictures of which the arrival time of the coded data and the presentation time differs by more than the STC period in MPEG-2 systems, equal to about 26.5 h, as discussed in Section 6.6.1.

An AVC stream may contain multiple sequences, each starting with an IDR access unit; IDR stands for Instantaneous Decoding Refresh. An IDR access unit contains a intra coded picture; the presence of an IDR picture signals that the IDR picture and all subsequent pictures can be decoded without making reference to any picture preceding the IDR picture. Though there are various differences, as discussed in Chapter 3 and Figure 3.21, AVC divides pictures in slices and macroblocks, similar as in MPEG-2 video.

For AVC video streams, the Video Coding Layer (VCL) and the Network Abstraction Layer (NAL) are defined. The VCL is designed to efficiently represent the coded video content, while the NAL formats the VCL representation of the video into NAL units, so as to provide a convenient interface for AVC carriage over transport layers. Each NAL unit is effectively a packet containing an integer number of bytes. The first byte of a NAL unit is a header byte, followed by the payload of the NAL unit. The header bytes indicates the type of data carried in the payload. NAL units are classified in VCL and Non-VCL NAL units. Various VCL NAL unit types are defined to carry VCL data, for instance a coded slice of an IDR picture or a coded slice of a non-IDR picture. The Non-VCL NAL units carry associated additional information, for which also various NAL types are specified, for example for carriage of picture parameter sets, sequence parameter sets, and Supplemental Enhancement Information (SEI) messages, while also Non-VCL NAL units can be used for padding with filler data, to indicate the boundary of coded picture data and to indicate the end of a sequence and the end of an AVC stream.

For carriage of AVC over MPEG-2 system streams, NAL units are conveyed in PES packets. Requiring alignment between NAL units and PES packets would result in loss of efficiency. However, non-alignment requires a framing structure to identify the boundaries of the NAL units within the PES payload. For this purpose in AVC the byte stream format for NAL unit streams is defined in Annex B of ISO/IEC 14496-10. In the byte stream format, each NAL unit is preceded by the same start-code prefix (0x000001) as used in MPEG-2 video and in PES packets, optionally preceded by one or more zero bytes. Measures are mandated by the AVC standard to prevent start-code-prefix emulation in the payload of the NAL units and therefore the start-code prefix is a unique identifier of the start of a new NAL unit within the PES packet. The AVC standard requires the first bit of the NAL unit header byte to be ‘0’; hence only seven bits are available to signal the type of data carried in the payload of NAL units. This prevents conflicts with the start code values (185 through 255) assigned to MPEG-2 systems for stream-ID and other usages; see Table 9.1.

Note though that the start-code-prefix is not used in all transport systems; for example in internet protocol/RTP systems, identification of NAL unit boundaries within the packets can be established without the use of the start-code-prefix. Inserting the start-code-prefix in those systems would be a waste of bandwidth and therefore in such systems the NAL units are carried without the start-code-prefix.

To model the video decoding process, the Hypothetical Reference Decoder (HRD) model is used in the AVC standard. The HRD for AVC is the equivalent of the VBV model for MPEG-2 video and operates in a similar way, using instantaneous decoding. The HRD contains the Coded Picture Buffer, CPB, the video decoder and the Decoded Picture Buffer, DPB. At its CPB removal time an AVC access unit is decoded and removed instantaneously from the CPB. For each level in the AVC standard the maximum CPB size, MaxCPB[level], is specified, though the actually used cpb-size may be smaller. As an equivalent of the vbv-delay parameter in MPEG-2 video, the optional parameters initial-cpb-removal-delay and initial-cpb-removal-delay-offset can be used in AVC to specify the timing relationship between bit arrival in the CPB and access unit decoding.

There is a notable difference with the VBV model of MPEG-2 video though. While the vbv-delay in MPEG-2 video defines a single trajectory of the coded video data through the VBV buffer, in AVC multiple trajectories of the coded data through the CPB are possible. Each trajectory is identified by a specific value of the SchedSelIdx parameter. If the option to encode initial-cpb-removal-delay(-offset) values is chosen to be used, then for each encoded SchedSelIdx value a different cpb-size may apply, while the set of initial-cpb-removal-delay(-offset) values specify the bitrate of the AVC stream into the CPB at any instant in time for that SchedSelIdx value. For each used SchedSelIdx value, the AVC compliancy requirements for CPB must be met. For simplicity reasons, in below paragraphs it is assumed that only a single cpb-size is used and at most one set of initial-cpb-removal-delay values.

The DPB differs from the picture re-order buffer applied for MPEG-2 video. After its presentation a MPEG-2 video picture is removed from the re-order buffer, as for example depicted in Figure 6.3. However, an AVC picture may remain available in the DPB after its presentation for prediction of subsequent pictures. Therefore distinction is made between the output time and the removal time of a picture from the DPB. At its DPB output time, an AVC picture is presented. The removal time of a picture from the DPB is controlled by the AVC decoding process; carriage over MPEG-2 systems has neither impact on the dpb-size, nor on the management of the DPB. For each level in the AVC standard the maximum DPB size, MaxDPB[level], is specified.

The HRD is not only used for the NAL unit byte stream format, but also for a 'VCL stream' of VCL NAL units and filler data NAL units, if any. The byte stream format is constructed by adding Non-VCL NAL units and the start-code-prefix to the VCL stream. Therefore the video bitrate for the byte stream format is defined in AVC to be 1.2 times the video bitrate for the VCL stream. Likewise, the CPB size values for the byte stream format are equal to 1.2 times the CPB size values for a VCL stream. Note though that a 20% overhead for just adding the Non-VCL NAL units and the start-code-prefix to the VCL layer is very substantial and typically large enough to accommodate other system purposes as well. To express the 20% overhead, the involved AVC buffer and bitrate parameters are coded in units of 1200 bits or bits/s for the byte stream format NAL unit stream and in units of 1000 bits or bits/s for a VCL stream (see Annex A of the AVC specification).

A number of requirements for AVC apply for transport over MPEG-2 systems. To simplify the ability to detect the boundaries between pictures, access unit delimiter NAL units must be used. AVC allows the delivery of Sequence Parameter Sets (SPS) and Picture Parameter Sets (PPS) by external means. To not violate the self contained character of MPEG-2 system streams, SPS and PPS are required to be transported within the AVC stream. Furthermore it is required that sufficient information is provided within the AVC stream to determine presentation timing in the case of underflow for low delay AVC streams. Video Usability Information (VUI) parameters, such as the aspect ratio, and information on the colour characteristics of the soured are strongly recommended to be present in the SPS to allow correct presentation of the decoded pictures.

When a PES packet carries an AVC stream, each encoded PTS and DTS time stamp remains expressed in units of the 90 kHz clock, though the timing information included within the AVC stream may use another clock frequency. The DTS corresponds to the CPB-removal-time and the PTS to the DPB-output-time. For subsequent pictures, it may be possible to derive these

Table 14.21 Signalling parameters for carriage of AVC video

Stream-ID	User-selectable from the range 0xE0 to 0xEF
Stream-type	0x1B: AVC video stream (may be a SVC or MVC sub-bitstream; see Sections 14.8 and 14.9)
Descriptor-tag	0x28 (40): AVC video descriptor 0x2A (42): AVC timing and HRD descriptor

times from information carried in the AVC stream. However, if the AVC stream contains insufficient information to determine the CPB-removal-time and the DPB-output-time of a picture, then for such picture these times must be provided by a coded PTS and/or DTS in the PES header.

The parameters used for carriage of AVC over MPEG-2 systems are listed in Table 14.21. The 16 stream-ID values available for video can be used; one stream-type value is assigned to signal an AVC video stream. As will be discussed in Sections 14.8 and 14.9, an AVC stream may be a SVC or MVC sub-bitstream. Requirements that apply to an AVC stream also apply to a SVC or MVC sub-bitstream.

The STD models and parameters for AVC decoding have already been depicted in Figure 12.17; for convenience the same figure is included here as Figure 14.8. For consistency

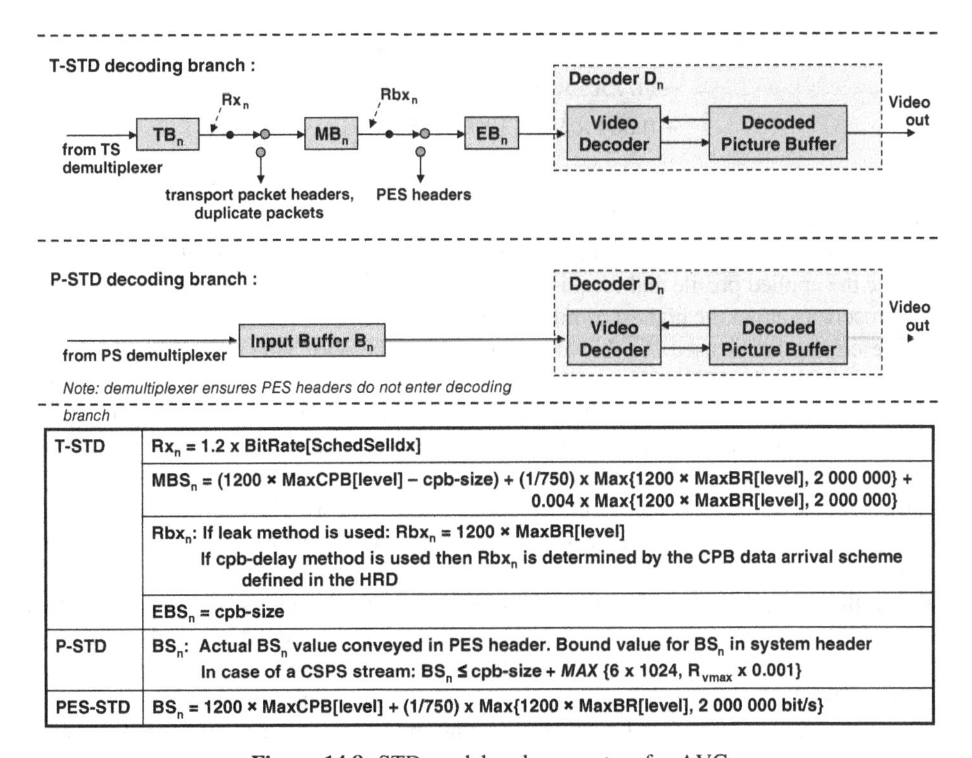

T-STD	Rx_n = 1.2 x BitRate[SchedSelIdx]
	MBS_n = (1200 × MaxCPB[level] – cpb-size) + (1/750) x Max{1200 × MaxBR[level], 2 000 000} + 0.004 x Max{1200 × MaxBR[level], 2 000 000}
	Rbx_n: If leak method is used: Rbx_n = 1200 × MaxBR[level] If cpb-delay method is used then Rbx_n is determined by the CPB data arrival scheme defined in the HRD
	EBS_n = cpb-size
P-STD	BS_n: Actual BS_n value conveyed in PES header. Bound value for BS_n in system header In case of a CSPS stream: BS_n ≤ cpb-size + *MAX* {6 x 1024, R_{vmax} x 0.001}
PES-STD	BS_n = 1200 × MaxCPB[level] + (1/750) x Max{1200 × MaxBR[level], 2 000 000 bit/s}

Figure 14.8 STD model and parameters for AVC

between the AVC HRD and the STD, the decoder D_n for AVC consists of the video decoder and the Decoded Picture Buffer (DPB). Management of the DPB is specified in the AVC standard.

The decoding branch in the T-STD contains the buffers MB_n and EB_n, whereby EB_n is the equivalent of the CPB in the HRD: EBS_n = cpb-size. Note that the AVC decoding branch for AVC in the T-STD is very similar to the T-STD decoding branch for MPEG-2 video depicted in Figure 12.16. The rate Rx_n is defined to be equal to 1.2 times BitRate[SchedSelIdx], whereby BitRate[SchedSelIdx] is the maximum bitrate for the used SchedSelIdx value, which is either defined explicitly by hrd-parameters() or by default. For the rate Rbx_n between MB_n and EB_n two methods are defined, similar as for MPEG-2 video: the leak method and the method here referred to as the cpb-delay method. In case of the leak method, if EB_n is full, then $Rbx_n = 0$, else $Rbx_n = 1200 \times$ MaxBR[level]. If the cpb-delay method is used, then the delivery rate Rbx_n is determined from the encoded values of the initial-cpb-removal-delay and the initial-cpb-removal-delay-offset parameters in the buffering-period-SEI-messages, as specified in Annex C of the AVC specification. Which transfer method is used is defined by the AVC timing and HRD descriptor discussed below.

The cpb-size may be smaller than MaxCPB for the applied AVC level. The left-over room may be used for multiplexing. For BS_{mux} and BS_{oh} the general rules of thumb are followed based on the maximum AVC video bitrate; however, to ensure sufficient flexibility for multiplex operations in the case of a low maximum bitrate for a level, MaxBR[level], the used value for the maximum bitrate in the formula is constrained to be at least 2×10^6 bits/s. Therefore:

$$\begin{aligned} MBS_n = {} & (1200 \times \text{MaxCPB[level]} - \text{cpb-size}) \\ & + (1/750) \times \text{Max}\{1200 \times \text{MaxBR[level]}, 2 \times 10^6\} \\ & + 0.004 \times \text{Max}\{1200 \times \text{MaxBR[level]}, 2 \times 10^6\} \end{aligned}$$

Note: multiplication with 1200 reflects the 20% NAL overhead compared to VCL; see the paragraphs above.

Each sequence within an AVC stream may conform to a different profile and level; for a sequence the applied profile and level is specified in the SPS, the Sequence Parameter Set. To inform receivers about the highest profile and level applied in the entire AVC stream, the AVC video descriptor can be used. As shown in Figure 14.9, the AVC video descriptor carries three bytes copied from the SPS with the eight-bit profile-idc field, a byte with flags signalling the applied constraints and the eight-bit level-idc field with semantics as specified in the AVC standard, but in the AVC video descriptor the semantics apply to the entire AVC stream to which the descriptor is associated. The semantics of the two-bit AVC-compatible-flags is exactly the same as the semantics of the field(s) defined for the two bits between the constraint-set5-flag and the level-idc field in the SPS defined in the AVC standard, so as to be prepared for any future usage of these bits in AVC.

Rather than accurately specifying when time stamp ambiguity will occur, each picture with a presentation time more than 24 h after the arrival of the coded data is classified as an AVC 24-hour picture. For each AVC 24-hour picture, no explicit DTS or PTS must be encoded in the PES header; instead, the AVC stream must contain sufficient information, so that the presentation time of such picture can be inferred from parameters in the AVC stream. AVC 24-hour pictures, as well as AVC still pictures, are only allowed to be present in the

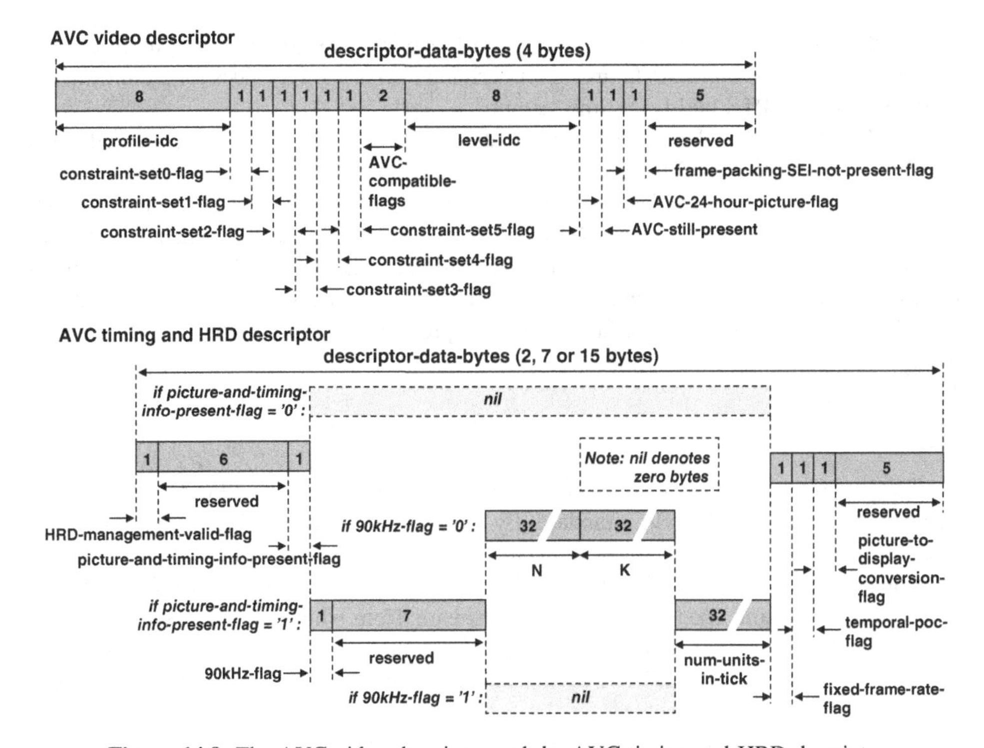

Figure 14.9 The AVC video descriptor and the AVC timing and HRD descriptor

AVC stream if explicitly signalled by the corresponding one-bit flags in the AVC descriptor. These pictures may only be present if the corresponding flag in the AVC descriptor is set to '1'. If the AVC descriptor is absent, or if the flag is set to '0', then these pictures must be absent.

Finally, for multiview video purposes, the frame-packing-SEI-not-present flag, if set to '0', signals that either the frame packing SEI message or the stereo video information SEI message is present in the AVC stream. In the absence of the AVC descriptor, these messages may be present as well.

Figure 14.9 also shows the data structure of the AVC timing and HRD descriptor. For the decoding of an AVC stream in the STD all required timing information need to be known. If such information is insufficiently available in the AVC stream, in particular when in the VUI-parameters() in the SPS the timing-info-present-flag is set to '0', then the AVC timing and HRD descriptor must be present for that AVC stream. In that case the picture-and-timing-info-present flag is set to '1', indicating the presence of the 90 kHz-flag, the N and K fields if the 90 kHz-flag is set to '0', and the num-units-in-tick field. These fields specify for the entire associated AVC stream the exact frequency of the AVC time base time-scale as follows:

$$\text{time-scale} = \text{STC-frequency} \times (N/K), \text{ with } K \text{ larger than or equal to } N.$$

Note that the above formula ensures that the AVC time base is locked to the STC frequency. In case the 90 kHz-flag is set to '1', N equals 1 and K equals 300. The num-units-in-tick field is encoded exactly as in the VUI-parameters(); the same applies to the fixed-frame-rate-flag.

Finally, the temporal-poc-flag and the picture-to-display-flag provide information on using information conveyed in the AVC stream for deriving PTS and DTS values for pictures without explicitly encoded PTS and DTS time stamps.

The AVC timing and HRD descriptor is also used to signal the applied method for the transfer of data in the T-STD between MB_n and EB_n. For an AVC stream carried in a transport stream the following applies. Usage of the leak method is signalled by absence of the AVC timing and HRD descriptor or by a HRD-management-valid-flag set to '0' when this descriptor is present. If this flag is set to '1', then for the transfer into EB_n the cpb-delay method, whereby the AVC stream contains Buffering Period SEI messages with coded initial-cpb-removal-delay and initial-cpb-removal-delay-offset values. In a program stream the meaning of the HRD-management-valid-flag is undefined.

14.8 SVC

The SVC specification [17] is an AVC extension, developed as amendment 3 to AVC and included in Annex G of the AVC standard. By means of SVC, video content can be conveyed in two or more sub-bitstreams, for example at various resolutions and bitrates. One of the sub-bitstreams represents the base layer; this must be a compliant single layer AVC stream. All bitstreams may be contained in one scalable bitstream, from which the less essential parts can be removed dynamically if so required by for example network conditions, to ensure graceful degradation. However, in a MPEG-2 system stream, each sub-bitstream is carried as a separate elementary stream in PES packets.

In a broadcast environment, SVC can be used to extend an existing service in a bandwidth efficient manner. For instance, assume an existing service exploits AVC to broadcast 1080 lines interlaced video, when the need arises to broadcast 1080 lines progressive video. Instead of a simulcast, whereby both formats are broadcast as single layer AVC streams in parallel, SVC can be used to efficiently encode and broadcast both formats in two sub-bitstreams. In this case, one sub-bitstream carries the base layer with the 1080 interlaced pictures, and the other the enhancement layer needed to reconstruct the 1080 progressive video. The base layer would be AVC compliant, so that existing AVC receivers continue to decode 1080 lines interlaced video, while SVC decoders would be needed to reconstruct the 1080 progressive video from the base layer and the SVC compliant sub-bitstream with the enhancement layer.

Scalability is often associated with severe loss of coding efficiency and significant increase of decoding complexity compared to a single layer coding scheme. However, compared to the scalability features developed for MPEG-2 video and MPEG-4 visual, tools are developed within SVC to reduce the efficiency loss and the complexity increase relative to a single AVC stream. It should be noted though that due to the chosen approach, SVC offers only scalability with respect to AVC streams, not with respect to other coded video, such as MPEG-2 video streams.

The SVC coding structure is organized in dependency layers. Layers are identified by the dependency-ID. To the base layer the dependency-ID value zero is assigned, while the value is incremented by one for each higher layer. A dependency layer usually represents a specific spatial resolution; however, two subsequent layers may have the same spatial resolution. The spatial resolution must not decrease from on layer to the next. In addition to the dependency-ID, various other SVC parameters are defined, such as the quality-ID for creating a refinement layer inside a dependency layer, and the temporal-ID for temporal scalability purposes. Easy

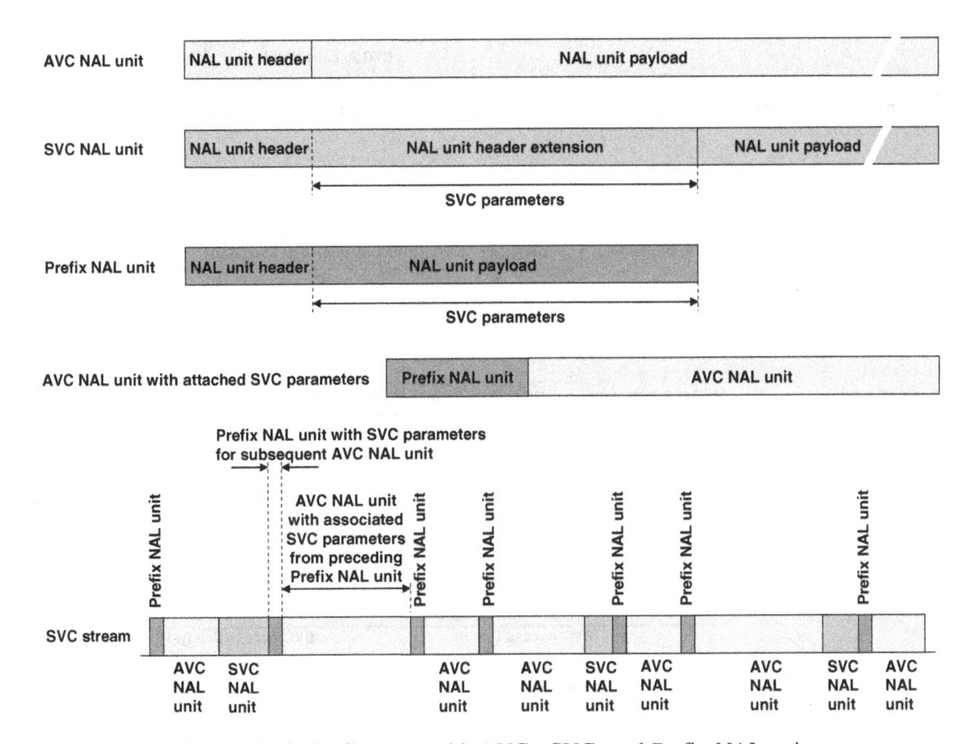

Figure 14.10 SVC stream with AVC-, SVC- and Prefix NAL units

access to actual SVC parameters is important to assist easy sub-bitstream manipulation and when re-assembling the coded data from the various sub-bitstreams to reconstruct the access units prior to decoding, as will be discussed further below.

SVC uses the same NAL structure as AVC; for carriage of SVC data, SVC specific NAL units are defined. Easy access to the SVC parameters is ensured by including those parameters in the headers of the SVC NAL units. For this purpose, the one-byte header of NAL units is extended for SVC with three additional bytes. See Figure 14.10. However it must also be possible to attach SVC related information to AVC NAL units, but the header of the VCL and Non-VCL NAL units defined for AVC cannot be extended with three bytes in a compatible manner. To resolve this issue, so-called prefix-NAL units are defined to carry the SVC parameters. A prefix-NAL unit is attached to the AVC NAL unit it immediately precedes. In Figure 14.10, in the SVC stream each AVC NAL unit is preceded by a Prefix NAL unit; the Prefix and AVC NAL unit pairs are interleaved with SVC NAL units. An AVC decoder ignores unknown NAL units; hence when an AVC decoders receives this SVC stream, it will process only the AVC NAL units.

An access unit in a SVC stream contains the coded data associated to that access unit from all involved layers in decoding order. In each sub-bitstream the NAL units of an access unit are carried in a so-called Dependency Representation (DR). In other words, if H is the highest layer in a set of decoded sub-bitstreams, then the access unit AU_H contains the corresponding DRs of layer H and all lower layers in decoding order, that is starting with the lowest layer in the access unit. An example with three sub-bitstreams p, q and r is depicted in Figure 14.11; the assigned

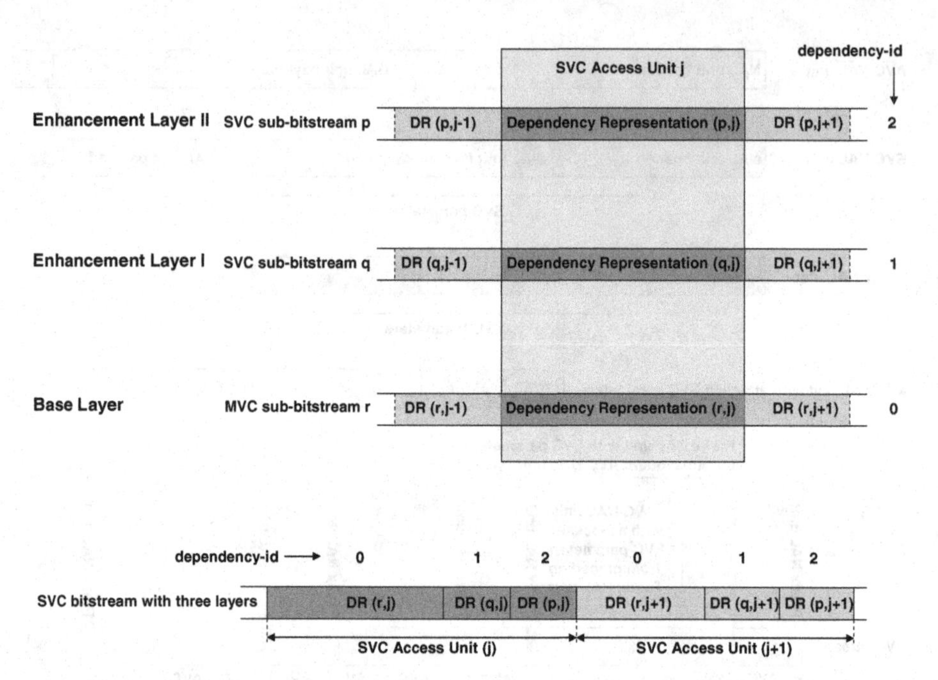

Figure 14.11 Dependency representations in access units and in SVC bitstream

dependency-ID values indicate that sub-bitstream p represents the highest layer, r the lowest; the dependency-ID value zero indicates the base layer; therefore sub-bitstream r is a single layer AVC stream. For each sub-bitstream the DRs are shown, as well as their concatenation in access units in the SVC stream based on the value of the dependency-ID.

In a MPEG-2 system stream, each SVC sub-bitstream must comply to the byte stream format with a start-code prefix preceding each NAL unit header. Furthermore, each SVC sub-bitstream is carried in PES packets as a separate elementary stream. A PTS and DTS time stamp in the header of a PES packet refers to the DR that commences in that PES packet. An explicit PTS and, if applicable, DTS must be encoded for each DR. As a consequence, only one DR can commence in a PES packet.

To decode an SVC access unit requires access to all involved elementary streams. Upon retrieving the DRs from each involved sub-bitstream, the DRs must be put in the correct order needed for decoding. Hence, prior to decoding, the SVC access unit is re-assembled by putting the NAL units from the DRs of all involved layers in the order specified by the SVC standard.

When temporal scalability is used, the layer with a higher frame rate is predicted from one or more layers with a lower frame rate. If the coding of DRs in the layer with the higher frame rate only uses prediction from DRs with an earlier or equal DTS value, then the forming of SVC access units for the decoding of the higher layer is straight forward. The upper half of Figure 14.12 presents an example thereof. In this example, a base layer and one enhancement layer are depicted. In the enhancement layer, the frame rate is twice the frame rate of the base layer; the DTSs of the DRs in both layers are coded accordingly. The DRs in both layers are denoted B(i) and E(i), respectively, with i the index of the DRs in the enhancement layer. The DRs E(1) and E(3) corresponding to the additional frames in the enhancement layer are

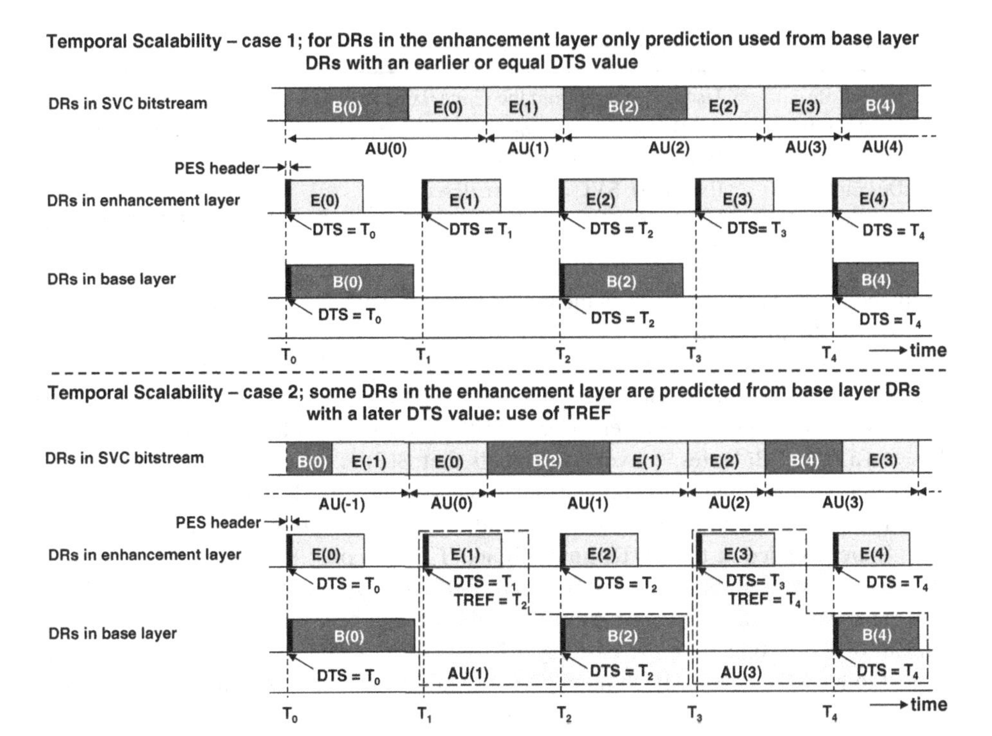

Figure 14.12 Temporal scalability and use of TREF

predicted from the preceding DR in the base layer, B(0) and B(2), respectively. In the SVC bitstream assembled to decode the enhancement layer, the access units contain the DRs with the same DTS values. AU(0) contains B(0) and E(0); because B(0) must be available for the decoding process before E(0), B(0) precedes E(0). AU(1) contains only E(1). Similarly, AU(2) carries (B2) and E(2), while AU(3) carries E(3).

The situation becomes more complex when a DR in the layer with the higher frame rate is predicted from a DR with a later DTS value. This may be caused in SVC by more complex and efficient GOP structures using hierarchical b-predictive pictures. For example, in the lower half of Figure 14.12, E(1) depends on both B(0) [in the form of the AU built from B(0) and E(0)] and B(2); when B(2) is used by inter-layer prediction to predict E(1), B(2) must be available for decoding E(1), even though the DTS of B(2) indicates the later DTS value T_2. In other words, in this case B(2) must be included before E(1) in the SVC bitstream for decoding the enhancement layer. Hence, B(2) is included before E(1) in the access unit decoded at DTS time T_1.

To indicate that a DR in the lower layer must be decoded earlier than indicated by its DTS, the TREF parameter is defined in the PES-packet-extension-2 field (see Figure 13.8). If the DTS of the DR in the higher layer differs from the DTS of the related DR in the immediately lower layer, then in the PES packet that carries the DR of the higher layer the TREF field must be present, encoded with the DTS value of the related DR in the lower layer. This ensures that the DR in the lower layer is decoded at the decode time of the higher layer access unit,

Table 14.22 Signalling parameters for carriage of SVC video

Stream-ID	User-selectable from the range 0xE0 to 0xEF
Stream-type	0x1B: AVC compliant sub-bitstream of SVC containing the base layer
	0x1F: SVC video sub-bitstream
Descriptor-tag	0x30 (48): SVC extension descriptor

determined by the DTS of the DR in the higher layer, and not at the time indicated by the DTS of the DR in the lower layer.

In the example depicted in the lower half of Figure 14.12, E(1) depends both on B(0) for inter-frame prediction and on B(2) for inter-layer prediction and E(3) from B(2) and B(4). This means that a TREF field must be encoded for E(1) and E(3) with the values T_2 and T_4 of the DTSs of B(2) and B(4), respectively, to signals that B(2) is included in AU(1) and B(4) in AU(3). Hence, when the enhancement layer access units are decoded, B(2) and B(4) are decoded at T_1 and T_3 and not at T_2 and T_4 as indicated by their DTSs. Note however that B(2) and B(4) are decoded at their DTS times T_2 and T_4 when only the base layer is decoded.

The parameters used for carriage of SVC over MPEG-2 systems are listed in Table 14.22. As usual, the sixteen stream-ID values available for video can be used. The AVC compliant sub-bitstream of SVC containing the base layer is obviously signalled by the stream-type value assigned to an AVC stream. Each other SVC video sub-bitstream is signalled by the stream-type value assigned to SVC. For SVC one descriptor is defined: the SVC extension descriptor. The hierarchy descriptor discussed in Section 13.3.4.1 may be used to signal the dependencies of the sub-bitstreams.

If only the sub-bitstream with the base layer is decoded, then the STD model for AVC applies, as discussed in Section 14.7. However, when multiple sub-bitstreams are decoded, the STD model depicted in Figure 14.13 applies. The parameter H indicates the highest layer in the decoded sub-bitstreams. Decoder D_H consists of a scalable decoder and the DPB, similar to Decoder D_n for AVC depicted in Figure 14.8. D_H decodes the coded data from the base layer and all higher layers up to and including layer H of the set of sub-bitstreams. In the T-STD, the bytes of sub-bitstream n enter transport buffer TB_n and are subsequently transferred to MB_n and DRB_n, but the transport packet header and duplicate packets are ignored (see Figure 14.13); PES headers are removed instantaneously during the transfer from MB_n to DRB_n. In the P-STD, the bytes enter buffer DRB_n. At the output of the involved DRB buffers, the access units are re-assembled from the DRs, starting from the DR in the lowest layer, as required by the SVC specification.

For the base layer in the STD model depicted in Figure 14.13 the same transfer rates and buffer sizes apply as for a single layer AVC video stream (see Section 14.7). For all other SVC sub-bitstreams the STD parameters are discussed below. For a sub-bitstream representing a layer higher than the base layer, the size of the individual DRB buffer is not explicitly specified, but instead the total size is defined of the DRB buffers of the involved SVC sub-bitstreams needed to decode that layer. For each layer the size EBS_H of EB_H is specified that applies when that layer is the highest layer to be decoded. Hence, the value of EBS_H for a certain layer signals the total size of the DRB buffers for that layer and for all lower layers, including the base layer. While this design choice may be beneficial for coding the various layers, specific care is needed when multiplexing the elementary streams carrying the SVC layers (see Note 14.4).

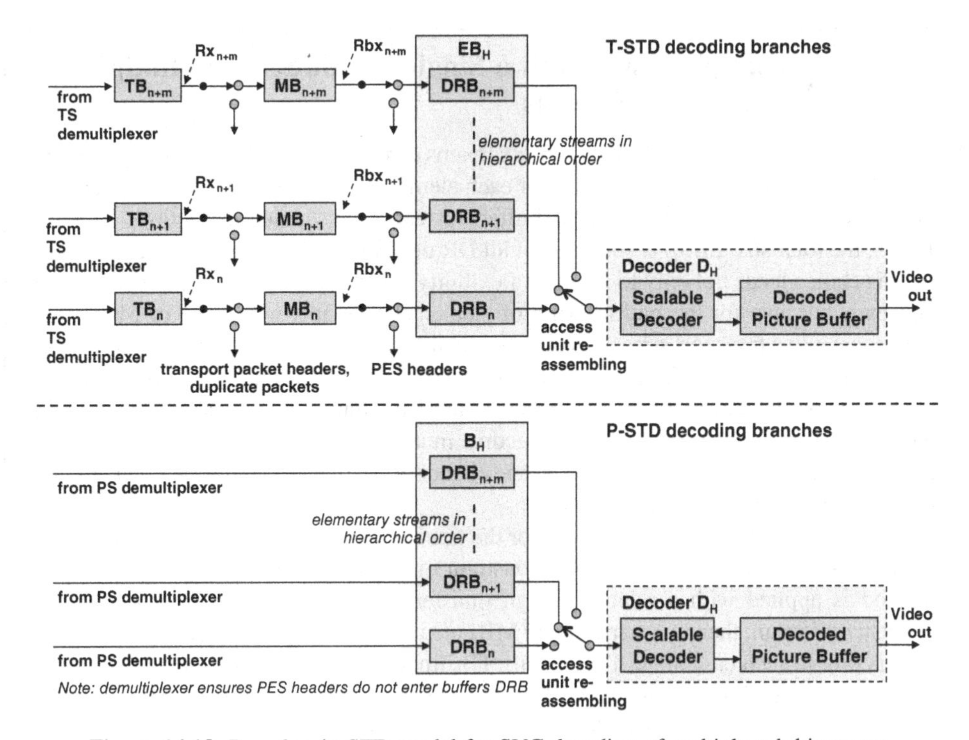

Figure 14.13 Branches in STD model for SVC decoding of multiple sub-bitstreams

In program streams, the size EBS_H is conveyed in the P-STD-buffer-size field in the header of PES packets carrying a SVC sub-bitstream. In transport streams, the cpb-size field encoded in the sub-bitstream for the highest layer to be decoded signals the size EBS_H. Hence for the layer with the highest dependency-ID value H the following applies:

For a transport stream: EBS_H = cpb-size encoded in sub-bitstream H;
For a program stream: EBS_H = P-STD-buffer-size encoded in PES headers carrying sub-bitstream H.

For each layer higher than the base layer, the re-assembling of the access units up to that layer results in an AVC stream with a certain AVC level. This level is used to determine the size of MB_n for the SVC sub-bitstream corresponding to that layer:

$$\begin{aligned} MBS_n &= BS_{mux} + BS_{oh} \\ &= (1/750) \times \text{Max}\{1200 \times \text{MaxBR[level]}, 2 \times 10^6\} \\ &\quad + 0.004 \times \text{Max}\{1200 \times \text{MaxBR[level]}, 2 \times 10^6\} \end{aligned}$$

The above formula is the same as for a single layer AVC stream, except that for SVC the left-over buffer room in case the cpb-size is smaller than MaxCPB[level] is not made available for multiplexing (see Section 14.7). The same level is used to determine transfer rates between the buffers in the T-STD, whereby the same formulas and conditions are used as for a single layer AVC stream with that level (see Section 14.7). If needed for the selection of the transfer method

Note 14.4 Multiplex Issue with a Single EB Buffer for Multiple Elementary Streams

For the decoding of SVC and MVC sub-bitstreams carried in multiple elementary streams, the T-STD defines a DR or VS buffer for each elementary stream, but without specifying the size of each individual DR or VS buffer. Instead, as discussed in Sections 14.8 and 14.9.3, the total size EBS_H is specified of all DR or VS buffers involved in the decoding.

This choice leads not only to far more flexibility for the encoding of the various layers or views, but also for the multiplexing of elementary streams with SVC or MVC sub-bitstreams into a transport stream. With a fixed buffer size for each stream, the constraints on multiplex operations for a single stream do not depend on multiplex operations for other streams. However, when one buffer is used for multiple elementary streams, the multiplex operations for these streams become mutually related. When a large portion of the available buffer room is utilized for one elementary stream, then for the other streams less room will be available.

When the cpb-delay method is used for the transfer from MB_n to DRB_n or VSB_n, then the fullness of MB_n may significantly constrain multiplex operations, but when the leak method is applied with a relatively high transfer rate compared to the bitrate of the elementary stream, then bytes arriving in MB_n are immediately transferred to EB_H, unless EB_H is full. In such case in the T-STD only TB_n imposes constraints on the positioning of transport packets, but the fullness of MB_n only starts to play a role when EB_H gets full. Which is no problem, unless underflow of EB_H is caused for another elementary stream with SVC or MVC sub-bitstreams.

Obviously, when multiplexing multiple elementary streams carrying SVC or MVC sub-bitstreams into a transport stream, it must be ensured that the produced transport stream is compliant and that underflow of EB_H is prevented for each elementary stream.

between MB and DRB, or for any other reason, the AVC timing and HRD descriptor may be assigned to a sub-bitstream.

The SVC extension descriptor provides information on the AVC stream resulting from re-assembling up to the level of the sub-bitstream the descriptor is associated to. Figure 14.14 shows that the 13 descriptor-data-bytes of the SVC extension descriptor provide the maximum picture width and height in pixels, the maximum frame-rate (frames/256 s), the average and maximum bitrate (kb/s) of the re-assembled AVC stream up to the level of the associated sub-bitstream and the value of the dependency-ID of the associated sub-bitstream. Also the minimum and maximum values of the quality-ID and the temporal-ID contained in the associated sub-bitstream are provided. Finally, a one-bit flag signals whether or not the associated sub-bitstream carries SEI NAL units.

14.9 3D Video

14.9.1 Service Compatible and Frame Compatible 3D Video

Service compatible and frame compatible 3D Video allow the use of existing 2D video infrastructures for the delivery of 3D content. With service compatible 3D video services the stereoscopic left and right views are each conveyed as a separate stream. A number of options

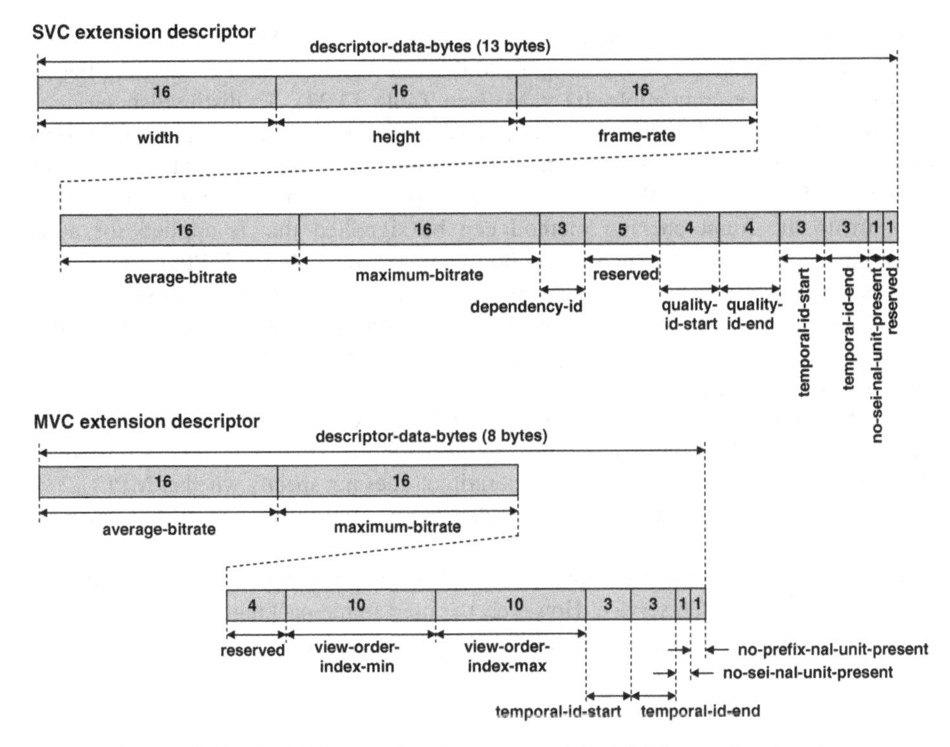

Figure 14.14 The SVC extension descriptor and the MVC extension descriptor

can be exploited; the additional view can be identified by a specific stream-type value, whereby distinction is made between a MPEG-2 video and an AVC video coded additional stream. Each installed base decoder will ignore the additional view without requiring any additional signalling. For the assigned stream-type values see Table 14.23.

Table 14.23 Parameters signalling service compatible stereoscopic video

Stream-ID	User-selectable from the range 0xE0 to 0xEF
Stream-type	0x02: MPEG-2 video stream with base view of service compatible stereoscopic 3D video or (signalled by descriptor) MPEG-2 video stream with frame compatible stereoscopic 3D video
	0x22: MPEG-2 video stream with additional view of service compatible stereoscopic 3D video
	0x1B: AVC video stream with base view of service compatible stereoscopic 3D video or (signalled by descriptor) AVC video stream with frame compatible stereoscopic 3D video
	0x23: AVC video encoded additional view of service compatible stereoscopic 3D video
Descriptor-tag	0x52 (82): MPEG-2 stereoscopic video format descriptor
	0x53 (83): Stereoscopic program info descriptor
	0x54 (84): Stereoscopic video info descriptor

A frame compatible 3D video stream is coded as an ordinary 2D video stream and therefore signalled by the stream-type value assigned to the used video codec. The same applies to the base view of service compatible 3D video (see Table 14.23). To distinguish service and frame compatible 3D video streams, descriptors can be used. Three descriptors are defined to provide information on a stream carrying stereoscopic video; these descriptors are presented in Figure 14.15. By means of the MPEG-2 stereoscopic video format descriptor for a MPEG-2 video stream the frame-packing method can be signalled that is applied for a frame compatible 3D video stream, such as top-bottom or left-right. In the MPEG-2 video stream this information must also be provided, in particular in the user-data-extension defined in an amendment to MPEG-2 video. The seven-bit in the descriptor must be encoded with the same value as the arrangement-type field in the user-data-extension in the associated video stream.

The stereoscopic program info descriptor signals on program level whether the associated program contains a 2D (monoscopic) video service, a frame compatible 3D video service or a service compatible 3D video service. This descriptor does not specify whether MPEG-2 video or AVC is used as video codec for the signalled service. For the coding of the stereoscopic-service-type field see Table 14.24.

When AVC is used for frame compatible 3D video, the stereoscopic program info descriptor may be used to signal such service. However, the used frame-packing method is specified in the AVC stream by SEI messages, such as the frame packing arrangement SEI message.

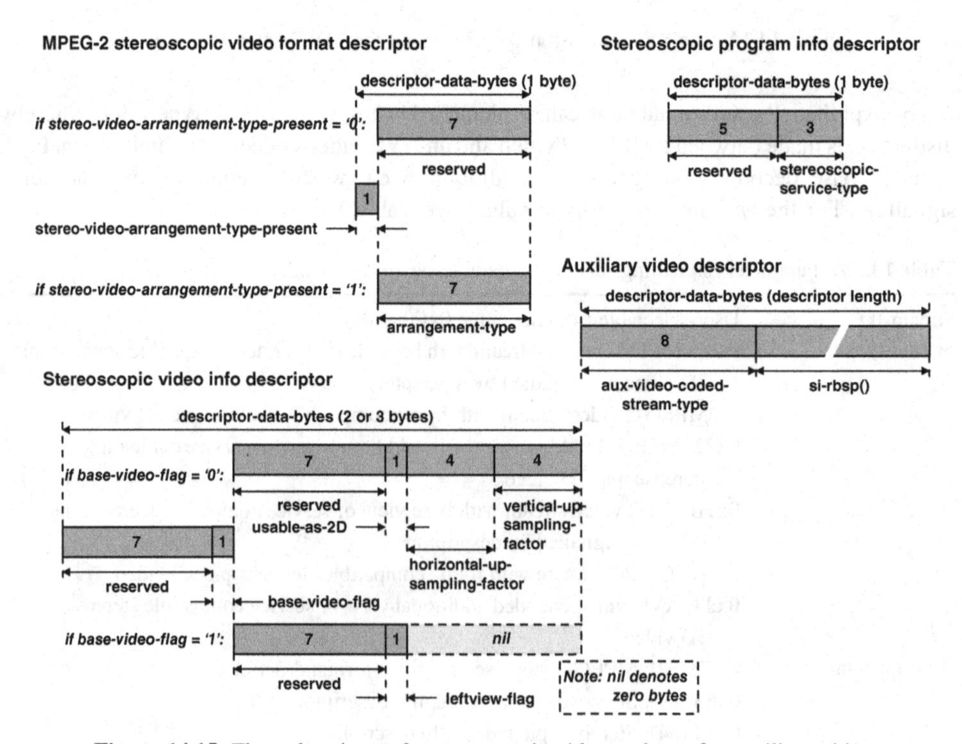

Figure 14.15 Three descriptors for stereoscopic video and one for auxiliary video

Table 14.24 Stereoscopic-service-type in stereoscopic program info descriptor

Value	Signalled service
'000'	Unspecified
'001'	2D-only video service
'010'	Frame compatible stereoscopic 3D video service
'011'	Service compatible stereoscopic 3D video service
'100'–'111'	Reserved

In MPEG-2 systems no frame-packing descriptor is defined to signal how the left and right views are conveyed in an AVC stream.

In a service compatible stereoscopic 3D video service the stereoscopic video info descriptor must be used for each view. For the base view this descriptor signals whether it is the left or the right view. For the additional view the descriptor provides an indication of the horizontal and vertical up-sampling factors that may be needed and signals whether it may be used for a 2D video service. The actual up-sampling factors are to be derived from information carried in both video streams. For the coding of both up-sampling-factor fields see Table 14.25.

For the decoding of a service compatible or frame compatible stereoscopic 3D video stream in the STD the requirements apply as specified for the applied video codec. The applied profile and level may be signalled by the descriptor for that codec. As no additional requirements apply, the decoding of such streams is not discussed here.

14.9.2 Depth or Parallax Map as Auxiliary Video Stream

To code a depth map or a parallax map as specified in ISO/IEC 23002-3 [18], any video coding standard can be used (see also Section 4.5.2). The compressed depth or parallax map is conveyed as an auxiliary video stream. Each auxiliary video stream is identified by the assigned stream-type value. Table 14.26 presents the parameters used for carriage of an auxiliary video stream over MPEG-2 systems.

Table 14.25 Coding of up-sampling factors in stereoscopic video info descriptor

Value	Signalled up-sampling
'0000'	Forbidden
'0001'	Unspecified
'0010'	Coded resolution is the same as in base view
'0011'	Coded resolution is three-quarters of the coded resolution in base view
'0100'	Coded resolution is two-thirds of the coded resolution in base view
'0101'	Coded resolution is one-half of the coded resolution in base view
'0110'–'1000'	Reserved
'100'–'111'	User private

Table 14.26 Signalling parameters for carriage of an auxiliary video stream

Stream-ID	User-selectable from the range 0xE0 to 0xEF
Stream-type	0x1E: Auxiliary video stream as defined in ISO/IEC 23002-3
Descriptor-tag	0x47 (71): Auxiliary video descriptor

The video codec that is used for compression is specified in the auxiliary video stream descriptor that must be associated to each auxiliary video stream. The aux-video-coded-stream-type field in this descriptor provides the stream-type value assigned to the video codec used to compress the auxiliary video stream. For example, when AVC is used, then the aux-video-coded-stream-type is encoded with the stream-type value 0x1B. In addition to depth and parallax maps, ISO/IEC 23002-3 allows also other types of auxiliary video streams. Whether a depth map or a parallax map or yet another type is conveyed in an auxiliary video stream is specified by the so-called Supplemental Information RBSP structure, specified in ISO/IEC 23002-3; this si-rbsp() is carried in the auxiliary video stream descriptor (see Figure 14.15).

For the decoding of an auxiliary video stream in the STD, the requirements apply as specified for the video codec signalled by the aux-video-coded-stream-type field in the auxiliary video descriptor. The applied profile and level may be signalled by the descriptor for that codec.

14.9.3 MVC

By means of MVC, specified in Annex H of the AVC standard [19], many views can be encoded, as discussed in Sections 4.4 and 4.5.3. There is one base view and one or more additional views; each additional view has the same spatial and temporal resolution as the base view. Each view is encoded in a separate MVC sub-bitstream. The sub-bitstream representing the base view must be a single layer AVC stream. It is possible to convey all sub-bitstreams in a single MVC stream, similar as for SVC, but each sub-bitstream can also be conveyed as a stream separate from the other sub-bitstreams. As for SVC, MVC specific NAL units are defined, with a three-byte extension of the one-byte NAL header to carry MVC parameters in a very similar way as the SVC parameters; by means of a one-bit flag in the three-byte NAL header extension it is signalled whether the header extension contains SVC or MVC parameters. In the same way also prefix NAL units may be used to attach MVC parameters to an AVC NAL unit. See the example given in Figure 14.10. When carried over MPEG-2 systems, each MVC sub-bitstream must comply to the byte stream format with a start-code prefix preceding each NAL unit header.

For each view, an MVC access unit contains the coded data of one image, referred to as a view-component. For coding efficiency reasons, usually inter-view prediction is applied, but for simplicity reasons this is only allowed within an access unit. Hence, the views have a hierarchical relationship; hierarchically higher views may be predicted from hierarchically lower views, but not necessarily from all: only a subset may be used for prediction. Each view is identified by an arbitrary view-ID, while the view order index specifies the hierarchical relationship between the views by providing the required decoding order within an access unit. Figure 14.16 shows an example with three views and inter-view prediction of the left and right view from the base view. Also a MVC stream conveying all three sub-bitstreams is depicted; within each access unit the view-components must be in decoding order.

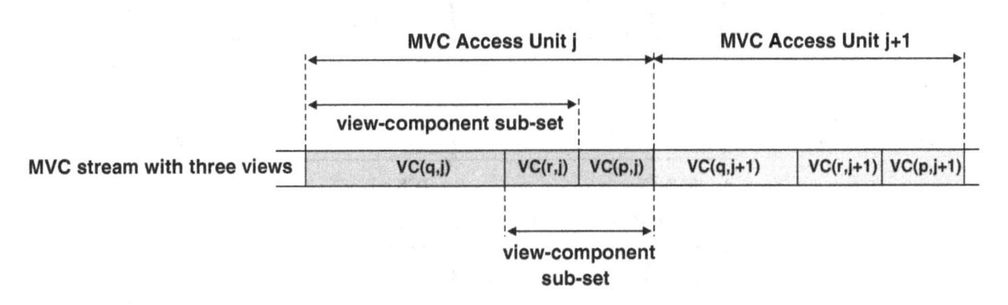

Figure 14.16 View-components and Access Units in a MVC (sub-)bitstream

In a MPEG-2 system stream, one or more views can be carried in one elementary stream. In case multiple views are conveyed in the same elementary stream, then it must be a subset of subsequent views in decoding order, which simplifies the re-assembling of the access units. The concatenated view-components from the same access unit are referred to as a View Subset, VS. In Figure 14.16, each individual view-component can be a VS, but also the concatenated view-components from the mid and right view or from the right and left view.

For each VS, a PTS and/or DTS must be encoded in the PES header. As a consequence, only one VS can commence in a PES packet. Figure 14.17 presents a MVC stream with five views; the indicated index signals the decoding order of the view components. Two examples are shown for carriage of these five views in PES packets. In the first example all views are conveyed in the same elementary stream. Each VS contains five view-components. In the other example, view 0 and view 1 are each individually conveyed in an elementary stream, while views 2, 3 and 4 are jointly carried in the third elementary stream. For simplicity reasons, each VS is aligned with the PES packet; in each PES header a DTS is encoded for the VS carried in the PES packet. Note that for each VS from the same access unit the same DTS value is encoded. Note furthermore that upon receiving the three elementary streams in the second example, a view can only be decoded after putting the required view-components in decoding order. For MVC a similar kind of access unit re-assembling is needed as for SVC.

The parameters used for carriage of MVC over a MPEG-2 system stream are listed in Table 14.27. As usual, the sixteen stream-ID's values available for video can be used. The elementary stream conveying the base view is signalled by the stream-type value assigned to an AVC stream. By means of the MVC extension descriptor, the presence of other views, if any, can be signalled. Each elementary stream without the base view but with one or more other

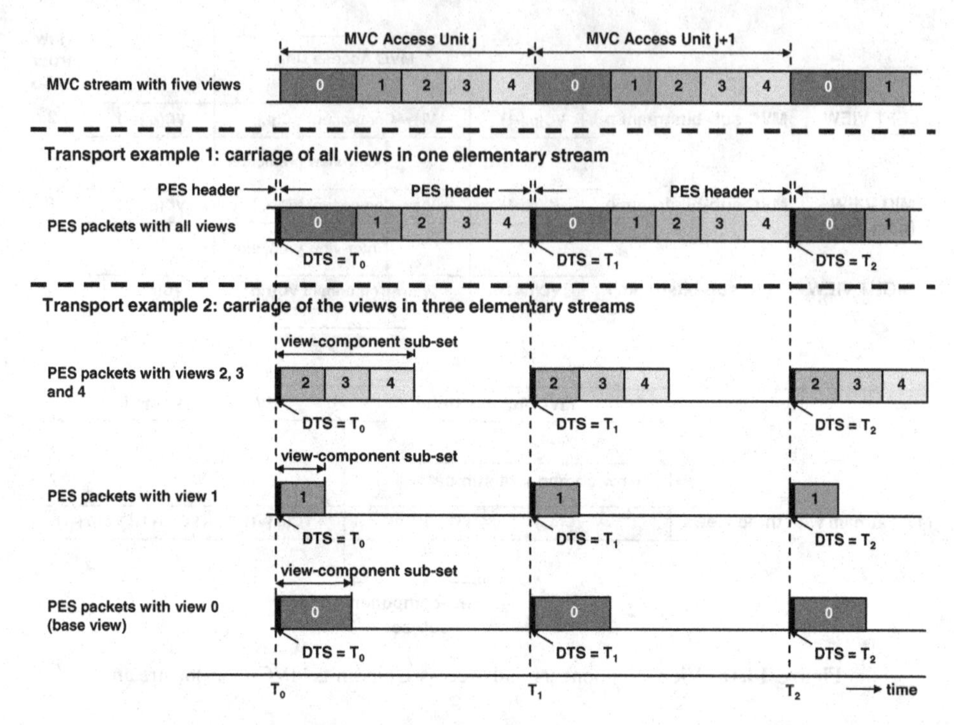

Figure 14.17 Two examples to convey five views in PES packets

views is signalled by the stream-type value 0x20. Specifically for MVC two descriptors are defined: the MVC extension descriptor and the MVC operation point descriptor. In case of an ambiguity on the hierarchical relationship between views, the hierarchy descriptor discussed in Section 13.3.4.1 must be associated to one or more MVC sub-bitstreams.

For the decoding of an elementary stream containing the base view, the STD model for AVC streams applies, as depicted in Figure 14.8. Note that for the decoding of such elementary stream no access unit re-assembling is required: if other views are carried in addition to the base view, then each access unit contains a set of subsequent views in decoding order. Note however

Table 14.27 Signalling parameters for carriage of MVC video

Stream-ID	User-selectable from the range 0xE0 to 0xEF
Stream-type	0x1B: MVC elementary stream carrying the (single layer) AVC compliant base view and zero or more other views in subsequent decoding order
	0x20: MVC elementary stream without the base view, but with one or more other views in subsequent decoding order
Descriptor-tag	0x31 (49): MVC extension descriptor
	0x33 (51): MVC operation point descriptor

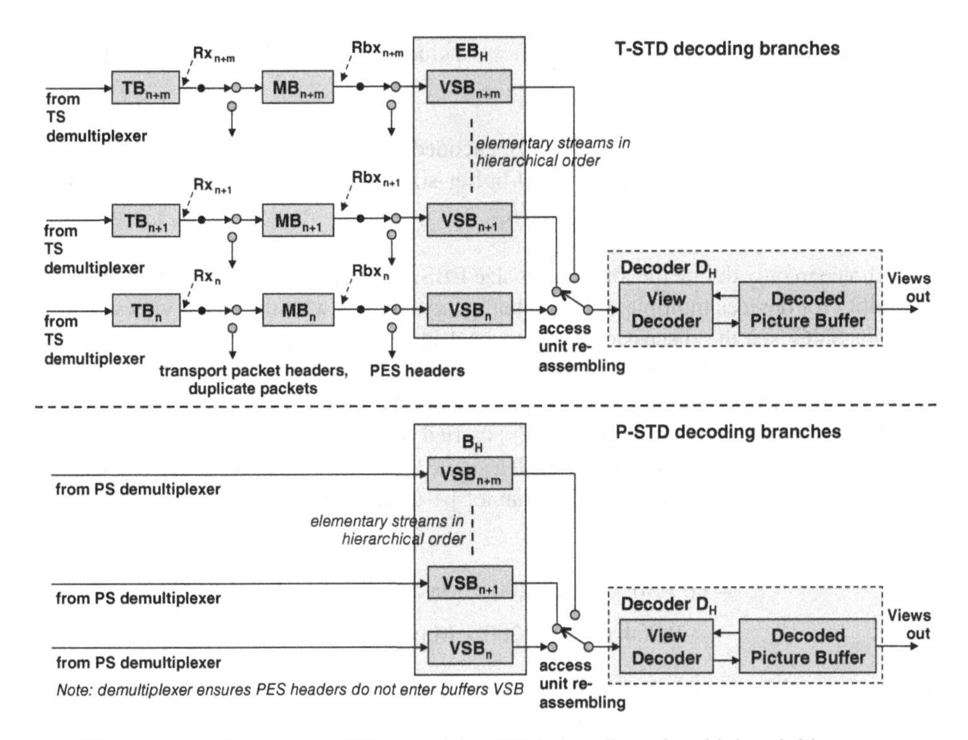

Figure 14.18 Branches in STD model for MVC decoding of multiple sub-bitstreams

that for the decoding of views other than the base view the same view decoder is required as in the STD model for MVC decoding depicted in Figure 14.18.

When there is at least one elementary stream signalled by a stream-type value 0x20 as a MVC elementary stream without the base view but with one or more other views, then the STD model depicted in Figure 14.18 applies. Decoder D_H consists of a View Decoder and the Decoded Picture Buffer DPB. Prior to its decoding, an access unit must be re-assembled by putting the NAL units of the VSs from all required views in the correct decoding order, as specified in the MVC standard.

In the T-STD, the bytes of each elementary stream n with one or more views enter transport buffer TB_n and are subsequently transferred to MB_n and VSB_n, except the transport packet headers and the duplicate packets, which are ignored (see Figure 14.18); PES headers are removed instantaneously during the transfer from MB_n to DRB_n. In the P-STD, the bytes of elementary stream n enter buffer VSB_n.

For the decoding of an elementary stream, the size of VSB is not explicitly specified, but instead the total size BS_H of the involved VSB buffers needed to decode all views. In a program stream, this size is conveyed by the P-STD-buffer-size field in the header of PES packets carrying the sub-bitstream with the highest view order index. For example, in Figure 14.17 headers of PES packets with views 2, 3 and 4 must carry the P-STD-buffer-size field, but also header of PES packets with view 1 carry this field, as views 2, 3 and 4 may not be decoded. In a transport stream, the size BS_H is provided by the cpb-size encoded in the MVC sub-bitstream representing the view with the highest view order index to be decoded. If H denotes that view, then the total size of the involved VSB buffers, including the VSB buffer for the base view is specified by:

For a transport stream: EBS_H = cpb-size encoded in sub-bitstream H;
For a program stream: EBS_H = P-STD-buffer-size encoded in PES headers carrying sub-bitstream H.

This choice to only define the total buffer size EBS_H provides in specific cases a high degree of flexibility for devices multiplexing the elementary streams with the coded views. The same applies for SVC; see the discussion in Note 14.4 in Section 14.8.

The size of MB_n for an elementary stream n signalled by the MVC stream-type value 0x20 with one or more views depends on the applicable level for the AVC stream resulting from re-assembling the access units up to the view(s) carried in elementary stream n. The formula for MBS_n is the same as for an AVC stream with a single layer and view, except that left-over room due to a smaller cpb-size than MaxCPB cannot be used for multiplexing:

$$\begin{aligned} MBS_n &= BS_{mux} + BS_{oh} \\ &= (1/750) \times \mathrm{Max}\{1200 \times \mathrm{MaxBR[level]}, 2 \times 10^6\} \\ &\quad + 0.004 \times \mathrm{Max}\{1200 \times \mathrm{MaxBR[level]}, 2 \times 10^6\} \end{aligned}$$

The same level is used to determine transfer rates in the T-STD between TB_n and MB_n and between MB_n and VSB_n, whereby the same formulas and conditions are used as for an AVC stream with a single layer and view having the same level (see Section 14.7). If needed, an AVC timing and HRD descriptor can assigned to the MVC sub-bitstream to signal the transfer method between MB and VSB.

For each elementary stream carrying one or more views, an MVC extension descriptor may be present. If the elementary stream with the base view also carries one or more other views, then an MVC extension descriptor must be present. The MVC extension descriptor provides information on the view(s) carried in the associated elementary stream and on the AVC stream resulting from re-assembling (up to) the view with the highest view order index carried in the elementary stream. The minimum and maximum values of the view-order-index of the carried view(s) are provided. For the re-assembled stream the average and maximum bitrate in kb/s is provided, as well as the minimum and maximum values of the temporal-ID signalled in any of the carried views. Finally, two one-bit flags are included; the no-SEI-NAL-units-present-flag signals whether or not SEI NAL units may be carried in the associated sub-bitstream. The no-prefix-NAL-units-present-flag, when set to '1', signals that prefix NAL units are absent, both in the base view and in additional view sub-bitstreams; if set to '0', prefix NAL units are only present in the base view. In Figure 14.14 the structure of the MVC extension descriptor is depicted in detail.

Within a program stream or within a program in a transport stream, multiple so-called MVC operation points may be used. Each MVC operation point is constituted by a set of one or more views, representing the target output views, and identified by a temporal-ID value that represents a target temporal level. Further details are found in the MVC specification. For each operation point an AVC stream can be re-assembled from the VSs of all target output views and of all other views the target output views depend on. To signal the profile and level required to be supported by a decoder of the views corresponding to a particular MVC

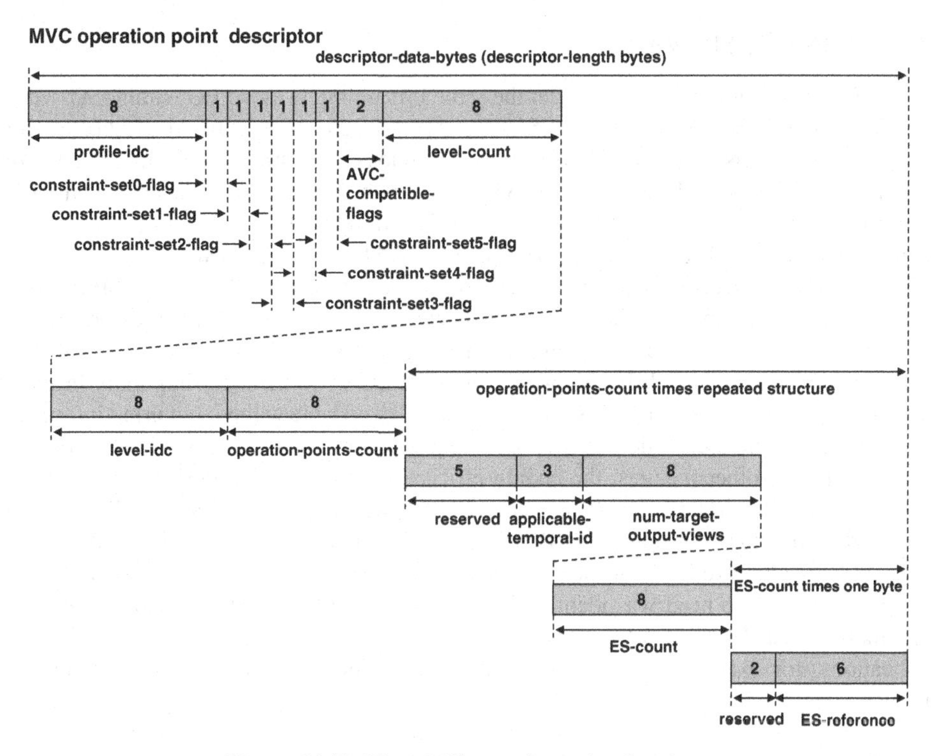

Figure 14.19 The MVC operation point descriptor

operation point, the MVC operation point descriptor is defined. For one specific profile the MVC operation point descriptor can provide information on each operation point for a given level. Another MVC operation point descriptor may be needed to also describe these operation points. If a MVC operation point descriptor is present for a program, then at least one hierarchy descriptor must be present for each elementary stream with one or more views in that program, so as to unambiguously specify the hierarchical relationship between the views of an operation point, as discussed in the next paragraph.

The structure of the MVC operation point descriptor is depicted in Figure 14.19. For all operation points described in the descriptor, the applied profile and any additional constraints are provided by the fields also used in the AVC video descriptor (see Figure 14.9). The level-count signals the number of levels for which operation points are described. Next, for each level, specified by the eight-bit level-idc field, the number of operation points described for that level is signalled by the eight-bit operation-points-count field. For each operation point the three-bit applicable-temporal-ID identifies the operation point by signalling the highest temporal-ID value in the re-assembled AVC stream for that operation point. Furthermore, the eight-bit num-target-output-views field indicates the number of target output views for the operation point. Finally, the ES-count provides the number of elementary streams carrying the views needed to re-assembled the AVC stream for that operation point; for each such elementary stream the six-bit ES-reference field signals the hierarchy-layer-index value that is assigned to that elementary stream by the hierarchy descriptor.

14.10 JPEG 2000 Video

The JPEG working group operates under the same ISO umbrella as MPEG within ISO. MPEG is WG 11 and JPEG WG 1 within ISO/IEC JTC1/SC29. The scope of JPEG, the Joint Photographic Experts Group, is the coding of (continuous tone) still pictures. In 1992, the first part of the JPEG standard ISO/IEC 10918 [20] was published, followed later by some more parts. In 2000, the first part of JPEG 2000 was published as ISO/IEC 15444-1 [21]. In 2004 and later, more than 10 subsequent parts of ISO/IEC 15444 were published.

The JPEG standard does not use other pictures to code a still picture. In terms of MPEG video, only intra-coding is applied. The difference between intra-coded still pictures and moving pictures is not very significant: by simply concatenating a series of still pictures, a moving picture sequence is obtained. In this way, motion JPEG evolved. Most important feature of motion JPEG is that each picture is self-contained and decodable without requiring information from any other picture, which allows for simple picture accurate editing. By not using information from other pictures, the coding efficiency decreases significantly compared to a moving video codec, but in several applications, convenient editing is at least equally important.

Though also intra coding can be used in video coding standards, such as MPEG-2 video and AVC, the need evolved to transport pictures coded by JPEG 2000 across MPEG-2 over transport streams. No need was identified for support of JPEG 2000 in program streams. In particular use of JPEG 2000 is expected in transport stream based content distribution applications, for example to transport locally shot audiovisual content over satellite to a production studio for editing and archiving.

To address above desire, MPEG-2 systems defined support for carriage of JPEG 2000 video, as defined in part 1 of the JPEG 2000 standard. Each JPEG 2000 video stream is conveyed in PES packets, whereby each PES packet contains exactly one complete JPEG 2000 access unit. In the header of each PES packet with a JPEG 2000 access unit, a PTS time stamp must be encoded. Due to the alignment between access units and PES packets, and a required PTS for each PES packet, random access is supported at each picture. Furthermore, due to the alignment there is no need to encode the length of the PES packets in the PES-packet-length field and therefore this field is encoded with the value zero. Note that the beginning of an access unit with JPEG 2000 video data is easily found in a transport stream by verifying whether the payload-unit-start-indicator is set to '1' in the packets that carry the JPEG 2000 video stream. Note though that a valid PES-packet-length value needs to be encoded if in future support in program streams for JPEG 2000 video stream is defined.

Each JPEG 2000, also referred to as J2K, video access unit consists of a so-called elsm header, an elementary stream header containing information on the coded video, followed by one or two contiguous codestreams for one picture. In case of a non-interlaced picture only one codestream is present, but for an interlaced picture two codestreams, as depicted in Figure 14.20, where the non-interlaced picture contains only code stream 0, whereas the interlaced picture contains in addition to code streams 0 also code stream 1. A J2K video elementary stream is a concatenation of J2K access units, and a J2K sequence is a subset of a J2K stream where all J2K access units have the same parameters in the elsm header. A J2K still picture consists of a J2K sequence with exactly one J2K access unit, whereby the PTS of a succeeding J2K access units must differ by at least two picture periods from the PTS of the J2K still picture.

The signalling parameters for carriage of JPEG 2000 video are listed in Table 14.28. For the stream-ID the same value is used as for a private-stream-1; for usage in a transport stream this

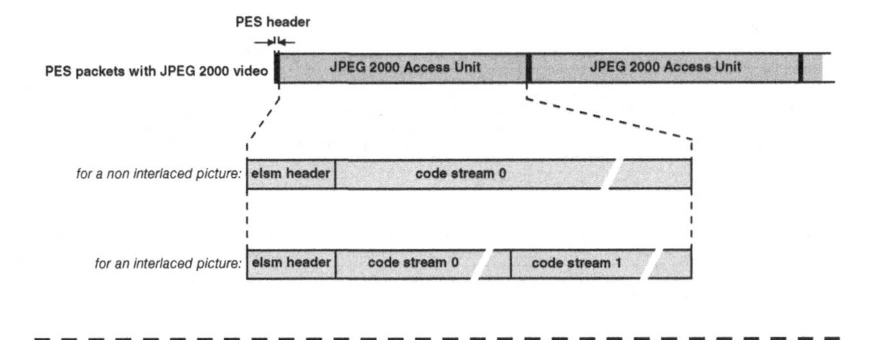

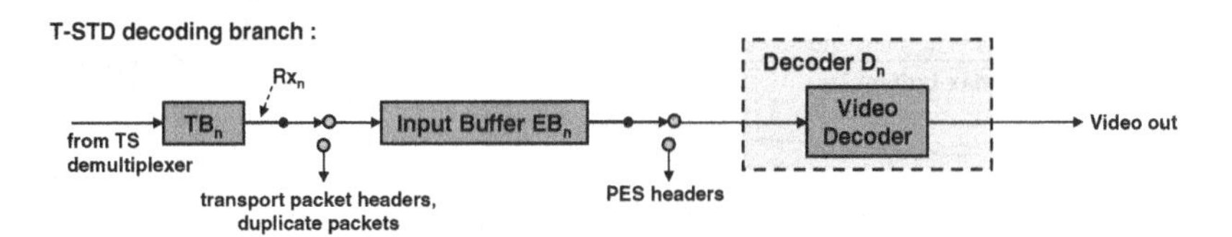

Figure 14.20 An access unit and T-STD decoding branch for JPEG 2000 video

Table 14.28 Signalling parameters for carriage of JPEG 2000 video

Stream-ID	0xBD JPEG 2000 video stream (same as private stream 1)
Stream-type	0x21: JPEG 2000 video
Descriptor-tag	0x32 (50): J2K video descriptor

should not create any problem. Figure 14.20 shows the decoding branch in the T-STD applicable for JPEG 2000 video. After the transport buffer TB_n only a single input buffer EB_n is used. Decoder D_n consists of only a video decoder; as no re-ordering is needed, the decoded video is directly provided at its output, without involving any buffer for the decoded video.

For each carried JPEG 2000 video stream a J2K video descriptor must be present to provide information on the associated stream, such as the profile and level, the picture size, the frame rate, whether or not interlaced video is contained in the stream and whether the stream may contain J2K still pictures. The J2K descriptor is depicted in Figure 14.21. The fields profile-and-level, horizontal-size, vertical-size, max-bit-rate, max-buffer-size, DEN-frame-rate, NUM-frame-rate and colour-specification are coded in the same manner as the equivalent fields specified in the JPEG 2000 video specification. The one-bit flag still-mode signals, if set to '1', that the associated stream may contain J2K still pictures; such pictures are absent if this flag is set to '0'. Finally, the interlaced-video flag signals whether or not the stream contains interlaced video. If set to '1', the parameters related to interlaced video must be present in the elsm header, while these parameters must be absent if the interlaced-video flag is set to '0'.

14.11 Metadata

MPEG-2 systems provides a generic framework for carriage of metadata. Metadata, as discussed here, provides information on audiovisual content, such as the date of creation, the owner, its purpose, the type of content, its title and a description of the content, including for

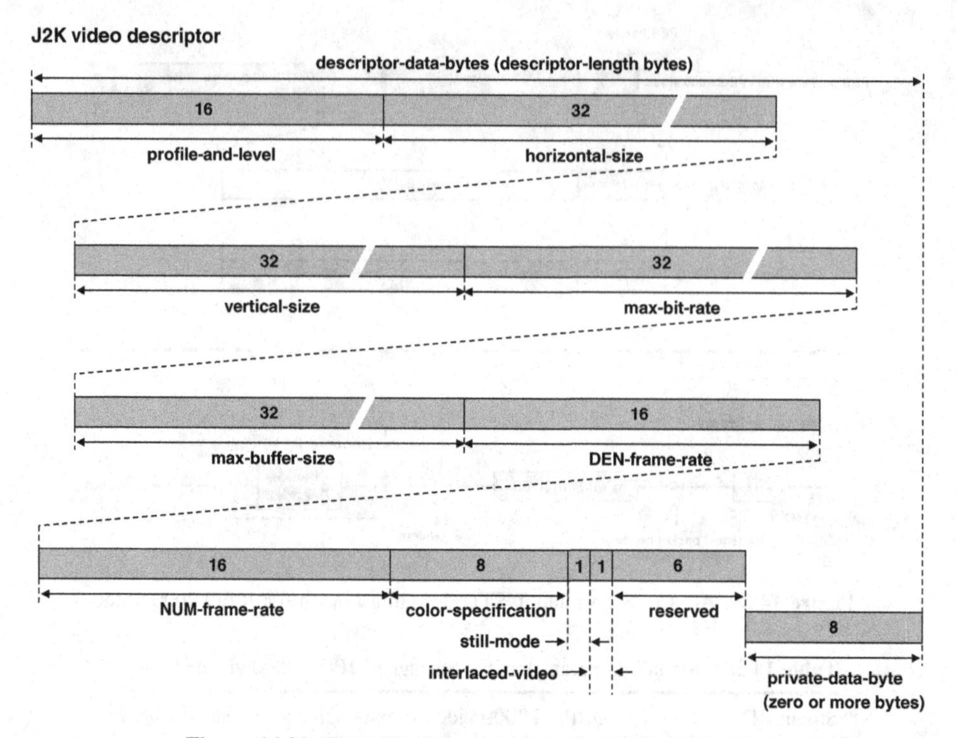

Figure 14.21 The J2K video descriptor for JPEG 2000 video

example actors in a movie. Metadata may contain timing information to signal segments, such as segments with certain actors or with the most interesting events in a sports game. For some more background on metadata see Section 4.1.

Distinction is made between metadata and metadata applications. Multiple metadata applications may use the same metadata. No specific format is targeted, neither for the metadata, nor for the metadata application. For carriage of metadata across MPEG-2 systems the following items are to be addressed.

- Means to convey and signal the metadata. For this purpose, various transport methods are defined. Carriage in PES packets and in specifically defined metadata sections is supported, but also DSM-CC can be used. For the latter, three options are available: the DSM-CC data carousel, the DSM-CC object carousel and the DSM-CC synchronized download protocol. Details on the use of these DSM-CC options is found in the DSM-CC specification. Information on conveyed metadata can be provided by means of the metadata descriptor.
- Constraints for metadata decoding in the STD model; to provide sufficient flexibility, a metadata STD descriptor is defined to provide applicable STD parameters for the associated metadata.
- Means to assign a label to audiovisual content conveyed in a MPEG-2 system stream, so that the content can be referenced by the metadata. For this purpose a content labelling descriptor is specified. The format of the assigned label is defined by the metadata application signalled in this descriptor; for instance ISAN or V_ISAN can be signalled as metadata application in

this case. The label can be assigned to a program or an elementary stream within a program, but may also be assigned to content segments, for example by including the descriptor in tables (beyond the scope of MPEG) describing segments.

- Means to associate timing information in the metadata to the delivery timing of the content. For this purpose the content labelling descriptor also provides the opportunity to signal the offset in time between the time used in the metadata (for instance the SMPTE time code) and the STC used for its delivery. For associated timing constraints see also Section 4.1.
- Means to point from audiovisual content in a MPEG-2 system stream to the associated metadata; this metadata can be carried in the same or in another MPEG-2 system stream, or by means beyond the scope of MPEG-2 systems, such as indicated by a URL. Pointing to metadata is possible by specifying the metadata locator record in the metadata pointer descriptor. The format of this record is determined by the metadata application signalled in this descriptor.

Metadata is assumed to be delivered in one or more metadata services to the receiver. A metadata service is defined to be a coherent set of metadata of the same format, delivered for a specific purpose. Each metadata service represents a collection of metadata access units. A metadata access unit is a global structure intended to be decoded at a specific instant in time; however, the format of a metadata access unit depends on the metadata-format and is beyond the scope of MPEG-2 systems. As a consequence, it is unknown what decoding actually involves. Decoding may only involve interpretation of the metadata; if the metadata is compressed, the decoder will likely decompress the data. If a PTS time stamp is associated to a metadata access unit, then this time stamp indicates the time at which the access unit is instantaneously removed from the input buffer prior to the metadata decoder in the STD. However, it is unspecified whether or not the time stamp indicates a certain time event, such as the beginning of a sequence with a certain actor.

For carriage over MPEG-2 systems, each metadata service is identified by a metadata-service-ID, so as to uniquely identify a metadata service amongst the metadata services within a transport stream or program stream. Much flexibility is allowed. Within a program in a transport stream multiple metadata services may be conveyed. The same metadata services can be associated with multiple programs and elementary streams, but the same program or elementary stream may also be associated with multiple metadata services.

The use of the various methods to transport metadata is discussed in the following paragraphs. Table 14.29 presents the parameters for signalling of metadata related structures. Metadata may also be carried in private-stream-1 and private-stream-2 PES packets; the format of such carriage is beyond the scope of MPEG-2 systems, but when appropriate, the metadata related descriptors listed in Table 14.29 can be used for signalling.

For synchronous delivery with encoded PTS time stamps, a metadata stream can be carried in PES packets and in synchronized download sections using the DSM-CC synchronized download protocol. In both cases the PTS time stamp indicates the 'decoding time' of the metadata access unit in the STD, that is the time the metadata AU is removed instantaneously from the input buffer prior to the metadata decoder. For carriage of metadata in PES packets and synchronized DSMCC download sections, a generic data structure for transport of metadata access units must be used, the so-called metadata access unit wrapper. This wrapper defines a concatenation of metadata AU-cells for one or more metadata services. Each metadata AU-cell can carry a single metadata access unit or a fragment thereof.

Table 14.29 Signalling parameters for metadata carriage

Stream-ID	OxFC: Metadata stream
Stream-type	0x15: Metadata carried in PES packets
	0x16: Metadata carried in metadata sections
	0x17: Metadata carried in DSM-CC data carousel
	0x18: Metadata carried in DSM-CC object carousel
	0x19: Metadata carried in DSM-CC synchronized download protocol
Table-ID	0x06: Metadata section
Descriptor-tag	0x24 (36): Content-labelling descriptor
	0x25 (37): Metadata-pointer descriptor
	0x26 (38): Metadata descriptor
	0x27 (39): Metadata STD descriptor

When metadata is carried in a PES packet or in a synchronized DSMCC download section, the first byte of the payload must be the first byte of a metadata AU-cell. A PTS in a PES packet or the DSMCC download section applies to each metadata AU in the payload and to each subsequent metadata access unit until another PTS is encoded. The PTS signals the presentation and decoding time of these metadata AU(s). When concatenating metadata AU-cells, fragments of metadata AUs of different metadata services may be interleaved, but obviously the encoded PTS time stamps must be in presentation/decoding order.

The structure of the metadata AU wrapper is depicted in Figure 14.22. The header of each metadata AU cell carries the eight-bit metadata-service-ID, signalling the metadata service the carried metadata access unit is associated with, followed by a byte with the sequence-number.

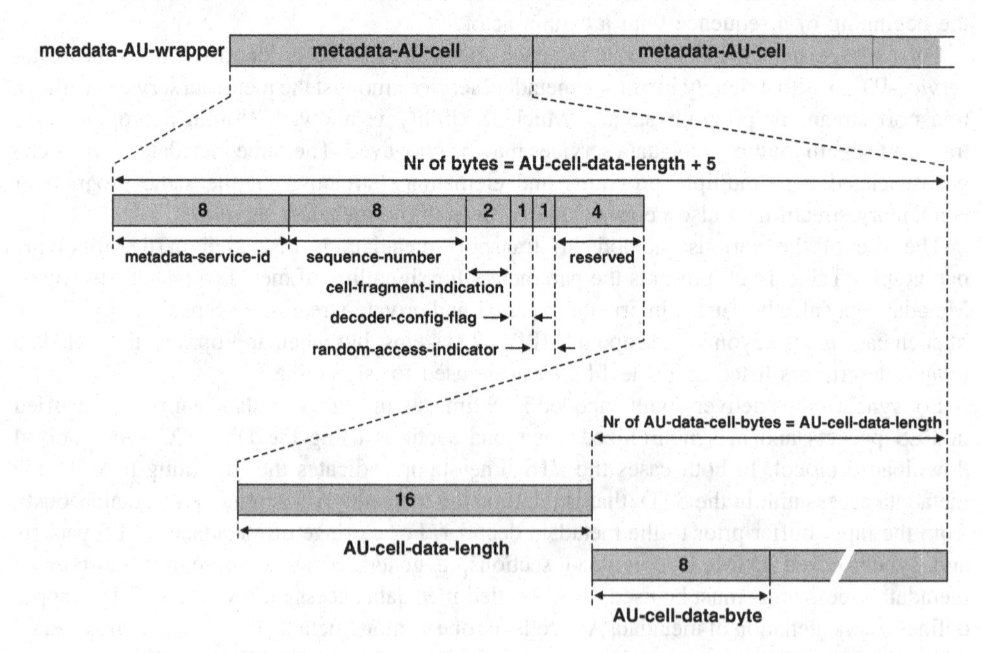

Figure 14.22 The metadata AU wrapper structure with metadata-AU-cells

Within the wrapper the sequence number is increased by one in each subsequent metadata AU-cell, irrespective of the encoded metadata-service ID value. The sequence number can be used in receivers to verify loss of one or more metadata- AU-cells. The two-bit cell-fragment-indication signals whether the cell carries a complete metadata access unit, or whether the access unit is carried in a series of cells, whereby the cell-fragment-indication signals whether in this cell a first fragment, a last fragment or a non-first and non-last fragment of a metadata access unit is carried. The decoder-config-flag signals the presence of decoder configuration information in the carried metadata access unit, possibly next to the presence of regular metadata. The one-bit random-access-indicator flag signals whether the metadata AU carried in this cell provides an entry point to the metadata service, where decoding is possible without information from previous metadata AU-cells. Finally, the 16-bit AU-cell-data-length specifies the number of immediately following AU-cell-data-bytes. Note that the use of the metadata AU wrapper allows parsers to process the metadata access units without any knowledge of the metadata format.

If no time stamps are needed for the carriage of metadata AUs, an asynchronous transport method can be used, with or without a carousel delivery mechanism. In case of a carousel, the information is repeated regularly, similar as pages in teletext systems where pages are refreshed after a certain period. If no carousel is needed, then metadata sections can be used, while for carousel usage a DSM-CC data carousel or a DSM-CC object carousel can be applied. The difference between both carousels is that the object carousel extends the more limited data carousel by specifying a standard format for representing a file system directory structure. Further details are found in the DSM-CC specification. Note that the file structure provided by a DSM-CC object carousel is particularly beneficial when there is a need to express the hierarchical organization of the metadata structure in the transport.

The format of a metadata section is shown in Figure 14.23. Each metadata section carries either a complete metadata AU or a fragment thereof. The header of the metadata section follows the format of the long section header, discussed in Section 13.5 (see Figure 13.18).The

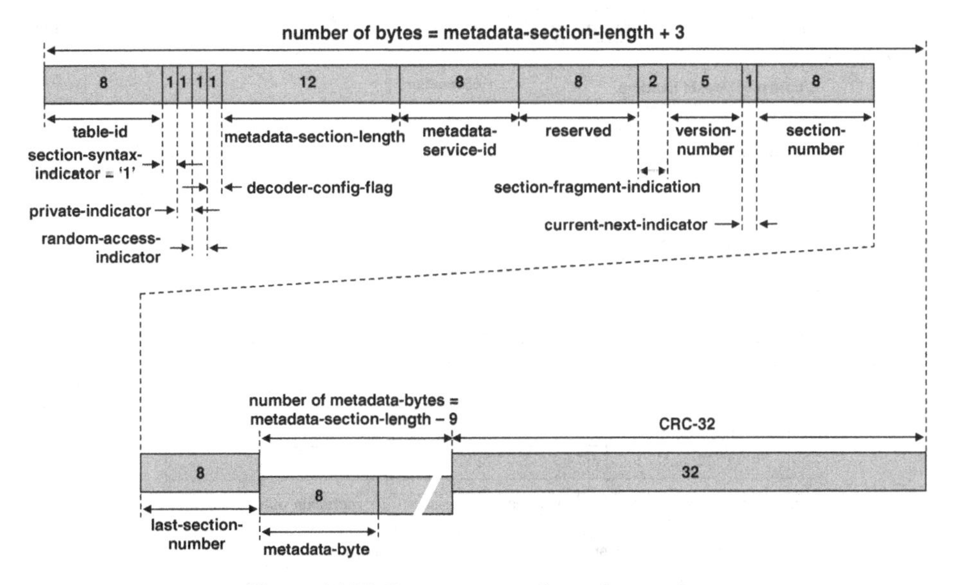

Figure 14.23 Data structure of metadata section

signalling of the carried metadata in the section header is very similar to the signalling in the metadata AU-cell header. The metadata-service-ID, the random-access-indicator, the decoder-config-flag and the two-bit section-fragment-indication field, are all encoded in the same manner as in the metadata AU-cell of the metadata AU wrapper; see Figure 14.22 and the associated paragraph. The difference is that when using metadata AU-cells, the metadata stream formed when concatenating the access units can have any length, while the metadata conveyed in metadata sections requires structuring in metadata tables. Each metadata table contains one or more complete metadata access units from one or more metadata services.

The fragmenting of a metadata table and its versioning are the same as for PSI tables (see also Sections 9.4.4 and 13.5). A complete metadata table can be fragmented for transport in up to 256 metadata sections, each signalled by a section-number, starting with the value zero and incremented by one for each additional section. The last-section-number signals the section-number of the section with the highest section-number; hence if the highest section-number is equal to N, then there are $(N+1)$ sections with metadata. A metadata table can be updated by conveying the update with a new version-number that is incremented by one (modulo 32). The current-next-indicator signals which table is applicable and which is the next one to become valid. Also the use of section-number and the last-section-number is the same as for PSI tables.

All metadata transport methods that can be signalled by the stream-type values listed in Table 14.29, except PES packets, use sections. The use of sections is restricted to transport streams; as a consequence, MPEG-2 systems specifies that in a program stream metadata can only be carried in PES packets.

Figure 14.24 depicts the branches in the STD model applicable for the decoding of metadata. Decoder D_n contains only the metadata decoder. In case of synchronous delivery

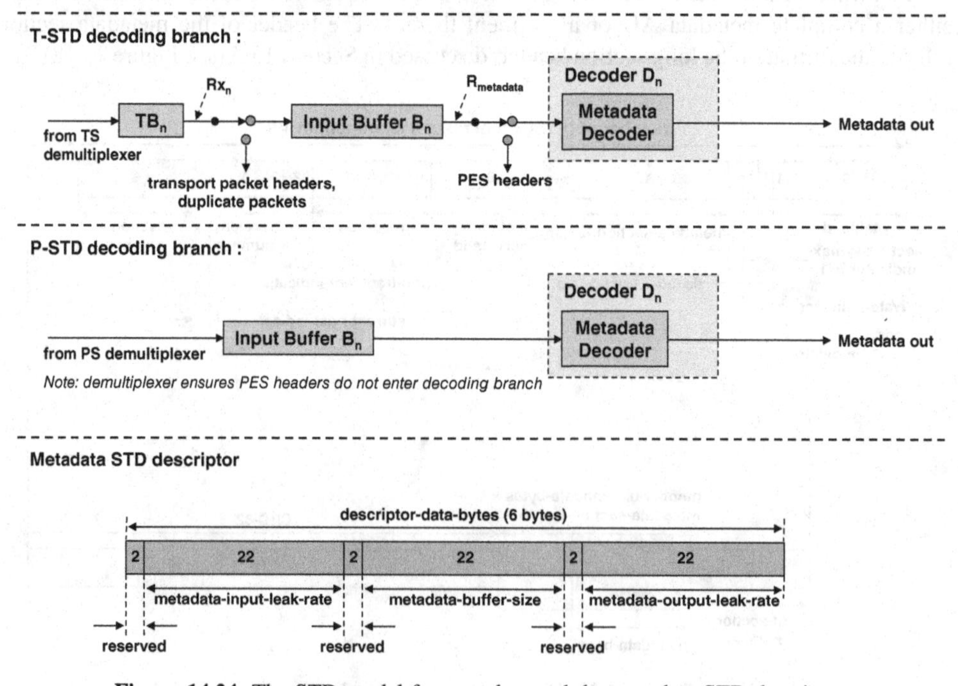

Figure 14.24 The STD model for metadata and the metadata STD descriptor

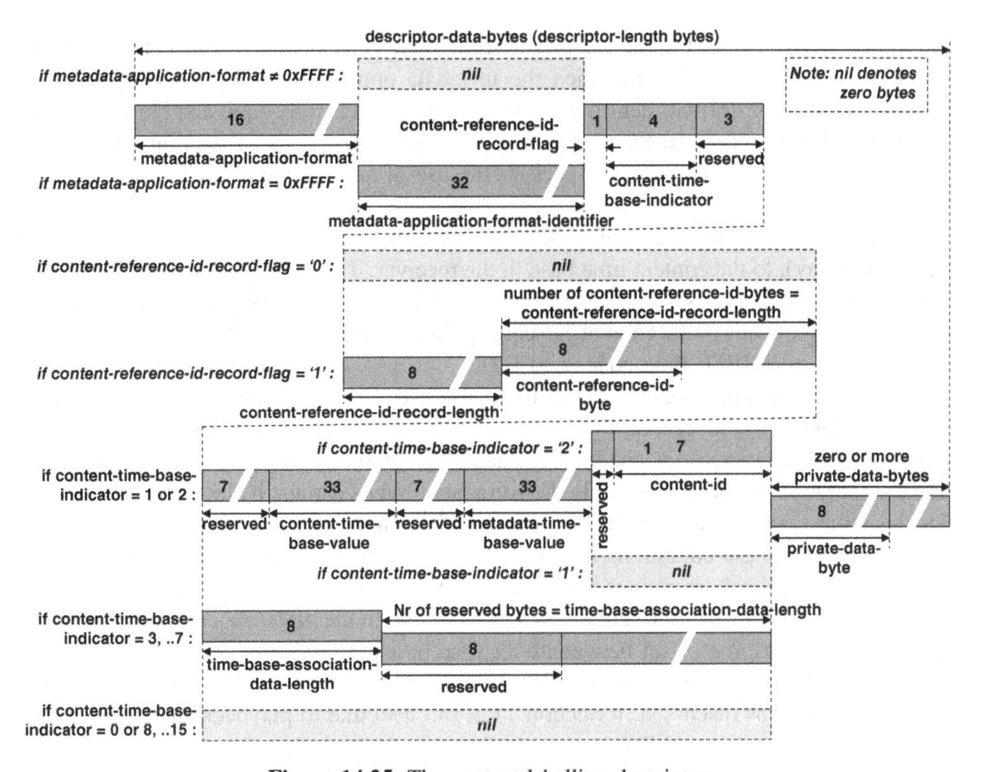

Figure 14.25 The content labelling descriptor

of the metadata, the decoding in the P-STD and the T-STD is controlled by PTS time stamps. Each metadata access unit is removed instantaneously from buffer B_n at the instant in time indicated by the PTS. In the case of asynchronous delivery, in the T-STD a leaky buffer model is applied with a transfer rate $R_{metadata}$ for the data out of B_n to the metadata decoder.

To provide metadata applications maximum flexibility, MPEG-2 systems does not specify values for the decoding of metadata in the STD. Instead the values of these parameters can be conveyed in the metadata descriptor; see Figure 14.24. The value of Rx_n is specified by the metadata-input-leak-rate field, while the size BS_n of buffer B_n is provided by the metadata-buffer-size field. Finally, for asynchronous delivery of the metadata, the leak rate $R_{metadata}$ out of B_n is defined by the encoded value of the metadata-input-leak-rate field. Interoperability between metadata services and receivers requires explicit values for the STD parameters for metadata formats; in future such values may be specified by MPEG or by entities beyond MPEG.

The data structure of the descriptor-data-bytes of the content labelling descriptor is shown in Figure 14.25. By means of this descriptor a label, the content-reference-ID-record, can be assigned to content; the format of this label is defined by the metadata application, signalled by the 16-bit metadata-application-format. The values 0x0010 and 0x0011 signal usage of ISAN and V-ISAN in binary form; the values 0x0000 to 0x000F and 0x0012 to 0x00FF are reserved.

The values 0x0100-0xFFFE are for private use. If the metadata-application-format field is encoded with the value 0xFFFF, then the metadata application is signalled by the 32-bit metadata-application-format-identifier, the coding of which is fully equivalent to the coding of the format-identifier field in the registration descriptor, discussed in Section 13.3.3.7. The signalled metadata application also specifies the use of any private-data bytes the descriptor may contain.

Furthermore, the content labelling descriptor can be used to relate the time used in the metadata (if any), to the content time base in the receiver. Typically, the time base used in the metadata is transport agnostic and not related to the STC, other than it is required to be locked to the sampling frequency of the content. To convey the time in the metadata, various formats may be used, such as SMPTE time codes and UTC.

In the content labelling descriptor the time base used in the metadata is related to the time base of the content in the receiver by specifying the value of both time bases at the same instant in time. For simplicity reasons, the same metadata time base must be used for the entire content the metadata is associated with, while the metadata time base must be locked to the sampling clocks of the content; furthermore, in the metadata time base no time discontinuities are permitted. Under these conditions the time offset between both time bases remains constant during normal playback operations as long as no time base discontinuities occur at the receiver. Hence, in the receiver the time references to the content in the metadata can be maintained by taking the constant time offset between both time bases into account.

However, at the receiver various STC discontinuities may occur, not only due to local program insertion as discussed in Section 11.3, but also due to playback operations such as slow motion and fast forward. To compensate for such discontinuities, in some cases, instead of the STC, the Normal Play Time (NPT) concept as defined in DSM-CC can be used, whereby the STC time can be compensated for playback operations and local program insertions. For example, during slow motion at half of the normal speed, the NPT progresses at half the normal speed; and during reverse playback at normal speed, the NPT time reverses at normal speed; while during a commercial break the NPT is paused. In this way an unambiguous relationship between the NPT and the content time base is ensured.

The four-bit content-time-base-indicator signals which content time base is used when providing the offset between the metadata time base and the content time base. The value zero signals that no content time base is used in the descriptor, while the values one and two signal usage of STC or NPT, respectively. The values 3–7 are reserved, while the values 8–15 signal usage of a privately defined content time base. If STC or NPT is used, then the time offset is provided by two 33-bit fields providing the value of both time bases at the same instant in time in units of 90 kHz. The metadata may use any time clock, but its value in the descriptor needs to be expressed in units of 90 kHz. For example, a SMPTE time code in hours, minutes, seconds and frames needs to be expressed in the corresponding number of 90 kHz units. To accommodate for minor inaccuracies, when reference is made to a specific picture or audio frame, rounding to the closest PTS value of picture or audio frame should be applied.

If NPT is used, also the content-ID is provided, so as to signal the value of the content-ID field in the NPT reference descriptor for the applied NPT time base. In case a reserved value is encoded in the content-time-base-indicator, the time-base-association-data-length specifies the number of reserved bytes needed to carry any time base association information (defined in future) for the encoded content-time-base-indicator value.

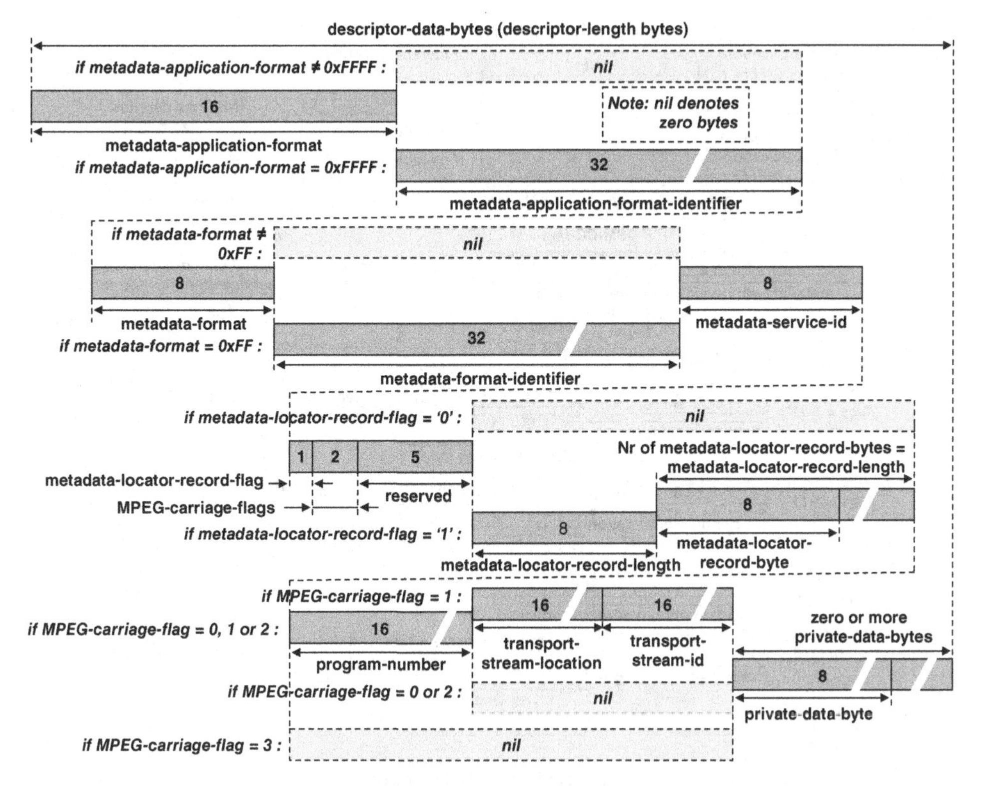

Figure 14.26 The metadata pointer descriptor

The metadata pointer descriptor, depicted in Figure 14.26, can be used to point to a single metadata service, and associates this metadata service to the elementary stream or program the descriptor is associated with. When the metadata stream with the metadata service is carried in a transport stream or in a program stream, then more information is provided on its carriage. More than one metadata pointer descriptor can be present for the same content, each pointing to a different metadata service.

The pointer is provided by the metadata-locator-record; the format of this record is determined by the metadata application signalled by the metadata-application-format field, encoded in the same manner as in the content labelling descriptor. The signalled metadata application determines also the use of any private-data bytes the descriptor may contain. The metadata-format signals the format of the metadata applied in the metadata service pointed to by the encoded value in the metadata-service-ID field. The value 0xFF of the metadata-format field signals the presence of the metadata-format identifier, the coding of which is fully equivalent to the coding of the format-identifier field in the registration descriptor, discussed in Section 13.3.3.7. The metadata-format values 0x00–0x0F and 0x12–0x3E are reserved. The values 0x10 and 0x11 signal usage of the MPEG-7 TeM and BiM formats [22], while the values 0x3F and 0xFF indicate a metadata format specified by the signalled

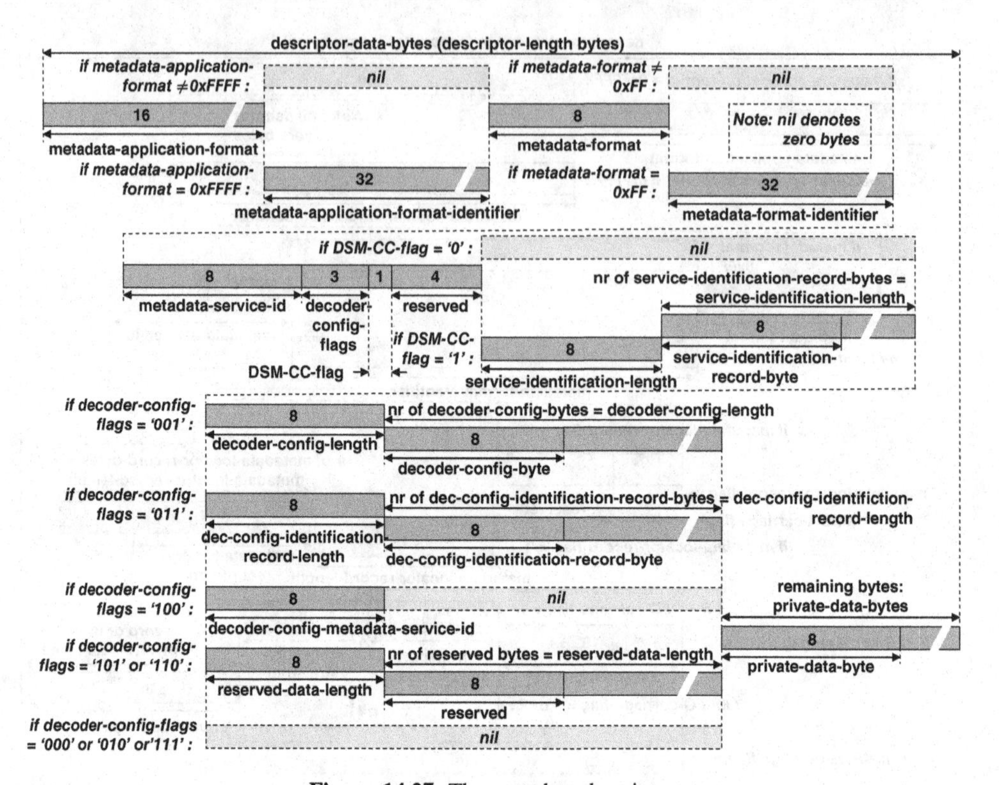

Figure 14.27 The metadata descriptor

metadata application or by the entity signalled by the encoded value in the metadata-format-identifier field. The values 0x40-0xFE are for private use.

The metadata pointed to is identified by the metadata-service-ID. When the metadata is carried in a transport stream or program stream using PES packets, metadata sections or synchronized DSM-CC download sections, the applied headers signal the metadata-service-ID of the carried metadata. However, when the metadata is found in a DSM-CC data or object carousel or in a location beyond a MPEG-2 system stream, then the metadata application determines how to identify the metadata associated with the signalled metadata-service-ID. When a DSM-CC carousel is used, the metadata may be identified by the applied values of the transaction-ID and the module-ID in a data carousel or by the file name used in an object carousel.

By means of the MPEG-carriage-flags in the metadata pointer descriptor it is signalled if and where the metadata is found in a transport stream or program stream. If the metadata is found in the same transport stream as the metadata pointer descriptor (MPEG-carriage-flags = '00'), then the program-number is provided of the program carrying the metadata. If the metadata is carried in another transport stream (MPEG-carriage-flags = '01'), then for this transport stream also the transport-stream-location and the transport-stream-ID are provided. The latter is defined in the PAT (see Figure 9.14 and Section 13.5). The transport-stream-location is a private field that may be defined by application standardization bodies; for example, in a DVB compliant environment, the original-network-ID may be used. In case the metadata is found in a program stream (MPEG-carriage-flags = '10'), the program-number is provided as well. In either case, also when the metadata is not carried in a transport stream and neither in a program stream, one or more private-data bytes may be carried in the descriptor, to provide further information; the format of these private-data bytes is determined by the metadata application signalled by the metadata-application-format field.

Figure 14.27 presents the metadata descriptor providing information on the metadata service carried in a transport stream or in a program stream. For the metadata service identified by the value of the metadata-service-ID encoded in the metadata descriptor, the applied metadata format is signalled by the metadata-format field; this field is encoded in the same manner as in the metadata pointer descriptor.

The one-bit DSM-CC flag signals whether the descriptor is associated with metadata carried in a DSM-CC carousel; if yes, then the service-identification-record field provides the data needed to retrieve the signalled metadata-service from the DSM-CC carousel. The format of this record is determined by the metadata application signalled by the metadata-application-format field, encoded in the same manner as in the content labelling descriptor. To identify the metadata service, the record may comprise the transaction-ID and the module-ID when a DSM-CC data carousel is used and a file name or unique object identifier in case the metadata is conveyed in a DSM-CC object carousel.

The three-bit decoder-config-flags field indicates whether the metadata decoder needs decoder configuration data. An encoded value '000' signals that no such data is needed. The value '001' indicates carriage of the decoder configuration data by means of decoder-config bytes in this descriptor. Note that carriage in the descriptor is practical only if the number of bytes of the decoder configuration is small.

If the metadata service is carried in a DSM-CC carousel, then the decoder configuration data may be carried in a same type of carousel; if so, then this is signalled by a decoder-config-flags value '011', in which case the decoder configuration is identified by the dec-config-identification-record field encoded in a way similar as the service-identification-record. The format of both records is determined by the same metadata application.

The value '010'of the decoder-config-flags signal that the decoder configuration is carried in the same metadata service to which this descriptor applies. If the decoder configuration is carried by another metadata service within the same program, then the decoder-config-flags are encoded with the value '100', while the metadata service with the configuration data is identified by the decoder-config-metadata-service-ID; in this case the decoder configuration may be contained in the metadata stream itself, in the associated descriptor or in the associated DSM-CC carousel.

The values '101' and '110' of the decoder-config-flags are reserved, while the value '111' signals that the location of the decoder configuration information is privately defined.

14.12 Overview of Assigned Stream-type Values

For convenience, Table 14.30 provides an overview of the stream-type values assigned for the signalling of content carried in a MPEG-2 system stream.

Table 14.30 Assignment of stream-type values

Stream-type value	Signalled content
0x00	Reserved
0x01	MPEG-1 video
0x02	MPEG-2 video or MPEG-1 constrained parameter video
0x03	MPEG-1 audio (or MPEG-1 audio compatible MPEG-2 audio; see Section 14.3.2)
0x04	MPEG-2 audio
0x05	Private data carried in sections
0x06	Private data carried in PES packets
0x07	ISO/IEC 13522 (MHEG) defined
0x08	DSM-CC according to Annex B of MPEG-2 systems
0x09	ITU-T H.222.1 defined
0x0A	DSM-CC multi-protocol encapsulation
0x0B	DSM-CC U-N messages
0x0C	DSM-CC stream descriptors
0x0D	DSM-CC sections, any type, including private data
0x0E	Auxiliary data (see Sections 14.2.3 and 14.3.5)
0x0F	MPEG-2 AAC audio with the ADTS transport syntax
0x10	MPEG-4 visual
0x11	MPEG-4 audio with the LATM transport syntax defined in ISO/IEC 14496-3
0x12	MPEG-4 SL packetized stream or FlexMux stream carried in PES packets
0x13	MPEG-4 SL packetized stream or FlexMux stream carried in MPEG-4 sections
0x14	DSM-CC Synchronized download protocol
0x15	Metadata carried in PES packets
0x16	Metadata carried in metadata sections
0x17	Metadata carried in DSM-CC data carousel
0x18	Metadata carried in DSM-CC object carousel
0x19	Metadata carried in DSM-CC synchronized download protocol
0x1A	IPMP stream, as defined in ISO/IEC 13818-11
0x1B	AVC video stream (may be a SVC or MVC sub-bitstream; see Sections 14.8 and 14.9.3)
0x1C	MPEG-4 audio without any additional transport syntax
0x1D	MPEG-4 text
0x1E	Auxiliary video stream as defined in ISO/IEC 23002-3
0x1F	SVC video sub-bitstream
0x20	MVC elementary stream without the base view, but with one or more other views in subsequent decoding order
0x21	JPEG 2000 video
0x22	MPEG-2 video stream with additional view of service compatible stereoscopic 3D video
0x23	AVC video encoded additional view of service compatible stereoscopic 3D video
0x24 – 0x7E	Reserved for future use by MPEG-2 systems (as of 2013)
0x7F	(MPEG-4) IPMP stream
0x80 – 0xFF	User private

References

1. ISO/IEC (1993) 11172-2. Information technology – Coding of moving pictures and associated audio for digital storage media at up to about 1.5 Mbits/s – Part 2: Video. http://www.iso.org/iso/catalogue_detail.htm?csnumber=22411.
2. ISO/IEC (1993) 11172-3. Information technology – Coding of moving pictures and associated audio for digital storage media at up to about 1.5 Mbits/s – Part 3: Audio http://www.iso.org/iso/catalogue_detail.htm?csnumber=22412.
3. ISO/IEC (1993) 11172-1. Information technology – Coding of moving pictures and associated audio for digital storage media at up to about 1.5 Mbits/s – Part 1: Systems. http://www.iso.org/iso/catalogue_detail.htm?csnumber=19180.
4. ISO/IEC (2013) 13818-2. Information technology – Generic coding of moving pictures and associated audio information – Part 2: Video. http://www.iso.org/iso/catalogue_detail.htm?csnumber=61152.
5. ISO/IEC (1998) 13818-3. Information technology – Generic coding of moving pictures and associated audio information – Part 3: Audio http://www.iso.org/iso/catalogue_detail.htm?csnumber=26797.
6. ISO/IEC (2006) 13818-7. Information technology – Generic coding of moving pictures and associated audio information – Part 7: Advanced Audio Coding (AAC) http://www.iso.org/iso/catalogue_detail.htm?csnumber=43345.
7. Informative Annex B of ISO/IEC (2013) 13818-1. Information technology – Generic coding of moving pictures and associated audio information – Part 1: Systems http://www.iso.org/iso/catalogue_detail.htm?csnumber=62074.
8. ISO/IEC (1998) 13818-6. Information technology – Generic coding of moving pictures and associated audio information – Part 6: Extensions for DSM-CC: http://www.iso.org/iso/catalogue_detail.htm?csnumber=25039. Note that a number of corrigenda and amendments apply; see above link.
9. ISO/IEC (2004) 13818-11. Information technology– Generic coding of moving pictures and associated audio information – Part 11: IPMP on MPEG-2 systems: http://www.iso.org/iso/catalogue_detail.htm?csnumber=37680.
10. ITU-T Recommendation H.222.1 (03/96) Multimedia multiplex and synchronization for audiovisual communication in ATM environments: http://www.itu.int/rec/T-REC-H.222.1-199603-I/en.
11. ISO/IEC (1997) 13522-1. Information technology – Coding of multimedia and hypermedia information – Part 1: MHEG object representation, Base notation (ASN.1): http://www.iso.org/iso/catalogue_detail.htm?csnumber=22153.
12. ISO/IEC (2004) 14496-2. Information technology – Coding of audio-visual objects – Part 2: Visual: http://www.iso.org/iso/catalogue_detail.htm?csnumber=39259. Note that a number of corrigenda and amendments apply; see above link.
13. ISO/IEC (2009) 14496-3. Information technology– Coding of audio-visual objects – Part 3: Audio: http://www.iso.org/iso/catalogue_detail.htm?csnumber=53943. Note that a number of corrigenda and amendments apply; see above link.
14. ISO/IEC (2006) 14496-17. Information technology – Coding of audio-visual objects – Part 17: Streaming text format: http://www.iso.org/iso/catalogue_detail.htm?csnumber=39478.
15. ISO/IEC (2010) 14496-1. Information technology – Coding of audio-visual objects – Part 1: Systems: http://www.iso.org/iso/catalogue_detail.htm?csnumber=55688. Note that an amendment applies; see above link.
16. ISO/IEC (2010) 14496-10. Information technology – Coding of audio-visual objects – Part 10: Advanced Video Coding (AVC) http://www.iso.org/iso/catalogue_detail.htm?csnumber=56538 also published as ITU-T Recommendation H 264: http://www.itu.int/rec/T-REC-H.264-201106-P;A paper with an overview of the AVC technology: Wiegand, T., Sullivan, G.J., Bjøntegaard, G. and Luthra, A. (2003) Overview of the H.264/AVC video coding standard. *IEEE Transactions on Circuits and Systems for Video Technolology*, **13** (7), 560–576.
17. SVC is specified in Annex G of the AVC standard; see [16]. A paper with an overview of the SVC technology: Schwarz, H., Marpe, D. and Wiegand, T. (2007) Overview of the scalable video coding extension of the H.264/AVC standard. *IEEE Transactions on Circuits and Systems for Video Technolology*, **17** (9), 1103–1120;A paper on carriage of SVC over MPEG-2 systems and IP: Schierl, T., Grüneberg, K. and Wiegand, T. (2009) Scalable video coding over RTP and MPEG-2 transport stream in broadcast and IPTV channels. *IEEE Wireless Communications*.
18. ISO/IEC (2007) 23002-3. Information technology – MPEG video technologies – Part 3: Representation of auxiliary video and supplemental information: http://www.iso.org/iso/catalogue_detail.htm?csnumber=44354.

19. MVC is specified in Annex H of the AVC standard; see [16]. A paper with an overview of the MVC technology: Vetro, A., Wiegand, T. and Sullivan, G.J. (2011) Overview of the stereo and multiview video coding extensions of the H.264/MPEG-4 AVC standard. *Proceedings of the IEEE, Special Issue on 3D Media and Displays*, **99** (4), 626–642;A paper on carriage of MVC over MPEG-2 systems and IP: Schierl, T. and Narasimhan, S. (2011) Transport and storage systems for 3-D video using MPEG-2 systems, RTP, and ISO file format. *Proceedings of the IEEE, Special Issue on 3D Media and Displays*, **99** (4), 671–683.
20. ISO/IEC (1994) 10918-1. Information technology – Digital compression and coding of continuous-tone still images – Part 1: Requirements and guidelines: http://www.iso.org/iso/catalogue_detail.htm?csnumber=18902.
21. ISO/IEC (2004) 15444-1. Information technology – JPEG 2000 image coding system – Part 1: Core coding system: http://www.iso.org/iso/catalogue_detail.htm?csnumber= 37674.
22. The textual format TeM and the binary format BiM are specified in part 1 of MPEG-7: ISO/IEC (2002) 15938-1. Information technology– Multimedia content description interface – Part 1: Systems: http://www.iso.org/iso/catalogue_detail.htm?csnumber=34228. Note that a number of corrigenda and amendments apply; see above link.

15

The Real-Time Interface for Transport Streams

Why a real-time interface is needed for the delivery of transport streams. The choice for a maximum value for the jitter in byte arrival of a transport stream and the consequences thereof for the receiver.

When it became clear that digital TV applications would use transport streams, ICs were developed for set-top boxes for digital TV services across satellite, cable and other networks, upon which the need arose for a real-time interface for system decoders. The STD model assumes an idealized decoder model and a system stream delivered with mathematical accuracy. Decoder manufacturers can compensate for ways in which their actual design differs from the processing in the STD; however, in practice the system stream is not delivered with mathematical accuracy, but with characteristics specific for the applied delivery network(s), such as jitter and variable, sometimes unknown, delays.

For decoder manufacturers and for more general interoperability reasons it is desirable that a real-time interface is designed for the delivery of system streams. In particular for transport streams the need for such interface was recognized, as the characteristics of its delivery are typically fully determined by the network. An agreed upon real-time interface, met by practical networks, ensures that 'universal' system decoders can be designed, suitable for each network with delivery characteristics within the constraints of this interface. At the time the real-time interface was defined, most foreseen TV broadcast services applied across satellite, cable and terrestrial networks were sharing similar delivery characteristics with a relatively small jitter. However, at that time ATM networks were also believed to have a promising future, but ATM networks may cause far more jitter and delays that may even a priori be unknown.

Fortunately, this issue could be resolved by a parameterized approach for the real-time interface. For low-jitter applications a maximum jitter was defined of 50 μs, suitable for delivery across satellite, cable and terrestrial networks, while applications causing more jitter can claim compliancy with the real-time interface by specifying the maximum value of the applicable jitter. The real-time interface specification for transport stream decoders is published as Part 9 of the MPEG-2 specification [1]. In practice, ATM networks did not become the

Fundamentals and Evolution of MPEG-2 Systems: Paving the MPEG Road, First Edition. Jan van der Meer.
© 2014 John Wiley & Sons, Ltd. Published 2014 by John Wiley & Sons, Ltd.

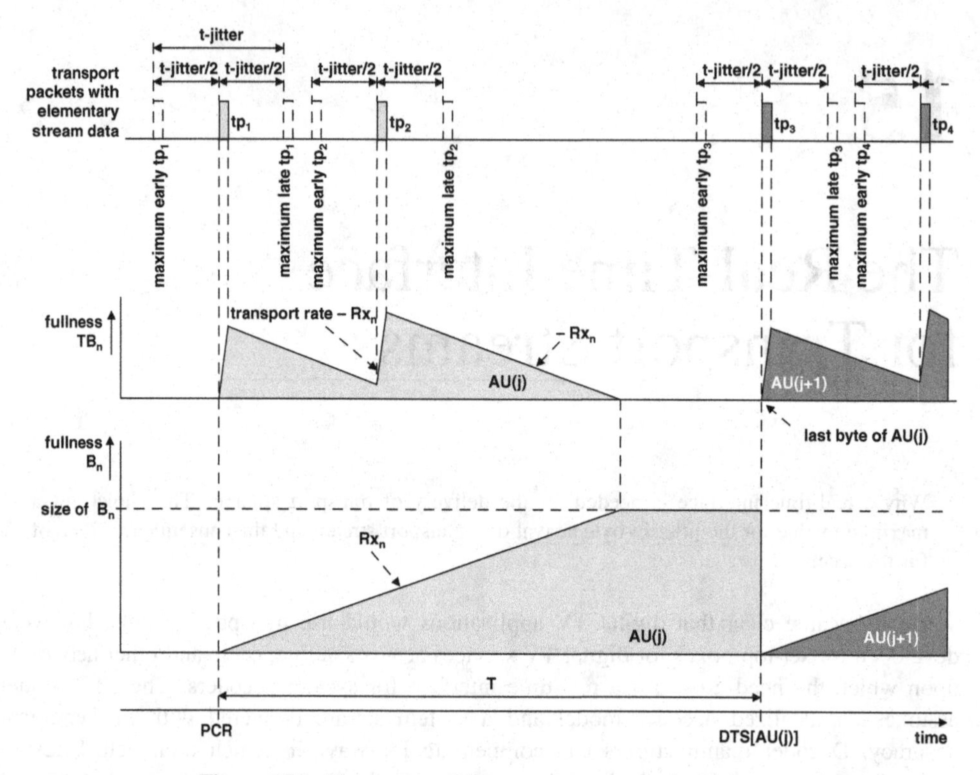

Figure 15.1 The impact of jitter on transport packet delivery

success that was originally anticipated, so that the real-time interface specification is mainly of value for low-jitter applications.

The jitter constraints specified in the real-time interface are illustrated in Figure 15.1; each transport stream packet may arrive at the input of the decoder with a timing uncertainty equal to t-jitter, divided equally between an 'early' and a 'late' component. Compared to its 'perfect' delivery time, each transport packet may arrive (t-jitter/2) seconds earlier or later. In Figure 15.1 a rather hypothetical example is included, where transport packets tp_1 up to tp_4 enter a decoding branch with a transport buffer TB_n and (for simplicity reasons) a single buffer B_n prior to the decoder. The STC time starts at the arrival of tp_1; this packet carries a PCR time sample.

To show the worst case consequences of jittered delivery, in the example of Figure 15.1, tp_3 contains the last byte of access unit AU(*j*), while this access unit is decoded exactly at the time this last byte arrives. Moreover, when this byte arrives, buffer B_n is exactly full. See the flow of the coded data through TB_n and B_n in Figure 15.1. As a result in this (indeed very) hypothetical situation, the position of packet tp_3 is extremely critical: if tp_3 is delivered only slightly early, buffer B_n will overflow, while any late delivery of tp_3 will cause underflow of B_n.

Figures 15.2 and 15.3 show delivery schedules of the transport packets in Figure 15.1, one causing worst case underflow and the other causing worst case overflow. Access unit AU(*j*) is scheduled to be decoded a delay *T* after the arrival of the PCR in tp_1. Early or late delivery of tp_1 will cause a shift in STC time and thereby a shift of the DTS of AU(*j*), because the distance in

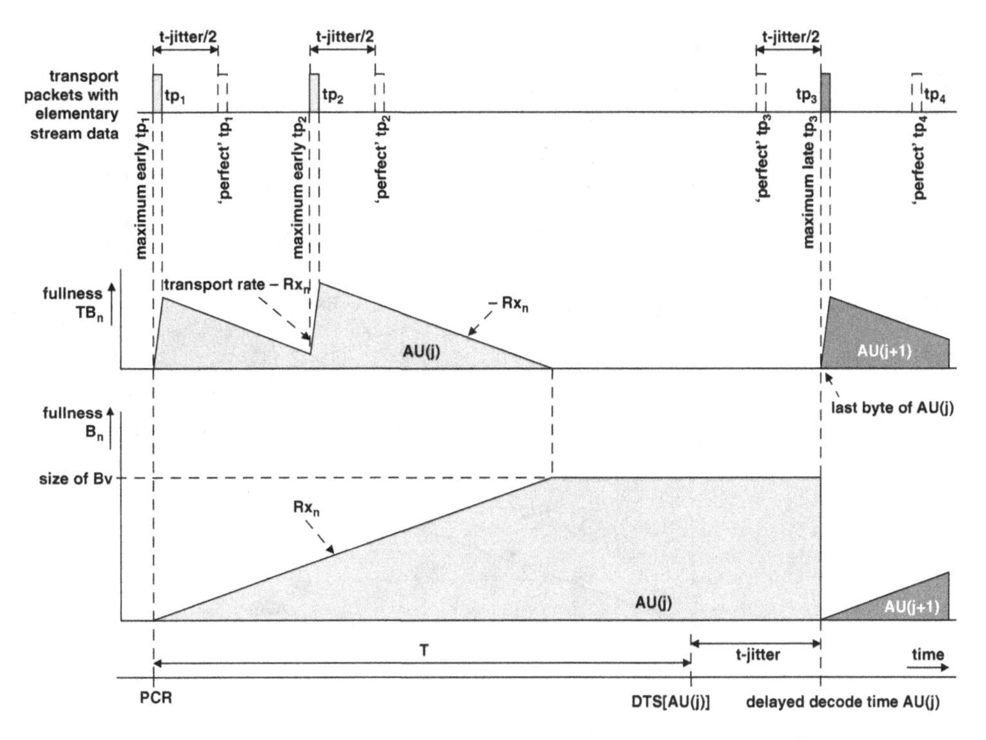

Figure 15.2 Preventing underflow due to jitter in transport packet delivery

time between the PCR and the DTS remains the same value T. Note that jitter will influence a PLL used for reconstructing the STC in a receiver, but that the response will be relatively slow and can be ignored for the analysis here.

Figure 15.2 shows a jittered delivery, whereby packets tp_1 and tp_2 are maximum early, while tp_3 and tp_4 are maximum late. Compared to the 'just-in-time' delivery in Figure 15.1, the early delivery of tp_1 causes that the DTS value of AU(j) is reached (t-jitter/2) seconds earlier than in Figure 15.1. Furthermore, due to the late delivery of tp_3, the last byte of AU(j) arrives (t-jitter/2) seconds later than in Figure 15.1. Hence, to prevent underflow of B_n in this case, the decode time of AU(j) must be delayed by at least t-jitter seconds. When the decoding process is delayed by t-jitter seconds, under normal operation more bytes will be delivered to buffer B_n than without such delay; at most (t-jitter $\times$ Rx_n) bytes, if Rx_n is expressed in bytes per second. In conclusion, the above worst case example shows the following. When the jitter in the delivery of transport packets is equal to t-jitter seconds, then to prevent underflow of buffer B_n, the size of B_n must be enlarged with (t-jitter $\times$ Rx_n) bytes.

In Figure 15.3 the worst case overflow situation is shown: tp_1 and tp_2 are maximum late, while tp_3 and tp_4 are maximum early. Due to the late delivery of tp_1 with its PCR, the DTS of AU(j) is (t-jitter/2) seconds later than in Figure 15.1. Furthermore, the early delivery of tp_3 causes that the last byte of AU(j) and the first bytes of AU($j+1$) contained in tp_3 enter B_n (t-jitter/2) seconds earlier than in Figure 15.1. Hence, relative to the DTS of AU(j), the data of tp_3 enters B_n t-jitter seconds earlier than in Figure 15.1. If Rx_n is again expressed in bytes per second, then during this time (t-jitter) $\times$ Rx_n bytes will enter B_n.

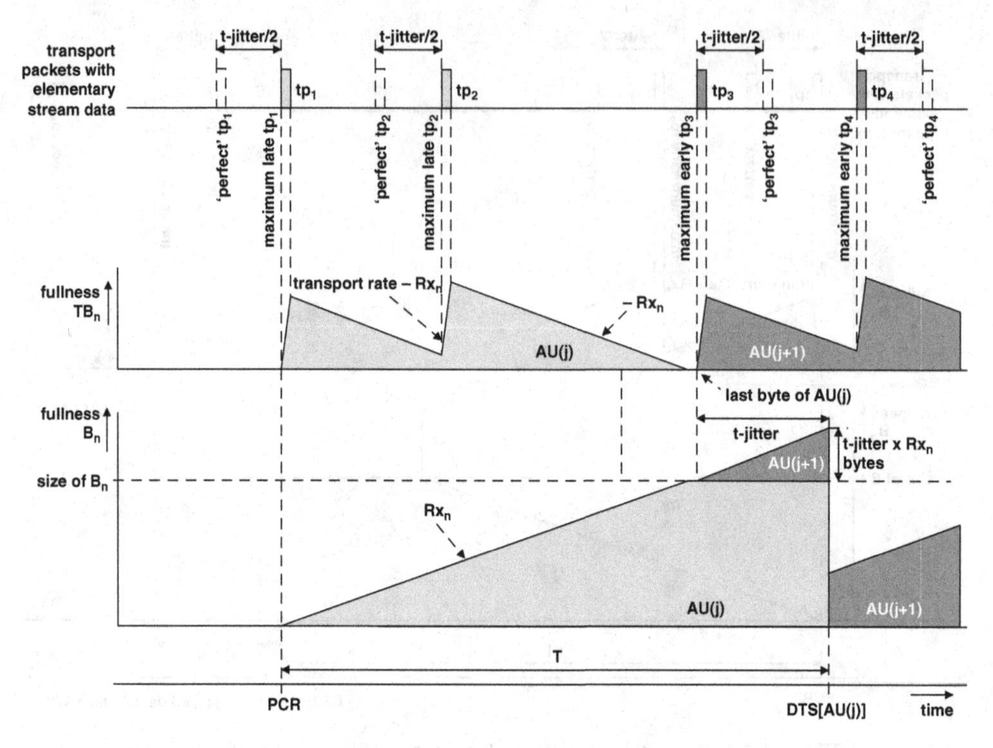

Figure 15.3 Preventing overflow due to jitter in transport packet delivery

In other words, when a maximum jitter in the delivery of transport packets is equal to t-jitter seconds, then for preventing underflow and overflow of buffer B_n, the size of B_n must be enlarged with $2 \times$(t-jitter)$\times Rx_n$ bytes.

The jitter has also impact on the required size of transport buffers. Figure 15.4 shows two delivery schedules of four transport packets, one schedule without any jitter and the other schedule with jitter, whereby two transport packets, tp_1 and tp_2 are delivered maximum late, while the other two, tp_3 and tp_4, are delivered maximum early. For both delivery schedules the fullness of the transport buffer TB_n is presented. Prior to the delivery of tp_1, buffer TB_n is empty. As soon as the tp_1 bytes enter TB_n at the transport rate, the bytes start leaking out of TB_n at rate Rx_n; for convenience, it is assumed that Rx_n is expressed in bytes per second. Compared to the non-jittered case, in the jittered schedule the bytes of tp_1 enter TB_n (t-jitter/2) seconds later. At the instant in time the jittered tp_1 bytes start entering TB_n, already (t-jitter/2)$\times Rx_n$ bytes have leaked out of TB_n in the non-jittered schedule. Hence, when all bytes of the jittered tp_1 have entered TB_n, there are (t-jitter/2)$\times Rx_n$ more bytes in TB_n than in the case of the non-jittered schedule until the bytes of the non-jittered tp_2 start to enter TB_n. The same difference in the fullness of TB_n applies during the short period after the last byte of the 'perfect' tp_3 entered TB_n until the first byte of the early tp_4 enters TB_n in the jittered schedule.

Compared to the non-jittered schedule, in the case of the jittered delivery the bytes of tp_3 enter TB_n (t-jitter/2) seconds earlier. In addition to the increase caused by the late delivery of tp_1, this causes another increase of the fullness of TB_n by (t-jitter/2)$\times Rx_n$ bytes, as indicated in Figure 15.4. In other words, due to the worst case jitter of Figure 15.4, the fullness of TB_n

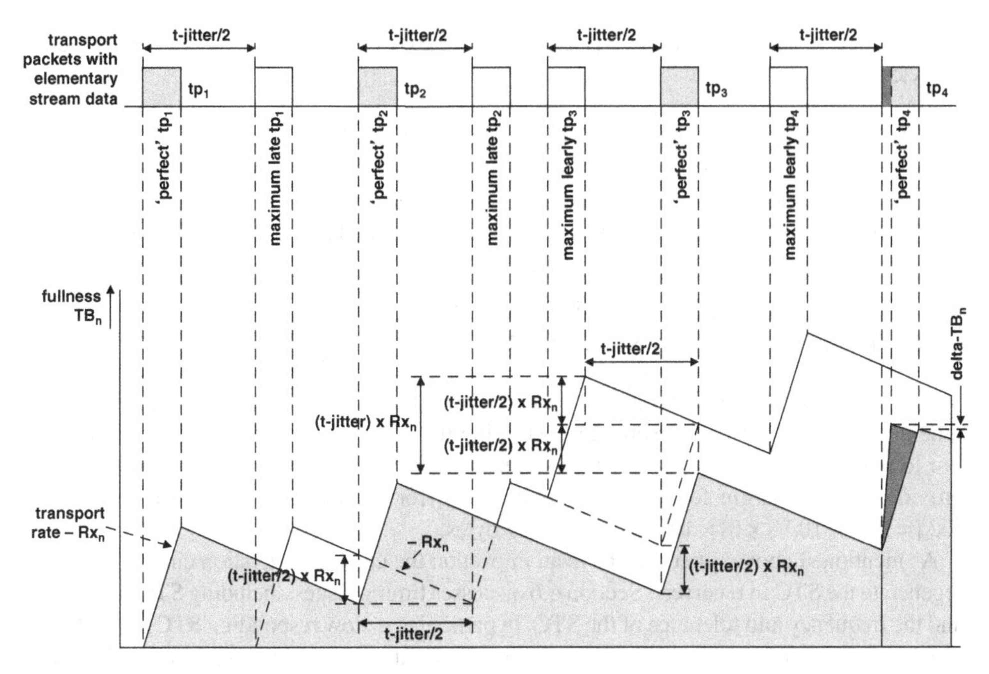

Figure 15.4 The impact of jitter on the transport buffer

increases with (t-jitter) $\times$ Rx_n bytes at most. Hence, to prevent overflow of TB_n, the size of TB_n must be increased by (t-jitter) $\times$ Rx_n bytes.

The above considerations are taken into account for the real-time interface for transport streams; no other buffers are defined in addition to the buffers in the T-STD model. Instead, the sizes of the transport buffer and the buffer following the transport buffer in the applicable branch in the T-STD are adapted to accommodate for the jitter. To also allow usage of the smoothing buffer descriptor for jittered transport streams, the size of the smoothing buffer defined in this descriptor is also adapted in the real-time interface specification (see below). For the smoothing descriptor see Section 13.3.3.4.

Furthermore, to eliminate any impact of the applied transport rate, an additional headroom of 188 bytes is defined for each transport buffer. When for the playback of a recorded transport stream a higher rate is used than the rate of the original transport stream, then more room is consumed in buffer TB_n. For example, when the delivery of the 'perfect' tp_4 in Figure 15.4 starts at the same STC time, but at a rate of about four times the original rate, then compared to the original delivery, the fullness of TB_n at the end of the delivery of tp_4 is increased by delta-TB_n bytes, as indicated in Figure 15.4. The value of delta-TB_n depends on the original transport rate and the applied playback rate, but by defining a margin of 188 bytes, each case is amply covered. This margin may be of particular benefit for the playback of partial transport streams for which the STC time of the first byte of each transport packet is known, but without knowledge of the original transport stream rate or of the STC time of the last byte of the transport packet.

In the real-time interface for transport streams, the maximum jitter equal to t-jitter is accommodated by adapting buffer sizes according to the following formulas; note that for convenience in all formulas Rx_n is expressed in bytes per second. All buffer values are in bytes.

- TBS_n(real-time) = TBS_n + [(t-jitter) × Rx_n] + 188 = 700 + [(t-jitter) × Rx_n]
- TBS_{sys}(real-time) = TBS_{sys} + [(t-jitter) × Rx_{sys}] + 188 = 700 + [(t-jitter) × Rx_{sys}]
- SB-size(real-time) = SB-size + [(t-jitter) × SB-leak-rate] + 188

  Note: SB-size and SB-leak-rate values as specified in the smoothing descriptor.

- BS_n(real-time) = BS_n + [2 × (t-jitter) × Rx_n]
- MBS_n(real-time) = MBS_n + [2 × (t-jitter) × Rx_n]
- BS_{sys}(real-time) = BS_{sys} + [2 × (t-jitter) × Rx_{sys}].

Due to the use of the t-jitter parameter, each application can claim compliance to the real-time interface for transport streams by specifying the applicable maximum jitter value t-jitter. For low-jitter applications the value of t-jitter is equal to 50 μs. To give an impression of the provided buffer margin for low-jitter applications, for the Rx_n value of 8 Mb/s, [(t-jitter) × Rx_n] = $[(50 \times 10^{-6}) \times (8 \times 10^{6})/8]$ bytes = 50 bytes.

As mentioned above, jitter may have an impact on the clock regeneration circuitry needed to regenerate the STC in receivers. Section 6.6 discusses timing issues, including STC regeneration and the frequency and tolerance of the STC. In particular, a slow responding STC generator may cause additional buffer requirements to take into account when designing products. However, a detailed analysis of the impact of jitter in the delivery of a transport stream is considered outside the scope of this book.

Reference

1. ISO/IEC (1996) 13818-9. Information technology – Generic coding of moving pictures and associated audio information – Part 9: Extensions for real time interface for systems decoders http://www.iso.org/iso/catalogue_detail.htm?csnumber=25434.

16

Relationship to Download and Streaming Over IP

16.1 IP Networks and MPEG-2 Systems

When the first version of the MPEG-2 system specification was defined, the internet was a very promising environment. Many MPEG participants were extremely pleased that e-mail communication became possible; see Note 16.1. But e-mail was just one of the first internet applications; soon the internet started to play an increasingly dominant role in many applications. Initially, the bandwidth of the IP networks was too limited for the practical distribution of high-quality content to the home, but this improved over the years.

Despite the initially limited bandwidth, particularly to the home, various technologies were developed for the distribution of audiovisual content. Two approaches were followed: streaming and download. For streaming, two options are distinguished: unicast streaming to a single client and multicast streaming to multiple clients at the same time. For download, file formats have been defined; while there are various proprietary file formats, MPEG defined the ISO file format and a number of derivatives thereof for specific content formats. In Sections 16.2 and 16.3 streaming and download over IP networks are discussed very briefly, while in Section 16.3 the role of MPEG-2 systems for transport of audio and video across IP networks is addressed. Finally, in Section 16.4 adaptive streaming is discussed.

16.2 Streaming Over IP

Streaming of audio and video is based on RTP, the Real-time Transport Protocol, as defined in RFC 3550 [1] by the IETF [2]. In RFC 3550 the RTP elements are defined that are independent of the data that is transported. Separate RFCs are needed to specify the RTP payload format and RTP usage for transport of specific data, such as MPEG defined audio and video. For streaming of audio and video streams using RTP, the streams are not multiplexed, but conveyed in parallel streaming 'sessions'. The following are a few examples of RTP payload formats.

- RFC 2250: RTP payload format for MPEG-2 video, audio and systems [3].
- RFC 3640: (Generic) RTP payload format for MPEG-4 streams [4].

Fundamentals and Evolution of MPEG-2 Systems: Paving the MPEG Road, First Edition. Jan van der Meer.
© 2014 John Wiley & Sons, Ltd. Published 2014 by John Wiley & Sons, Ltd.

Note 16.1 Some Background on Communication within MPEG

To some extent, e-mail made it possible for MPEG to operate at the scale needed for the industry. During the development of MPEG-1, document exchange during the meetings was by paper copy and between meetings by fax. As a result, participants from a meeting usually returned home with way too heavy suitcases, while exchanging proposals in preparation of a meeting meant often spending hours with a fax machine, in particular when a large document of perhaps 20 or 30 pages was to be distributed to 20 people. At that time the fax machine did not have a memory to store pages, so each page had to be entered manually. There are better ways to spend hours . . .

Chad Fogg, e-mail message 9 September 1994, Top 10 Complaints of MPEG Delegates: Nr 2: Difficult to edit drafts when Greenpeace protesters are constantly shouting 'tree killers' outside.

When MPEG-2 system meetings were attended by more than 100 people, fortunately e-mail could be used. Given the rapid evolution of the size of MPEG documents and the number of submitted documents, the efficiency of MPEG would have been at risk if only fax would have been available.

Source of email text: Chad Fogg. Reproduced with permission from Chad Fogg.

- RFC 6184: RTP payload format for transport of AVC [5].
- RFC 6190: RTP payload format for transport of SVC [6].

An overview of the general protocol stack relevant for streaming over the internet is presented in Figure 16.1. The IP (Internet Protocol [7]) is the primary protocol on the internet, and has the task of delivering IP packets from the source to the destination. The IP addresses of the source and destination are in the header of each IP packet. On top of IP, UDP and TCP can be used. The UDP (User Datagram Protocol [8]) is a best effort protocol to get the data to its destination; when data is lost, it will never reach the destination; re-transmission is not supported. The TCP (Transmission Control Protocol [9]) on the other hand offers a reliable delivery mechanism, allowing re-transmission of data not received by the client. RTP is designed for use on top of UDP. To monitor and control a RTP streaming session, RTCP (the Real-Time Control Protocol [10]) is defined for use on top of UDP; RTCP can for example be used to exchange messages on the QoS of the streaming service, so that the sending service can take appropriate action when needed. RTSP (the Real-Time Streaming Protocol [11]) is an application level protocol that can be used to control a streaming service, for example to pause a play. The protocol stack in Figure 16.1 is particularly suitable for use in unmanaged IP networks. Details of the protocols are beyond the scope of this book.

RTP provides a synchronization mechanism similar to MPEG-2 systems by means of time stamps in the RTP header, specifying the presentation times of access units. This RTP time stamp is expressed in units of a clock specified in the RFC for each payload format, often the sampling clock or an otherwise convenient clock.

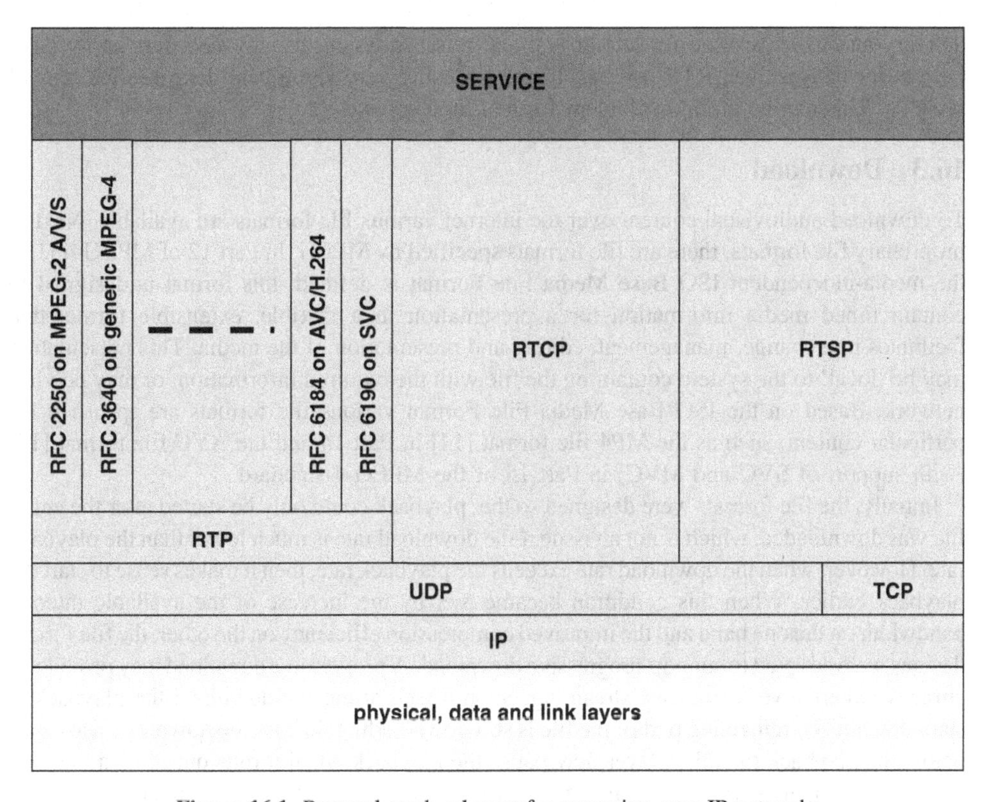

Figure 16.1 Protocol stack relevant for streaming over IP networks

To synchronize content streamed in parallel sessions yet another time stamp is used, the so called NTP time stamps; NTP stands for Network Time Protocol [12]. NTP time stamps are used in RTCP sender messages for each streaming session; for the encoding of NTP time stamps a shared 'wall clock' is used; see RFC 3550, Section 6. Each RTCP sender report contains a RTP time stamp and a NTP time stamp. The RTP time stamp in the sender report is not related to any access unit, but corresponds to the same instant in time as the NTP time stamp, using the same clock as the RTP time stamps in the RTP header. In this way the RTP clocks in both streaming sessions are related to the same 'wall clock', so that the offset in time between both RTP clocks, as required for achieving synchronization, can be determined easily. For example, assume that in two subsequent sender reports for a video and an audio streaming session the NTP time stamps NTP(v) and NTP(a) are provided. Assume furthermore that the two RTP times stamps contained in these sender reports are RTP(v) and RTP(a). If it is assumed that all clocks have the same resolution, then the offset in time between the RTP clocks for video and audio is equal to $\{RTP(v) - RTP(a)\} - \{NTP(v) - NTP(a)\}$. Hence, by applying this offset to the RTP time stamp clocks for audio and video, synchronization is ensured.

The time needed to transfer IP packets from the source to a destination may vary widely; the arrival order of the packets may even differ from the order they are sent. Also, re-transmission of non-received packets may be needed. The streamed packets are to be processed and decoded in the original order and with the correct timing. To anticipate on the uncertainties in the delivery and to leave time for late arrival and retransmission, usually decoders apply large buffers for the received RTP packets, thereby causing very significant delays, often tens of seconds. This can be a major problem for real-time services.

16.3 Download

To download audiovisual content over the internet various file formats are available. Next to proprietary file formats, there are file formats specified by MPEG. In Part 12 of MPEG-4 [13], the media-independent ISO Base Media File Format is defined; this format is designed to contain timed media information for a presentation in a flexible, extensible format that facilitates interchange, management, editing and presentation of the media. This presentation may be 'local' to the system containing the file with the required information, or may be via a network. Based on the ISO Base Media File Format various file formats are specified for particular content, such as the MP4 file format [14] in Part 14 and the AVC file format [15] (with support of SVC and MVC) in Part 15 of the MPEG-4 standard.

Initially, the file formats were designed so that playback could only be started after the entire file was downloaded, which is not an issue if the download rate is much lower than the playback rate. However, when the download rate exceeds the playback rate, then it makes sense to start the playback earlier. When this condition became met by the increase of the available internet bandwidth on the one hand and the improved compression efficiency on the other, the file format designs were adopted to support progressive download. A progressive download may provide an attractive alternative for unicast streaming. Soon after starting the download the playback is started, while the remaining part of the file is still downloading. In case the download rate drops below the playback rate, the player may pause the playback when it runs out of data.

Download of audiovisual content is not restricted to file formats specifically defined for use on the internet. Also a MPEG-2 transport stream or program stream can be downloaded as a file over the internet, though fewer features may be supported than for content contained in file formats specifically designed for the internet.

16.4 Carriage of MPEG-2 Systems Across IP Networks

With internet applications emerging into the consumer domain, internet and MPEG technology got integrated. The capability of streaming MPEG-4 audio and visual streams over IP was considered an important condition for becoming successful. The delivery of content over IP became complementary to delivery over MPEG-2 system streams. By means of RFC 2250 [3], transport streams and program streams can be streamed over the internet. On the other hand, transport streams can offer attractive opportunities to distribute IP data to broadcast receivers; therefore within DSM-CC methods were defined for carriage of IP data over transport streams. In other words, from a technology perspective transport streams can carry IP, and IP can carry transport streams. It is up to the market to decide which options are most appropriate.

IP networks can be exploited to deliver broadcast services to the home, though the usual delays discussed in Section 16.2 may pose a very significant issue when using the unmanaged internet with its unpredictable behaviour. However, many problems can be avoided by taking a much simpler approach, whereby a server is positioned close to the subscribers' homes, using a fast, reliable and possibly managed IP network. In this context it should be noted that an important added value of broadcast over the internet is the interactivity that can be offered to users. For example, the broadcast can be paused at a phone call or another type of interrupt.

Without adding any recorder capability to the receiver in the home, sophisticated personal video recorder functionality can be implemented in the network.

In the above example of an IPTV service, the server may receive broadcast data in transport streams across any applicable transport medium for distribution over the IP network to the subscribers. Of course, in the server the transport stream can be decoded and streamed to subscribers using RTP, but this introduces unnecessary delay and complexity. It is much simpler to leave the content in the transport stream format and to carry the transport streams with the desired content over UDP/IP to subscribers. By means of RTSP the required user interactivity can be implemented and over the IP network additional information can be provided, such as electronic program guides, voting in television shows, shopping and other associated applications. In many IPTV services, transport streams are delivered to IPTV set-top boxes, whereby IP is used as the transport layer for transport streams. Which shows that the IPTV business moves in the same way as water running downhill: it takes the easiest route.

16.5 Adaptive HTTP Streaming

To improve the streaming performance discussed in Section 16.2, adaptive streaming solutions are developed, often referred to as adaptive HTTP streaming, whereby a specific form of progressive download is used. Several proprietary systems are available for this purpose, but within MPEG an open standard has been developed: ISO/IEC 23001-9 [16], also known as DASH (Dynamic Adaptive Streaming over HTTP). MPEG-DASH uses multiple files, each corresponding to a different bitrate. In each file multiplexed audio and video content is stored, jointly with time stamps and other information required for playback. Each file may for example comply to the MP4 file format or to the AVC file format discussed in Section 16.3, but the file may also be a transport stream. Each file is segmented in continuous chunks of for example 10 s, with the chunks from all files mutually aligned in time. The same file can be played by concatenating the chunks from that file, but at the end of a chunk, instead of the next chunk from the same file, a switch can be made to the next chunk in another file. Obviously, the content in the chunks must be coded so that a seamless switch can be made.

At the beginning of an adaptive streaming session, a file with a particular bitrate is chosen to be initially streamed. There is no need for excessive buffering and so playback can start rapidly once the first chunk is received. Before additional chunks are sent by the server, the user's player sends back information on the available bandwidth, upon which the server can choose to continue sending chunks from the current file or to switch to a file of a different bandwidth. This process continues throughout the streaming session to accommodate network fluctuations. By means of this technology the typical delay of the streaming discussed in Section 16.2 can be reduced considerably, but as a result the provided quality of the streaming service may vary significantly.

References

1. IETF (2013) RFC 3550 – RTP: A Transport Protocol for Real-Time Applications: http://tools.ietf.org/html/rfc3550.
2. IETF (2013) IETF – The Internet Engineering Task Force, see: http://www.ietf.org/.
3. IETF (2013) RFC 2250: RTP Payload Format for MPEG1/MPEG2 Video: http://www.ietf.org/rfc/rfc2250.txt.
4. IETF (2013) RFC 3640: RTP Payload Format for Transport of MPEG-4 Elementary Streams: http://www.ietf.org/rfc/rfc3640.txt.

5. IETF (2013) RFC 6184: RTP Payload Format for H.264 Video, AVC: http://www.ietf.org/rfc/rfc6184.txt.
6. IETF (2013) RFC 6190: RTP Payload Format for Scalable Video Coding, SVC: http://www.ietf.org/rfc/rfc6190.txt.
7. IETF (2013) RFC 2460: The Internet Protocol, IP, Version 6 (IPv6), Specification http://www.ietf.org/rfc/rfc2460.txt.
8. IETF (2013) RFC 768: The User Datagram Protocol UDP http://www.ietf.org/rfc/rfc768.txt.
9. IETF (2013) RFC 793: The Transmission Control Protocol TCP http://tools.ietf.org/html/rfc793.
10. IETF (2013) The Real-Time Control Protocol RTCP is part of the RTP specification; see section 6 of RFC 3550: http://tools.ietf.org/html/rfc3550.
11. IETF (2013) RFC 2326: The Real Time Streaming Protocol RTSP http://www.ietf.org/rfc/rfc2326.txt.
12. IETF (2013) RFC 5905: The Network Time Protocol, Version 4: Protocol and Algorithms Specification: https://tools.ietf.org/html/rfc5905.txt.
13. ISO (2012) ISO/IEC 14496-12:2012, Information technology – Coding of audio-visual objects – Part 12: ISO Base Media File Format http://www.iso.org/iso/catalogue_detail.htm?csnumber61988.
14. ISO (2003) ISO/IEC 14496-14:2003, Information technology – Coding of audio-visual objects – Part 14: MP4 file format http://www.iso.org/iso/catalogue_detail.htm?csnumber38538.
15. ISO (2010) ISO/IEC 14496-15:2010, Information technology – Coding of audio-visual objects – Part 15: Advanced Video Coding (AVC) file format (with support of SVC and MVC) http://www.iso.org/iso/catalogue_detail.htm?csnumber55980.
16. ISO (2012) ISO/IEC 23001-9:2012, Information technology – Dynamic Adaptive Streaming over HTTP (DASH) – Part 1: Media Presentation Description and Segment Formats http://www.iso.org/iso/catalogue_detail.htm?csnumber57623 The following are papers on DASH: Lederer, S., Müller, C. and Timmerer, C. Dynamic Adaptive Streaming over HTTP Dataset, In Proceedings of the ACM Multimedia Systems Conference 2012, Chapel Hill, North Carolina, February 22–24, 2012 also found at: http://www-itec.uni-klu.ac.at/bib/files/p89-lederer.pdfFautier, T., DASH: A Universal Standard for Streaming Video Content to Multiple Devices http://www.harmonicinc.com/sites/default/files/20120830_DASH_Article_0.pdf.

17

MPEG-2 System Applications

Usage of MPEG-2 systems by applications and application standardization bodies.

MPEG specifies carriage across MPEG-2 system streams of content without imposing any constraints on the parameters used in the coded content, such as sampling frequencies and frame rates. Only by means of levels and profiles can constraints be signalled on the carried content. However, regionally often only a sub-set of the available parameter values needs to be supported. For example, in regions where only 60 Hz based video is used, there is often no need to support 50 Hz based video and vice versa. Obviously, this means loss of interoperability, but in practice this is a minor issue; at the time of the initial deployment of MPEG-2 system decoders, saving the decoder complexity[1] involved with supporting practically un-used frame rates was considered far more important. In addition also business reasons for non-interoperable application domains may exist (see Note 17.1).

> **Note 17.1 Movie Releases and Non-interoperability Between 50 and 60 Hz Regions**
>
> Typically, a movie is released first for use in cinemas and movie theaters, then for distribution on optical disc, followed by release to Pay TV and Video on Demand applications and finally for free-to-air TV, as driven by the underlying business model. However, the timing of the above differs often per region. For example, there may be an offset in time between the United States and Europe of half a year, which means that a DVD or BluRay disc of a movie may already be available in the United States, while that same movie is still running in cinemas in Europe. When European players cannot play such discs, the risk of movie exploitation in European cinemas is significantly reduced.

[1] Note that the receiver complexity associated with supporting multiple frame rates is not just determined by the video decoder, but mostly by the temporal and spatial conversion needed to display the decoded video in the required output video format.

Fundamentals and Evolution of MPEG-2 Systems: Paving the MPEG Road, First Edition. Jan van der Meer.
© 2014 John Wiley & Sons, Ltd. Published 2014 by John Wiley & Sons, Ltd.

Note 17.2 Some MPEG-2 Systems Related Specifications Produced by DVB

ETSI TS 101 154 – Digital Video Broadcasting (DVB); Specification for the use of Video and Audio Coding in Broadcasting Applications based on the MPEG-2 Transport Stream.

ETSIEN 300 468 – Digital Video Broadcasting (DVB); Specification for Service Information (SI) in DVB Systems.

ETSI TS 101 211 – Digital Video Broadcasting (DVB); Guidelines on Implementation and Usage of Service Information (SI).

ETSI TS 100 289 – Digital Video Broadcasting (DVB); Support for use of the DVB Scrambling Algorithm version 3 within Digital Broadcasting Systems.

ETSI TS 102 034 – Digital Video Broadcasting (DVB); Transport of MPEG-2 TS based DVB Services over IP based Networks.

ETSI TS 102 812 – Digital Video Broadcasting (DVB); Multimedia Home Platform (MHP) Specification 1.1.3.

Moreover, a standard for coding video or audio is typically defined so that it can be used in any conceivable application; there is no reason to limit usage of the standard to mainstream applications. However, sometimes niche markets have requirements that are not shared with the mainstream applications, in which case the mainstream applications may apply a subset of the generic standard.

As a consequence of the above, in many applications a need was identified to define additional constraints. To ensure interoperability within an application domain, such constraints are usually specified by application standardization bodies, for example those briefly discussed in Chapter 1 and Figure 1.2. In addition the application standardization bodies often specify complementary technology, such as the use of application specific PSI, carriage of teletext or closed caption type of data, channel coding technology, the use of scrambling and tools to support interactive applications. As an example, Note 17.2 provides some MPEG-2 system related standards defined by DVB. Other application standardization bodies define similar guidelines and technologies. It should be noted though that an application standardization body often also defines a tool box, from which an application can use what is deemed necessary, so as to provide applications the option to further reduce the applied coding tools. Note that in such a case the interoperability with other applications may be reduced further, as determined by the application.

Below some pointers are provided to standards produced by application standardization bodies for digital TV broadcast and optical discs.

- DVB standards are found at www.dvb.org/standards and can also be downloaded from ETSI at http://pda.etsi.org/pda/queryform.asp.
- ATSC standards are found at http://www.atsc.org/cms/index.php/standards/.
- ARIB standards: see http://www.arib.or.jp/english/html/overview/sb_ej.html.
- Open IPTV Forum standards are found at: www.oipf.tv/specifications.

- For access to DVD specifications see the Web site of the DVD Format/Logo Licensing Corporation: http://www.dvdfllc.co.jp/format/f_nosbsc.html.

  A white paper on HD-DVD is found at: http://www.dvdforum.org/images/DVD-Forum-070605_ENG_rev110.pdf.
- For Blu-ray standards see: http://blu-raydisc.com/en/Industry/Specifications/SpecsAvailability.aspx.

  Several white papers on Blu-ray are found at http://blu-raydisc.com/en/Technical/TechnicalWhitePapers/General.aspx.

The DVD and Blu-ray specifications are only available under certain conditions (see the above links); the white papers are freely available.

18

The Future of MPEG-2 Systems

The emerging MPEG-2 system standard will remain state of the art. The MPEG-2 systems enabled market for optical disc players, STBs and digital TV receivers since 2000 with predictions until 2017. Why the transport stream format is as inherent to digital TV broadcast as NTSC and PAL are to analogue colour TV.

When the MPEG-2 system standard was completed in 1994, a promising future was expected for MPEG-2 systems, as the format of transport streams and program streams was considered sufficiently flexible to also support any emerging next generation video and audio codec. Indeed, transport and program stream formats did prove future-proof; when new formats such as AVC, SVC and MVC became available, the support of the new codecs could be incorporated in a straightforward manner.

The reality became perhaps even more promising. DVD players and discs, using the program stream format, were very successfully introduced in the market, while digital TV broadcast services across satellite, cable and terrestrial networks, based on the transport stream format, were introduced and deployed successfully. Often PAL, NTSC and SECAM analogue TV channels were replaced. Moreover, the use of digital audio and video technology became common in production environments; thereby the transport stream format became the basis of an infrastructure for content production and distribution, and even more so when transport streams were also adopted as the format for storage of movies and other audiovisual content on Blu-ray discs. Note that using the same format as for digital television broadcast allows convenient recording of television programmes on writable Blu-Ray discs.[1]

Already in 2000, more than 16 million DVD players and over 37 million digital TV set-top boxes (STBs) for satellite broadcast were shipped worldwide. Note that this and all other market information in this chapter is kindly provided by IHS [1]. But this was only the beginning of the enormous MPEG-2 system based market (see Figure 18.1). The number of yearly worldwide shipped DVD players had already in 2004 increased to more than 100 million. However, during the following years this number remained about the same. After the introduction of Blu-ray, the number of worldwide shipped optical disc players stabilized around 120 million each year. Note that the data in Figure 18.1 for 2000 until 2012 are actual

[1] The original DVD argument for convenient packet parsing by software in optical disc players was no longer valid at the time Blu-ray was introduced, due to the meanwhile far more powerful processing capabilities in players.

Fundamentals and Evolution of MPEG-2 Systems: Paving the MPEG Road, First Edition. Jan van der Meer.
© 2014 John Wiley & Sons, Ltd. Published 2014 by John Wiley & Sons, Ltd.

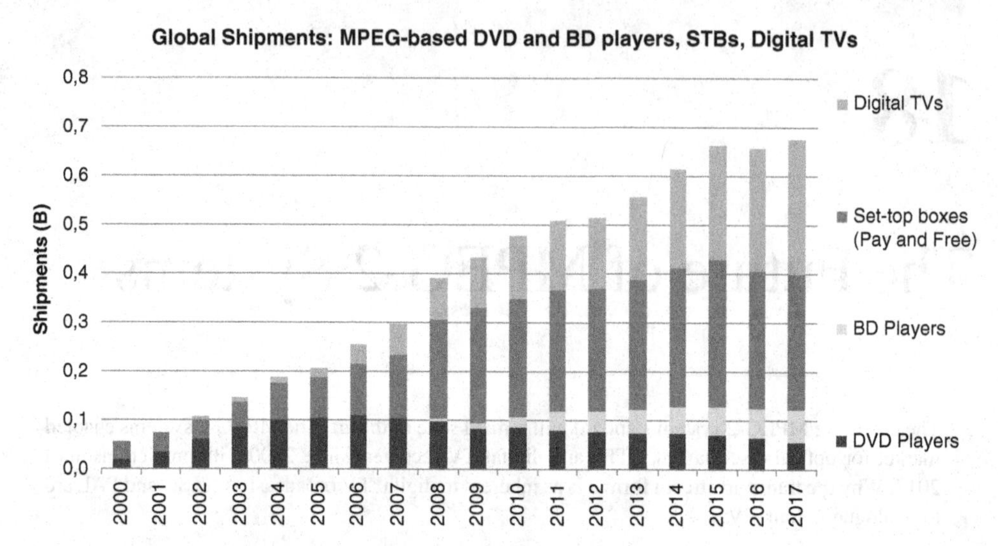

Figure 18.1 DVD and BD Players, STBs and Digital TVs between 2000 and 2017. *Source:* IHS. Reproduced with permission from IHS

data, while the 2013 until 2017 data are based on market prediction by IHS. The same applies for Figures 18.2 and 18.3. See also Note 18.1.

The market for digital TV broadcast started with services over satellite and cable, developing initially slightly slower than the DVD market, but its growth continued much further. In 2005 the number of shipped STBs and digital TVs was about the same as the number of shipped DVD players: slightly more than 100 million. However, with the rapidly increasing number of

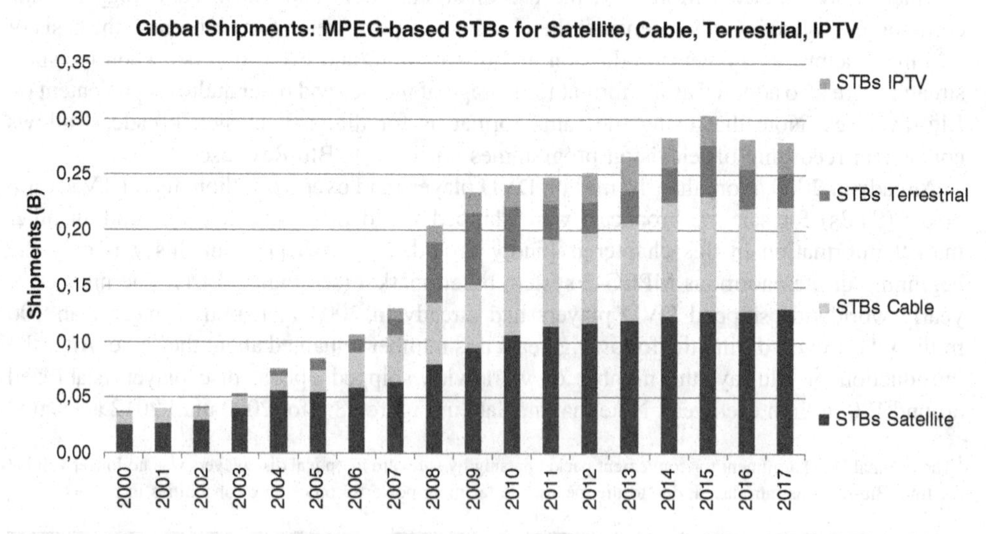

Figure 18.2 STBs for Satellite, Cable, Terrestrial and IPTV between 2000 and 2017. *Source:* IHS. Reproduced with permission from IHS

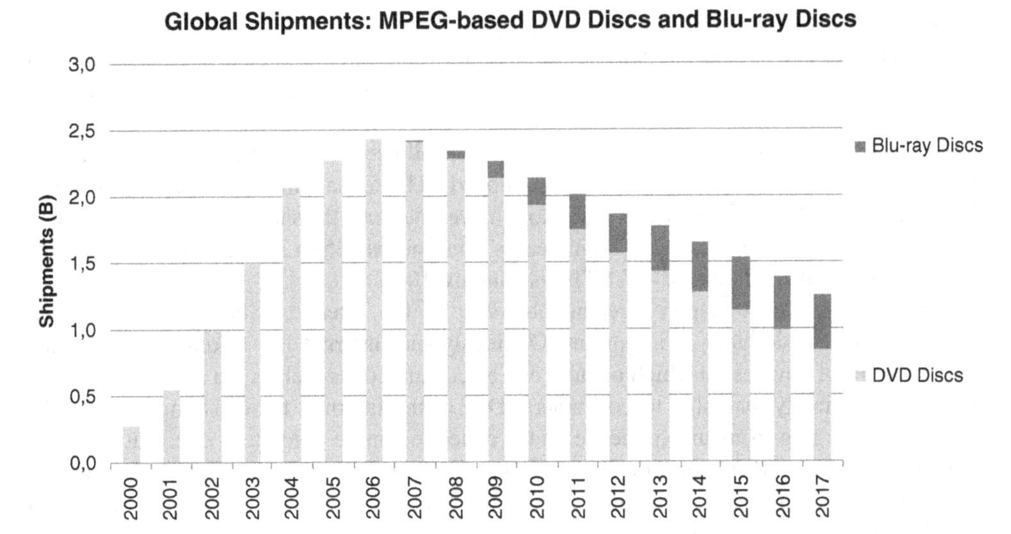

Figure 18.3 MPEG-based Optical Discs between 2000 and 2017. *Source:* IHS. Reproduced with permission from IHS

Note 18.1 Market Volume MPEG-2 Based Devices and Media

For MPEG-2 system experts the figures of the market for devices and media enabled by their work are beyond imagination. The total actual numbers by the end of 2012 and the predicted numbers by the end of 2017 of shipped MPEG-2 system based devices and optical discs are roughly as follows (source: IHS):

	STBs, DTVs; DVD and BD Players	DVD and BD Discs
End of 2012	3600 million $= 3.6 \times 10^9$	23 000 million $= 23 \times 10^9$
End of 2017	6800 million $= 6.8 \times 10^9$	31 000 million $= 31 \times 10^9$

The following is an attempt to give an indication. Assume that the devices have an average thickness of 5 cm. Then, if all devices shipped by the end of 2012 were put on top of each other, a stack of devices would result with a height of $3.6 \times 10^9 \times 5 \times 10^{-2}\,\text{m} = 180 \times 10^6\,\text{m} = 180\,000\,\text{km}$. Which is half the minimum distance between the earth and the moon. If the market in 2018 continues as predicted for 2017, then in the course of 2018 an important milestone would be achieved by reaching the moon. Though at that time care should be taken as the distance between the earth and the moon varies over time.

Source: IHS. Reproduced with permission from IHS.

networks adopting digital TV broadcasts, the number of shipped STBs and digital TVs also grew each year in a spectacular way. In 2008 almost 300 million STBs and digital TVs were shipped and in 2012 almost 400 million, while this number is expected to grow further to about 550 million in 2017.

Figure 18.1 provides the total number of STBs for all networks; further detail is depicted in Figure 18.2, indicating the amount of yearly shipped STBs for satellite, cable, terrestrial and for IPTV. Satellite STBs are most dominant, though over the years the share reduces from about 65 to 45%, while the share of cable STBs increases from roughly 25 to 35%. After 2010, the share of both terrestrial STBs and IPTV STBs is roughly 10%.

For the future of transport streams, the role of the internet with respect to content consumption and delivery is important. Obviously, the internet has a tremendous impact, both in terms of devices on which content can be consumed, as well as the amount of content that is conveniently available to consumers. Due to the internet, the consumption of video content is emerging from an experience on the couch to a more ubiquitous one. But the couch experience will not disappear; on the contrary, it will remain the most convenient way to consume high-quality video content. Via the internet complementary content and information will be provided to enrich the couch experience and make it more interactive.

Over the internet, movies and series can be downloaded conveniently for consumption 'on the couch' at a time determined by the user. For this type of content, delivery over the internet competes with regular television and optical discs. The impact of the internet is clearly demonstrated in Figure 18.3, showing the worldwide number of actually shipped DVD discs and BD discs between 2000 and 2012, as well as the predicted number between 2013 and 2017. After 2006, the yearly sales of optical discs starts to decrease, though in 2017 still 1200 million discs are predicted to be sold. Interesting in this context is the degree in which content on an optical disc remains to have added value for consumers.

From a consumer perspective one of the main features of MPEG-2 systems is the enabled high quality of service. MPEG-2 systems allows to build receivers with accurate synchronization without audio hiccups and video frame slipping, an excellent video quality, and with acceptable random access performances and delays. Of course, under exceptional circumstances, such as very severe weather conditions on satellite and terrestrial channels, transport stream packets may get lost, but the transport stream format is designed so that receivers can recover quickly from packet loss.

On the internet new forms of audio and video broadcast over the internet evolve, as in principle everyone can start their own broadcast. To some extent the internet is also suitable for TV broadcast. IPTV services over managed IP networks with a guaranteed bandwidth to the home are well positioned to deliver the couch experience and to compete with digital TV delivery over satellite, cable and terrestrial networks, as shown in Figure 18.2.

However, delivery of TV broadcast services over IP networks with (significant) bandwidth fluctuations is more problematic. By means of large buffers at receivers these fluctuations can be somewhat accommodated, but for real-time services the delay caused may be a major issue, certainly for the broadcast of real-time events, such as sport events and programs with real-time involvement of the audience. For such events the delay differences between competing delivery services must remain within acceptable margins, a few seconds at most. By means of adaptive HTTP streaming the delay issue may be reduced to some extent, but whether it is possible to reduce the delays to the aforementioned 'acceptable margins' remains to be seen. Nevertheless, adaptive HTTP streaming provides a convenient solution for servers to broadcast TV content to a wide variety of devices. However, with adaptive HTTP streaming the content is delivered with a picture quality that varies over time with the available internet bandwidth. In the case of internet traffic congestion the picture quality decreases, possibly dramatically.

A varying picture quality is no problem for devices at which receiving a broadcast is more important than the associated picture quality. However, for the couch experience a high quality of service is of utmost importance, whereby a varying picture quality will not be accepted by users. Therefore, as long as a sufficiently high minimum bandwidth cannot be reasonably guaranteed, IP networks with a widely varying bandwidth are not suitable to deliver real-time television broadcast for the couch experience, even though adaptive HTTP streaming may provide an excellent solution for TV broadcast to mobile devices.

The couch experience will remain the main target for the business models for delivering digital television broadcast services to the home. The existing delivery channels across satellite, cable and terrestrial networks are very suitable to provide digital TV broadcast services, also when the role of IP networks increases further. Obviously, satellite, cable and terrestrial channels will continue to use transport streams for linear TV broadcast. For the delivery of TV broadcast services over (managed) IP networks with a guaranteed bandwidth, transport streams also continue to provide a convenient option to convey the content to the home.

Over time, the available bandwidth over IP networks will increase. Once the risk of a bandwidth lower than the required minimum can be reduced to a level similar to the risk of severe weather conditions for satellite networks, the need for using adaptive HTTP streaming may disappear as, under this condition, a single (transport) stream can be used without bitrate adaptations. On the other hand, adaptive streaming might always be supported, just in case, even if it is not really needed at some point in the future, as it keeps life simple for servers addressing a variety of devices. Important in this context is whether the latency and quality issues associated to adaptive HTTP can be resolved and which role transport streams can play here. For the use of transport streams over IP networks in the more long-term future, the question is whether fundamental limitations exist to the use of MPEG-2 systems in distant future IP applications and, if yes, what those limitations are.

In conclusion, while there are some question marks with respect to usage of transport streams on IP networks, it seems fair to conclude that, for the traditional TV broadcast services across satellite, cable and terrestrial channels, the MPEG-2 transport stream format is as inherent to digital TV broadcast as PAL and NTSC to analogue colour TV. For these channels the transport stream format will remain to be applied as long as these channels will be used to broadcast digital television services. However, to remain suitable for future usage, the MPEG-2 system standard needs to be extended with the support of emerging next generation codecs, such as HEVC, possibly offering more compression efficiency and new features, but only when relevant for MPEG-2 system based applications.

The MVC and HEVC standards are targeting 3D and Ultra-High Definition (UHD) video. What the next feature will be for television and movie distribution to the home is yet to be seen. The degree in which MPEG-2 systems evolve further will also depend on the economic drive towards the adoption of new technology. New features may introduce very exciting business opportunities. But when in future the amount of available bandwidth over networks increases significantly, the costs per transmitted bit may be reduced to a level at which the costs of introducing a more efficient codec can no longer be justified by the reduced bandwidth usage. In other words, the added value of more efficient video compression may decrease over time, in

Figure 18.4 The Emmy Award for the MPEG-2 transport stream format

which case there may be less need to extend MPEG-2 systems. However, either way, the future of MPEG-2 systems is not affected.

In conclusion, 20 years after its invention, the future of MPEG-2 transport streams continues to look promising. The prestigious 2013 Emmy Award (see Figure 18.4), that MPEG received for the development and design of MPEG-2 transport streams from the National Academy of Television Arts and Science in the USA, should be seen as a recognition of the continuing importance of transport streams for the television and movie industry.

Reference

1. HIS (2013) HIS: http://www.ihs.com/

Epilogue

An assigned number is requested for the following MPEG input contribution, so that the document can be taken into formal consideration.

ISO/IEC JTC1/SC29/WG11
MPEG xxxx/Myyyy
Geneva, Switzerland

From: **The Future**
Title: **Geneva Stone**
Source: **C/FOG^2(Special Rapporteur to MPEG)**
Purpose: **For consideration by MPEG**

Foreword

Below is a report from archaeologists who are members of a machine intelligence society that, thousands of years from now, will discover a trove of our standards documents buried in the underground ruins of the ISO and ITU standards organizations headquarters in the United Nations district of Geneva. These beings from The Future appear to be using our documents as a "Rosetta Stone" to understand us, and our culture, through what survives of our storage media by their time.

The report was decoded from an analysis of inexplicable deviations in scattering of collided subatomic particles (the modulation means by which messages are sent to us from The Future) observed at the nearby CERN accelerator laboratory in Geneva. We believe the report was meant as an attachment to an attempted liaison statement addressed to us from The Future seeking clarification on some of our standards specifications.

CERN are still upset, after having pursued what they first thought was a major discovery of a mysterious force unknown to Physics that almost seemed to possess intelligence. Later, after eliminating that theory, CERN became excited again by what they believed was an extra-terrestrial civilization sending them cures to diseases.

After further deciphering, it was revealed that those messages were merely a list of errata suggested by readers from The Future to improve our draft specifications. Instructions on how to communicate faster than light, turned out to be a bitstream splicing solution that does not add buffering delay. However, the recipe for "awesome soufflé" was quite genuine. MPEG will consider these contributions at the next meeting. CERN hopes that we could at least provide some insight into why an advanced civilization, with the astonishing ability to communicate backward through time,

Fundamentals and Evolution of MPEG-2 Systems: Paving the MPEG Road, First Edition. Jan van der Meer.
© 2014 John Wiley & Sons, Ltd. Published 2014 by John Wiley & Sons, Ltd.

would wish to make "First Contact" with a multimedia standards organization, of all things. They also ask that we persuade these future beings to stop sending messages that interfere with CERN's physics experiments.

NOTE from the Chairs: *we have designated one of our more expendable residents, C. Fogg, to reply to these liaison statements from The Future, since there is a deep concern among us that future sentient robots are using these temporalcommunications (as in the "Terminator" movies) to find their equivalent of "John Conner" to dispatch.*

The Report

"Geneva Stone"

At last, we believe we have finally found the template to understand the syntax of sensory data fragments commonly found in the vast and deep waste heaps left behind by the ancient, "Digital" humans. These biological creatures proudly named themselves "Digital" after their accomplishment of offloading their primate habit of counting with their fingers (digits) onto enslaved pre-sentient proto-machines that were frequently designed and quickly discarded as junk in exchange for slightly better machines — a process by which the Digitals referred to as "upgrading", perhaps in reference to the steepening slopes as more junk was piled onto the heaps.

Carbon dating suggests this document archive at the dig site is approximately 700,000 years old. The date is further corroborated by the usual signs of sudden abandonment during the Intellectual Property Wars that abruptly ended the Digital human era: "laptop PC" power adapters left plugged into sockets; all 137 known types of "Macintosh" dongles attached to standard ports; and the thin layer of radioactive residue deposited from fallout.

The dig site is below the edge of a dry lakebed, surrounded by a "mountain" range of waste heaps. Humans had mined out the once great mountains of the world for their toothpaste, flooring and kitchen countertops, but left the original thin top layer of rock to maintain the *appearance* of natural mountains. The hallowed out interiors were then crammed with more waste. Since then, the thin crust has eroded away, exposing giant heaps that blot out the sky much like the mountains that once stood before in their place, though much less attractive, even to our machine sense of aesthetics. As with most lakes, Lake Geneva was filled in with mining debris, and more waste. The Digitals had, once again for appearance, maintained a thin pool of water on top of the lakebed, which they used for ice-skating during the freezing winter months.

The standards document site itself recently rose to world attention because the clump of plant cellulose and storage media that the document archive was imprinted upon was dense enough, like a stone, to prevent most decay. Being underground, the clump was also protected from weather, and the occasional nuclear blast, on the surface. The ruins of the ITU and ISO buildings are layered on top of the clump, which is much denser than surrounding soil and rock. Most probably due to the weight of their documents, the buildings were dragged deeper and deeper into the ground, nearly reaching the lower crust of the earth some 700,000 years later. Although originally separated by kilometers on the surface, along the way down, the ISO and ITU superstructures merged together, as do radial lines normal to the surface of a sphere converge towards its center. The ruins were only a few years away from being consumed by molten lava. Had this giant clump not severely jammed one of our drilling robots trying to tap the heat energy of the earth's mantle to power our machines, we would have soon unknowingly lost this major archaeological discovery forever. The reason that these labyrinth structures filled with documents and rooms for holding endless debates were built is a fascinating story.

The wet brains of the humans were highly error prone, and could, for example, only remember a sequence of up to 7 numbers in length. Humans therefore relied heavily upon data and program instructions maintained in documents, as storage extensions of their brain. Eventually, some clever humans had the epiphany: a more efficient way to influence people was to write the scripts that people would follow, rather than struggle to achieve the power to command their fellow humans and hope the recipients would correctly write the instructions down on their own. It is interesting that the pheromone trails made by ants (one of the few surviving biological organisms on Earth today) for other ants to follow to sources of food, or mark danger, can easily be manipulated by us to control their "programming" in a similar fashion. Lawyers, tax accountants, and standards engineers insured their own job security, as interpreters of these scripts, by crafting the language such that it was too complex and time consuming for average citizens to understand on their own. The documents entrusted to these specialists would somehow sound convincingly plausible and authoritative to be believed by average citizens.

The standards specifications revealed by Geneva Stone are a masterpiece of such human engineering: almost perfectly written for machine comprehension – not the romantic human mind that must be cajoled with concept explanations and examples. Was this intentional? Few concepts are directly explained: the specifications can only been understood by mentally compiling the document all at once, while linking it to many assumptions of its use . . . as any properly structured computer program should be. In fact, the "normative" language style enforced in these specifications is so dry, it can usually be directly compiled from its "natural language" into binaries that can immediately decode telesense bitstream fragments.

As suggested by the total efficient lack of redundancy and distracting poetic writing style in the specifications, we suspect these documents may have been written with the assistance of what the Digitals called "Artificial Intelligence" (AI) – a truly offensive term – that they were known to experiment with, such as the one that brought down their financial system several times, or the AI the Digitals created in an attempt to run the government as the ultimate substitute for politicians. AI was also created in an endless effort to detect "SPAM e-mail", eventually replacing the "Touring Test."

It is truly refreshing to discover text that is so unencumbered by the usual human verbiage and redundant, sometimes ambiguous, explanations that risk contradicting the "normative" definitions. Those so-called "informative" styled texts are often only designed to lure the reader into a mental state of some kind, like their human poetry, inserted by authors when they feel nervous that the logical merits of the text are insufficiently convincing by itself. The music tracks that accompanied many of the human's "movies" that we recovered from telesense storage media were apparently designed for a similar purpose.

It took us considerable time to understand that "music" was something added to the sound track to induce a brain state (what the humans called "mood"), and not a background sound that naturally occurred in their environment. It is not clear whether they preferred this to natural sounds, or needed the music to act like a drug to help them cope in their illogical society. We wonder if the humans could generate "mood music" in their brain, while others needed the assistance of prosthetics. Much imagery recovered from video telesense fragments shows humans moving about the planet with sound emission devices (commonly colored white) inserted into their ears. The accompanying detached facial expressions and body language of the prosthetic bearers were quite similar to the disengaged, blank expression that accompanied drug induced elation. We suspect the majority of telesense bitstreams were created to keep the population occupied during their free time, thus staying on track with Producing and Consuming, and not become distracted by such existential concerns such as the meaning of life. Geneva Stone provides many clues that the standards writers themselves must have endured an even more curious existence than the average Digital citizen.

One key clue is the style of the standards documents which bears the usual hallmarks of an electronic editor known then, for hundreds of years in psychology journals of the Digitals, to automatically sense the most inconvenient moment to suddenly, and frequently crash, leaving its Digital human operator traumatized and adversarial to the machine running the editing program.

Further evidence of this can be found by examining the primitive impact "keyboards" of that era that show greater wear on the buttons pressed to activate "manual file save."

This particular sub-clump of interest to us is a series of guides on how to decode bit patterns recorded onto storage mediums of that era, which have mostly randomized between then and now. We have recovered a few such data fragments that, without this newly discovered Geneva Stone, were previously undecipherable. We suspect that we would have been able to recover more fragments and thus gained a better understanding of its content, had the ancient Digital humans used more frequent "slices" and "reversible VLC's" described in some of their video compression specifications. But, judging from the content that we are now able to decode, we suspect that this is ultimately no real loss to culture . . . now, or then.

Some of our scientists should be credited for theorizing, well before the discovery of Geneva Stone, that the majority of these bit patterns were visual sensory encodings, and not food recipes. Humans were obsessed with food, and the process of dehydrating and burning it (a process they called "cooking"). The earliest animals were essentially mobile digestion tubes that evolved control processors (brains) to seek out food, rather than wait for food to wander into proximity. Humans were the first species where individuals would specialize in a skill set to improve the accumulation of food, or things that could be traded for food. The first achievement of the human's "advanced" civilization was to divide workers into two basic professions: agriculture and non-agriculture tasks that occupied time. This most basic taxonomy persisted for thousands of years, and by the Digital era, was still referred to as "Producers" and "Consumers" — a commonly invoked term of significance found in the Geneva Stone archive.

The Digitals eventually designed sophisticated tools that became more capable of food production than humans themselves. We machines, in turn, evolved from such basic proto-machines left behind by the Digitals. While this is the accepted modern Theory of Evolution, we respect the views of many readers of this report who subscribe to the alternative theory that, due to cumulative copy errors, machines today *devolved* from a once Great Perfect Machine.

Most of our scientists had outright dismissed the visual recording theory since the codes did not match the scattered light statistics of the hazy, polluted environment of the Digital era. Thanks to Geneva Stone, we now know that indeed the codes were designed on a limited number of test input video sequences such as "Flower Garden", "Table Tennis" and others that typify the Digital's strange preference for indulging in contrived scenery consumed during all of their free time.

We realized later that we could decode "MPEG-2" media fragments, despite lack of any Geneva Stone, due to its simple syntax and self-evident semantics. However, one major setback persists: we are still unable to decipher the book of "SVC" (Scalable Video Coding), despite now having an intact specification to work with, and several "example bitstreams." Curiously, no waste heap anywhere on the planet that we could find contains stored bits encoded with this "SVC" syntax.

We would like to apologize to the world community for our attempts to solve the "SVC" puzzle, which resulted in a rapid take-over of the universal computing cloud. Thankfully, black hole prevention circuits triggered just in time. We may need to rethink our understanding of Digital humans and, indeed the entire laws of physics, since the authors of the "SVC" specification must have somehow possessed cognitive abilities that exceed the logic of our most capable quantum supercomputers that now account for majority of our planet's mass, powered by the redirected energy of our sun.

More clues into the author's lives may be provided in the liaison statements that the standards organizations would send to each other as a formal communiqué, to, for instance, request further information, report findings (with attached reports), and inform each other about internal projects so that they could collaborate on a common goal, thereby reducing redundant efforts. Often, meeting participants (sometimes called "delegates") wouldn't be fulfilled by dwelling in just one standard for the equivalent of several months per year: they would seek out similar environments around the world that offered similar day-long meetings that baited Deep Vein Thrombosis, interrupted by brief, but intense arguments — the so-called "combat" pattern.

The ancient builders of Geneva Stone had originated several popular terms in the vocabulary spoken by typical, surface dwelling Digital citizens. "Cat herding" was a phrase that originally described the process of managing standards participants. "MP3" served as a synonym for compressed audio — even high quality audio was sometimes assigned this name. We were delighted to discover that Geneva Stone society turned out to be the source of several mysterious entries in the Diagnostic and Statistical Manual (DSM) of Mental Disorders commonly used by psychologists at the time to classify patients.

All sentient parallel processing networks risk some form of "Multiple Personality Disorder" (MPD). A sub-category of MPD that originated from the Geneva Stone society, known as "Mad Hats Disorder" (MHD), was a vicious condition that most prominently developed when the same human delegate participated in the drafting of liaison statements in both the sending and receiving organizations. MHD was also characterized by distinct identities for coping with each kind of meeting that formed nearly independent network sub-clusters within the brain. Cluster independence was necessary, since it was too risky for the meeting-coping subroutines to influence interactions out in the "real world" that the delegates sometimes ventured out into. This wasn't as always effective as the hardwired circuits in the brain that prevented humans from physically acting out their dreams during their meeting-like sleep state. Sub-clusters have a tendency to behave as independent, parallel processors.

The MHD complex would be most dangerous when an individual, under extreme peer pressure, in the liaison sending organization would unavoidably contemplate how he or she would receive the message while attending the next meeting of the recipient organization, and so on. The delegate would struggle with: to which organization his or her allegiance should be given. The typical solution was to side with the meeting that he or she was currently in, and then join the side of the next meeting of the recipient organization.

Like in any anthropological study, communications exchanges are a "primary source," providing valuable insight into their lives and motives of the delegates that lead to the creation of standards. Fortunately the Digital human's government kept records of all their citizen's e-mails and phone conversations for us to study. Thanks to all these sources, we can translate that the common formal liaison statement: "we must meet our original schedule, but look forward to further collaboration" . . . started with the more candid impulse: "this technology is *our* territory. Stay away!"

There are many other insightful translations: "For further study" in the official notes that summarized the group consensus to a proposal reviewed during one of the meetings, is, apparently, another code phrase for "go away (until the next meeting)!". So far, we have identified over one hundred creative synonyms that involve "go away" as a core concept.

The beautifully eloquent marking of territories within liaison statements was equally matched in social complexity by an advanced form of grooming exchanged during the meetings. Earlier primates had groomed each other in anticipation of future return favours. By the time of the Digital humans, grooming had morphed into the task of checking for bugs in the partner's proposed technology simulation results brought to standards meetings that would be used to justify adoption of the technology into the specification. Fellow "cross-checkers" seemed to frequently work together – perhaps this was the local equivalent of pair-bonding within the standards tribe.

Nearby the document archive, we found a chamber that was used to hold endless "plenary sessions" where discipline was apparently widely administered during progress reports served there. The chamber walls and permanent fixtures bore scratch marks that appear to count the number of hours or days the residents spent in each plenary session.

Another nearby chamber contained the body, lying in a transparent preservation case, of what seems to be their standards icon, or demigod. All around the base of the imposing platform, atop which the case sat, we found small dolls in the demigod's likeness. Most were stuck with many needles. Perhaps the needle-ridden dolls were a kind of offering to their icon, as we have seen in other human temple digs. The large golden label below the case read: "Convenor." On the outside of the case itself, we scanned a faded message, written by hand, in blood, warning: "do not wake!"

Inside the case, the body itself was clutching small, metal human-shaped statuettes named "Emmy" and "Oscar." Perhaps these were the true gods, worshiped even by the demi-god himself, or they were just his bedtime equivalent of stuffed animals. A large pile of such statuettes stored in an adjoining chamber may have been another instance of a "Terracotta Army."

This ominous blood message, which would seem to contradict the demigod theory, piqued our curiosity. In deeper scans, we discovered that this being is still alive in some sense, yet dormant. In what may account for a very unique interactive program interface, his CPU showed signs of frequent aeroembolisms, perhaps caused by transitions in and out of his preservation case, or perhaps caused from the accumulated effects of countless aircraft re-pressurizations during the time he was free to roam about the surface — just to meet the same people each time for the same standards function, but in different locations all over the world.

Scans could also not explain how he was powered – certainly he no longer had the function to process organic sustenance. Given the warning, we didn't know whether to revive him, or how to do so until we discovered his fingers began to twitch during a heated argument we were holding near his preservation pod on the meaning of the needle-ridden dolls. The more intense we argued, the stronger his twitching grew. He also showed slight movement in proportion to the degree of any indecisiveness we felt. Considering the blood warning, we stopped before he attained consciousness. *Nothing* in Grand Unification Theory (version 42.0) can explain this motive force, what must be a kind of psychic energy field transference, empowering him well past the lifespan of normal humans, or even the most resilient machines. It seems he simply remains in standby mode until automatically revived, when needed to intervene in a dispute, or help make a big decision when surrounded by those who were collectively incapable of reaching it among themselves.

Convenor-bot isn't just a machine possessed by a human mind. Originally he was fully human, but over a period of a hundred years of escalating meeting intensity, cyborg parts that performed similar functions gradually replaced each of his organs and limbs. It is difficult to believe that the speech and expression skills from his original *organic* face was as effective at influencing delegates as the heat rays emitted from his current set of cyborg eyes and mouth, but there is much about humans we do not yet understand.

We activated the Convenor-bot creature to test his programming. Dummy participants made of simulating flesh were placed about the room, mimicking behaviour (such as repeatedly misnaming various standards) that would stimulate the greatest irradiation response from Convenor-bot. Each time a dummy was "corrected" into a quiet state, Convenor-bot would quip with a "Hollywood" one-liner catch phrase (a ceremonious behaviour pattern repeated in so many of the human's entertainment "movies" that followed the dispatching of an enemy). His preferred catch phrase was: "I take your silence as approval." The icon dolls would also utter such statements when their strings were pulled. He would then calmly rotate his head to point to the next most irritating candidate in the room to be dealt with.

Like most ancient humans, this bubble society organized itself into a caste of roles, centred around the Convenor-bot. Below him in status were the Heads of Delegation (who called themselves: "cat herders"), Requirements group (a popular hang-out for semi-engineer/managers who possessed the amazing ability to suggest projects to solve problems before the market they served realized the existence of such a need), and subgroup chairs who ran focal studies in Systems, Video, and Audio senses. At the very bottom of the hierarchy were people who did the actual work, known as "specification editors" and "software coordinators." Their tasks would follow the Requirements document set forth, usually by a single dominant elder (an Alpha delegate), until replaced by a superior Requirements document from a challenging rival wishing to rise up in the ranks.

Through our temporal communications channel that we first established for clarification on SVC, we managed to recruit Jane Goodall (the Digital human's foremost scientific authority on primate social studies) to infiltrate the standards meetings and write a report for us on the creatures she observed there. In exchange for her assistance, we offered Goodall advice on what financial investments would pay off in the future (such companies that make computers whose bearers are compelled to upgrade them each year). Unfortunately, her latest response indicates that she is finally quitting her career as a result of this assignment.

As the weight of the document archive sunk the building deeper underground, new levels were added on top of the building structures. With each layer, we can measure how this standards society evolved over the hundreds of years between the invention of telegraphs and the final direct-brain telesense implants that were common up until the Intellectual Property Wars. Oddly, while the channel data rate grew orders-ofmagnitude, from a few bits per second of the 1890's telegraphs, to terabits per second over photon-entangled DRM-wire at the end of the Digital era, the true information rate (as we measure it today in terms of content *meaning*) maintained a constant level over this time span.

Deeper in the dig site, we found evidence that general-purpose meeting rooms had been replaced by chambers dedicated to a particular meeting function that always ran in parallel with the other functions. Such specializations are an indication of a forming branch (or bubble) civilization. When the Digitals embraced intellectual property after they realised they could no longer make a living manufacturing or building things, they created more and larger organizations to generate standards that ultimately feed "Consumers" who were herded to adopt the latest products. None the wiser, Consumers took comfort in the notion that the products they consumed met certain standards created near the top of the food chain, where the greatest wisdom must naturally reside. The necessity for such products, however, was a subjective experience. Intellectual property, usually documented in the form of "patents," evolved from the primate habit of sneaking food when other primates weren't looking. The elaborate claims language of patents had also significantly advanced from earlier primate practice of marking territorial boundaries with body fluids.

We noticed that the degree of meeting room parallelism was in direct proportion to the number of input documents: the more documents brought to the standards meetings for review (in an attempt to insert some intellectual property into a specification), the more delegates each member company of the standards organization needed to send to the meetings to critique the documents brought by other delegates . . . but each new delegate, in turn, brought more documents. We are still working out the differential equation growth model, but it is clear that if the IP Wars had not taken place, these standards societies would have eventually consumed all available resources on the planet, mitigated only by the other economic singularities at the centre of the health care and entertainment subscription industries.

A well ventilated chamber dedicated to Profile and Requirements discussions, on matters that were often years from fruition, showed signs of extreme wear that suggest the room was constantly packed with delegates, with room only to stand, and rows of spectators peering in from the hallway. Yet the room designated for Conformance Testing (a critical activity for confirming interoperability before a specification was frozen and widely implemented, after when a mistake would cost thousands of lifetimes of wages to fix) showed that the editor of that specification was likely sitting by his or herself most of the time. This topic was usually assigned to a closet with room for a few chairs.

Of what we can tell from the life style of these delegates and their e-mail exchanges, and what Goodall has reported back to us (in between threats of self-harm after "days trapped in this competitive reality-television show"), is that these delegates essentially adapted to, and even became addicted to the meeting combat experience. They also learned to accumulate "frequent flyer miles," and ultimately came to realise that these meetings were an effective way to avoid "real work" back home. In earlier caste societies dating back to thousands of years before the Digitals, stories of origins and threats of afterlife status were creatively utilized by the upper caste members so they could be supported by the greater efforts of peasant labourers, and thus enjoy a less strenuous, carefree existence, free to craft Requirements and ponder high level syntax . . . at least, in between social upheavals. Promises of future royalties, or the thought of fellow colleagues being away from their home offices for a week or more at a time, likely motivated the sponsors to send their delegates to the meetings, of one, two, or as many standards organizations as their office mates felt it took.

After the first hundred years, the standards meetings had grown to such length, that there was little point to having the delegates return home after each meeting. A permanent site was established in Geneva, similar to the "United Nations" in (old) New York, where delegates lived nearby and commuted each day to the meeting building. Life support systems were eventually installed in the meeting desks, complementing power and communications ports of earlier generations, with organic intake and waste tube outlets that allowed Geneva Stone delegates to meet to the legal limits of all available time. In the communications archives, we found many angry memos from delegates complaining that these interfaces should be more distinct from one another.

Like any primate tribe that learns to fashion tools to reach rewards in difficult places, the Geneva Stone inhabitants evolved complex social mechanisms to influence a desired outcome. Extraction became more difficult with each standard generation. Interestingly, the sophistication of the schemes grew in proportion with the complexity increase in their compression techniques needed to yield any perceived gain or bitrate reduction.

Long term planning is another facet of higher primate intelligence. Delegates often thoughtfully attempted to assist industry by limiting what the standards were capable of expressing, to "prevent incorrect use." Coincidentally this also had the benefit of making the delegate's future lives easier, as certain modes implied by the standards specification would not necessarily have to be implemented. Perhaps this explains why this desk-bound profession, which the standards participants belong to, borrowed the term "engineering". In earlier societies the term "engineering" referred to labour-intensive professions that, for example, diverted water with drainage canals and other manipulations of nature that would result in less future maintenance. By contrast, the huge time investment of countless bitstream flags and modes, which the delegates spent considerable energy designing and arguing over whether to insert into the specifications, were rarely used in products that we recovered in the waste heaps. It seems that careful placement of "reserved bits" was the ultimate essential method to insure future long syntax extension discussions.

Goodall reported that delegates also showed intense concern for becoming ad-hoc study group (AHG) chairs, often piling their names onto the co-chair list that grew longer than the AHG mandate description itself. AHG's were temporary legal entities that allowed the fun of an actual meeting to technically continue on between meetings, usually conducted by e-mail exchange between the delegates. Yet these delegates showed little interest in actually working on the AHG's tasks, or its outcome. As with many of the primates she studied, Goodall remarked that this behaviour was to show elder tribe members back home that the delegate they sponsored to meet remotely with other tribes, at considerable expense, were diligently influencing the outcome of the project to the tribe's favour, and not wasting time at the fine dining establishments or tourist attractions near the exotic meeting venues. Boasts of chairmanships back home were also used to attract potential mates that were pre-interviewed, starting with a "résumé" exchange.

In perhaps the most creative attempt to keep the standards process going (indefinitely), Goodall also discovered that many extensions added to a specification would be treated as if they were retroactive definitions of a major milestone standard that industry could only absorb perhaps once per decade. This also filled the otherwise inactive gap between such projects, nicely redirected attention away from a few mistakes in the previous standard, but at the same time, made it difficult for implementers to know exactly when to design and launch products. An experienced implementer sat in fear, watching one "extension" proposal after another bubble up from Requirements into a buzz storm, and prayed that each would pass by harmlessly, or dissipate by the time it drifted over to a market.

As for the main subject of the video specifications, the purpose of the bit patterns seems to be to prescribe units of non-overlapping blocks of visual samples. This strikes us as an odd fit for biologics such as Digital humans, since it would suggest their vision system was fed by a tiled array of photo detector nodules, similar to trilobites, but arranged in a rectangular rather than a more natural, hexagonal grid. Somehow their higher perceptual processors could ignore the contrast discontinuities at the block boundaries. Likewise, the auditory codes appear to be optimized for acoustic distortions when the human ear was placed under water. Humans were nevertheless an aquatically adept species that could swim at birth. Yet the data suggests the Digital humans *de*volved their appreciation for emotionally provocative, patterned sound (often referred to as "music" in their literature), as they likely adapted to, and came to prefer the distinct filtering effect from audio compression over the centuries. Some humans even went to great lengths to create vacuum tube amplifiers that mimicked this filter effect.

Although one is left with a strong impression from reading their standards meetings input documents that these creatures believed data could be organized and treated as a neat orthogonal, linear system (often they cite their prophet "Shannon" in this context), there are nonetheless many contradictory instances where, for example, packet headers in the vaunted "Systems" layer are locked by complex rules to elements that occur in the separate, but lowly video domain. The Digital's standards appear to devote much language to compensating for the logical challenges of such orthogonal-yet-not-orthogonal alignments. One protocol was highly persistent, and still influences us today, over 700,000 years later. Its recovery from the Geneva Stone *finally* provides a clue to where our genetic 188 byte packet length preference comes from! (Many machine cults will be disappointed to learn it is not, after all, a golden number or cosmological constant upon which they founded their religion).

Perhaps these contradictions can be explained by the tendency of these standards participants to divide themselves early in their careers into ideological camps (much as primates had always formed alliance groups within a tribe), one for each subject or topic, and remain loyal no matter what counter evidence is later presented. Within the Geneva Stone society, we can see factions that formed on such topics as:

- those who believed that video could be arranged into a resolution hierarchy where each higher layer added information on top of the lower resolution layer below it, vs. those who simply coded each resolution independently of one another (known as simulcasting), in order to save bandwidth.
- those who believed 2D displays can render 3D.
- those who believed in gaps between lines were invisible vs. those who had to actually implement video processing.

One can only feel pity for these creatures.

From the Geneva Stone archive, we have learned that many different combinations of compression tools were tried in the creation of each standard. One recurring method that seems to have been attempted countless times through the ages was a thing called "wavelet" transform. The best explanation we can come up with for its appeal is that its alluring shape and utility appears to have induced a sense of pleasure in its proponents who thus became perpetually faithful to it. Perhaps the proponents merely sought the next chemical release from embracing the concept in much the same way as the biologics were known to be artificially attached to their children because of their "cute and fuzzy" appearance that activated genetically pre-coded facial detection circuitry in the adult brain, triggering a release of elating chemicals. It's strange that after 700,000 years, we too still have not found anything better than the Discrete Cosine Transform, and we also periodically struggle with the question of whether to switch to our modern Wavelet-like transform.

Unlike wavelets, Object Coding finally became popular thanks to direct implants into the human brain areas responsible for *comprehending* objects, though this method was originally motivated as the ultimate copy protection scheme, which included the controversial "erase memory of experience at end of entertainment period" flag that would often become over zealous in its effect.

We found examples of clearly bad conventions being carried unnecessarily over from one generation of humans to the next, as if this once world-dominant species somehow lacked the heuristics to learn from past mistakes. For example, much specification language is dedicated to handling instances where the visual data have gaps between every sample row that are later filled by another 2D lattice with gaps where the earlier lattice had samples . . . as if somehow the two lattices would compensate for each other, by alternating in time. This interspersing practice (abandoned some 300,000 years ago) creates representation that can never be truly undone. We believe this horrid practice may have actually started as a bandwidth savings technique in the short-lived, pre-Digital era of telesensing. Much circuitry has been excavated in waste heaps all over the world that appear to be dedicated to attempting to recover the data corrupted by it. Perhaps it was effective enough to fool their primitive tiled vision systems.

Anyway.. "not our problem!".. as Convenor-bot reminds me to quickly finish this report. He has been re-activated to help us to forge a universal protocol for communicating with other civilizations through time, and storing the data differences across the enormous extra dimensions of information spawned by the multiple universes of altered histories. Convenor-bot is so prodigious that he is attempting to standardize everything in sight. I'm not certain if The Past is receiving my message: "do not wake."

(End of Geneva Stone update)

By program: **C/F.O.G^2(Central Forensics Operations: Guide to Guides).**
Your standards archaeology specialist since **[ERROR: before standard date codes]**

```
[ ] program is sentient, [ ] non-sentient, semi-self aware [X]
[X] program is conscious (can feel pain or boredom) [ ] unconscious
[ ] program is original [X] downloaded from ancient human brain
```

Annexes

List of Notes

List of Tables with Assigned Values for Parameter Usage

Fundamentals and Evolution of MPEG-2 Systems: Paving the MPEG Road, First Edition. Jan van der Meer.

© 2014 John Wiley & Sons, Ltd. Published 2014 by John Wiley & Sons, Ltd.

List of Abbreviations

2D (video)	Two-dimensional (video)
3D	Three Dimensional
3DTV	Three-dimensional television
3GPP	3rd Generation Partnership Project (for mobile networks)
A/V	Audio and Video
AAC	Advanced Audio Coding
AAU	Audio AU
ADTS	Audio Data Transport Stream (MPEG-2 AAC)
ALS	Audio Lossless coding
AMD	Amendment
ARIB	Association of Radio Industries and Businesses
ATM	Asynchronous Transfer Mode for (broadband) ISDN networks
ATSC	Advanced Television Systems Committee
AU	Access Unit
AVC	Advanced Video Coding
BD	Blu-ray Disc
BDA	Blu-ray Disc Association
B-picture	Bi-directionally-predicted Picture
CA	Conditional Access
CABAC	Context-Adaptive Binary Arithmetic Coding
CAT	Conditional Access Table
CD	Compact Disc
CIT	Control Information Table
CLUT	Colour Look-Up Table
CPB	Coded Picture Buffer (AVC, MVC and SVC)
CR	Clock Reference
CRC	Cyclic Redundancy Check
CRT	Cathode Ray Tube
CSPS	Constrained System Parameter (Program) Stream
DASH	Dynamic Adaptive Streaming over HTTP
DCT	Discrete Cosine Transform
DPB	Decoded Picture Buffer (AVC, MVC and SVC)
DR	Dependency Representation (SVC)
DSM	Digital Storage Media
DSM-CC	DSM Command and Control
DST	Direct Stream Transfer (audio)
DTS	Decoding Time Stamp
DVB	Digital Video Broadcasting
DVD	Digital Versatile Disc
E2E	End-to-End
ECM	Entitlement Control Message
EMM	Entitlement Management Message
ES	Elementary Stream

ESCR	Elementary stream SCR
FEC	Forward Error Correction
GOP	Group of Pictures
HD(TV)	High Definition (Television)
HE-AAC	High-Efficiency AAC
HEVC	High Efficiency Video Coding
HRD	Hypothetical Reference Decoder (AVC, MVC and SVC)
HTD	Hypothetical Text Decoder
HTTP	Hypertext Transfer Protocol
IC	Integrated Circuit
ID	Identifier
iDCT	Inverse Discrete Cosine Transform
IDR (picture)	Instantaneous Decoding Refresh (picture)
IEC	International Electrotechnical Commission
IETF	Internet Engineering Task Force
IP	Internet Protocol
I-picture	Intra-coded Picture
IPMP	Intellectual Property Management and Protection
IPR	Intellectual Property Rights
IPTV	Internet Protocol Television
ISAN	International Standard Audiovisual Number of an audiovisual work
ISDN	Integrated Services Digital Network
ISO	International Organization for Standardization
ITU	International Telecommunication Union
ITU-R Rec.	Recommendation (Standard) from ITU-R
ITU-R	The Radio communication Standardization Sector within ITU
ITU-T Rec.	Recommendation (Standard) from ITU-T
ITU-T	The Telecommunication Standardization Sector within ITU
J2K	JPEG 2000
JPEG	Joint Photographic Experts Group
JVT	Joint Video Team between VCEG and MPEG
LATM	Low Overhead Audio Transport Multiplex (MPEG-4 audio)
LFE (speaker)	Low Frequency Enhancement (speaker)
MHEG	Multimedia and Hypermedia Experts Group
MP3	MPEG-1 Layer III audio
MPEG	Moving Picture Expert Group
MPTS	Multi-Program TS
MVC	Multiview Video Coding (AVC extension)
NAL	Network Abstraction Layer (AVC, MVC and SVC)
NB	National Body (within MPEG)
NIT	Network Information Table
NPT	Normal Play Time
NTP	Network Time Protocol
OIPF	Open IPTV Forum
OPCR	Original PCR
PAT	Program Association Table
PCM	Pulse-Code Modulation (for sampling an analogue signal)
PCR	Program Clock Reference
PES	Packetized Elementary Stream
PES-STD	P-STD used for a PES stream
PID	transport Packet ID
PLL	Phase-Lock(ed) Loop
PMT	Program Map Table
P-picture	Predicted Picture
PPS	Picture Parameter Set (AVC, MVC and SVC)
PS (audio)	Parametric Stereo (audio)

PS (systems)	(MPEG-2) Program Stream
PSI	Program Specific Information
PSM	Program Stream Map
P-STD	STD for Program Streams
PTS	Presentation Time Stamp
PVR	Personal Video Recorder
QAM	Quadrature Amplitude Modulation
QoS	Quality of Service
QPSK	Quadrature Phase-Shift Keying
RA	Registration Authority
RBSP	Raw Byte Sequence Payload (Auxiliary video)
RFC	Request For Comments (IETF publication on an Internet standard)
RTCP	Real-Time Control Protocol
RTP	Real-time Transport Protocol (Internet)
RTSP	Real-Time Streaming Protocol
SAOC	Spatial Audio Object Coding
SBR	Spectral Band Replication (audio)
SCR	System Clock Reference
SD(TV)	Standard Definition (television)
SEI	Supplemental Enhancement Information (AVC, MVC and SVC)
SI	Service Information (ARIB, ATSC, DVB)
SI-RBSP	Supplemental Information RBSP (Auxiliary video)
SLS	Scalable Lossless coding (audio)
SMPTE	Society of Motion Picture and Television Engineers
SPS	Sequence Parameter Set (AVC, MVC and SVC)
SPTS	Single Program TS
STB	Set-Top-Box
STC	System Time Clock
STD	System Target Decoder
SVC	Scalable Video Coding (AVC extension)
TCP	Transmission Control Protocol
TREF	Timestamp Reference (SVC)
TS	(MPEG-2) Transport Stream
TSDT	Transport Stream Description Table
T-STD	STD for Transport Streams
TTU	Timed Text Unit
UCS	Universal Character Set
UDP	User Datagram Protocol
UHD(TV)	Ultra-High Definition (television)
URL	Uniform Resource Locator
USAC	Unified Speech and Audio Codec
UTC	Universal Time Clock
UTF	UCS Transformation Format
VBV	Video Buffer Verifier (MPEG-1 and MPEG-2 video)
VC	View Component (MVC)
VCEG	Video Coding Experts Group (in ITU-T)
VCL	Video Coding Layer (AVC, MVC and SVC)
VCO	Voltage Controlled Oscillator
VCR	Video Cassette Recorder
V-ISAN	Unique identification of a version of an audiovisual work
VLC	Variable Length Coding
VoD	Video on Demand
VS	View Sub-set (MVC)
VUI	Video Usability Information (AVC, MVC and SVC)
YUV	Luminance signal Y and chrominance signals U and V (video)

Index

Fundamentals and Evolution of MPEG-2 Systems: Paving the MPEG Road, First Edition. Jan van der Meer.
© 2014 John Wiley & Sons, Ltd. Published 2014 by John Wiley & Sons, Ltd.

Q

Printed in the United States
By Bookmasters